A TREATISE ON THE DIFFERENTIAL GEOMETRY OF CURVES AND SURFACES

LUTHER PFAHLER EISENHART

DOVER PUBLICATIONS, INC.
MINEOLA, NEW YORK

DOVER PHOENIX EDITIONS

Bibliographical Note

This Dover edition, first published in 1960 and reprinted in 2004, is an unabridged and unaltered republication of the work first published in 1909 by Ginn and Company, Boston.

International Standard Book Number: 0-486-43820-1

Manufactured in the United States of America
Dover Publications, Inc., 31 East 2nd Street, Mineola, N.Y. 11501

PREFACE

This book is a development from courses which I have given in Princeton for a number of years. During this time I have come to feel that more would be accomplished by my students if they had an introductory treatise written in English and otherwise adapted to the use of men beginning their graduate work.

Chapter I is devoted to the theory of twisted curves, the method in general being that which is usually followed in discussions of this subject. But in addition I have introduced the idea of moving axes, and have derived the formulas pertaining thereto from the previously obtained Frenet-Serret formulas. In this way the student is made familiar with a method which is similar to that used by Darboux in the first volume of his *Leçons*, and to that of Cesaro in his *Geometria Intrinseca*. This method is not only of great advantage in the treatment of certain topics and in the solution of problems, but it is valuable in developing geometrical thinking.

The remainder of the book may be divided into three parts. The first, consisting of Chapters II–VI, deals with the geometry of a surface in the neighborhood of a point and the developments therefrom, such as curves and systems of curves defined by differential equations. To a large extent the method is that of Gauss, by which the properties of a surface are derived from the discussion of two quadratic differential forms. However, little or no space is given to the algebraic treatment of differential forms and their invariants. In addition, the method of moving axes, as defined in the first chapter, has been extended so as to be applicable to an investigation of the properties of surfaces and groups of surfaces. The extent of the theory concerning ordinary points is so great that no attempt has been made to consider the exceptional problems. For a discussion of such questions as the existence of integrals of differential equations and boundary conditions the reader must consult the treatises which deal particularly with these subjects.

In Chapters VII and VIII the theory previously developed is applied to several groups of surfaces, such as the quadrics, ruled surfaces, minimal surfaces, surfaces of constant total curvature, and surfaces with plane and spherical lines of curvature.

The idea of applicability of surfaces is introduced in Chapter III as a particular case of conformal representation, and throughout the book attention is called to examples of applicable surfaces. However, the general problems concerned with the applicability of surfaces are discussed in Chapters IX and X, the latter of which deals entirely with the recent method of Weingarten and its developments. The remaining four chapters are devoted to a discussion of infinitesimal deformation of surfaces, congruences of straight lines and of circles, and triply orthogonal systems of surfaces.

It will be noticed that the book contains many examples, and the student will find that whereas certain of them are merely direct applications of the formulas, others constitute extensions of the theory which might properly be included as portions of a more extensive treatise. At first I felt constrained to give such references as would enable the reader to consult the journals and treatises from which some of these problems were taken, but finally it seemed best to furnish no such key, only to remark that the *Encyklopädie der mathematischen Wissenschaften* may be of assistance. And the same may be said about references to the sources of the subject-matter of the book. Many important citations have been made, but there has not been an attempt to give every reference. However, I desire to acknowledge my indebtedness to the treatises of Darboux, Bianchi, and Scheffers. But the difficulty is that for many years I have consulted these authors so freely that now it is impossible for me to say, except in certain cases, what specific debts I owe to each.

In its present form, the material of the first eight chapters has been given to beginning classes in each of the last two years; and the remainder of the book, with certain enlargements, has constituted an advanced course which has been followed several times. It is impossible for me to give suitable credit for the suggestions made and the assistance rendered by my students during these years, but I am conscious of helpful suggestions made by my colleagues, Professors Veblen, MacInnes, and Swift, and by my former colleague, Professor Bliss of Chicago. I wish also to thank Mr. A. K. Krause for making the drawings for the figures.

It remains for me to express my appreciation of the courtesy shown by Ginn and Company, and of the assistance given by them during the printing of this book.

LUTHER PFAHLER EISENHART

CONTENTS

CHAPTER I

CURVES IN SPACE

CHAPTER II

CURVILINEAR COÖRDINATES ON A SURFACE. ENVELOPES

CHAPTER III

LINEAR ELEMENT OF A SURFACE. DIFFERENTIAL PARAMETERS. CONFORMAL REPRESENTATION

CHAPTER IV

GEOMETRY OF A SURFACE IN THE NEIGHBORHOOD OF A POINT

CHAPTER V

FUNDAMENTAL EQUATIONS. THE MOVING TRIHEDRAL

CHAPTER VI

SYSTEMS OF CURVES. GEODESICS

CHAPTER VII

QUADRICS. RULED SURFACES. MINIMAL SURFACES

CHAPTER VIII

SURFACES OF CONSTANT TOTAL CURVATURE. *W*-SURFACES. SURFACES WITH PLANE OR SPHERICAL LINES OF CURVATURE

CHAPTER IX

DEFORMATION OF SURFACES

CHAPTER X

DEFORMATION OF SURFACES. THE METHOD OF WEINGARTEN

CHAPTER XI

INFINITESIMAL DEFORMATION OF SURFACES

CHAPTER XII

RECTILINEAR CONGRUENCES

CHAPTER XIII

CYCLIC SYSTEMS

CHAPTER XIV

TRIPLY ORTHOGONAL SYSTEMS OF SURFACES

NOTES AND CORRECTIONS

Page 18, line 6. After "vanish" insert "or $\rho_{\cdot} = 0$."

Page 36, Ex. 3. The curve C is plane and P is a point of the plane.

Page 49, Ex. 3. "Point of a plane" should read "point of a non-isotropic plane."

Page 95, line 7. After "that" insert "a real function."

Page 104. In the statement of the first property at the foot of the page, after "*straight line*" insert "*not through the origin.*"

Page 117, § 49. In displayed equation preceding (16) after

$$\frac{\partial^2 x}{\partial v^2}\left(\frac{dv}{ds}\right)^2 \text{ add } + \frac{\partial x}{\partial u}\frac{d^2 u}{ds^2} + \frac{\partial x}{\partial v}\frac{d^2 v}{ds^2}.$$

Page 123. The last paragraph is incorrect. The position of the surface relative to the tangent plane depends upon the character of the terms of third and higher orders.

Page 131, line 1. In place of "that is, . . . positive" read "which may be the case or not, when the curvature of S_1 is negative."

Page 177, Ex. 4. After "system" insert "which is parametric."

Page 188, Ex. 23. In place of "a single parameter" read "either u or v alone."

Page 195, Ex. 1. The following should be substituted for the printed form: "1. If an asymptotic line is a plane curve, not a straight line, each point of the curve is a parabolic point of the surface."

Page 197, § 81. If the curves

$$x_1 = U_1, \quad y_1 = U_2, \quad z_1 = U_3; \qquad x_2 = V_1, \quad y_2 = V_2, \quad z_2 = V_3$$

do not lie on the surface (32), the surface may be generated by curves congruent to them after the manner described in this section.

Page 198, § 82. If a surface of positive curvature is referred to real parametric curves, it follows from (IV, 40) that the parameters of the asymptotic lines are conjugate-imaginary. The second quadratic form is real, as follows from (IV, 4). Consequently λ is real.

Page 201, illustrative example. It is not possible to change the parameter v so that $\phi(v)$ may be made equal to unity. However, if $\sqrt{G}$ is replaced by $\phi\sqrt{G}$ in the last equation of page 201 and the first equation of page 202, the resulting equations are correct, and likewise the theorem.

Page 217, line 25. After "integral" insert "with an arbitrary constant."

Page 219, Ex. 6. It is understood that a is an arbitrary constant. Replace "and the curves . . . trajectories" by "and $\theta_2 =$ const. are geodesic parallels."

Page 257, line 7; also page 259, lines 3 and 26. Replace "*congruent*" by "*superposable by a translation.*"

Page 269, Ex. 18 (g). The section of the surface by each of the planes $x = 0$, $y = 0$ consists of a line of curvature and a double cubic; the double cubics are the locus of the double points of the lines of curvature.

Page 282, line 9. In order to prove that the parametric curves on S_1 are lines of curvature, it is necessary and sufficient to show that the tangents to the parametric curves on S_1 and on its spherical representation at corresponding points are parallel (cf. § 61). The expressions on this page for the displacements of M_1 and of its spherical representation satisfy this condition.

Page 284. In displayed type (34) at end of first line add "$+ \cos\theta\, d\lambda$" and at end of second line add "$+ \sin\theta\, d\lambda$."

Page 341, Ex. 3. Before "surface of revolution" insert "non-developable."

Page 360, line 22. This theorem cannot be stated as true for any surface as a consequence of (29), since it has not been shown that (29) hold for any surface. If x, y, z are the cartesian coördinates of any surface, and we write

$$x = u, \quad y + iz = v, \quad y - iz = 2\psi,$$

then the linear element is

$$ds^2 = du^2 + 2\,dv d\psi = du^2 + 2\frac{\partial\psi}{\partial u}dudv + 2\frac{\partial\psi}{\partial v}dv^2,$$

which is of the form (41). Hence the linear element of any surface can be given this form. Furthermore, the steps of § 148 are reversible, and as a consequence equations (29) hold for any surface. Thus the method of § 148 can be applied to the determination of surfaces having a given linear element when one such surface is known.

Page 400, line 9. Replace "ω_1, ω_2" by "$\omega_1 + \frac{\pi}{2}, \omega_2 + \frac{\pi}{2}$."

Page 412, Ex. 9. When this condition is satisfied, there exists a unique congruence with the given spherical representation of its developables; it is determined by

$$\frac{\partial \log \rho}{\partial u} + \begin{Bmatrix}12\\2\end{Bmatrix}' = 0, \quad \frac{\partial \log \rho}{\partial v} + \begin{Bmatrix}12\\1\end{Bmatrix}' = 0.$$

But every congruence with this representation does not have degenerate focal surfaces.

Page 436, lines 18–20. This condition is sufficient except when the surfaces S_1 are spheres or minimal surfaces (§ 61). However, from the expressions (34) and those preceding (32) it follows that the tangents to the parametric curves on S_1 and on its spherical representation at corresponding points are parallel. Consequently the parametric curves on the surfaces S_1 are their lines of curvature (§ 61).

Page 441, line 27. After "zero" insert "(in fact, L vanishes identically)."

DIFFERENTIAL GEOMETRY

CHAPTER I

CURVES IN SPACE

1. Parametric equations of a curve. Consider space referred to fixed rectangular axes, and let (x, y, z) denote as usual the coördinates of a point with respect to these axes. In the plane $z = 0$ draw a circle of radius r and center (a, b). The coördinates of a point P on the circle can be expressed in the form

$$x = a + r\cos u, \quad y = b + r\sin u, \quad z = 0, \tag{1}$$

where u denotes the angle which the radius to P makes with the positive x-axis. As u varies from $0°$ to $360°$, the point P describes the circle. The quantities a, b, r determine the position and size of the circle; whereas u determines the position of a point upon it. In this sense it is a *variable* or *parameter* for the circle. And equations (1) are called *parametric equations* of the circle.

A straight line in space is determined by a point on it, $P_0(a, b, c)$, and its direction-cosines α, β, γ. The latter fix also the sense of the line. Let P be another point on the line, and let the distance P_0P be denoted by u, which is positive or negative. The rectangular coördinates of P are then expressible in the form

$$x = a + u\alpha, \quad y = b + u\beta, \quad z = c + u\gamma. \tag{2}$$

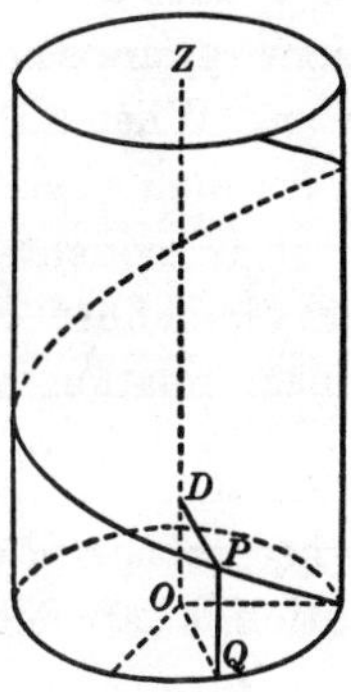

FIG. 1

To each value of u there corresponds a point on the line, and the coördinates of any point on the line are expressible as in (2). These equations are consequently *parametric equations* of the straight line.

When, as in fig. 1, a line segment PD, of constant length a, perpendicular to a line OZ at D, revolves uniformly about OZ as axis, and at the same time D moves along it with uniform velocity, the locus of P is called a *circular helix.* If the line OZ be taken for the z-axis, the initial position of PD for the positive x-axis, and the angle between the latter and a subsequent position of PD be denoted by u, the equations of the helix can be written in the parametric form

$$x = a \cos u, \quad y = a \sin u, \quad z = bu, \tag{3}$$

where the constant b is determined by the velocity of rotation of PD and of translation of D. Thus, as the line PD describes a radian, D moves the distance b along OZ.

In all of the above equations u is the variable or parameter. Hence, with reference to the locus under consideration, the coördinates are functions of u alone. We indicate this by writing these equations

$$x = f_1(u), \quad y = f_2(u), \quad z = f_3(u). \tag{4}$$

The functions f_1, f_2, f_3 have definite forms when the locus is a circle, straight line or circular helix. But we proceed to the general case and consider equations (4), when f_1, f_2, f_3 are any functions whatever, analytic for all values of u, or at least for a certain domain.* The locus of the point whose coördinates are given by (4), as u takes all values in the domain considered, is a *curve.* Equations (4) are said to be the *equations of the curve in the parametric form.* When all the points of the curve do not lie in the same plane it is called a *space curve* or a *twisted curve*; otherwise, a *plane curve.*

It is evident that a necessary and sufficient condition that a curve, defined by equations (4), be plane, is that there exist a linear relation between the functions, such as

$$af_1 + bf_2 + cf_3 + d = 0, \tag{5}$$

where a, b, c, d denote constants not all equal to zero. This condition is satisfied by equations (1) and (2), but not by (3).

If u in (4) be replaced by any function of v, say

$$v = \phi(u), \tag{6}$$

equations (4) assume a new form,

$$x = F_1(v), \quad y = F_2(v), \quad z = F_3(v). \tag{7}$$

* E.g. in case u is supposed to be real, it lies on a segment between two fixed values; when it is complex, it lies within a closed region in the plane of the complex variable.

It is evident that the values of x, y, z, given by (7) for a value of v, are equal to those given by (4) for the corresponding value of u obtained from (6). Consequently equations (4) and (7) define the same curve, u and v being the respective parameters. Since there is no restriction upon the function ϕ, except that it be analytic, it follows that a curve can be given parametric representation in an infinity of ways.

2. Other forms of the equations of a curve. If the first of equations (4) be solved for u, giving $u = \phi(x)$, then, in terms of x as parameter, equations (7) are

$$x = x, \quad y = F_2(x), \quad z = F_3(x). \tag{8}$$

In this form the curve is really defined by the last two equations, or, if it be a plane curve in the xy-plane, its equation is in the customary form

$$y = f(x). \tag{9}$$

The points in space whose coördinates satisfy the equation $y = F_2(x)$ lie on the cylinder whose elements are parallel to the z-axis and whose cross section by the xy-plane is the curve $y = F_2(x)$. In like manner, the equation $z = F_3(x)$ defines a cylinder whose elements are parallel to the y-axis. Hence the curve with the equations (8) is the locus of points common to two cylinders with perpendicular axes. Conversely, if lines are drawn through the points of a space curve normal to two planes perpendicular to one another, we obtain two such cylinders whose intersection is the given curve. Hence equations (8) furnish a perfectly general definition of a space curve.

In general, the parameter u can be eliminated from equations (4) in such a way that there result two equations, each of which involves all three rectangular coördinates. Thus,

$$\Phi_1(x, y, z) = 0, \qquad \Phi_2(x, y, z) = 0. \tag{10}$$

Moreover, if two equations of this kind be solved for y and z as functions of x, we get equations of the form (8), and, in turn, of the form (4), by replacing x by an arbitrary function of u. Hence equations (10) also are the general equations of a curve. It will be seen later that each of these equations defines a surface.

It should be remarked, however, that when a curve is defined as the intersection of two cylinders (8), or of two surfaces (10), it may happen that these curves of intersection consist of several parts, so that the new equations define more than the original ones.

For example, the curve defined by the parametric equations

$$x = u, \qquad y = u^2, \qquad z = u^3, \tag{i}$$

is a twisted cubic, for every plane meets the curve in three points. Thus, the plane

$$ax + by + cz + d = 0$$

meets the curve in the three points whose parametric values are the roots of the equation

$$cu^3 + bu^2 + au + d = 0.$$

This cubic lies upon the three cylinders

$$y = x^2, \qquad z = x^3, \qquad y^3 = z^2.$$

The intersection of the first and second cylinders is a curve of the sixth degree, of the first and third it is of the sixth degree, whereas the last two intersect in a curve of the ninth degree. Hence in every case the given cubic is only a part of the curve of intersection — that part which lies on all three cylinders.

Again, we may eliminate u from equations (i), thus

$$xy = z, \qquad y^2 = xz, \tag{ii}$$

of which the first defines a hyperbolic paraboloid and the second a hyperbolic-parabolic cone. The straight line $y = 0$, $z = 0$ lies on both of these surfaces, but not on the cylinder $y = x^2$. Hence the intersection of the surfaces (ii) consists of this line and the cubic. The generators of the paraboloid are defined by

$$x = a, \qquad z = ay; \qquad y = b, \qquad z = bx;$$

for all values of the constants a and b. From (i) we see that the cubic meets each generator of the first family in one point and of the second family in two points.

3. Linear element. By definition the *length of an arc* of a curve is the limit, when it exists, toward which the perimeter of an inscribed polygon tends as the number of sides increases and their lengths uniformly approach zero. Curves for which such a limit does not exist will be excluded from the subsequent discussion.

Consider the arc of a curve whose end points m_0, m_a, are deter mined by the parametric values u_0 and a, and let $m_1, m_2, \cdots$, be intermediate points with parametric values $u_1, u_2, \cdots$. The length l_k of the chord $\overline{m_k m_{k+1}}$ is

$$\begin{aligned} l_k &= \sqrt{(x_{k+1} - x_k)^2 + (y_{k+1} - y_k)^2 + (z_{k+1} - z_k)^2} \\ &= \sqrt{\Sigma_i [f_i(u_{k+1}) - f_i(u_k)]^2}. \qquad i = 1, 2, 3 \end{aligned}$$

By the mean value theorem of the differential calculus this is equal to

$$l_k = \sqrt{f_1'^2(\xi_1) + f_2'^2(\xi_2) + f_3'^2(\xi_3)}\cdot (u_{k+1} - u_k),$$

where
$$\xi_i = u_k + \theta_i(u_{k+1} - u_k), \qquad 0 < \theta_i < 1$$

and the primes indicate differentiation.

As defined, the length of the arc $m_o m_a$ is the limit of Σl_k, as the lengths $\overline{m_k m_{k+1}}$ tend to zero. From the definition of a definite integral this limit is equal to

$$\int_{u_0}^{a} \sqrt{f_1'^2(u) + f_2'^2(u) + f_3'^2(u)}\, du.$$

Hence, if s denotes the length of the arc from a fixed point (u_o) to a variable point (u), we have

(11) $$s = \int_{u_0}^{u} \sqrt{f_1'^2 + f_2'^2 + f_3'^2}\, du.$$

This equation gives s as a function of u. We write it

(12) $$s = \phi(u),$$

and from (11) it follows that

(13) $$\frac{ds}{du} = \sqrt{f_1'^2 + f_2'^2 + f_3'^2},$$

which we may write in the form

(14) $$ds^2 = dx^2 + dy^2 + dz^2.$$

As thus expressed ds is called the *element of length*, or *linear element*, of the curve.

In the preceding discussion we have tacitly assumed that u is real. When it is complex we take equation (11) as the definition of the length of the arc.

If equation (12) be solved for u in terms of s, and the result be substituted in (4), the resulting equations also define the curve, and s is the parameter. From (11) follows the theorem:

A necessary and sufficient condition that the parameter u be the arc measured from the point $u = u_0$ is

(15) $$f_1'^2 + f_2'^2 + f_3'^2 = 1.$$

An exceptional case should be noted here, namely,

(16) $$f_1'^2 + f_2'^2 + f_3'^2 = 0.$$

Unless f_1', f_2', f_3' be zero and the curve reduce to a point, at least one of the coördinates must be imaginary. For this case s is zero. Hence these imaginary curves are called *curves of length zero*, or *minimal curves*. For the present they will be excluded from the discussion.

Let the arc be the parameter of a given curve and s and $s+e$ its values for two points $M(x, y, z)$ and $M_1(x_1, y_1, z_1)$. By Taylor's theorem we have

$$(17)\qquad \begin{cases} x_1 = x + x'e + x''\dfrac{e^2}{2} + \cdots, \\ y_1 = y + y'e + y''\dfrac{e^2}{2} + \cdots, \\ z_1 = z + z'e + z''\dfrac{e^2}{2} + \cdots, \end{cases}$$

where an accent indicates differentiation with respect to s.

Unless x', y', z' are all zero, that is, unless the locus is a point and not a curve, one at least of the lengths $x_1 - x$, $y_1 - y$, $z_1 - z$ is of the order of magnitude of e. If these lengths be denoted by δx, δy, δz, and e by δs, then we have

$$\sqrt{\delta x^2 + \delta y^2 + \delta z^2} = \delta s + l_2,$$

where l_2 denotes the aggregate of terms of the second and higher orders in δs. Hence, as M_1 approaches M the ratio of the lengths of the chord and the arc MM_1 approaches unity; and in the limit we have $ds^2 = dx^2 + dy^2 + dz^2$.

4. Tangent to a curve. The *tangent to a curve* at a point M is the limiting position of the secant through M and a point M_1 of the curve as the latter approaches M as a limit.

In order to find the equation of the tangent we take s for parameter and write the expressions for the coördinates of M_1 in the form (17). The equations of the secant through M and M_1 are

$$\frac{X - x}{x_1 - x} = \frac{Y - y}{y_1 - y} = \frac{Z - z}{z_1 - z}.$$

If each member of these equations be multiplied by e and the denominators be replaced by their values from (17), we have in the limit as M_1 approaches M

$$(18)\qquad \frac{X - x}{x'} = \frac{Y - y}{y'} = \frac{Z - z}{z'}.$$

If α, β, γ denote the direction-cosines of the tangent in consequence of (15), we may take

$$(19) \qquad \alpha = x', \quad \beta = y', \quad \gamma = z'.$$

When the parameter u is any whatever, these equations are *

$$(20) \quad \alpha = \frac{f_1'}{\sqrt{f_1'^2 + f_2'^2 + f_3'^2}}, \quad \beta = \frac{f_2'}{\sqrt{f_1'^2 + f_2'^2 + f_3'^2}}, \quad \gamma = \frac{f_3'}{\sqrt{f_1'^2 + f_2'^2 + f_3'^2}}.$$

They may also be written thus:

$$(21) \qquad \alpha = \frac{dx}{ds}, \quad \beta = \frac{dy}{ds}, \quad \gamma = \frac{dz}{ds}.$$

From these equations it follows that, if the convention be made that the positive direction on the curve is that in which the parameter increases, the positive direction upon the tangent is the same as upon the curve.

A fundamental property of the tangent is discovered by considering the expression for the distance from the point M_1, with the coördinates (17), to any line through M. We write the equation of such a line in the form

$$(22) \qquad \frac{X - x}{a} = \frac{Y - y}{b} = \frac{Z - z}{c},$$

where a, b, c are the direction-cosines.

The distance from M_1 to this line is equal to

$$(23) \qquad \{[(bx' - ay')e + \tfrac{1}{2}(bx'' - ay'')e^2 + \cdots]^2 + [(cy' - bz')e + \cdots]^2 + [(az' - cx')e + \cdots]^2\}^{\frac{1}{2}}.$$

Hence, if MM_1 be considered an infinitesimal of the first order, this distance also is of the first order unless

$$\frac{a}{x'} = \frac{b}{y'} = \frac{c}{z'},$$

in which case it is of the second order at least. But when these equations are satisfied, equations (22) define the tangent at M. Therefore, of all the lines through a point of a curve the tangent is nearest to the curve.

* Whenever the functions x', y', z' appear in a formula it is understood that the arc s is the parameter; otherwise we use f_1', f_2', f_3', indicating by accents derivatives with respect to the argument u.

5. Order of contact. Normal plane. When the curve is such that there are points for which

$$(24) \qquad \frac{x''}{x'} = \frac{y''}{y'} = \frac{z''}{z'},$$

the distance from M_1 to the tangent is of the third order at least. In this case the tangent is said to have *contact of the second order*, whereas, ordinarily, the *contact is of the first order*. And, in general, the tangent to a curve has contact of the nth order at a point, if the following conditions are satisfied for $n = 2, \cdots, n-1$, and n:

$$(25) \qquad \frac{x^{(n)}}{x^{(n-1)}} = \frac{y^{(n)}}{y^{(n-1)}} = \frac{z^{(n)}}{z^{(n-1)}}.$$

When the parameter of the curve is any whatever, equations (24), (25) are reducible to the respective equations

$$\frac{f_1''}{f_1'} = \frac{f_2''}{f_2'} = \frac{f_3''}{f_3'}; \qquad \frac{f_1^{(n)}}{f_1^{(n-1)}} = \frac{f_2^{(n)}}{f_2^{(n-1)}} = \frac{f_3^{(n)}}{f_3^{(n-1)}}.$$

The plane normal to the tangent to a curve at the point of contact is called the *normal plane* at the point. Its equation is

$$(26) \qquad (X-x)\alpha + (Y-y)\beta + (Z-z)\gamma = 0,$$

where α, β, γ have the values (20).

EXAMPLES

1. Put the equations of the circular helix (3) in the form (8).

2. Express the equations of the circular helix in terms of the arc measured from a point of the curve, and show that the tangents to the curve meet the elements of the circular cylinder under constant angle.

3. Show that if at every point of a curve the tangency is of the second order, the curve is a straight line.

4. Prove that a necessary and sufficient condition that at the point (x_0, y_0) of the plane curve $y = f(x)$ the tangent has contact of the nth order is $f''(x_0) = f'''(x_0) = \cdots = f^{(n)}(x_0) = 0$; also, that according as n is even or odd the tangent crosses the curve at the point or does not.

5. Prove the following properties of the twisted cubic:

(*a*) Of all the planes through a point of the cubic one and only one meets the cubic in three coincident points; its equation is $3u^2x - 3uy + z - u^3 = 0$.

(*b*) There are no double points, but the orthogonal projection on a plane has a double point.

(*c*) Four planes determined by a variable chord of the cubic and by each of four fixed points of the curve are in constant cross-ratio.

6. Curvature. Radius of first curvature. Let M, M_1 be two points of a curve, Δs the length of the arc between these points, and $\Delta\theta$ the angle between the tangents. The limiting value of $\Delta\theta/\Delta s$ as M_1 approaches M, namely $d\theta/ds$, measures the rate of change of the direction of the tangent at M as the point of contact moves along the curve. This limiting value is called the *first curvature* of the curve at M, and its reciprocal the ***radius of first curvature***; the latter will be denoted by ρ.

In order to find an expression for ρ in terms of the quantities defining the curve, we introduce the idea of spherical representation as follows. We take the sphere * of unit radius with center at the origin and draw radii parallel to the positive directions of the tangents to the curve, or such a portion of it that no two tangents are parallel. The locus of the extremities is a curve upon the sphere, which is in one-to-one correspondence with the given curve. In this sense we have a ***spherical representation***, or *spherical indicatrix*, of the curve.

The angle $\Delta\theta$ between the tangents to the curve at the points M, M_1 is measured by the arc of the great circle between their representative points m, m_1 on the sphere. If $\Delta\sigma$ denotes the length of the arc of the spherical indicatrix between m and m_1, then by the result at the close of § 3,

$$\frac{d\theta}{d\sigma} = \lim \frac{\Delta\theta}{\Delta\sigma} = 1.$$

Hence we have

$$\frac{1}{\rho} = \frac{d\sigma}{ds}, \tag{27}$$

where $d\sigma$ is the linear element of the spherical indicatrix.

The coördinates of m are the direction-cosines α, β, γ of the tangent at M; consequently

$$\frac{1}{\rho^2} = \left(\frac{d\sigma}{ds}\right)^2 = \left(\frac{d\alpha}{ds}\right)^2 + \left(\frac{d\beta}{ds}\right)^2 + \left(\frac{d\gamma}{ds}\right)^2. \tag{27'}$$

When the arc s is the parameter, this formula becomes

$$\frac{1}{\rho^2} = x''^2 + y''^2 + z''^2. \tag{28}$$

* Hereafter we refer to this as the *unit sphere*.

However, when the parameter is any whatever, u, we have from (12), (13), (20),

$$\frac{d\alpha}{ds}=\frac{1}{\phi'}\frac{d}{du}\left(\frac{f_1'}{\phi'}\right)=\frac{1}{\phi'^3}\left(\phi' f_1''-\phi'' f_1'\right), \tag{29}$$

and

$$\phi'\phi''=f_1'f_1''+f_2'f_2''+f_3'f_3''.$$

Hence we find by substitution

$$\frac{1}{\rho^2}=\frac{f_1''^2+f_2''^2+f_3''^2-\phi''^2}{\phi'^4}, \tag{30}$$

which sometimes is written thus:

$$\frac{1}{\rho^2}=\frac{(d^2x)^2+(d^2y)^2+(d^2z)^2-(d^2s)^2}{ds^4}. \tag{31}$$

The sign of ρ is not determined by these formulas. We make the convention that it is always positive and thus fix the sense of a displacement on the spherical indicatrix.

7. Osculating plane. Consider the plane through the tangent to a curve at a point M and through a point M_1 of the curve. The limiting position of this plane as M_1 approaches M is called the *osculating plane at M.* In deriving its equation and thus establishing its existence we assume that the arc s is the parameter, and take the coördinates of M_1 in the form (17).

The equation of a plane through $M\ (x, y, z)$ is of the form

$$(X-x)\,a+(Y-y)\,b+(Z-z)\,c=0, \tag{32}$$

X, Y, Z being the current coördinates. When the plane passes through the tangent at M, the coefficients a, b, c are such that

$$x'a+y'b+z'c=0. \tag{33}$$

If the values (17) for x_1, y_1, z_1 be substituted in (32) for X, Y, Z, and the resulting equation be divided by $\frac{e^2}{2}$, we get

$$(x''a+y''b+z''c)+\eta=0,$$

where η represents the aggregate of the terms of first and higher orders in e. As M_1 approaches M, η approaches zero, and in the limit we have

$$x''a+y''b+z''c=0. \tag{34}$$

Eliminating a, b, c from equations (32), (33), (34) we obtain, as the equation of the osculating plane,

$$(35)\qquad \begin{vmatrix} X-x & Y-y & Z-z \\ x' & y' & z' \\ x'' & y'' & z'' \end{vmatrix} = 0.$$

From this we find that when the curve is defined by equations (4) in terms of a general parameter u, the equation of the osculating plane is

$$(36)\qquad \begin{vmatrix} X-x & Y-y & Z-z \\ f_1' & f_2' & f_3' \\ f_1'' & f_2'' & f_3'' \end{vmatrix} = 0.$$

The plane defined by either of these equations is unique except when the tangent at the point has contact of an order higher than the first. In the latter case equations (33), (34) are not independent, as follows from (24); and if the contact of the tangent is of the nth order, the equations

$$x^{(r)}a + y^{(r)}b + z^{(r)}c = 0,$$

for all values of r up to and including n are not independent of one another. But for $r = n+1$, this equation and (33) are independent, and we have as the equation of the osculating plane at this singular point,

$$\begin{vmatrix} X-x & Y-y & Z-z \\ x' & y' & z' \\ x^{(n+1)} & y^{(n+1)} & z^{(n+1)} \end{vmatrix} = 0.$$

When a curve is plane, and its plane is taken for the xy-plane, the equation (35) reduces to $Z = 0$. Hence the osculating plane of a plane curve is the plane of the latter, and consequently is the same for all points of the curve. Conversely, when the osculating plane of a curve is the same for all its points, the curve is plane, for all the points of the curve lie in the fixed osculating plane.

The equation of the osculating plane of the twisted cubic (§ 2) is readily reducible to

$$\text{(i)}\qquad 3u^2X - 3uY + Z - u^3 = 0,$$

where X, Y, Z are current coördinates. From the definition of the osculating plane and the fact that the curve is a cubic, it follows that the osculating plane meets the curve only at the point of osculation. As equation (i) is a cubic in u, it follows that through a point (x_0, y_0, z_0) not on the curve there pass three planes which osculate the cubic. Let u_1, u_2, u_3 denote the parameter values of these points. Then from (i) we have

$$u_1 + u_2 + u_3 = 3x_0, \quad u_1u_2 + u_2u_3 + u_3u_1 = 3y_0, \quad u_1u_2u_3 = z_0.$$

By means of these relations the equation of the plane through the corresponding three points on the cubic is reducible to

$$(X - x_0)\,3\,y_0 - (Y - y_0)\,3\,x_0 + (Z - z_0) = 0.$$

This plane passes through the point (x_0, y_0, z_0); hence we have the theorems:

The points of contact of the three osculating planes of a twisted cubic through a point not on the curve lie in a plane through the point.

The osculating planes at three points of a twisted cubic meet in a point which lies in the plane of the three points.

By means of these theorems we can establish a dual relation in space by making a point correspond to the plane through the points of osculation of the three osculating planes through the point, and a plane to the point of intersection of the three planes which osculate the cubic at the points where it is met by the plane. In particular, to a point on the cubic corresponds the osculating plane at the point, and vice versa.

8. Principal normal and binormal. Evidently there are an infinity of normals to a curve at a point. Two of these are of particular interest: the normal, which lies in the osculating plane at the point, called the *principal normal*; and the normal, which is perpendicular to this plane, called the *binormal*.

If the direction-cosines of the binormal be denoted by λ, μ, ν, we have from (35)

$$\lambda : \mu : \nu = (y'z'' - z'y'') : (z'x'' - x'z'') : (x'y'' - y'x'').$$

In consequence of the identity

$$\Sigma(y'z'' - z'y'')^2 = \Sigma x'^2 \cdot \Sigma x''^2 - (\Sigma x'x'')^2,$$

the value of the common ratio is reducible by means of (19) and (28) to $\pm\rho$.* We take the positive direction of the binormal to be such that this ratio shall be $+\rho$; then

$$(37)\qquad \lambda = \rho\,(y'z'' - z'y''),\quad \mu = \rho\,(z'x'' - x'z''),\quad \nu = \rho\,(x'y'' - x''y').$$

When the parameter u is general, these formulas are

$$(38)\ \lambda = \frac{\rho}{\phi'^3}(f_2'f_3'' - f_3'f_2''),\ \mu = \frac{\rho}{\phi'^3}(f_3'f_1'' - f_1'f_3''),\ \nu = \frac{\rho}{\phi'^3}(f_1'f_2'' - f_2'f_1''),$$

or in other form:

$$(38')\ \lambda = \rho\,\frac{dy\,d^2z - dz\,d^2y}{ds^3},\quad \mu = \rho\,\frac{dz\,d^2x - dx\,d^2z}{ds^3},\quad \nu = \rho\,\frac{dx\,d^2y - dy\,d^2x}{ds^3}.$$

* For $\Sigma x'x'' = 0$, as is seen by differentiating $\Sigma x'^2 = 1$ with respect to s.

By definition the principal normal is perpendicular to both the tangent and binormal. We make the convention that its positive direction is such that the positive directions of the tangent, principal normal and binormal at a point have the same mutual orientation as the positive directions of the x-, y-, z-axes respectively. These directions are represented in fig. 2 by the lines MT, MC, MB. Hence, if l, m, n, denote the direction-cosines of the principal normal, we have *

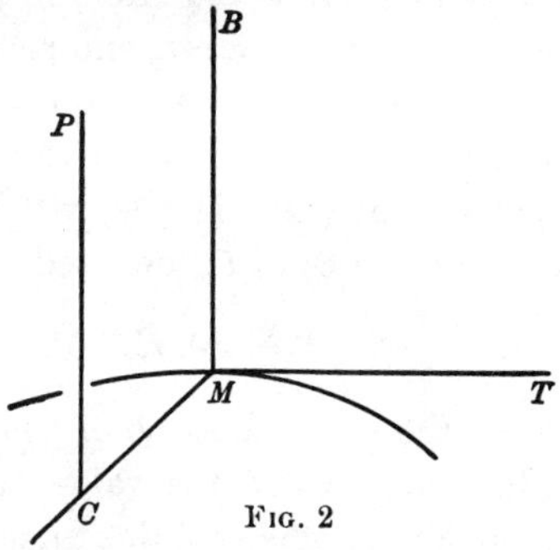

FIG. 2

$$(39)\qquad \begin{vmatrix} \alpha & \beta & \gamma \\ l & m & n \\ \lambda & \mu & \nu \end{vmatrix} = +1,$$

from which it follows that

$$(40)\qquad \begin{cases} \alpha = m\nu - n\mu, & \beta = n\lambda - l\nu, \quad \gamma = l\mu - m\lambda, \\ l = \mu\gamma - \nu\beta, & m = \nu\alpha - \lambda\gamma, \quad n = \lambda\beta - \mu\alpha, \\ \lambda = \beta n - \gamma m, & \mu = \gamma l - \alpha n, \quad \nu = \alpha m - \beta l. \end{cases}$$

Substituting the values of α, β, γ; λ, μ, ν from (19) and (37) in the expressions for l, m, n, the resulting equations are reducible to

$$(41)\qquad l = \rho x'', \quad m = \rho y'', \quad n = \rho z''.$$

Hence, when the parameter u is general, we have

$$(42)\quad l = \frac{\rho}{\phi'^3}(\phi' f_1'' - \phi'' f_1'), \quad m = \frac{\rho}{\phi'^3}(\phi' f_2'' - \phi'' f_2'), \quad n = \frac{\rho}{\phi'^3}(\phi' f_3'' - \phi'' f_3'),$$

or in other form,

$$(42')\quad l = \frac{ds\,d^2x - dx\,d^2s}{ds^3}, \quad m = \frac{ds\,d^2y - dy\,d^2s}{ds^3}, \quad n = \frac{ds\,d^2z - dz\,d^2s}{ds^3}.$$

In consequence of (29) equations (42) may be written:

$$(43)\qquad l = \rho\frac{d\alpha}{ds}, \quad m = \rho\frac{d\beta}{ds}, \quad n = \rho\frac{d\gamma}{ds},$$

or by means of (27),

$$l = \frac{d\alpha}{d\sigma}, \quad m = \frac{d\beta}{d\sigma}, \quad n = \frac{d\gamma}{d\sigma}.$$

Hence the tangent to the spherical indicatrix of a curve is parallel to the principal normal to the curve and has the same sense.

* C. Smith, *Solid Geometry*, 11th ed., p. 31.

9. Osculating circle. Center of first curvature. We have defined the osculating plane to a curve at a point M to be the limiting position of the plane determined by the tangent at M and by a point M_1 of the curve, as the latter approaches M along the curve. We consider now the circle in this plane which has the same tangent at M as the curve, and passes through M_1. The limiting position of this circle, as M_1 approaches M, is called the *osculating circle* to the curve at M. It is evident that its center C_0 is on the principal normal at M. Hence, with reference to any fixed axes in space, the coördinates of C_0, denoted by X_0, Y_0, Z_0, are of the form

$$X_0 = x + rl, \quad Y_0 = y + rm, \quad Z_0 = z + rn,$$

where the absolute value of r is the radius of the osculating circle.

In order to find the value of r, we return to the consideration of the circle, when M_1 does not have its limiting position, and we let X, Y, Z; l_1, m_1, n_1; r_1 denote respectively coördinates of the center of the circle, the direction-cosines of the diameter through M and the radius. If x_1, y_1, z_1 be the coördinates of M_1, they have the values (17), and since M_1 is on the circle, we have

$$r_1^2 = \Sigma(X - x_1)^2 = \Sigma(r_1 l_1 - ex' - \tfrac{1}{2} e^2 x'' \cdots)^2.$$

If we notice that $\Sigma x' l_1 = 0$, and after reducing the above equation divide through by e^2, we have

$$1 - r_1 \Sigma l_1 x'' + \eta = 0,$$

where η involves terms of the first and higher orders in e. In the limit r_1 becomes r, $\Sigma x'' l_1$ becomes $\Sigma x'' l$, that is $\frac{1}{\rho}$, and this equation reduces to

$$1 - \frac{r}{\rho} = 0,$$

so that r is equal to the radius of curvature. On this account the osculating circle is called the *circle of curvature* and its center *the center of first curvature* for the point. Since r is positive the center of curvature is on the positive half of the principal normal, and consequently its coördinates are

$$X_0 = x + \rho l, \quad Y_0 = y + \rho m, \quad Z_0 = z + \rho n. \tag{44}$$

The line normal to the osculating plane at the center of curvature is called the *polar line* or *polar* of the curve for the corresponding point. Its equations are

$$\frac{X - x - \rho l}{\lambda} = \frac{Y - y - \rho m}{\mu} = \frac{Z - z - \rho n}{\nu}. \tag{45}$$

In fig. 2 C represents the center of curvature and CP the polar line for M.

A curve may be looked upon as the path of a point moving under the action of a system of forces. From this point of view it is convenient to take for parameter the time which has elapsed since the point passed a given position. Let t denote this parameter. As t is a function of s, we have

$$\frac{dx}{dt} = \frac{dx}{ds}\frac{ds}{dt} = \alpha\frac{ds}{dt}, \quad \frac{dy}{dt} = \beta\frac{ds}{dt}, \quad \frac{dz}{dt} = \gamma\frac{ds}{dt}.$$

Hence the rate of change of the position of the point with the time, or its *velocity*, may be represented by the length $\frac{ds}{dt}$ laid off on the tangent to the curve. In like manner, by means of (41), we have

$$\frac{d^2x}{dt^2} = \alpha\frac{d^2s}{dt^2} + \frac{l}{\rho}\left(\frac{ds}{dt}\right)^2, \quad \frac{d^2y}{dt^2} = \beta\frac{d^2s}{dt^2} + \frac{m}{\rho}\left(\frac{ds}{dt}\right)^2, \quad \frac{d^2z}{dt^2} = \gamma\frac{d^2s}{dt^2} + \frac{n}{\rho}\left(\frac{ds}{dt}\right)^2.$$

From this it is seen that the rate of change of the velocity at a point, or the *acceleration*, may be represented by a vector in the osculating plane at the point, through the latter and whose components on the tangent and principal normal are $\frac{d^2s}{dt^2}$ and $\frac{1}{\rho}\left(\frac{ds}{dt}\right)^2$.

EXAMPLES

1. Prove that the curvature of a plane curve defined by the equation $M(x, y)\,dx + N(x, y)\,dy = 0$ is

$$\frac{1}{\rho} = \frac{MN\left(\frac{\partial M}{\partial y} + \frac{\partial N}{\partial x}\right) - N^2\frac{\partial M}{\partial x} - M^2\frac{\partial N}{\partial y}}{(M^2 + N^2)^{\frac{3}{2}}}.$$

2. Show that the normal planes to the curve,

$$x = a\sin^2 u, \quad y = a\sin u\cos u, \quad z = a\cos u,$$

pass through the origin, and find the spherical indicatrix of the curve.

3. The straight line is the only real curve of zero curvature at every point.

4. Derive the following properties of the twisted cubic:

(*a*) In any plane there is one line, and only one, through which two osculating planes can be drawn.

(*b*) Four fixed osculating planes are cut by the line of intersection of any two osculating planes in four points whose cross-ratio is constant.

(*c*) Four planes through a variable tangent and four fixed points of the curve are in constant cross-ratio.

(*d*) What is the dual of (*c*) by the results of § 7?

5. Determine the form of the function ϕ so that the principal normals to the curve $x = u$, $y = \sin u$, $z = \phi(u)$ are parallel to the yz-plane.

6. Find the osculating plane and radius of first curvature of

$$x = a\cos u + b\sin u, \quad y = a\sin u + b\cos u, \quad z = c\sin 2u.$$

10. Torsion. Frenet-Serret formulas. It has been seen that, unless a curve be plane, the osculating plane varies as the point moves along the curve. The change in the direction depends evidently upon the form of the curve. The ratio of the angle $\Delta\theta_1$ between the binormals at two points of the curve and their curvilinear distance Δs expresses our idea of the mean change in the direction of the osculating plane. And so we take the limit of this ratio, as one point approaches the other, as the measure of the rate of this change at the latter point. This limit is called the *second curvature*, or *torsion*, of the curve, and its inverse the *radius of second curvature*, or the *radius of torsion*. The latter will be denoted by τ.

In order to establish the existence of this limit and to find an expression for it in terms of the functions defining the curve, we draw radii of the unit sphere parallel to the positive binormals of the curve and take the locus of the end points of these radii as a second spherical representation of the curve. The coördinates of points of this representative curve on the sphere are λ, μ, ν. Proceeding in a manner similar to that in § 6, we obtain the equation

$$\frac{1}{\tau^2} = \frac{d\sigma_1^2}{ds^2} = \left(\frac{d\lambda}{ds}\right)^2 + \left(\frac{d\mu}{ds}\right)^2 + \left(\frac{d\nu}{ds}\right)^2, \tag{46}$$

where $d\sigma_1$ is the linear element of the spherical indicatrix of the binormals.

In order that a real curve have zero torsion at every point, the cosines λ, μ, ν must be constant. By a change of the fixed axes, which evidently has no effect upon the form of the curve, the cosines can be given the values $\lambda = 1$, $\mu = \nu = 0$. It follows from (40) that $\alpha = 0$, and consequently $x = \text{const.}$ Hence a necessary and sufficient condition that the torsion of a real curve be zero at every point is that the curve be plane.

In the subsequent discussion we shall need the derivatives with respect to s of the direction-cosines α, β, γ; l, m, n; λ, μ, ν. We deduce them now. From (41) we have

$$\alpha' = \frac{l}{\rho}, \quad \beta' = \frac{m}{\rho}, \quad \gamma' = \frac{n}{\rho}. \tag{47}$$

In order to find the values of λ', μ', ν', we differentiate with respect to s the identities,

$$\lambda^2+\mu^2+\nu^2=1, \qquad \alpha\lambda+\beta\mu+\gamma\nu=0,$$

and, in consequence of (47), obtain

$$\lambda\lambda'+\mu\mu'+\nu\nu'=0, \qquad \alpha\lambda'+\beta\mu'+\gamma\nu'=0.$$

From these, by (40), follows the proportion

$$\lambda':\mu':\nu'=l:m:n,$$

and the factor of proportionality is $\pm 1/\tau$, as is seen from (46). The algebraic sign of τ is not determined by the latter equation. We fix its sign by writing the above proportion thus:

$$\lambda'=\frac{l}{\tau}, \quad \mu'=\frac{m}{\tau}, \quad \nu'=\frac{n}{\tau}. \tag{48}$$

If the identity $l=\mu\gamma-\nu\beta$ be differentiated with respect to s the result is reducible by (40), (47), and (48) to

$$l'=-\left(\frac{\alpha}{\rho}+\frac{\lambda}{\tau}\right). \tag{49}$$

Similar expressions can be found for m' and n'. Gathering together these results, we have the following formulas fundamental in the theory of twisted curves, and called the Frenet-Serret formulas:

$$\left\{\begin{array}{l} \alpha'=\dfrac{l}{\rho}, \quad \beta'=\dfrac{m}{\rho}, \quad \gamma'=\dfrac{n}{\rho}, \\ l'=-\left(\dfrac{\alpha}{\rho}+\dfrac{\lambda}{\tau}\right), \quad m'=-\left(\dfrac{\beta}{\rho}+\dfrac{\mu}{\tau}\right), \quad n'=-\left(\dfrac{\gamma}{\rho}+\dfrac{\nu}{\tau}\right), \\ \lambda'=\dfrac{l}{\tau}, \quad \mu'=\dfrac{m}{\tau}, \quad \nu'=\dfrac{n}{\tau}. \end{array}\right. \tag{50}$$

As an example, we derive another expression for the torsion. If the equation $\qquad \lambda=\rho(y'z''-z'y'')$

be differentiated with respect to s, the result may be written

$$\frac{l}{\tau}=\frac{\rho'}{\rho}\lambda+\rho(y'z'''-z'y''').$$

If this equation and similar ones for m/τ, n/τ be multiplied by l, m, n respectively and added, we have, in consequence of (50) and (41),

$$\frac{1}{\tau}=-\rho^2\begin{vmatrix} x' & y' & z' \\ x'' & y'' & z'' \\ x''' & y''' & z''' \end{vmatrix}. \tag{51}$$

The last three of equations (50) give the rate of change of the direction-cosines of the osculating plane of a curve as the point of osculation moves along the curve. From these equations it follows that a necessary and sufficient condition that this rate of change at a point be zero is that the values of s for the point make the determinant in equation (51) vanish. At such a point the osculating plane is said to be *stationary*.

11. Form of curve in the neighborhood of a point. The sign of torsion. We have made the convention that the positive directions of the tangent, principal normal, and binormal shall have the same relative orientation as the fixed x-, y-, z-axes respectively. When we take these lines at a point M_0 for axes, the equations of the curve can be put in a very convenient form. If the coördinates be expressed in terms of the arc measured from M_0, we have from (19) and (41) that for $s = 0$

$$x' = 1, \quad y' = z' = 0; \qquad x'' = 0, \quad y'' = \frac{1}{\rho}, \quad z'' = 0.$$

When the values of l and λ from (41) and (37) are substituted in the fourth of equations (50), we obtain

$$x''' = -\frac{x'}{\rho^2} - \frac{1}{\tau}(y'z'' - z'y'') - \frac{\rho'}{\rho}x''. \tag{52}$$

From this and similar expressions for y''' and z''' we find that for $s = 0$

$$x''' = -\frac{1}{\rho^2}, \quad y''' = -\frac{\rho'}{\rho^2}, \quad z''' = -\frac{1}{\rho\tau}.$$

Hence, by Maclaurin's theorem, the coördinates x, y, z can be expressed in the form

$$\left\{\begin{aligned} x &= s - \frac{1}{6\rho^2}s^3 + \cdots, \\ y &= \frac{s^2}{2\rho} - \frac{1}{6}\frac{\rho'}{\rho^2}s^3 + \cdots, \\ z &= -\frac{1}{6\rho\tau}s^3 + \cdots, \end{aligned}\right. \tag{53}$$

where ρ and τ are the radii of first and second curvature at the point $s = 0$, and the unwritten terms are of the fourth and higher powers in s.

From the last of these equations it is seen that for sufficiently small values of s the sign of z changes with the sign of s unless

$1/\tau = 0$ at M_0. Hence, unless the osculating plane is stationary at a point, the curve crosses the plane at the point.* Furthermore, when a point moves along a curve in the positive direction, it passes from the positive to the negative side of the osculating plane at a point, or vice versa, according as the torsion at the latter is positive or negative. In the former case the curve is said to be *sinistrorsum*, in the latter *dextrorsum*.

As another consequence of this equation, we remark that as a variable point M on the curve approaches M_0, the distance from M to the osculating plane at M_0 is of the third order of magnitude in comparison with MM_0. By means of the other equations (53) we find that the distance to any other plane through M_0 is of the second order at most. Hence we have the theorem:

The osculating plane to a twisted curve at an ordinary point is crossed by the curve, and of all the planes through the point it lies nearest to the curve.

From the second of (53) it is seen that y is positive for sufficiently small values of s, positive or negative. Hence, in the neighborhood of an ordinary point, the curve lies entirely on one side of the plane determined by the tangent and binormal — on the side of the positive direction of the principal normal.

These properties of a twisted curve are discovered, likewise, from a consideration of the projections upon the coördinate planes of the approximate curve, whose equations consist of the first terms in (53). The projection on the osculating plane is the parabola $x = s,\ y = s^2/2\rho$, whose axis is the principal normal to the curve. On the plane of the tangent and binormal it is the cubic $x = s,\ z = -s^3/6\rho\tau$, which has the tangent to the curve for an inflectional tangent. And the curve projects upon the plane of the binormal and principal normal into the semicubical parabola $y = s^2/2\rho,\ z = -s^3/6\rho\tau$, with the latter for cuspidal tangent.

These results are represented by the following figures, which picture the projection of the curve upon the osculating plane, normal plane, and the plane of the tangent and binormal. In the third figure the heavy line corresponds to the case where τ is positive and the dotted line to the case where τ is negative.

* This result can be derived readily by geometrical considerations.

The preceding results serve also to give a means of determining the variation in the osculating plane as the point moves along the curve. By means of (50) the direction-cosines λ, μ, ν can be given the form

$$\lambda = \lambda_0 + \frac{l_0}{\tau_0} s + \cdots, \quad \mu = \mu_0 + \frac{m_0}{\tau_0} s + \cdots, \quad \nu = \nu_0 + \frac{n_0}{\tau_0} s + \cdots,$$

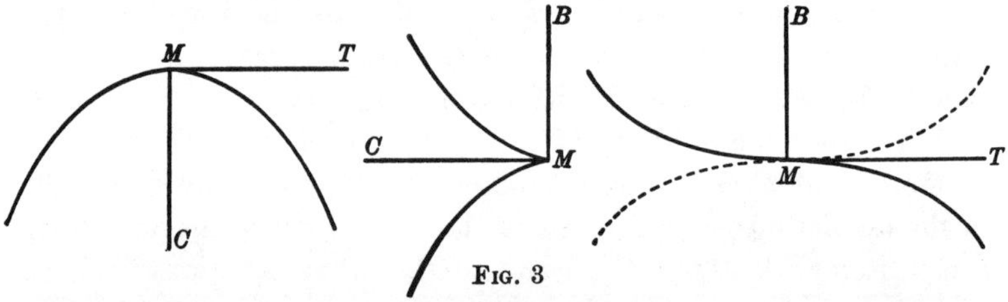

FIG. 3

where the subscript null indicates the value of a function for $s = 0$ and the unwritten terms are of the second and higher terms in s. If the coördinate axes are those which lead to (53), the values of λ, μ, ν for the point of parameter δs are

$$\lambda = 0, \quad \mu = \frac{\delta s}{\tau_0}, \quad \nu = 1,$$

to within terms of higher order, and consequently the equation of this osculating plane at this point M_1 is

$$Y\frac{\delta s}{\tau_0} + Z = 0.$$

If we put $Y = \rho_0$, we get the z-coördinate of the point in which this plane is cut by the polar line for the point $s = 0$; it is $-\rho_0 \delta s / \tau_0$. Hence, according as τ_0 is positive or negative at M, the osculating plane at the near-by point M_1 cuts the polar line for M on the negative or positive side of the osculating plane at M.

12. Cylindrical helices.

As another example of the use of formulas (50) we derive several properties of cylindrical helices. By definition, a cylindrical helix is a curve which lies upon a cylinder and cuts the elements of the cylinder under constant angle. If the axis of z be taken parallel to the elements of the cylinder, we have $\gamma = \text{const.}$ Hence, from (50),

$$n = 0, \quad \frac{\gamma}{\rho} + \frac{\nu}{\tau} = 0, \quad \nu' = 0,$$

from which it follows that the cylindrical helices have the following properties:

The principal normal is perpendicular to the element of the cylinder at the point, and consequently coincides with the normal to the cylinder at the point (§ 22).

The radii of first and second curvature are in constant ratio.

Bertrand has established the converse theorem: *Every curve whose radii of first and second curvature are in constant ratio is a cylindrical helix.* In order to prove it, we put $\tau = \kappa\rho$, and remark from (50) that

$$\frac{d\alpha}{ds} = \kappa\frac{d\lambda}{ds}, \quad \frac{d\beta}{ds} = \kappa\frac{d\mu}{ds}, \quad \frac{d\gamma}{ds} = \kappa\frac{d\nu}{ds},$$

from which we get $\quad \alpha = \kappa\lambda + a, \quad \beta = \kappa\mu + b, \quad \gamma = \kappa\nu + c,$

where a, b, c are constants. From these equations we find

$$a^2 + b^2 + c^2 = 1 + \kappa^2, \qquad a\alpha + b\beta + c\gamma = 1.$$

Hence the tangents to the curve make the constant angle $\cos^{-1}\dfrac{1}{\sqrt{1+\kappa^2}}$ with the lines whose direction-cosines are $\dfrac{a,\, b,\, c}{\sqrt{1+\kappa^2}}$. Consequently the curve is a cylindrical helix, and the elements of the helix have the above direction.

EXAMPLES

1. Find the length of the curve $x = a(u - \sin u)$, $y = a\cos u$, between the points for which u has the values $-\pi$ and π; show that the locus of the center of curvature is of the same form as the given curve.

2. Find the coördinates of the center of curvature of

$$x = a\cos u, \quad y = a\sin u, \quad z = a\cos 2u.$$

3. Find the radii of curvature and torsion of

$$x = a(u - \sin u), \quad y = a(1 - \cos u), \quad z = bu.$$

4. If the principal normals of a curve are parallel to a fixed plane, the curve is a cylindrical helix.

5. Show that the curve $x = e^u$, $y = e^{-u}$, $z = \sqrt{2}\,u$ is a cylindrical helix and that the right section of the cylinder is a catenary; also that the curve lies upon a cylinder whose right section is an equilateral hyperbola. Express the coördinates in terms of the arc and find the radii of first and second curvature.

6. Show that if θ and ϕ denote the angles which the tangent and binormal to a curve make with a fixed line in space, then $\dfrac{\sin\theta\, d\theta}{\sin\phi\, d\phi} = \dfrac{\tau}{\rho}$.

7. When two curves are symmetric with respect to the origin, their radii of first curvature are equal and their radii of torsion differ only in sign.

8. The osculating circle at an ordinary point of a curve has contact of the second order with the latter; and all other circles which lie in the osculating plane and are tangent to the curve at the point have contact of the first order.

9. A necessary and sufficient condition that the osculating circle at a point have contact of the third order is that $\rho' = 0$ and $1/\tau = 0$ at the point; at such a point the circle is said to *superosculate* the curve.

10. Show that any twisted curve may be defined by equations of the form

$$x = s - \frac{1}{6\rho^2}s^3 + \frac{1}{8}\frac{\rho'}{\rho^3}s^4 + \cdots,$$

$$y = \frac{s^2}{2\rho} - \frac{1}{6}\frac{\rho'}{\rho^2}s^3 + \frac{1}{24}\left[\left(\frac{1}{\rho}\right)'' - \frac{1}{\rho^3} - \frac{1}{\rho\tau^2}\right]s^4 + \cdots,$$

$$z = -\frac{1}{6\rho\tau}s^3 - \frac{1}{24}\left[\left(\frac{1}{\rho\tau}\right)' + \frac{1}{\tau}\left(\frac{1}{\rho}\right)'\right]s^4 + \cdots,$$

where ρ and τ are the radii of first and second curvature at the point $s = 0$.

11. When the equations of a curve are in the form (4), the torsion is given by

$$\frac{1}{\tau} = -\frac{\rho^2}{\phi'^6}\begin{vmatrix} f_1' & f_2' & f_3' \\ f_1'' & f_2'' & f_3'' \\ f_1''' & f_2''' & f_3''' \end{vmatrix},$$

where ϕ has the significance of equation (12).

12. The locus of the centers of curvature of a twisted curve of constant first curvature is a curve of the same kind.

13. When all the osculating planes of a curve pass through a fixed point, the curve is plane.

14. Determine $f(u)$ so that the curve $x = a\cos u$, $y = a\sin u$, $z = f(u)$ shall be plane. What is the form of the curve?

13. Intrinsic equations. Fundamental theorem. Let C_1 and C_2 be two curves defined in terms of their respective arcs s, and let points upon each with the same values of s correspond. We assume, furthermore, that at corresponding points the radii of first curvature have the same value, and also the radii of second curvature. We shall show that C_1 and C_2 are congruent.

By a motion in space the points of the two curves for which $s = 0$ can be made to coincide in such a way that the tangents, principal normals, and binormals to them at the point coincide also. Hence if we use the notation of the preceding sections and indicate by subscripts 1 and 2 the functions of C_1 and C_2, we have, when $s = 0$,

$$x_1 = x_2, \quad \alpha_1 = \alpha_2, \quad l_1 = l_2, \quad \lambda_1 = \lambda_2, \tag{54}$$

and other similar equations.

The Frenet-Serret formulas for the two curves are

$$\frac{d\alpha_1}{ds} = \frac{l_1}{\rho}, \quad \frac{dl_1}{ds} = -\left(\frac{\alpha_1}{\rho} + \frac{\lambda_1}{\tau}\right), \quad \frac{d\lambda_1}{ds} = \frac{l_1}{\tau},$$

$$\frac{d\alpha_2}{ds} = \frac{l_2}{\rho}, \quad \frac{dl_2}{ds} = -\left(\frac{\alpha_2}{\rho} + \frac{\lambda_2}{\tau}\right), \quad \frac{d\lambda_2}{ds} = \frac{l_2}{\tau},$$

the functions without subscripts being the same for both curves. If the equations of the first row be multiplied by α_2, l_2, λ_2 respectively, and of the second row by α_1, l_1, λ_1, and all added, we have

$$\frac{d}{ds}(\alpha_1\alpha_2 + l_1 l_2 + \lambda_1\lambda_2) = 0, \tag{55}$$

and consequently $\qquad \alpha_1\alpha_2 + l_1 l_2 + \lambda_1\lambda_2 = \text{const.}$

This constant is equal to unity for $s = 0$, as is seen from (54), and hence for all values of s we have

$$\alpha_1\alpha_2 + l_1 l_2 + \lambda_1\lambda_2 = 1.$$

Combining this equation with the identities

$$\alpha_1^2+l_1^2+\lambda_1^2=1, \qquad \alpha_2^2+l_2^2+\lambda_2^2=1,$$

we obtain $$(\alpha_1-\alpha_2)^2+(l_1-l_2)^2+(\lambda_1-\lambda_2)^2=0.$$

Hence $\alpha_1=\alpha_2$, $l_1=l_2$, $\lambda_1=\lambda_2$. Moreover, since in like manner $\beta_1=\beta_2$, $\gamma_1=\gamma_2$, we have

$$\frac{d}{ds}(x_1-x_2)=0, \quad \frac{d}{ds}(y_1-y_2)=0, \quad \frac{d}{ds}(z_1-z_2)=0.$$

Consequently the differences x_1-x_2, y_1-y_2, z_1-z_2 are constant. But for $s=0$ they are zero, and so we have the theorem:

Two curves whose radii of first and second curvature are the same functions of the arc are congruent.

From this it follows that a curve is determined, to within its position in space, by the expressions for the radii of first and second curvature in terms of the arc. And so the equations of a curve may be written in the form

(56) $$\rho=f_1(s), \qquad \tau=f_2(s).$$

They are called its *intrinsic equations.*

We inquire, conversely, whether two equations (56), in which f_1 and f_2 are any functions whatever of a parameter s, are intrinsic equations of a curve for which s is the length of arc.

In answering this question we show, in the first place, that the equations

(57) $$\frac{du}{ds}=\frac{v}{\rho}, \quad \frac{dv}{ds}=-\left(\frac{u}{\rho}+\frac{w}{\tau}\right), \quad \frac{dw}{ds}=\frac{v}{\tau}$$

admit of three sets of solutions, namely:

(58) $u=\alpha,\ v=l,\ w=\lambda;\ u=\beta,\ v=m,\ w=\mu;\ u=\gamma,\ v=n,\ w=\nu;$

which are such that for each value of s the quantities α, β, γ; l, m, n; λ, μ, ν are the direction-cosines of three mutually perpendicular lines. In fact, we know * that a system (57) admits of a unique set of solutions whose values for $s=0$ are given arbitrarily. Consequently these equations admit of three sets of solutions

* Picard, *Traité d'Analyse*, Vol. II, p. 313; Goursat, *Cours d'Analyse Mathématique*, Vol. II, p. 356.

whose values for $s=0$ are 1, 0, 0; 0, 1, 0; 0, 0, 1 respectively. By an argument similar to that applied to equation (55) we prove that for all values of s the solutions (58) satisfy the conditions

$$\alpha\beta + lm + \lambda\mu = 0, \quad \beta\gamma + mn + \mu\nu = 0, \quad \gamma\alpha + nl + \nu\lambda = 0. \tag{59}$$

In like manner, since it follows from (57) that

$$u\frac{du}{ds} + v\frac{dv}{ds} + w\frac{dw}{ds} = 0,$$

we prove that these solutions satisfy the conditions

$$\alpha^2 + l^2 + \lambda^2 = 1, \quad \beta^2 + m^2 + \mu^2 = 1, \quad \gamma^2 + n^2 + \nu^2 = 1. \tag{60}$$

But the conditions (59), (60) are equivalent to (40), and consequently the three sets of functions α, β, γ; l, m, n; λ, μ, ν are the direction-cosines of three mutually perpendicular lines for all values of s.

Suppose we have such a set of solutions. For the curve

$$x = \int \alpha\, ds, \quad y = \int \beta\, ds, \quad z = \int \gamma\, ds, \tag{61}$$

the functions α, β, γ are the direction-cosines of the tangent, and since $ds^2 = dx^2 + dy^2 + dz^2$, s measures the arc of the curve. From (61) and the first of (57) we get

$$\frac{d^2x}{ds^2} = \frac{l}{\rho}, \quad \frac{d^2y}{ds^2} = \frac{m}{\rho}, \quad \frac{d^2z}{ds^2} = \frac{n}{\rho}; \quad \left(\frac{d^2x}{ds^2}\right)^2 + \left(\frac{d^2y}{ds^2}\right)^2 + \left(\frac{d^2z}{ds^2}\right)^2 = \frac{1}{\rho^2}.$$

Hence if ρ be positive for all values of s, it is the radius of curvature of the curve (61), and l, m, n are the direction-cosines of the principal normal in the positive sense. In consequence of (40) the functions λ, μ, ν are the direction-cosines of the binormal; hence from (50) and the third of (57) it follows that τ is the radius of torsion of the curve. Therefore we have the following theorem fundamental in the theory of curves:

Given any two analytic functions, $f_1(s), f_2(s)$, of which the former is positive for all values of s within a certain domain; there exists a curve for which $\rho = f_1(s)$, $\tau = f_2(s)$, and s is the arc, for values of s in the given domain. The determination of the curve reduces to the finding of three sets of solutions of equations (57), *satisfying the conditions* (59), (60), *and to quadratures.*

We proceed now to the integration of equations (57). Since each set of integrals of the desired kind must satisfy the relation

$$u^2 + v^2 + w^2 = 1, \tag{62}$$

we introduce with Darboux * two functions σ and ω, defined by

$$\begin{cases} \dfrac{u+iv}{1-w} = \dfrac{1+w}{u-iv} = \sigma, \\ \dfrac{u-iv}{1-w} = \dfrac{1+w}{u+iv} = -\dfrac{1}{\omega}. \end{cases} \tag{63}$$

It is evident that the functions σ and $-\dfrac{1}{\omega}$ are conjugate imaginaries.

Solving for u, v, w, we get

$$u = \frac{1-\sigma\omega}{\sigma-\omega}, \quad v = i\frac{1+\sigma\omega}{\sigma-\omega}, \quad w = \frac{\sigma+\omega}{\sigma-\omega}. \tag{64}$$

If these values be substituted in equations (57), it is found that the functions σ and ω are solutions of the equation

$$\frac{d\theta}{ds} = \frac{i}{2\tau} - \frac{i}{\rho}\theta - \frac{i}{2\tau}\theta^2. \tag{65}$$

And conversely, any two different solutions of (65), when substituted in (64), lead to a set of solutions of equations (57) satisfying the relation (62). Our problem reduces then to the integration of equation (65).

14. Riccati equations. Equation (65) may be written

$$\frac{d\theta}{ds} = L + 2M\theta + N\theta^2, \tag{66}$$

where L, M, N are functions of s. This equation is a generalized form of an equation first studied by Riccati,† and consequently is named for him. As Riccati equations occur frequently in the theory of curves and surfaces, we shall establish several of their properties.

Theorem. *When a particular integral of a Riccati equation is known, the general integral can be obtained by two quadratures.*

* *Leçons sur la Théorie Générale des Surfaces*, Vol. I, p. 22. We shall refer to this treatise frequently, and for brevity give our references the form Darboux, I, 22.

† Cf. Forsyth, *Differential Equations*, chap. v; also Cohen, *Differential Equations*, pp. 173–177.

Let θ_1 be a particular integral of (66). If we put $\theta = 1/\phi + \theta_1$, the equation for the determination of ϕ is

$$\frac{d\phi}{ds} + 2(M + N\theta_1)\phi + N = 0. \tag{67}$$

As this equation is linear and of the first order, it can be solved by two quadratures. Since the general integral of (67) is of the form $\phi = f_1(s) + af_2(s)$, where a denotes the constant of integration, the general integral of equation (66) is of the form

$$\theta = \frac{aP + Q}{aR + S}, \tag{68}$$

where P, Q, R, S are functions of s.

Theorem. *When two particular integrals of a Riccati equation are known, the general integral can be found by one quadrature.*

Let θ_1 and θ_2 be two solutions of equation (66). If we effect the substitution $\theta = \dfrac{1}{\psi} + \theta_2$, the equation in ψ is

$$\frac{d\psi}{ds} + 2(M + N\theta_2)\psi + N = 0.$$

If this equation and (67) be multiplied by $1/\psi$ and $1/\phi$ respectively, and subtracted, the resulting equation is reducible to $\dfrac{d}{ds}(\psi/\phi) = N(\theta_1 - \theta_2)\psi/\phi$. Consequently the general integral of (66) is given by

$$\frac{\theta - \theta_1}{\theta - \theta_2} = \frac{\psi}{\phi} = ae^{\int N(\theta_1 - \theta_2)\,ds}, \tag{69}$$

where a is the constant of integration.

Since equation (68) may be looked upon as a linear fractional substitution upon a, four particular solutions θ_1, θ_2, θ_3, θ_4, corresponding to four values a_1, a_2, a_3, a_4 of a, are in the same cross-ratio as these constants. Hence we have the theorem:

The cross-ratio of any four particular integrals of a Riccati equation is constant.

From this it follows that if three particular integrals are known, the general integral can be obtained without quadrature.

15. The determination of the coördinates of a curve defined by its intrinsic equations. We return to the consideration of equation (65) and indicate by

$$\sigma_i = \frac{a_i P + Q}{a_i R + S}, \qquad \omega_i = \frac{b_i P + Q}{b_i R + S}, \qquad (i = 1, 2, 3) \tag{70}$$

six particular integrals of this equation. From these we obtain three sets of solutions of equations (57), namely

$$\alpha = \frac{1 - \sigma_1 \omega_1}{\sigma_1 - \omega_1}, \qquad l = i\frac{1 + \sigma_1 \omega_1}{\sigma_1 - \omega_1}, \qquad \lambda = \frac{\sigma_1 + \omega_1}{\sigma_1 - \omega_1}, \tag{71}$$

and similar expressions in σ_2, ω_2; σ_3, ω_3 respectively for β, m, μ; γ, n, ν. These expressions satisfy the conditions (60). In order that (59) also may be satisfied we must have

$$\frac{\sigma_i - \sigma_k}{\sigma_i - \omega_k} : \frac{\omega_i - \sigma_k}{\omega_i - \omega_k} = -1,$$

which is reducible to

$$\frac{a_i - a_k}{a_i - b_k} : \frac{b_i - a_k}{b_i - b_k} = -1. \qquad \left(\begin{matrix} i = 1, 2, 3 \\ k = 1, 2, 3 \end{matrix} \; i \neq k\right) \tag{72}$$

Hence each two of the three pairs of constants a_1, b_1; a_2, b_2; a_3, b_3, form a harmonic range.

When the values (70) for σ_i, ω_i are substituted in the expressions for α, β, γ, it is found that

$$\left\{\begin{aligned} \alpha &= \frac{1 - a_1 b_1}{a_1 - b_1} U + i\frac{1 + a_1 b_1}{a_1 - b_1} V + \frac{a_1 + b_1}{a_1 - b_1} W, \\ \beta &= \frac{1 - a_2 b_2}{a_2 - b_2} U + i\frac{1 + a_2 b_2}{a_2 - b_2} V + \frac{a_2 + b_2}{a_2 - b_2} W, \\ \gamma &= \frac{1 - a_3 b_3}{a_3 - b_3} U + i\frac{1 + a_3 b_3}{a_3 - b_3} V + \frac{a_3 + b_3}{a_3 - b_3} W, \end{aligned}\right. \tag{73}$$

where, for the sake of brevity, we have put

$$\left\{\begin{aligned} U &= \frac{(P^2 - R^2) - (Q^2 - S^2)}{2(PS - QR)}, \\ V &= i\frac{(P^2 - R^2) + (Q^2 - S^2)}{2(PS - QR)}, \\ W &= \frac{RS - PQ}{PS - QR}. \end{aligned}\right. \tag{74}$$

The coefficients of U, V, and W in (73) are of the same form as the expressions (71) for α, l, λ; β, m, μ; γ, n, ν. Moreover, the equations of condition (59) are equivalent to (72). Hence these coefficients are the direction-cosines of three fixed directions in space mutually perpendicular to one another. If lines through the origin of coördinates parallel to these three lines be taken for a new set of axes, the expressions for α, β, γ with reference to these axes reduce to U, V, W respectively.* These results may be stated thus:

If the general solution of equation (65) *be*

$$\theta = \frac{aP+Q}{aR+S}, \tag{68}$$

the curve whose radii of first and second curvature are ρ *and* τ *respectively is given by*

$$x = \int \frac{(P^2-R^2)-(Q^2-S^2)}{2(PS-QR)}\,ds, \quad y = i\int \frac{(P^2-R^2)+(Q^2-S^2)}{2(PS-QR)}\,ds,$$

$$z = \int \frac{RS-PQ}{PS-QR}\,ds.\dagger$$

It must be remarked that the new axes of coördinates are not necessarily real, so that when it is important to know whether the curves are real it will be advisable to consider the general formulas (73). An example of this will be given later.

We shall apply the preceding results to several problems.

When the curve is plane the torsion is zero, and conversely. For this case equation (65) reduces to $\frac{d\theta}{ds} = -\frac{i}{\rho}\theta$, of which the general integral is

$$\theta = ae^{-i\int\frac{ds}{\rho}} = ae^{-i\sigma},$$

where a is an arbitrary constant, and by (27) σ is the measure of the arc of the spherical indicatrix of the tangent. This solution is of the form (68), with

$$P = e^{-i\sigma}, \quad Q = R = 0, \quad S = 1.$$

Therefore the coördinates are given by

$$x = \int \cos\sigma\, ds, \quad y = \int \sin\sigma\, ds, \quad z = 0. \tag{75}$$

Hence the coördinates of any plane curve can be put in this form.

* This is the same thing as taking

$$a_1 = -b_1 = 1, \quad a_2 = -b_2 = i, \quad a_3 = \infty, \quad b_3 = 0.$$

† Scheffers, *Anwendung der Differential und Integral Rechnung auf Geometrie*, Vol. I, p. 219. Leipsic, 1902.

We have seen that *cylindrical helices* are characterized by the property that the radii of first and second curvature are in constant ratio. If we put $\tau = \rho c$, equation (65) may be written

$$\frac{d\theta}{ds} = \frac{i}{2\tau}(1 - 2c\theta - \theta^2).$$

Two particular integrals are the roots of the equation $\theta^2 + 2c\theta - 1 = 0$. These roots are real and unequal if c is real; we consider only this case, and put

$$(76) \qquad \theta_1 = -c - \sqrt{c^2+1}, \quad \theta_2 = -c + \sqrt{c^2+1}, \quad \theta_1\theta_2 = -1.$$

From (69) it follows that the general solution of the above equation is

$$(77) \qquad \theta = \frac{ae^{it}\theta_2 - \theta_1}{ae^{it} - 1},$$

where we have put

$$(78) \qquad t = \frac{\theta_2 - \theta_1}{2}\int\frac{ds}{\tau} = \frac{\sqrt{c^2+1}}{c}\int\frac{ds}{\rho}.$$

Since σ and $-\frac{1}{\omega}$ in (63) are conjugate imaginary, if we take

$$\sigma = \frac{ae^{it}\theta_2 - \theta_1}{ae^{it} - 1}, \qquad \omega = \frac{be^{it}\theta_2 - \theta_1}{be^{it} - 1},$$

then a and b must be such that

$$\frac{ae^{it}\theta_2 - \theta_1}{ae^{it} - 1} = -\frac{b_0e^{-it} - 1}{b_0e^{-it}\theta_2 - \theta_1},$$

where b_0 denotes the conjugate imaginary of b. This reduces, in consequence of (76), to

$$(79) \qquad ab_0 = -\frac{(1+\theta_1^2)}{(1+\theta_2^2)} = -\theta_1^2.$$

One solution of this is given by taking ∞ and 0 for a and b; we put $a_3 = \infty$, $b_3 = 0$. If these values be substituted in (72), we get $a_i + b_i = 0$, where $i = 1, 2$. So that equation (79) becomes $b_ib_{i0} = \theta_1^2$, where $i = 1, 2$. The solutions of this equation are $b_1 = \theta_1$, $b_2 = -i\theta_1$. From (77) $P = e^{it}\theta_2$, $Q = -\theta_1$, $R = e^{it}$, $S = -1$, so that

$$U = \frac{c}{2\sqrt{c^2+1}}(\theta_2e^{it} - \theta_1e^{-it}), \quad V = \frac{ic}{2\sqrt{c^2+1}}(\theta_2e^{it} + \theta_1e^{-it}), \quad W = \frac{1}{\sqrt{c^2+1}}.$$

When the foregoing values are substituted in (73), and the resulting values of α, β, γ in (61), we get

$$(80) \qquad x = \frac{c}{\sqrt{c^2+1}}\int\cos t\,ds, \quad y = \frac{c}{\sqrt{c^2+1}}\int\sin t\,ds, \quad z = \frac{s}{\sqrt{c^2+1}}.$$

From the last of these expressions we find that the tangent to the curve makes a constant angle with the z-axis — the direction of the elements of the cylinder. And the cross-section of the cylinder is defined by

$$x_1 = \int\cos t\,ds_1, \qquad y_1 = \int\sin t\,ds_1,$$

where s_1 denotes the arc of this section measured from a point of it. If ρ_1 denotes the radius of curvature of the right section, we find that $\rho c^2 = \rho_1(c^2+1)$.

EXAMPLES

1. Find the coördinates of the cylindrical helix whose intrinsic equations are $\rho = \tau = s$.

2. Show that the helix whose intrinsic equations are $\rho = \tau = (s^2 + 4)/\sqrt{2}$ lies upon a cylinder whose cross-section is a catenary.

3. Establish the following properties for the curve with the intrinsic equations $\rho = as$, $\tau = bs$, where a and b are constants:

(a) the Cartesian coördinates are reducible to $x = Ae^{ht}\cos t$, $y = Ae^{ht}\sin t$, $z = Be^{ht}$, where A, B, h are functions of a and b;

(b) the curve lies upon a circular cone whose axis coincides with the z-axis and cuts the elements of the cone under constant angle.

16. Moving trihedral. In § 11 we took for fixed axes of reference the tangent, principal normal, and binormal to a curve at a point M_0 of it, and expressed the coördinates of any other point of the curve with respect to these axes as power series in the arc s of the curve between the two points. Since M_0 is any point of the curve, there is a set of such axes for each of its points. Hence, instead of considering only the points whose locus is the curve, we may look upon the moving point as the intersection of three mutually perpendicular lines which move along with the point, the whole figure rotating so that in each position the lines coincide with the tangent, principal normal, and binormal at the point. We shall refer to such a configuration as the *moving trihedral.* In the solution of certain problems it is of advantage to refer the curve to this moving trihedral as axes. We proceed to the consideration of this idea.

With reference to the trihedral at a point M, the direction-cosines of the tangent, principal normal, and binormal at M have the values

$$\alpha = 1,\ \beta = \gamma = 0;\quad l = 0,\ m = 1,\ n = 0;\quad \lambda = \mu = 0,\ \nu = 1.$$

As the trihedral begins to move, the rates of change of these functions with s are found from the Frenet formulas (50) to have the values

$$\frac{d\alpha}{ds} = 0,\ \frac{d\beta}{ds} = \frac{1}{\rho},\ \frac{d\gamma}{ds} = 0;\qquad \frac{dl}{ds} = -\frac{1}{\rho},\ \frac{dm}{ds} = 0,$$

$$\frac{dn}{ds} = -\frac{1}{\tau};\quad \frac{d\lambda}{ds} = 0,\ \frac{d\mu}{ds} = \frac{1}{\tau},\ \frac{d\nu}{ds} = 0.$$

Let ξ, η, ζ denote coördinates referring to the axes at M, and ξ', η', ζ' those with reference to the axes at M', and let $MM' = \Delta s$ (see fig. 4). Since the rate of change of α is zero and $\alpha = 1$ at M, the cosine of the angle between the ξ- and ξ'-axes is 1 to within terms of higher than the first order in Δs. Likewise the cosine of the angle between the ξ- and η'-axes is $-\Delta s/\rho$. We calculate the cosines of the angles between all the axes, and the results may be tabulated as follows:

(81)

	ξ	η	ζ
ξ'	1	$\dfrac{\Delta s}{\rho}$	0
η'	$-\dfrac{\Delta s}{\rho}$	1	$-\dfrac{\Delta s}{\tau}$
ζ'	0	$\dfrac{\Delta s}{\tau}$	1

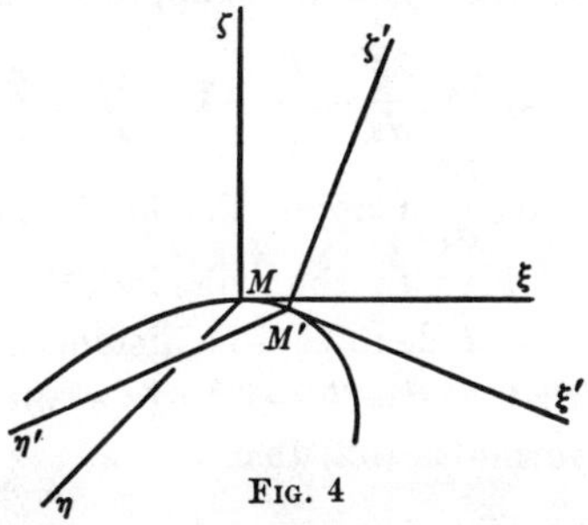

FIG. 4

Let P be a point whose coördinates with respect to the trihedral at M are ξ, η, ζ. Suppose that as M describes the given curve C, P describes a path Γ. It may happen that in this motion P is fixed relatively to the moving trihedral, but in general the change in the position of P will be due not only to the motion of the trihedral but also to a motion relative to it. In the latter general case, if P' denotes the point on Γ corresponding to M' on C, the coördinates of P' relative to the axes at M and M' may be written

$$\xi + \Delta_1\xi, \quad \eta + \Delta_1\eta, \quad \zeta + \Delta_1\zeta; \qquad \xi + \Delta_2\xi, \quad \eta + \Delta_2\eta, \quad \zeta + \Delta_2\zeta.$$

Thus $\Delta_2\theta$ indicates the variation of a function θ relative to the moving trihedral, and $\Delta_1\theta$ the variation due to the latter and to the motion of the trihedral.

To within terms of higher order the coördinates of M' are $(\Delta s, 0, 0)$ with respect to the axes at M, and with the aid of (81) the equations of the transformation of coördinates with respect to the two axes are expressible thus:

$$\begin{aligned}
\xi + \Delta_1\xi &= \Delta s + (\xi + \Delta_2\xi) - (\eta + \Delta_2\eta)\frac{\Delta s}{\rho}, \\
\eta + \Delta_1\eta &= \qquad (\xi + \Delta_2\xi)\frac{\Delta s}{\rho} + (\eta + \Delta_2\eta) + (\zeta + \Delta_2\zeta)\frac{\Delta s}{\tau}, \\
\zeta + \Delta_1\zeta &= \qquad -(\eta + \Delta_2\eta)\frac{\Delta s}{\tau} + (\zeta + \Delta_2\zeta).
\end{aligned}$$

These reduce to

$$\frac{\Delta_1\xi}{\Delta s}=\frac{\Delta_2\xi}{\Delta s}+1-\frac{\eta+\Delta_2\eta}{\rho},\qquad \frac{\Delta_1\zeta}{\Delta s}=\frac{\Delta_2\zeta}{\Delta s}-\frac{\eta+\Delta_2\eta}{\tau},$$

$$\frac{\Delta_1\eta}{\Delta s}=\frac{\Delta_2\eta}{\Delta s}+\frac{\xi+\Delta_2\xi}{\rho}+\frac{\zeta+\Delta_2\zeta}{\tau}.$$

In the limit as M' approaches M these equations become

$$\frac{\delta\xi}{ds}=\frac{d\xi}{ds}+1-\frac{\eta}{\rho},\quad \frac{\delta\eta}{ds}=\frac{d\eta}{ds}+\frac{\xi}{\rho}+\frac{\zeta}{\tau},\quad \frac{\delta\zeta}{ds}=\frac{d\zeta}{ds}-\frac{\eta}{\tau}; \tag{82}$$

thus $\frac{\delta\theta}{ds}$ denotes the absolute rate of change of θ, and $\frac{d\theta}{ds}$ that relative to the trihedral.*

If t denotes the distance between P and a point P_1 (ξ_1, η_1, ζ_1), that is $t^2=(\xi_1-\xi)^2+(\eta_1-\eta)^2+(\zeta_1-\zeta)^2$, we find by means of the formulas (82) that

$$\frac{dt}{ds}=\frac{\delta t}{ds}.$$

If a, b, c denote the direction-cosines of PP_1 with respect to the axes at M, then

$$\xi_1=\xi+at,\qquad \eta_1=\eta+bt,\qquad \zeta_1=\zeta+ct.$$

When we express the condition that ξ_1, η_1, ζ_1 as well as ξ, η, ζ satisfy equations (82), we are brought to the following fundamental relations between the variations of a, b, c:

$$\frac{\delta a}{ds}=\frac{da}{ds}-\frac{b}{\rho},\quad \frac{\delta b}{ds}=\frac{db}{ds}+\frac{a}{\rho}+\frac{c}{\tau},\quad \frac{\delta c}{ds}=\frac{dc}{ds}-\frac{b}{\tau}. \tag{83}$$

If the point P remains fixed in space as M moves along the curve, the left-hand members of equations (82) are zero and the equations reduce to

$$\frac{d\xi}{ds}=\frac{\eta}{\rho}-1,\qquad \frac{d\eta}{ds}=-\left(\frac{\xi}{\rho}+\frac{\zeta}{\tau}\right),\qquad \frac{d\zeta}{ds}=\frac{\eta}{\tau}. \tag{84}$$

Moreover, the direction-cosines of a line fixed in space satisfy the equations

$$\frac{da}{ds}=\frac{b}{\rho},\qquad \frac{db}{ds}=-\left(\frac{a}{\rho}+\frac{c}{\tau}\right),\qquad \frac{dc}{ds}=\frac{b}{\tau}. \tag{85}$$

* Cf. Cesaro, *Lezioni di Geometria Intrinseca*, pp. 122–128. Naples, 1896.

These are the Frenet-Serret formulas, as might have been expected.

We shall show that the solution of these equations carries with it the solution of (84). Suppose we have three sets of solutions of (85), α, l, λ; β, m, μ; γ, n, ν, whose values for $s=0$ are

$$1, \quad 0, \quad 0; \qquad 0, \quad 1, \quad 0; \qquad 0, \quad 0, \quad 1. \tag{86}$$

They are the direction-cosines, with respect to the moving trihedral with vertex M, of three fixed directions in space mutually perpendicular to one another. Let O be a fixed point, and through it draw the lines with the directions just found. Take these lines for coördinate axes and let x, y, z denote the coördinates of M with respect to them. If ξ, η, ζ denote the coördinates of O with respect to the moving trihedral, then $-\xi, -\eta, -\zeta$ are the coördinates of M with respect to the trihedral with vertex at O and edges parallel to the corresponding edges of the trihedral at M. Consequently we have

$$\begin{aligned} \xi &= -(\alpha x + \beta y + \gamma z), \\ \eta &= -(lx + my + nz), \\ \zeta &= -(\lambda x + \mu y + \nu z). \end{aligned} \tag{87}$$

If these values be substituted in (84) and we take account of (50) and (85), we find that the equations are identically satisfied. If ξ_0, η_0, ζ_0 denote the values of ξ, η, ζ for $s=0$, it follows from (86) and (87) that they differ only in sign from the initial values of x, y, z. Hence if we write, in conformity with (21),

$$x = \int_0^s \alpha\, ds - \xi_0, \quad y = \int_0^s \beta\, ds - \eta_0, \quad \zeta = \int_0^s \gamma\, ds - \zeta_0, \tag{88}$$

and substitute these values in (87), they become the general solution of equations (84). We have seen that the solution of equations (85) reduces to the integration of the Riccati equation (65).

17. Illustrative examples. As an example of the foregoing method we consider the curve which is the locus of a point on the tangent to a twisted curve C at a constant distance a from the point of contact.

The coördinates of the point M_1 of the curve with reference to the axes at M are a, 0, 0. In this case equations (82) reduce to

$$\frac{\delta \xi}{ds} = 1, \quad \frac{\delta \eta}{ds} = \frac{a}{\rho}, \quad \frac{\delta \zeta}{ds} = 0. \tag{i}$$

Hence if s_1 denotes the length of arc of C_1 from the point corresponding to $s = 0$ on C, we have

(ii) $$s_1 = \int_0^s \left(1 + \frac{a^2}{\rho^2}\right)^{\frac{1}{2}} ds,$$

and the direction-cosines of the tangent to C_1 with reference to the moving axes are given by

$$\alpha_1 = \frac{\rho}{\sqrt{a^2+\rho^2}}, \qquad \beta_1 = \frac{a}{\sqrt{a^2+\rho^2}}, \qquad \gamma_1 = 0.$$

Hence the tangent to C_1 is parallel to the osculating plane at the corresponding point of C.

By means of (83) we find

$$\frac{\delta\alpha_1}{ds} = \frac{d}{ds}\left(\frac{\rho}{\sqrt{a^2+\rho^2}}\right) - \frac{a}{\rho\sqrt{a^2+\rho^2}} = \frac{a^2\rho'}{(a^2+\rho^2)^{\frac{3}{2}}} - \frac{a}{\rho\sqrt{a^2+\rho^2}}.$$

Proceeding in like manner with β_1 and γ_1, and making use of (ii), we have

$$\frac{\delta\alpha_1}{\delta s_1} = \frac{a^2\rho\rho'}{(a^2+\rho^2)^2} - \frac{a}{a^2+\rho^2}, \qquad \frac{\delta\beta_1}{\delta s_1} = \frac{-a\rho^2\rho'}{(a^2+\rho^2)^2} + \frac{\rho}{a^2+\rho^2},$$

$$\frac{\delta\gamma_1}{\delta s_1} = -\frac{a\rho}{\tau(a^2+\rho^2)}.$$

From these expressions and (27′) we obtain the following expression for the square of the first curvature of C_1:

(iii) $$\frac{1}{\rho_1^2} = \frac{1}{a^2+\rho^2}\left(\frac{a\rho\rho'}{a^2+\rho^2} - 1\right)^2 + \frac{a^2\rho^2}{\tau^2(a^2+\rho^2)^2}.$$

The direction-cosines of the principal normal of C_1 are

$$l_1 = \rho_1\frac{\delta\alpha_1}{\delta s_1}, \qquad m_1 = \rho_1\frac{\delta\beta_1}{\delta s_1}, \qquad n_1 = \rho_1\frac{\delta\gamma_1}{\delta s_1}.$$

By means of (40) we derive the following expressions for the direction-cosines of the binormal:

$$\lambda_1 = -\frac{a^2\rho\rho_1}{\tau(a^2+\rho^2)^{\frac{3}{2}}}, \qquad \mu_1 = \frac{a\rho^2\rho_1}{\tau(a^2+\rho^2)^{\frac{3}{2}}}, \qquad \nu_1 = \left(1 - \frac{a\rho\rho'}{a^2+\rho^2}\right)\frac{\rho_1}{\sqrt{a^2+\rho^2}}.$$

In order to find the expression for τ_1, the radius of torsion of C_1, we have only to substitute the above values in the equation

$$\frac{l_1}{\tau_1} = \frac{\delta\lambda_1}{\delta s_1} = \frac{\rho}{\sqrt{a^2+\rho^2}}\left(\frac{d\lambda_1}{ds} - \frac{\mu_1}{\rho}\right).$$

We leave this calculation to the reader and proceed to an application of the preceding results.

We inquire whether there is a curve C such that C_1 is a straight line. The necessary and sufficient condition is that $1/\rho_1$ be zero (Ex. 3, p. 15). From (iii) it follows that we must have

$$\frac{a\rho\rho'}{a^2+\rho^2} - 1 = 0, \qquad \frac{1}{\tau} = 0.$$

From the second of these equations it follows that C must be plane, and from the former we get, by integration,

$$\log(a^2 + \rho^2) = \frac{2s}{a} + c,$$

where c is a constant of integration. If the point $s = 0$ be chosen so that we may take $c = \log a^2$, this equation reduces to

$$\rho = a\sqrt{e^{\frac{2s}{a}} - 1}.$$

If θ denotes the angle which the line C_1 makes with the ξ-axis, we have, from (i),

$$\tan\theta = \frac{\delta\eta}{\delta\xi} = \frac{a}{\rho} = \frac{1}{\sqrt{e^{\frac{2s}{a}} - 1}}.$$

Differentiating this equation with respect to s, we can put the result in the form

$$\frac{d\theta}{ds} = -\frac{1}{\rho};$$

consequently (§ 6),

(89) $$-\theta = \int \frac{ds}{\rho} = \sigma.$$

When these values are substituted in equations (75), we obtain the coördinates of C in the form

$$x = \int \sqrt{1 - e^{-\frac{2s}{a}}}\, ds, \qquad y = ae^{-\frac{s}{a}},$$

or, in terms of θ,

(90) $$x = -a\left[\log\tan\frac{\theta}{2} + \cos\theta\right], \qquad y = a\sin\theta.$$

The curve, with these equations, is called the *tractrix*. As just seen, it possesses the property that there is associated with it a straight line such that the segments of the tangents between the points of tangency and points of intersection with the given line are of constant length.

Theorem. *The orthogonal trajectories of the osculating plane of a twisted curve can be found by quadratures.*

With reference to the moving axes the coördinates of a point in the osculating plane are $(\xi, \eta, 0)$. The necessary and sufficient condition that this point describe an orthogonal trajectory of the osculating plane as M moves along the given curve is that $\frac{\delta\xi}{ds}$ and $\frac{\delta\eta}{ds}$ in (82) be zero. Hence we have for the determination of ξ and η the equations

$$\frac{d\xi}{d\sigma} - \eta + \rho = 0, \qquad \frac{d\eta}{d\sigma} + \xi = 0,$$

where σ is given by (89). Eliminating ξ, we have

$$\frac{d^2\eta}{d\sigma^2} + \eta = \rho.$$

Hence η can be found by quadratures as a function of σ, and consequently of s, and then ξ is given directly.

Problem. *Find a necessary and sufficient condition that a curve lie upon a sphere.*

If ξ, η, ζ denote the coördinates of the center, and R the radius of the sphere, we have $\xi^2 + \eta^2 + \zeta^2 = R^2$. Since the center is fixed, the derivatives of ξ, η, ζ are given by (84). Consequently, when we differentiate the above equation, the resulting equation reduces to $\xi = 0$, which shows that the normal plane to the curve at each point passes through the center of the sphere. If this equation be differentiated, we get $\eta = \rho$; hence the center of the sphere is on the polar line for each point. Another differentiation gives, together with the preceding, the following coördinates of the center of the sphere:

$$\xi = 0, \qquad \eta = \rho, \qquad \zeta = -\tau\rho'. \tag{91}$$

When the last of these equations is differentiated we obtain the desired condition

$$\frac{\rho}{\tau} + (\tau\rho')' = 0. \tag{92}$$

Conversely, when this condition is satisfied, the point with the coördinates (91) is fixed in space and at constant distance from points of the curve. A curve which lies upon a sphere is called a *spherical* curve. Hence equation (92) is a necessary and sufficient condition that a curve be spherical.

EXAMPLES

1. Let C be a plane curve and C_1 an orthogonal trajectory of the normals to C. Show that the segments of these normals between C and C_1 are of the same length.

2. Let C and C_1 be two curves in the same plane, and say that the points correspond in which the curves are met by a line through a fixed point P. Show that if the tangents at corresponding points are parallel, the two curves are similar and P is the center of similitude.

3. The locus of the point of projection of a fixed point P upon the tangent to a curve C is called the pedal curve of C with respect to P. Show that if r is the distance from P to a point M on C, and θ the angle which the line PM makes with the tangent to C at M, the arc s_1 and radius of curvature ρ_1 of the pedal curve are given by

$$s_1 = \int \frac{r}{\rho}\, ds, \qquad \rho_1 = \frac{r^2}{2\,r - \rho \sin\theta},$$

where s and ρ are the arc and the radius of curvature of C.

4. Find the intrinsic and parametric equations of a plane curve which is such that the segment on any tangent between the point of contact and the projection of a fixed point is of constant length.

5. Find the intrinsic equation of the plane curve which meets under constant angle all the lines passing through a fixed point.

6. The plane curve which is such that the locus of the mid-point of the segment of the normal between a point of the curve and the center of curvature is a straight line is the cycloid whose intrinsic equation is $\rho^2 + s^2 = a^2$.

7. Investigate the curve which is the locus of the point on the principal normal of a given curve and at constant distance from the latter.

18. Osculating sphere. Consider any curve whatever referred to its moving trihedral. The point whose coördinates have the values (91) lies on the normal to the osculating plane at the center of curvature, that is, on the polar line. Consequently the moving sphere whose center is at this point, and whose radius is $\sqrt{\rho^2+\tau^2\rho'^2}$, cuts the osculating plane in the osculating circle. This sphere is called the *osculating sphere* to the curve at the point. We shall derive the property of this sphere which accounts for its name.

When the tangent to a curve at a point M is tangent likewise to a sphere at this point, the center of the sphere lies in the normal plane to the curve at M. If R denotes its radius and the curve is referred to the trihedral at M, the coördinates of the center C of the sphere are of the form $(0, y_1, z_1)$ and $y_1^2+z_1^2=R^2$. Let $P(x, y, z)$ be a point of the curve near M, and Q the point in which the line CP cuts the sphere. If PQ be denoted by δ, we have, from (53),

$$(R+\delta)^2=\left(s-\frac{1}{6\rho^2}s^3+\cdots\right)^2+\left(y_1-\frac{1}{2\rho}s^2+\frac{1}{6\rho^2}\rho' s^3+\cdots\right)^2$$
$$+\left(z_1+\frac{1}{6\tau\rho}s^3+\cdots\right)^2,$$

which reduces to

$$2R\delta+\delta^2=s^2\left(1-\frac{y_1}{\rho}\right)+\frac{s^3}{3\rho}\left(\frac{\rho'}{\rho}y_1+\frac{z_1}{\tau}\right)+\cdots.$$

Hence δ is of the second order, in comparison with MP, unless $y_1=\rho$, that is, unless the center is on the polar line; then it is of the third order unless $z_1=-\rho'\tau$, in which case the sphere is the osculating sphere. Hence we have the theorem:

The osculating sphere to a curve at a point has contact with the curve of the third order; other spheres with their centers on the polar line, and tangent to the curve, have contact with the curve of the second order; all other spheres tangent to the curve at a point have contact of the first order.

The radius of the osculating sphere is given by

$$R^2=\rho^2+\tau^2\rho'^2, \tag{93}$$

and the coördinates of the center, referred to fixed axes in space, are

$$x_1=x+\rho l-\rho'\tau\lambda,\quad y_1=y+\rho m-\rho'\tau\mu,\quad z_1=z+\rho n-\rho'\tau\nu. \tag{94}$$

Hence when ρ is constant the centers of the osculating sphère and the osculating circle coincide. Then the radius of the sphere is necessarily constant. Conversely, it follows from the equation (93) that a necessary and sufficient condition that R be constant is

$$\rho'\tau\left[\frac{\rho}{\tau}+(\tau\rho')'\right]=0,$$

that is, either the curvature is constant, or the curve is spherical.

If equations (94) be differentiated with respect to s, we get

$$(95)\quad \begin{cases} x_1'=-\lambda\left[\dfrac{\rho}{\tau}+(\tau\rho')'\right], \\ y_1'=-\mu\left[\dfrac{\rho}{\tau}+(\tau\rho')'\right], \\ z_1'=-\nu\left[\dfrac{\rho}{\tau}+(\tau\rho')'\right]. \end{cases}$$

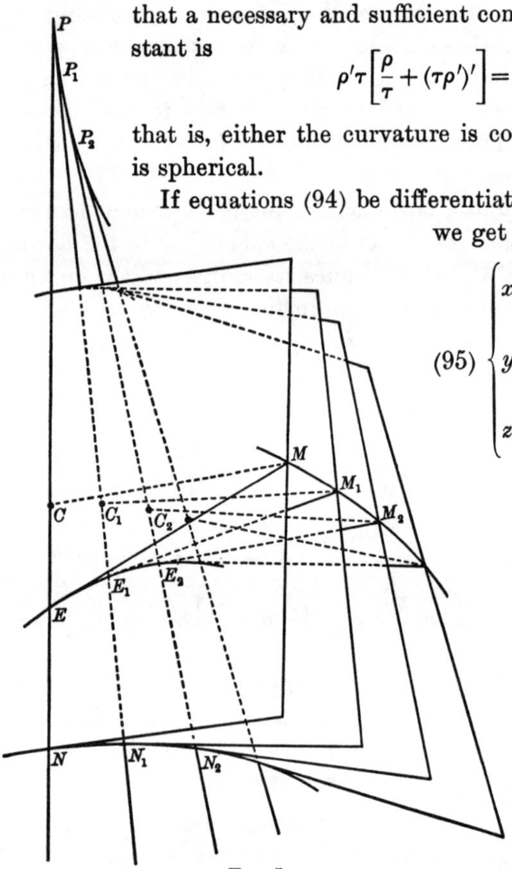

Fig. 5

From these expressions it is seen that the center of the osculating sphere is fixed only in case of spherical curves. Also, the tangent to the locus of the center is parallel to the binormal. Combining this result with a previous one, we have the theorem:

The polar line for a point on a curve is tangent to the locus of the center of the osculating sphere to the curve at the corresponding point.

This result is represented in fig. 5, in which the curve is the locus of the points M; the points $C, C_1, C_2, \cdots$ are the corresponding centers of curvature; the planes MCN, $M_1C_1N_1, \cdots$ are normal to the curve; the lines CP, $C_1P_1, \cdots$ are the polar lines; and the points $P, P_1, P_2, \cdots$ are the centers of the osculating spheres.

19. Bertrand curves. Bertrand proposed the following problem: *To determine the curves whose principal normals are the principal normals of another curve.* In solving this problem we make use of the moving trihedral. We must find the necessary and sufficient condition that the point $M_1(\xi = 0, \eta = k, \zeta = 0)$ generate a curve C_1 whose principal normal coincides with the η-axis of the moving trihedral. Since the point M_1 remains on the moving η-axis, we have $d\xi = d\zeta = 0$. And since M_1 tends to move at right angles to this axis, $\delta\eta = 0$. Now equations (82) reduce to

$$\text{(96)} \qquad \frac{\delta\xi}{ds} = 1 - \frac{k}{\rho}, \qquad \frac{dk}{ds} = 0, \qquad \frac{\delta\zeta}{ds} = -\frac{k}{\tau}.$$

From the second we see that k is a constant. Moreover, if ω denotes the angle which the tangent at M_1 makes with the tangent at M, we have, from the first and third of these equations,

$$\tan\omega = \frac{\delta\zeta}{\delta\xi} = \frac{k\rho}{\tau(k-\rho)},$$

or

$$\text{(97)} \qquad \frac{\sin\omega}{\rho} - \frac{\cos\omega}{\tau} = \frac{\sin\omega}{k}.$$

We have seen (§ 11) that according as τ is positive or negative, the osculating plane to a curve at a point M' near M cuts the polar line for M below or above the osculating plane at M. From these considerations it follows that when $\tau > 0$, ω is in the third, fourth, or first quadrants according as $k > \rho$, $0 < k < \rho$, or $k < 0$; and when $\tau < 0$, ω is in the second, first, or fourth quadrant, accordingly. It is readily found that these results are consistent with equation (97).

By means of (97) it is found from (96) that

$$\text{(98)} \qquad \frac{\delta s_1}{ds} = \left[\left(\frac{\delta\xi}{ds}\right)^2 + \left(\frac{\delta\zeta}{ds}\right)^2\right]^{\frac{1}{2}} = -\frac{k}{\tau\sin\omega},$$

the negative sign being taken so that the left-hand member may be positive.

Thus far we have expressed only the condition that the locus of M_1 cut the moving η-axis orthogonally, but not that this axis shall be the principal normal to the curve C_1 also. For this we consider the moving trihedral for C_1 and let a_1, b_1, c_1 denote the direction-cosines with respect to it of a fixed direction in space, as M_1D in fig. 6. They satisfy equations similar to (85), namely

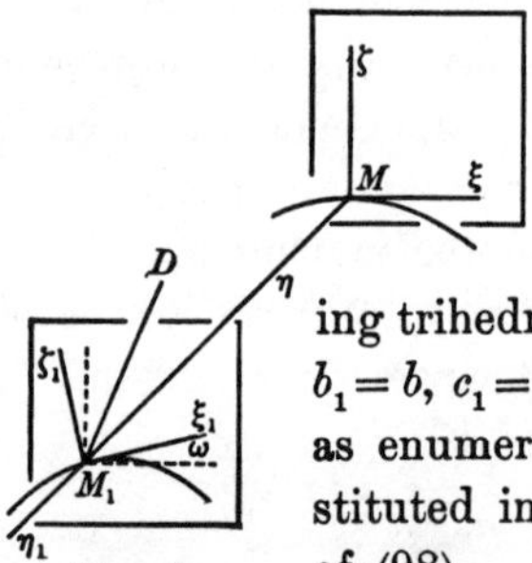

Fig. 6

$$\frac{da_1}{ds_1}=\frac{b_1}{\rho_1},\quad \frac{db_1}{ds_1}=-\left(\frac{a_1}{\rho_1}+\frac{c_1}{\tau_1}\right),\quad \frac{dc_1}{ds_1}=\frac{b_1}{\tau_1}. \tag{99}$$

If a, b, c are the direction-cosines of the same direction, with respect to the moving trihedral at M, we must have $a_1=a\cos\omega+c\sin\omega$, $b_1=b$, $c_1=-a\sin\omega+c\cos\omega$, for all possible cases, as enumerated above. When these values are substituted in the above equations, we get, by means of (98),

$$\left[\frac{\cos\omega}{\rho}+\frac{\sin\omega}{\tau}+\frac{k}{\rho_1\tau\sin\omega}\right]b+(c\cos\omega-a\sin\omega)\frac{d\omega}{ds}=0,$$

$$\left[\frac{\tau\sin\omega}{k\rho}+\frac{\cos\omega}{\rho_1}-\frac{\sin\omega}{\tau_1}\right]a+\left[\frac{\sin\omega}{k}+\frac{\sin\omega}{\rho_1}+\frac{\cos\omega}{\tau_1}\right]c=0,$$

$$\left[\frac{\sin\omega}{\rho}-\frac{\cos\omega}{\tau}-\frac{k}{\tau_1\tau\sin\omega}\right]b+(c\sin\omega+a\cos\omega)\frac{d\omega}{ds}=0.$$

Since these equations must be true for every fixed line, the coefficients of a, b, c in each of these equations must be zero. The resulting equations of condition reduce to

$$\begin{cases}\omega=\text{const.}, \qquad \tau\tau_1=\dfrac{k^2}{\sin^2\omega},\\[2mm] \dfrac{\sin\omega}{\rho_1}+\dfrac{\cos\omega}{\tau_1}+\dfrac{\sin\omega}{k}=0.\end{cases} \tag{100}$$

Since ω is a constant, equation (97) is a linear relation between the first and second curvatures of the curve C. And the last of equations (100) shows that a similar relation holds for the curve C_1.

Conversely, given a curve C whose first and second curvatures satisfy the relation

$$\frac{A}{\rho}+\frac{B}{\tau}=C, \tag{101}$$

where A, B, C are constants different from zero; if we take

$$k=\frac{A}{C}, \qquad \cot\omega=-\frac{B}{A},$$

and for ρ_1 and τ_1 the values given by (100), equations (99) are satisfied identically, and the point $(0, k, 0)$ on the principal normal generates the curve C_1, *conjugate* to C. We gather these results about the curves of Bertrand into the following theorem:

A necessary and sufficient condition that the principal normals of one curve be the principal normals of a second is that a linear relation exist between the first and second curvatures; the distance between corresponding points of the two curves is constant, the osculating planes at these points cut under constant angle, and the torsions of the two curves have the same sign.

We consider, finally, several particular cases, which we have excluded in the consideration of equation (101).

When $C = 0$ and $A \neq 0$, the ratio of ρ and τ is constant. Hence the curve is a helix and its conjugate is at infinity. When $A = 0$, that is, when the curve has constant torsion, the conjugate curve coincides with the original. When $A = C = 0$, k is indeterminate; hence plane curves admit of an infinity of conjugates, — they are the curves *parallel* to the given curve. The only other curve which has more than one conjugate is a circular helix, for since ρ and τ are constant, A/C can be given any value whatever; both the given helix and the circular helices conjugate to it are traced on circular cylinders with the same axis.

20. Tangent surface of a curve. For the further discussion of the properties of curves it is necessary to introduce certain curves and surfaces which can be associated with them. However, in considering these surfaces we limit our discussion to those properties which have to do with the associated curves, and leave other considerations to their proper places in later chapters.

The totality of all the points on the tangents to a twisted curve C constitute the *tangent surface of the curve*. As thus defined, the surface consists of an infinity of straight lines, which are called the *generators* of the surface. Any point P on this surface lies on one of these lines, and is determined by this line and the distance t from P to the point M where the line touches the curve, as is shown in fig. 7. If the coördinates x, y, z of M are expressed in terms of the arc s, the coördinates of P are given by

$$\xi = x + x't, \quad \eta = y + y't, \quad \zeta = z + z't, \tag{102}$$

where the accents denote differentiation with respect to s. When the equations of the curve have the general form

$$x = f_1(u), \quad y = f_2(u), \quad z = f_3(u),$$

the coördinates of P can be expressed thus:

$$(103) \quad \xi = f_1(u) + vf_1'(u), \quad \eta = f_2(u) + vf_2'(u), \quad \zeta = f_3(u) + vf_3'(u),$$

where
$$v = \frac{t}{[f_1'^2 + f_2'^2 + f_3'^2]^{\frac{1}{2}}}.$$

From this it is seen that v is equal to the distance MP only when s is the parameter.

As given by equations (102) or (103), the coördinates of a point on the tangent surface are functions of two parameters. A relation between these parameters, such as

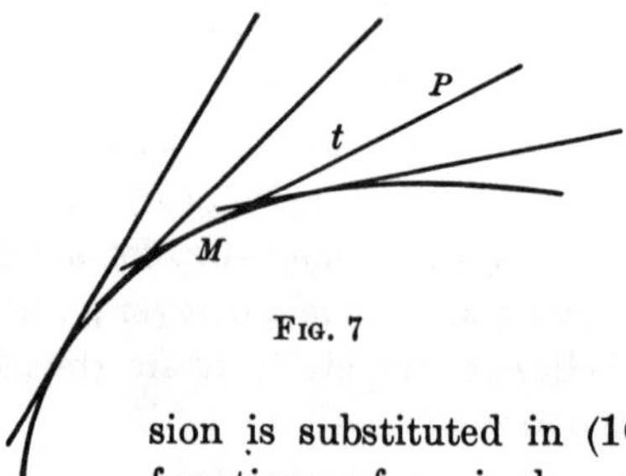

FIG. 7

$$(104) \qquad f(s, t) = 0,$$

defines a curve which lies upon the surface. For, when this equation is solved for t in terms of s and the resulting expression is substituted in (102), the coördinates ξ, η, ζ are functions of a single parameter, and consequently the locus of the point (ξ, η, ζ) is a curve (§ 1).

By definition, the element of arc of this curve is given by $d\sigma^2 = d\xi^2 + d\eta^2 + d\zeta^2$. This is expressible by means of (102) and (41) in the form

$$(105) \qquad d\sigma^2 = \left(1 + \frac{t^2}{\rho^2}\right) ds^2 + 2\, ds\, dt + dt^2,$$

where t is supposed to be the expression in s obtained from (104), and ρ is the radius of curvature of the curve C, of which the surface is the tangent surface. This result is true whatever be the relation (104). Hence equation (105) gives the element of length of any curve on the surface, and $d\sigma$ is called the *linear element of the surface.*

According as t in equations (102) has a positive or negative value, the point lies on the portion of the tangent drawn in the positive direction from the curve or in the opposite direction. It is now our purpose to get an idea of the form of the surface in the neighborhood of the curve.

In consequence of (53) equations (102) can be written

$$\xi=\left(s-\frac{1}{6\rho^2}s^3+\cdots\right)+\left(1-\frac{1}{2\rho^2}s^2+\cdots\right)t,$$

$$\eta=\left(\frac{s^2}{2\rho}-\frac{1}{6}\frac{\rho'}{\rho^2}s^3+\cdots\right)+\left(\frac{s}{\rho}-\frac{1}{2}\frac{\rho'}{\rho^2}s^2+\cdots\right)t,$$

$$\zeta=\left(-\frac{1}{6\rho\tau}s^3+\cdots\right)+\left(-\frac{1}{2\rho\tau}s^2+\cdots\right)t.$$

The plane $\xi=0$ cuts the surface in a curve Γ. The point M_0 of C, at which $s=0$, is also a point of Γ. From the above expression for ξ it is seen that for points of Γ near M_0 the parameters s and t differ only in sign. Hence, neglecting powers of s and t of higher orders, the equations of Γ in the neighborhood of M_0 are

$$\xi=0,\quad \eta=-\frac{t^2}{2\rho},\quad \zeta=-\frac{1}{3\rho\tau}t^3.$$

By eliminating t from the last two equations, we find that in the neighborhood of M_0 the curve Γ has the form of a semicubical parabola with the η-axis, that is the principal normal to C, for cuspidal tangent. Since any point of the curve C can be taken for M_0, we have the theorem:

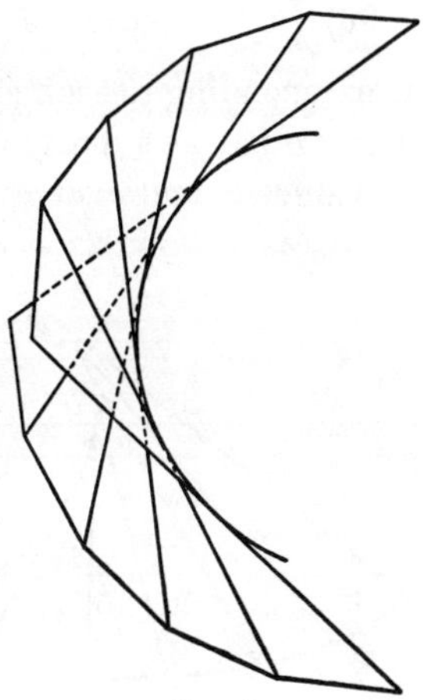

FIG. 8

The tangent surface of a curve consists of two sheets, corresponding respectively to positive and negative values of t, *which are tangent to one another along the curve, and thus form a sharp edge.*

On this account the curve is called the *edge of regression* of the surface. An idea of the form of the surface may be had from fig. 8.

21. Involutes and evolutes of a curve. When the tangents of a curve C are normal to a curve C_1, the latter is called an *involute* of C, and C is called an *evolute* of C_1. As thus defined, the involutes of a twisted curve lie upon its tangent surface, and those of a plane curve in its plane. The latter is only a particular case of the former, so that the problem of finding the involutes of a curve is that of finding the curves upon the tangent surface which cut the generators orthogonally.

We write the equations of the tangent surface in the form

$$\xi = x + \alpha t, \quad \eta = y + \beta t, \quad \zeta = z + \gamma t.$$

Assuming that s is the parameter of the curve, the problem reduces to the determination of a relation between t and s such that

$$\alpha\, d\xi + \beta\, d\eta + \gamma\, d\zeta = 0.$$

By means of (50) this reduces to $dt + ds = 0$, so that $t = c - s$, where c is an arbitrary constant. Hence the coördinates x_1, y_1, z_1 of an involute are expressible in the form

$$(106) \qquad x_1 = x + \alpha(c - s), \quad y_1 = y + \beta(c - s), \quad z_1 = z + \gamma(c - s).$$

Corresponding to each value of c there is an involute; consequently a curve has an infinity of involutes. If two involutes correspond to values c_1 and c_2 of c, the segment of each tangent between the curves is of length $c_1 - c_2$. Hence the involutes are said to form a system of *parallel* curves on the tangent surface.

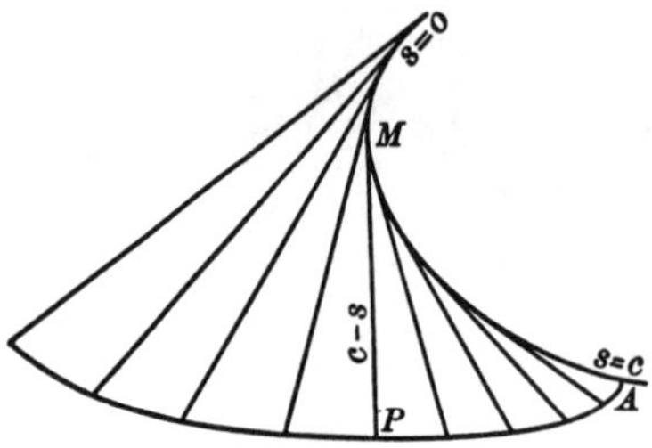

Fig. 9

When s is known the involutes are given directly by equations (106). Hence the complete determination of the involutes of a given curve requires one quadrature at most.

From the definition of t and its above value, an involute can be generated mechanically in the following manner, as represented in fig. 9. Take a string of length c and bring it into coincidence with the curve, with one end at the point $s = 0$; call the other end A. If the former point be fixed and the string be unwound gradually from the curve beginning at A, this point will trace out an involute on the tangent surface.

By differentiating equations (106), we get

$$dx_1 = \frac{l(c-s)}{\rho}\, ds, \quad dy_1 = \frac{m(c-s)}{\rho}\, ds, \quad dz_1 = \frac{n(c-s)}{\rho}\, ds.$$

Hence the tangent to an involute is parallel to the principal normal of the curve at the corresponding point, and consequently the tangents at these points are perpendicular to one another.

As an example of the foregoing theory, we determine the involutes of the circular helix, whose equations are

$$x = a \cos u, \quad y = a \sin u, \quad z = au \cot \theta,$$

where a is the radius of the cylinder and θ the constant angle which the tangent to the curve makes with axis of the cylinder. Now

$$s = a \operatorname{cosec} \theta \cdot u, \qquad \alpha, \beta, \gamma = \frac{-\sin u, \cos u, \cot \theta}{\operatorname{cosec} \theta}.$$

Hence the equations of the involutes are

$$x_1 = a \cos u + (au - c \sin \theta) \sin u, \quad y_1 = a \sin u - (au - c \sin \theta) \cos u, \quad z_1 = c \cos \theta.$$

From the last of these equations it follows that the involutes are plane curves whose planes are normal to the axis of the cylinder, and from the expressions for x_1 and y_1 it is seen that these curves are the involutes of the circular sections of the cylinder.

We proceed to the inverse problem:

Given a curve C, to find its evolutes.

The problem reduces to the determination of a succession of normals to C which are tangent to a curve C_0. If M_0 be the point on C_0 corresponding to M on C, it lies in the normal plane to C at M, and consequently its coördinates are of the form

$$x_0 = x + pl + q\lambda, \quad y_0 = y + pm + q\mu, \quad z_0 = z + pn + q\nu,$$

where p and q are the distances from M_0 to the binormal and principal normal respectively. These quantities p and q must be such that the line M_0M is tangent to the locus of M_0 at this point, that is, we must have

$$\frac{dx_0}{ds} = \kappa (x - x_0), \quad \frac{dy_0}{ds} = \kappa (y - y_0), \quad \frac{dz_0}{ds} = \kappa (z - z_0),$$

where κ denotes a factor of proportionality. When the above values are substituted in these equations, we get

$$\alpha\left(1 - \frac{p}{\rho}\right) + l\left(\frac{dp}{ds} + \frac{q}{\tau} + p\kappa\right) + \lambda\left(\frac{dq}{ds} - \frac{p}{\tau} + q\kappa\right) = 0,$$

and two other equations obtained by replacing α, l, λ by β, m, μ and γ, n, ν. Hence the expressions in parentheses vanish. From the first it follows that p is equal to ρ; consequently M_0 lies on the polar line of C at M. The other equations of condition can be written

$$\frac{d\rho}{ds}+\frac{q}{\tau}+\rho\kappa=0, \qquad \frac{dq}{ds}-\frac{\rho}{\tau}+q\kappa=0.$$

Eliminating κ, we get

$$\frac{ds}{\tau}=\frac{\rho\,dq-q\,d\rho}{\rho^2+q^2}=d\tan^{-1}\frac{q}{\rho}.$$

For the sake of convenience we put $\omega=\int\frac{ds}{\tau}$, and obtain by integration

$$\frac{q}{\rho}=\tan(\omega+c),$$

where c is the constant of integration. As c is arbitrary, there is an infinity of evolutes of the curve C; they are defined by the following equations, in which c is constant for an evolute but changes with it:

$$x_0=x+l\rho+\lambda\rho\tan(\omega+c), \qquad y_0=y+m\rho+\mu\rho\tan(\omega+c),$$
$$z_0=z+n\rho+\nu\rho\tan(\omega+c).$$

From the definition of q it follows that q/ρ is equal to the tangent of the angle which MM_0 makes with the principal normal to C at M. Calling this angle θ, we have $\theta=\omega+c$. The foregoing results give the following theorem:

A curve C admits of an infinity of evolutes; when each of the normals to C, which are tangent to one of its evolutes, is turned through the same angle in the corresponding normal plane to C, these new normals are tangent to another evolute of C.

In fig. 5 the locus of the points E is an evolute of the given curve.

Each system of normals to C which are tangent to an evolute C_0 constitute a tangent surface of which C_0 is the edge of regression. Hence the evolutes of C are the edges of regression of an infinity of tangent surfaces, all of which pass through C.

From the definition of ω it follows that ω is constant only when the curve C is plane. In this case we may take ω equal to zero. Then when $c=0$ we have the evolute C_0 in the plane of the curve. The other evolutes lie upon the right cylinder formed of the normals to the plane at points of C_0, and cut the elements of the cylinder under the constant angle $90° - c$, and consequently are helices. Hence we have the theorem:

The evolutes of a plane curve are the helices traced on the right cylinder whose base is the plane evolute. Conversely, *every cylindrical helix is the evolute of an infinity of plane curves.*

EXAMPLES

1. Find the coördinates of the center of the osculating sphere of the twisted cubic.

2. The angle between the radius of the osculating sphere for any curve and the locus of the center of the sphere is equal to the angle between the radius of the osculating circle and the locus of the center of curvature.

3. The locus of the center of curvature of a curve is an evolute only when the curve is plane.

4. Find the radii of first and second curvature of the curve $x = a \sin u \cos u$, $y = a \cos^2 u$, $z = a \sin u$. Show that the curve is spherical, and give a geometrical construction. Find its evolutes.

5. Derive the properties of Bertrand curves (§ 19) without the use of the moving trihedral.

6. Find the involutes and evolutes of the twisted cubic.

7. Determine whether there is a curve whose binormals are the binormals of a second curve.

8. Derive the results of § 21 by means of the moving trihedral.

22. Minimal curves. In the preceding discussion we have made exception of the curves, defined by

$$x = f_1(u), \qquad y = f_2(u), \qquad z = f_3(u),$$

when these functions satisfy the condition

$$f_1'^2 + f_2'^2 + f_3'^2 = 0. \tag{107}$$

As these imaginary curves are of interest in certain parts of the theory of surfaces, we devote this closing section to their discussion.

The equation of condition may be written in the form

$$\frac{f_1' + if_2'}{-f_3'} = \frac{f_3'}{f_1' - if_2'} = v,$$

where v is a constant or a function of u. These equations are equivalent to the following:

$$f_1' : f_2' : f_3' = \frac{1 - v^2}{2} : i\frac{(1 + v^2)}{2} : v. \tag{108}$$

At most, the common ratio is a function of u, say $f(u)$. And so if we disregard additive constants of integration, as they can be removed by a translation of the curve in space, we can replace the above equations by

$$(109)\qquad x=\int\frac{1-v^2}{2}f(u)du,\quad y=i\int\frac{1+v^2}{2}f(u)du,\quad z=\int vf(u)du.$$

We consider first the case when v is constant and call it a. If we change the parameter of the curve by replacing $\int f(u)\,du$ by a new parameter which we call u, we have, without loss of generality,

$$(110)\qquad x=\frac{1-a^2}{2}u,\qquad y=i\frac{1+a^2}{2}u,\qquad z=au.$$

For each value of a these are the equations of an imaginary straight line through the origin. Eliminating a, we find that the envelope of these lines is the imaginary cone, with vertex at the origin, whose equation is

$$(111)\qquad x^2+y^2+z^2=0.$$

Every point on the cone is at zero distance from the vertex, and from the equations of the lines it is seen that the distance between any two points on a line is zero. We call these generators of the cone *minimal straight lines*. Through any point in space there are an infinity of them; their direction-cosines are proportional to

$$\frac{1-a^2}{2},\qquad i\frac{1+a^2}{2},\qquad a,$$

where a is arbitrary. The locus of these lines is the cone whose vertex is at the point and whose generators pass through the circle at infinity. For, the equation in homogeneous coördinates of the sphere of unit radius and center at the origin is $x^2+y^2+z^2=w^2$, so that the equations of the circle at infinity are

$$x^2+y^2+z^2=0,\qquad w=0.$$

Hence the cone (111) passes through the circle at infinity.

We consider now the case where v in equations (109) is a function of u. If we take this function of u for a new parameter, and for convenience call it u, equations (109) may be written in the form

$$(112)\qquad x=\int\frac{1-u^2}{2}F(u)du,\quad y=i\int\frac{1+u^2}{2}F(u)\,du,\quad z=\int uF(u)du,$$

where, as is seen from (108), $F(u)$ can be any function of u different from zero.

If we replace $F(u)$ by the third derivative of a function $f(u)$, thus $F(u) = f'''(u)$, equations (112) can be integrated by parts and put in the form

$$(113) \qquad \begin{cases} x = \frac{1}{2}(1-u^2)f''(u) + uf'(u) - f(u), \\ y = \frac{i}{2}(1+u^2)f''(u) - iuf'(u) + if(u), \\ z = uf''(u) - f'(u). \end{cases}$$

Since F must be different from zero, $f(u)$ can have any form other than $c_1u^2 + c_2u + c_3$, where c_1, c_2, c_3 are arbitrary constants.

EXAMPLES

1. Show that the tangents to a minimal curve are minimal lines, and that a curve whose tangents are minimal lines is minimal.

2. Show that the osculating plane of a minimal curve can be written $(X-x)A + (Y-y)B + (Z-z)C = 0$, where $A^2 + B^2 + C^2 = 0$. A plane whose equation is of this sort is called an *isotropic plane.*

3. Show that through each point of a plane two minimal straight lines pass which lie in the latter.

4. Determine the order of the minimal curves for which the function f in (113) satisfies the condition $4f'''f^{\text{v}} - 5f^{\text{iv}2} = 0$.

5. Show that the equations of a minimal curve, for which f in (113) satisfies the condition $4f'''f^{\text{v}} - 5f^{\text{iv}2} = af'''^3$, where a is a constant, can be put in the form

$$x = \frac{8}{a}\cos t, \quad y = \frac{8}{a}\sin t, \quad z = \frac{8i}{a}t.$$

GENERAL EXAMPLES

1. Show that the equations of any plane curve can be put in the form

$$x = \int \cos\phi f(\phi)\,d\phi, \qquad y = \int \sin\phi f(\phi)\,d\phi,$$

and determine the geometrical significance of ϕ.

2. Prove that the necessary and sufficient condition that the parameter u in the equations $x = f_1(u)$, $y = f_2(u)$ have the significance of ϕ in Ex. 1 is

$$f_1'f_2'' - f_1''f_2' = f_1'^2 + f_2'^2.$$

3. Prove that the general projective transformation transforms an osculating plane of a curve into an osculating plane of the transform.

4. The principal normal to a curve is normal to the locus of the centers of curvature at the points where ρ is a maximum or minimum.

5. A certain plane curve possesses the property that if C be its center of curvature for a point P, Q the projection of P on the x-axis, and T the point where the tangent at P meets this axis, the area of the triangle CQT is constant. Find the equations of the curve in terms of the angle which the tangent forms with the x-axis.

6. The binormal at a point M of a curve is the limiting position of the common perpendicular to the tangents at M and M', as M' approaches M.

7. The tangents to the spherical indicatrices of the tangent and binormal of a twisted curve at corresponding points are parallel.

8. Any curve upon the unit sphere serves for the spherical indicatrix of the binormal of a curve of constant torsion. Find the coördinates of the curve.

9. The equations

$$x = a\int \frac{l dk - k dl}{h^2 + k^2 + l^2}, \qquad y = a\int \frac{h dl - l dh}{h^2 + k^2 + l^2}, \qquad z = a\int \frac{k dh - h dk}{h^2 + k^2 + l^2},$$

where a is constant and h, k, l are functions of a single parameter, define a curve whose radius of torsion is a.

10. If, in Ex. 9, we have

$$h = \cos\mu\theta - \sqrt{\frac{\mu}{\lambda}}\cos\lambda\theta, \quad k = \sin\mu\theta + \sqrt{\frac{\mu}{\lambda}}\sin\lambda\theta, \quad l = 2\left(\frac{\mu}{\lambda}\right)^{\frac{1}{2}}\cos\frac{\lambda+\mu}{2}\theta,$$

where λ and μ are constants whose ratio is commensurable, the integrands are expressible as linear homogeneous functions of sines and cosines of multiples of θ, and consequently the curve is algebraic.

11. Equations (1) define a family of circles, if a, b, r are functions of a parameter t. Show that the determination of their orthogonal trajectories requires the solution of the Riccati equation,

$$\frac{d\theta}{dt} = \frac{1}{r}\frac{da}{dt}\theta - \frac{1}{2r}\frac{db}{dt}(1 - \theta^2),$$

where $\theta = \tan u/2$.

12. Find the vector representing the rate of change of the acceleration of a moving point.

13. When a curve is spherical, the center of curvature for the point is the foot of the perpendicular upon the osculating plane at the point from the center of the sphere.

14. The radii of first and second curvature of a curve which lies upon a sphere and cuts the meridians under constant angle are in the relation $1 + a\tau + b\rho^2\tau = 0$, where a and b are constants.

15. An *epitrochoidal curve* is generated by a point in the plane of a circle which rolls, without slipping, on another circle, whose plane meets the plane of the first circle under constant angle. Find its equations and show that it is a spherical curve.

16. If two curves are in a one-to-one correspondence with the tangents at corresponding points parallel, the principal normals at these points are parallel and likewise the binormals; two curves so related are said to be deducible from one another by a *transformation of Combescure.*

17. If two curves are in a one-to-one correspondence and the osculating planes at corresponding points are parallel, either curve can be obtained from the other by a transformation of Combescure.

18. Show that the radius of the osculating sphere of a curve is given by $R^2 = \tau^2\rho^4[x'''^2 + y'''^2 + z'''^2] - \tau^2$, where the prime denotes differentiation with respect to the arc.

19. At corresponding points of a twisted curve and the locus of the center of its osculating sphere the principal normals are parallel, and the tangent to one curve is parallel to the binormal to the other; also the product of the radii of torsion of the two curves is equal to the product of the radii of first curvature, or to within the sign, according as the positive directions of the principal normals are different or the same.

20. Determine the twisted curves which are such that the centers of the spheres osculating the curve of centers of the osculating spheres of the given curve are points of the latter.

21. Show that the binormals to a curve do not constitute the tangent surface of another curve.

22. Determine the directions of the principal normal and binormal to an involute of a given curve.

23. Show that the equations

$$x = a\int\phi(u)\sin u\,du, \qquad y = a\int\phi(u)\cos u\,du, \qquad z = a\int\phi(u)\psi(u)\,du,$$

where $\phi(u) = (1+\psi^2+\psi'^2)^{\frac{1}{2}}(1+\psi^2)^{-\frac{3}{2}}$ and $\psi(u)$ is any function whatever, define a curve of constant curvature.

24. Find the curve for which $\psi(u) = \tan u$, in example 23.

25. Prove that in order that the principal normals of a curve be the binormals of another, the relation $a\left(\frac{1}{\rho^2}+\frac{1}{\tau^2}\right) = \frac{b}{\rho}$ must hold, where a and b are constants. Show that such curves are defined by equations of example 23 when

$$\phi = \frac{(1+\psi^2+\psi'^2)^3 + (1+\psi^2)^3(\psi''+\psi)^2}{(1+\psi^2)^{\frac{3}{2}}(1+\psi^2+\psi'^2)^{\frac{5}{2}}}.$$

26. Let λ_1, μ_1, ν_1 be the coördinates of a point on the unit sphere expressed as functions of the arc σ_1 of the curve. Show that the equations

$$x = ek\int\lambda_1 d\sigma_1 - k\cot\omega\int(\mu_1\nu_1' - \mu_1'\nu_1)\,d\sigma_1,$$

$$y = ek\int\mu_1 d\sigma_1 - k\cot\omega\int(\nu_1\lambda_1' - \nu_1'\lambda_1)\,d\sigma_1,$$

$$z = ek\int\nu_1 d\sigma_1 - k\cot\omega\int(\lambda_1\mu_1' - \lambda_1'\mu_1)\,d\sigma_1,$$

where k and ω are constant, $e = \pm 1$, and the primes indicate differentiation with respect to σ_1, define a Bertrand curve for which ρ and τ satisfy the relation (97); show also that λ_1, μ_1, ν_1 are the direction-cosines of the binormal to the conjugate curve.

CHAPTER II

CURVILINEAR COÖRDINATES ON A SURFACE. ENVELOPES

23. Parametric equations of a surface. In the preceding chapter we have seen that the coördinates of a point on the tangent surface of a curve are expressible in the form

$$(1)\quad x=f_1(u)+vf_1'(u),\quad y=f_2(u)+vf_2'(u),\quad z=f_3(u)+vf_3'(u),$$

where $$\xi=f_1(u),\qquad \eta=f_2(u),\qquad \zeta=f_3(u),$$

are the equations of the curve, and v is proportional to the distance between the points (ξ, η, ζ), (x, y, z) on the same generator. Since the coördinates of the surface are expressed by (1) as functions of two independent *parameters* u, v, the *equations of the surface* may be written

$$(2)\qquad \begin{cases} x=f_1(u, v),\\ y=f_2(u, v),\\ z=f_3(u, v).\end{cases}$$

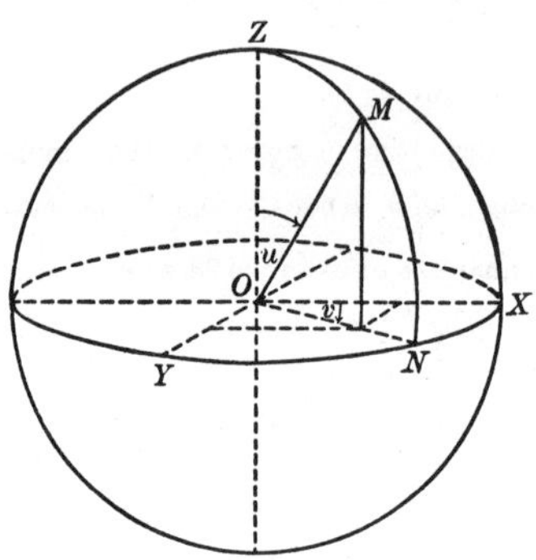

FIG. 10

Consider also a sphere of radius a whose center is at the origin O (fig. 10). If v denotes the angle, measured in the positive sense, which the plane through the z-axis and a point M of the sphere makes with the xz-plane, and u denotes the angle between the radius OM and the positive z-axis, the coördinates of M may be written

$$(3)\qquad x=a\sin u\cos v,\qquad y=a\sin u\sin v,\qquad z=a\cos u.$$

Here, again, the coördinates of any point on the sphere are expressible as functions of two parameters, and the *equations of the sphere* are of the form (2)*.

* Notice that in this case f_3 is a function of u alone.

In the two preceding cases the functions f_1, f_2, f_3 have particular forms. We consider the general case where f_1, f_2, f_3 are any functions of two independent parameters u, v, analytic for all values of u and v, or at least for values within a certain domain. The locus of the point whose coördinates are given by (2) for all values of u and v in the domain is called a *surface*. And equations (2) are called *parametric equations of the surface.*

It is to be understood that one or more of the functions f may involve a single parameter. For instance, any cylinder may be defined by equations of the form

$$x = f_1(u), \qquad y = f_2(u), \qquad z = f_3(u, v).$$

If we replace u and v in (2) by independent functions of two other parameters u_1, v_1, thus

$$(4) \qquad u = F_1(u_1, v_1), \qquad v = F_2(u_1, v_1),$$

the resulting equations may be written

$$(5) \qquad x = \phi_1(u_1, v_1), \qquad y = \phi_2(u_1, v_1), \qquad z = \phi_3(u_1, v_1).$$

If particular values of u_1 and v_1 be substituted in (4) and the resulting values of u and v be substituted in (2), we obtain the values of x, y, z given by (5), when u_1 and v_1 have been given the particular values. Hence equations (2) and (5) define the same surface, provided that F_1 and F_2 are of such a form that ϕ_1, ϕ_2, ϕ_3 satisfy the general conditions imposed upon the F's. Hence the equations of a surface may be expressed in parametric form in the number of ways of the generality of two arbitrary functions.

Suppose the first two of equations (2) solved for u and v in terms of x and y, and let $u = F_1(x, y)$, $v = F_2(x, y)$ be a set of solutions. When these equations are taken as equations (4), equations (5) become

$$x = x, \qquad y = y, \qquad z = f(x, y),$$

which may be replaced by the single relation,

$$(6) \qquad z = f(x, y).$$

If there is only one set of solutions of the first two of equations (2), equation (6) defines the surface as completely as (2). If, however, there are n sets of solutions, the surface would be defined by n equations, $z = f_i(x, y)$.

It may be said that equation (6) is obtained from equations (2) by eliminating u and v. This is a particular form of elimination, the more general giving an implicit relation between x, y, z, as

$$F(x, y, z) = 0. \tag{7}$$

If we have a locus of points whose coördinates satisfy a relation of the form (6), it is a surface in the above sense. For, if we take x and y equal to any analytic functions of u and v, namely f_1 and f_2, and substitute in (6), we obtain $z = f_3(u, v)$.

In like manner equation (7) may be solved for z, and one or more equations of the form (6) obtained, unless z does not appear in (7). In the latter case there is a relation between x and y alone, so that the surface is a cylinder whose elements are parallel to the z-axis, and its parametric equations are of the form

$$x = f_1(u), \qquad y = f_2(u), \qquad z = f_3(u, v).$$

Hence a surface can be defined analytically by equations (2), (6), or (7). Of these forms the last is the oldest. It was used exclusively until the time of Monge, who proposed the form (6); the latter has the advantage that many of the equations, which define properties of the surface, are simpler in form than when equation (7) is used. The parametric method of definition is due to Gauss. In many respects it is superior to both of the other methods. It will be used almost entirely in the following treatment.

24. Parametric curves. When the parameter u in equations (2) is put equal to a constant, the resulting equations define a curve on the surface for which v is the parameter. If we let u vary continuously, we get a continuous array of curves whose totality constitutes the surface. Hence a surface may be considered as generated by the motion of a curve. Thus the tangent surface of a curve is described by the tangent as the point of contact moves along the curve; and a sphere results from the revolution of a circle about a diameter.

We have just seen that upon a surface (2) there lie an infinity of curves whose equations are given by equations (2), when u is constant, each constant value of u determining a curve. We call them the curves $u = const.$ on the surface. In a similar way, there is an infinite family of curves $v = const.$* The curves of these two families are called the *parametric curves* for the given equations of the surface, and u and v are the *curvilinear coördinates* of a point upon the surface.† We say that the *positive direction* of a parametric curve is that in which the parameter increases.

If we replace v in equations (2) by a function of u, say

$$v = \phi(u), \tag{8}$$

the coördinates x, y, z are functions of a single parameter u, and consequently the locus of the point (x, y, z) is a curve. Hence equation (8) defines a curve on the surface (2). For example, the equation $v = au$ defines a helix on the cylinder

$$x = a \cos u, \qquad y = a \sin u, \qquad z = v.$$

Frequently equation (8) is written in the implicit form,

$$F(u, v) = 0. \tag{9}$$

Conversely, any curve upon the surface is defined by an equation of this form. For, if t be the parameter of the curve, both u and v in equations (2) are functions of t; thus $u = \phi_1(t)$, $v = \phi_2(t)$. Eliminating t between these equations, we get a relation such as (9).

We return to the consideration of the change of parameters, defined by equations (4). To a pair of values of u_1 and v_1 there correspond unique values of u and v. On the contrary, it may happen that another pair of values of u_1 and v_1 give the same values of u and v. But the values of x, y, z given by (5) will be the same in both cases; this follows from the manner in which these equations were derived. On this account when equations (4) are solved for u_1 and v_1 in terms of u and v, and there is more than one set of solutions, we must specify which solution will be used. We write the solution

$$u_1 = \Phi_1(u, v), \qquad v_1 = \Phi_2(u, v). \tag{10}$$

In terms of the original parameters, the parametric lines $u_1 = \text{const.}$ and $v_1 = \text{const.}$ have the equations,

$$\Phi_1(u, v) = a, \qquad \Phi_2(u, v) = b,$$

* On the sphere defined by equations (3) the curves $v = \text{const.}$ are meridians and $u = \text{const.}$ parallels.

† When a plane is referred to rectangular coördinates, the parametric lines are the two families of straight lines parallel to the coördinate axes.

where a and b denote constants. Unless u or v is absent from either of these equations the curves are necessarily distinct from the parametric curves $u = \text{const.}$ and $v = \text{const.}$ Suppose, now, that v does not appear in Φ_1; then u_1 is constant when u is constant, and vice versa. Consequently a curve $u_1 = \text{const.}$ is a member of the family of curves $u = \text{const.}$ Hence, when a transformation of parameters is made by means of equations of the form

$$u_1 = \Phi_1(u), \qquad v_1 = \Phi_2(v),$$

or

$$u_1 = \Phi_1(v), \qquad v_1 = \Phi_2(u),$$

the two systems of parametric curves are the same, the difference being in the value of the parameter which is constant along a curve.

EXAMPLES

1. A surface which is the locus of a family of straight lines, which meet another straight line orthogonally and are arranged according to a given law, is called a *right conoid*; its equations are of the form $x = u \cos v$, $y = u \sin v$, $z = \phi(v)$. Show that when $\phi(v) = a \cot v + b$ the conoid is a hyperbolic paraboloid.

2. Find the equations of the right conoid whose axis is the axis of z, and which passes through the ellipse $x = a$, $\dfrac{y^2}{b^2} + \dfrac{z^2}{c^2} = 1$.

3. When a sphere of radius a is defined by (3), find the relation between u and v along the curve of intersection of the sphere and the surface $x^4 + y^4 + z^4 = \frac{1}{2} a^4$. Show that the curves of intersection are four great circles.

4. Upon the surface $x = \sqrt{u^2 + \frac{1}{2}} \cos v$, $y = \sqrt{u^2 + \frac{1}{2}} \sin v$, $z = u$, determine the curves whose tangents make with the z-axis the angle $\tan^{-1}\sqrt{2}$. Show that two of these curves pass through every point, and find their radii of first and second curvature.

25. Tangent plane. A tangent line to a curve upon a surface is called a *tangent line to the surface* at the point of contact. It is evident that there are an infinity of tangent lines to a surface at a point. We shall show that all of these lines lie in a plane, which is called the *tangent plane* to the surface at the point.

To this end we consider a curve C upon a surface and let $M(x, y, z)$ be the point at which the tangent is drawn. The equations of the tangent are (§ 4)

$$\frac{\xi - x}{\dfrac{dx}{ds}} = \frac{\eta - y}{\dfrac{dy}{ds}} = \frac{\zeta - z}{\dfrac{dz}{ds}} = \lambda,$$

where ξ, η, ζ are the coördinates of a point on the line, depending for their values upon the parameter λ. If the equation in curvilinear coördinates of the curve C is $v = \phi(u)$, the above equations may be written

$$\xi - x = \lambda\left(\frac{\partial x}{\partial u} + \phi'\frac{\partial x}{\partial v}\right)\frac{du}{ds},$$

$$\eta - y = \lambda\left(\frac{\partial y}{\partial u} + \phi'\frac{\partial y}{\partial v}\right)\frac{du}{ds},$$

$$\zeta - z = \lambda\left(\frac{\partial z}{\partial u} + \phi'\frac{\partial z}{\partial v}\right)\frac{du}{ds},$$

where the prime indicates differentiation. In order to obtain the locus of these tangent lines, we eliminate ϕ' and λ from these equations. This gives

$$\text{(11)} \qquad \begin{vmatrix} \xi - x & \eta - y & \zeta - z \\ \dfrac{\partial x}{\partial u} & \dfrac{\partial y}{\partial u} & \dfrac{\partial z}{\partial u} \\ \dfrac{\partial x}{\partial v} & \dfrac{\partial y}{\partial v} & \dfrac{\partial z}{\partial v} \end{vmatrix} = 0,$$

which evidently is the equation of a plane through the point M. The normal to this plane at the point of contact is called the *normal* to the surface at the point.

As an example, we find the equation of the tangent plane to the tangent surface of a curve at any point. If the values from (1) be substituted in equation (11), the resulting equation is reducible to

$$\text{(12)} \qquad \begin{vmatrix} \xi - f_1 & \eta - f_2 & \zeta - f_3 \\ f_1' & f_2' & f_3' \\ f_1'' & f_2'' & f_3'' \end{vmatrix} = 0.$$

Hence the equation of the tangent plane is independent of v, and depends only upon u. In consequence of (I, 36)* we have the theorem:

The tangent plane to the tangent surface of a curve is the same at all points of a generator; it is the osculating plane of the curve at the point where the generator touches the curve.

When the surface is defined by an equation of the form $F(x, y, z) = 0$, we imagine that x, y, z are functions of u and v, and differentiate with respect to the latter. This gives

$$\text{(13)} \qquad \frac{\partial F}{\partial x}\frac{\partial x}{\partial u} + \frac{\partial F}{\partial y}\frac{\partial y}{\partial u} + \frac{\partial F}{\partial z}\frac{\partial z}{\partial u} = 0, \qquad \frac{\partial F}{\partial x}\frac{\partial x}{\partial v} + \frac{\partial F}{\partial y}\frac{\partial y}{\partial v} + \frac{\partial F}{\partial z}\frac{\partial z}{\partial v} = 0.$$

* In references of this sort the Roman numerals refer to the chapter.

By means of these equations the equation (11) of the tangent plane can be given the form

$$(\xi - x)\frac{\partial F}{\partial x} + (\eta - y)\frac{\partial F}{\partial y} + (\zeta - z)\frac{\partial F}{\partial z} = 0.$$

When the Monge form of the equation of a surface, namely $z = f(x, y)$, is used, it is customary to put

$$\frac{\partial z}{\partial x} = p, \qquad \frac{\partial z}{\partial y} = q. \tag{14}$$

Consequently the equation of the tangent plane is

$$(\xi - x)p + (\eta - y)q - (\zeta - z) = 0. \tag{15}$$

In the first chapter we found that a curve is defined by two equations of the form

$$F_1(x, y, z) = 0, \qquad F_2(x, y, z) = 0. \tag{16}$$

Hence a curve is the locus of the points common to two surfaces. The equations of the tangent to the curve are

$$\frac{\xi - x}{dx} = \frac{\eta - y}{dy} = \frac{\zeta - z}{dz},$$

where dx, dy, dz satisfy the relations

$$\frac{\partial F_1}{\partial x}dx + \frac{\partial F_1}{\partial y}dy + \frac{\partial F_1}{\partial z}dz = 0, \qquad \frac{\partial F_2}{\partial x}dx + \frac{\partial F_2}{\partial y}dy + \frac{\partial F_2}{\partial z}dz = 0.$$

Consequently the equations of the tangent can be put in the form

$$\frac{\xi - x}{\dfrac{\partial F_1}{\partial y}\dfrac{\partial F_2}{\partial z} - \dfrac{\partial F_1}{\partial z}\dfrac{\partial F_2}{\partial y}} = \frac{\eta - y}{\dfrac{\partial F_1}{\partial z}\dfrac{\partial F_2}{\partial x} - \dfrac{\partial F_1}{\partial x}\dfrac{\partial F_2}{\partial z}} = \frac{\zeta - z}{\dfrac{\partial F_1}{\partial x}\dfrac{\partial F_2}{\partial y} - \dfrac{\partial F_1}{\partial y}\dfrac{\partial F_2}{\partial x}}. \tag{17}$$

Comparing this result with (13), we see that the tangent line to a curve at a point M is the intersection of the tangent planes at M to two surfaces which intersect along the curve.

EXAMPLES

1. Show that the volume of the tetrahedron formed by the coördinate planes and the tangent plane at any point of the surface $x = u$, $y = v$, $z = a^3/uv$ is constant.

2. Show that the sum of the squares of the intercepts of the axes by the tangent plane to the surface

$$x = u^3 \sin^3 v, \qquad y = u^3 \cos^3 v, \qquad z = (a^2 - u^2)^{\frac{3}{2}},$$

at any point is constant.

3. Given the right conoid for which $\phi(v) = a \sin 2v$. Show that any tangent plane to the surface cuts it in an ellipse, and that if perpendiculars be drawn to the generators from any point the feet of the perpendiculars lie in a plane ellipse.

4. Show that the tangent planes, at points of a generator, to the right conoid for which $\phi(v) = a\sqrt{\tan v}$, meet the plane $z = 0$ in parallel lines.

5. Find the equations of the tangent to the curve whose equations are

$$ax^2 + by^2 + cz^2 = 1, \qquad bx^2 + cy^2 + az^2 = 1.$$

6. Find the equations of the tangent to the curve whose equations are

$$z(x+z)(x-a) = a^3, \qquad z(y+z)(y-a) = a^3,$$

and show that the curve is plane.

7. The distance from a point M' of a surface to the tangent plane at a near-by point M is of the second order when MM' is of the first order; and for other planes through M the distance from M' is ordinarily of the first order.

26. One-parameter families of surfaces. Envelopes. An equation of the form

$$F(x, y, z, a) = 0 \tag{18}$$

defines an infinity of surfaces, each surface being determined by a value of the parameter a. Such a system is called a *one-parameter family of surfaces*. For example, the tangent planes to the tangent surface of a twisted curve form such a family.

The two surfaces corresponding to values a and a' of the parameter meet in a curve whose equations may be written

$$F(x, y, z, a) = 0, \qquad \frac{F(x, y, z, a') - F(x, y, z, a)}{a' - a} = 0.$$

As a' approaches a, this curve approaches a limiting form whose equations are

$$F(x, y, z, a) = 0, \qquad \frac{\partial F(x, y, z, a)}{\partial a} = 0. \tag{19}$$

The curve thus defined is called the *characteristic* of the surface of parameter a. As a varies we have a family of these characteristics, and their locus, called the *envelope* of the family of surfaces, is a surface whose equation is obtained by eliminating a from the two equations (19). This elimination may be accomplished by solving the second of (19) for a, thus:

$$a = \phi(x, y, z),$$

and substituting in the first with the result

$$F(x, y, z, \phi) = 0.$$

The equation of the tangent plane to this surface is

$$\left(\frac{\partial F}{\partial x}+\frac{\partial F}{\partial \phi}\frac{\partial \phi}{\partial x}\right)(\xi-x)+\left(\frac{\partial F}{\partial y}+\frac{\partial F}{\partial \phi}\frac{\partial \phi}{\partial y}\right)(\eta-y)+\left(\frac{\partial F}{\partial z}+\frac{\partial F}{\partial \phi}\frac{\partial \phi}{\partial z}\right)(\zeta-z)=0. \tag{20}$$

For a particular value of a, say a_0, equations (19) define the curve in which the surface $F(x, y, z, a_0)=0$ meets the envelope; and from the second of (19) it follows that at all points of this curve equation (20) of the tangent plane to the envelope reduces to

$$\frac{\partial F}{\partial x}(\xi-x)+\frac{\partial F}{\partial y}(\eta-y)+\frac{\partial F}{\partial z}(\zeta-z)=0.$$

This, however, is the equation of the tangent plane to the surface $F(x, y, z, a_0)=0$. If we say that two surfaces with the same tangent plane at a common point are tangent to one another, we have:

The envelope of a family of surfaces of one parameter is tangent to each surface along the characteristic of the latter.

The equations of the characteristic of the surface of parameter a_1 are

$$F(x, y, z, a_1)=0, \qquad \left(\frac{\partial F}{\partial a}\right)_{a=a_1}=0. \tag{21}$$

This characteristic meets the characteristic (19) in the point whose coördinates satisfy (19) and (21), or, what is the same thing, equations (19) and

$$\frac{F(x, y, z, a_1)-F(x, y, z, a)}{a_1-a}=0, \qquad \frac{\left(\frac{\partial F}{\partial a}\right)_{a=a_1}-\left(\frac{\partial F}{\partial a}\right)_{a=a}}{a_1-a}=0.$$

As a_1 approaches a, this point of intersection approaches a limiting position whose coördinates satisfy the three equations

$$F=0, \qquad \frac{\partial F}{\partial a}=0, \qquad \frac{\partial^2 F}{\partial a^2}=0. \tag{22}$$

If these equations be solved for x, y, z, we have

$$x=f_1(a), \qquad y=f_2(a), \qquad z=f_3(a). \tag{23}$$

These are parametric equations of a curve, which is called the *edge of regression* of the envelope.

The direction-cosines of the tangent to the edge of regression are proportional to $\frac{dx}{da}, \frac{dy}{da}, \frac{dz}{da}$. If we imagine that x, y, z in (19) are replaced by the values (23), and we differentiate these equations with respect to a, we get, in consequence of (22),

$$\frac{\partial F}{\partial x}\frac{dx}{da} + \frac{\partial F}{\partial y}\frac{dy}{da} + \frac{\partial F}{\partial z}\frac{dz}{da} = 0,$$

$$\frac{\partial^2 F}{\partial a\,\partial x}\frac{dx}{da} + \frac{\partial^2 F}{\partial a\,\partial y}\frac{dy}{da} + \frac{\partial^2 F}{\partial a\,\partial z}\frac{dz}{da} = 0.$$

From these we obtain

$$\frac{dx}{da} : \frac{dy}{da} : \frac{dz}{da} = \left\| \begin{array}{ccc} \frac{\partial F}{\partial x} & \frac{\partial F}{\partial y} & \frac{\partial F}{\partial z} \\ \frac{\partial^2 F}{\partial a\,\partial x} & \frac{\partial^2 F}{\partial a\,\partial y} & \frac{\partial^2 F}{\partial a\,\partial z} \end{array} \right\|.$$

But from (17) it follows that the minors of the right-hand member are proportional to the direction-cosines of the tangent to the curve (19). Hence we have the theorem:

The characteristics of a family of surfaces of one parameter are tangent to the edge of regression.

27. Developable surfaces. Rectifying developable. A simple example of a family of surfaces of one parameter is afforded by a family of planes of one parameter. Their envelope is called a *developable surface*; the full significance of this term will be shown later (§ 43). The characteristics are straight lines which are tangent to a curve, the edge of regression. When the edge of regression is a point, the surface is a cone or cylinder, according as the point is at a finite or infinite distance. We exclude this case for the present and assume that the coördinates x, y, z of a point on the edge of regression are expressed in terms of the arc s.

We may write the equation of the plane

$$(24) \qquad (X - x)\,a + (Y - y)\,b + (Z - z)\,c = 0,$$

where a, b, c also are functions of s. The characteristics are defined by this equation and its derivative with respect to s, namely:

$$(25) \qquad (X - x)\,a' + (Y - y)\,b' + (Z - z)\,c' - ax' - by' - cz' = 0.$$

Since these equations define the tangent to the curve, they must be equivalent to the equations

$$\frac{X-x}{x'}=\frac{Y-y}{y'}=\frac{Z-z}{z'}.$$

Hence we must have

$$(26) \qquad ax'+by'+cz'=0, \qquad a'x'+b'y'+c'z'=0.$$

If the first of these equations be differentiated with respect to s, the resulting equation is reducible, in consequence of the second of (26), to

$$ax''+by''+cz''=0.$$

From this equation and (26) we find

$$a:b:c=(y'z''-z'y''):(z'x''-x'z''):(x'y''-y'x'').$$

Hence by (§ 7) we have the theorem:

On the envelope of a one-parameter family of planes the planes osculate the edge of regression.

We leave it to the reader to prove that the edge of regression of the osculating planes of a twisted curve is the curve itself.

The envelope of the plane normal to the principal normal to a curve at a point of the curve is called the *rectifying developable* of the latter. We shall find the equations of its edge of regression.

The equation of this plane is

$$(27) \qquad (X-x)l+(Y-y)m+(Z-z)n=0.$$

If we differentiate this equation with respect to the arc of the curve, and make use of the Frenet formulas (I, 50), we obtain

$$(28) \quad (X-x)\left(\frac{\alpha}{\rho}+\frac{\lambda}{\tau}\right)+(Y-y)\left(\frac{\beta}{\rho}+\frac{\mu}{\tau}\right)+(Z-z)\left(\frac{\gamma}{\rho}+\frac{\nu}{\tau}\right)=0.$$

From these equations we derive the equations of the characteristic in the form

$$X=x+\left(\frac{\alpha}{\tau}-\frac{\lambda}{\rho}\right)t, \quad Y=y+\left(\frac{\beta}{\tau}-\frac{\mu}{\rho}\right)t, \quad Z=z+\left(\frac{\gamma}{\tau}-\frac{\nu}{\rho}\right)t,$$

t being the parameter of points on the characteristic. In order to find the value of t corresponding to the point where the characteristic touches the edge of regression, we combine these equations with the derivative of (28) with respect to s, namely:

$$(X-x)\left[\frac{\alpha\rho'}{\rho^2}+\frac{\lambda\tau'}{\tau^2}\right]+(Y-y)\left[\frac{\beta\rho'}{\rho^2}+\frac{\mu\tau'}{\tau^2}\right]+(Z-z)\left[\frac{\gamma\rho'}{\rho^2}+\frac{\nu\tau'}{\tau^2}\right]+\frac{1}{\rho}=0,$$

and obtain
$$\left(\frac{\rho'}{\tau\rho^2}-\frac{\tau'}{\rho\tau^2}\right)t+\frac{1}{\rho}=0.$$

Hence the coördinates of the edge of regression of the rectifying developable are

$$(29)\quad \xi=x+\frac{\alpha\rho-\lambda\tau}{\tau'\rho-\rho'\tau}\tau,\quad \eta=y+\frac{\beta\rho-\mu\tau}{\tau'\rho-\rho'\tau}\tau,\quad \zeta=z+\frac{\gamma\rho-\nu\tau}{\tau'\rho-\rho'\tau}\tau.$$

Problem. Under what conditions does the equation $F(x, y, z) = 0$ define a developable surface?

We assume that x, y, z are functions of two parameters u, v, such that the curves $u =$ const. are the generators, and $v =$ const. are any other lines. The equation of the tangent plane is

$$\text{(i)}\qquad (X-x)\frac{\partial F}{\partial x}+(Y-y)\frac{\partial F}{\partial y}+(Z-z)\frac{\partial F}{\partial z}=0.$$

This equation should involve u and be independent of v. Its characteristic is given by (i) and

$$\text{(ii)}\qquad A\frac{\partial x}{\partial u}+B\frac{\partial y}{\partial u}+C\frac{\partial z}{\partial u}=0,$$

where we have put, for the sake of brevity,

$$A=(X-x)\frac{\partial^2 F}{\partial x^2}+(Y-y)\frac{\partial^2 F}{\partial x\,\partial y}+(Z-z)\frac{\partial^2 F}{\partial x\,\partial z}-\frac{\partial F}{\partial x},$$

$$B=(X-x)\frac{\partial^2 F}{\partial x\,\partial y}+(Y-y)\frac{\partial^2 F}{\partial y^2}+(Z-z)\frac{\partial^2 F}{\partial y\,\partial z}-\frac{\partial F}{\partial y};$$

$$C=(X-x)\frac{\partial^2 F}{\partial x\,\partial z}+(Y-y)\frac{\partial^2 F}{\partial y\,\partial z}+(Z-z)\frac{\partial^2 F}{\partial z^2}-\frac{\partial F}{\partial z}.$$

Since equation (i) is independent of v, we have

$$\text{(iii)}\qquad A\frac{\partial x}{\partial v}+B\frac{\partial y}{\partial v}+C\frac{\partial z}{\partial v}=0$$

Comparing equations (ii) and (iii) with (13) we see that

$$A-\lambda\frac{\partial F}{\partial x}=0,\qquad B-\lambda\frac{\partial F}{\partial y}=0,\qquad C-\lambda\frac{\partial F}{\partial z}=0,$$

where λ denotes a factor of proportionality. If we eliminate $X - x$, $Y - y$, $Z - z$, and λ from these equations and (i), we obtain the desired condition

$$\begin{vmatrix} \frac{\partial^2 F}{\partial x^2} & \frac{\partial^2 F}{\partial x\,\partial y} & \frac{\partial^2 F}{\partial x\,\partial z} & \frac{\partial F}{\partial x} \\ \frac{\partial^2 F}{\partial x\,\partial y} & \frac{\partial^2 F}{\partial y^2} & \frac{\partial^2 F}{\partial y\,\partial z} & \frac{\partial F}{\partial y} \\ \frac{\partial^2 F}{\partial x\,\partial z} & \frac{\partial^2 F}{\partial y\,\partial z} & \frac{\partial^2 F}{\partial z^2} & \frac{\partial F}{\partial z} \\ \frac{\partial F}{\partial x} & \frac{\partial F}{\partial y} & \frac{\partial F}{\partial z} & 0 \end{vmatrix} = 0.$$

EXAMPLES

1. Find the envelope and edge of regression of the family of planes normal to a given curve.

2. Find the rectifying developable of a cylindrical helix.

3. Prove that the rectifying developable of a curve is the polar developable of its involutes, and conversely.

4. Find the edge of regression of the envelope of the planes

$$x \sin u - y \cos u + z - au = 0.$$

5. Determine the envelope of a one-parameter family of planes parallel to a given line.

6. Given a one-parameter family of planes which cut the xy-plane under constant angle; the intersections of these planes with the latter plane envelop a curve C. Show that the edge of regression of the envelope of the planes is an evolute of C.

7. When a plane curve lies on a developable surface its plane meets the tangent planes to the surface in the tangent lines to the curve. Determine the developable surface which passes through a parabola and the circle, described in a perpendicular plane, on the latus rectum for diameter, and show that it is a cone.

8. Determine the developable surface which passes through the two parabolas $y^2 = 4\,ax$, $z = 0$; $x^2 = 4\,ay$, $z = b$, and show that its edge of regression lies on the surface $y^3 z = x^3 (b - z)$.

28. Applications of the moving trihedral. Problems concerning the envelope of a family of surfaces are sometimes more readily solved when the surfaces are referred to the moving trihedral of a curve, which is associated in some manner with the family of surfaces, the parameter of points on the curve being the parameter of the family.

Let

$$(30) \qquad F(\xi, \eta, \zeta, s) = 0$$

define such a family of surfaces. Since ξ, η, ζ are functions of s, the equations of the characteristics are (30) and

$$\frac{dF}{ds}=\frac{\partial F}{\partial \xi}\frac{d\xi}{ds}+\frac{\partial F}{\partial \eta}\frac{d\eta}{ds}+\frac{\partial F}{\partial \zeta}\frac{d\zeta}{ds}+\frac{\partial F}{\partial s}=0.$$

But the characteristics being fixed in space, we have (I, 84)

$$\text{(31)} \qquad \frac{d\xi}{ds}=\frac{\eta}{\rho}-1, \qquad \frac{d\eta}{ds}=-\left(\frac{\xi}{\rho}+\frac{\zeta}{\tau}\right), \qquad \frac{d\zeta}{ds}=\frac{\eta}{\tau}.$$

Hence the equations of the characteristics are

$$\text{(32)} \qquad F=0, \qquad \frac{\partial F}{\partial \xi}\left(\frac{\eta}{\rho}-1\right)-\frac{\partial F}{\partial \eta}\left(\frac{\xi}{\rho}+\frac{\zeta}{\tau}\right)+\frac{\partial F}{\partial \zeta}\frac{\eta}{\tau}+\frac{\partial F}{\partial s}=0.$$

If, for the sake of brevity, we let $\phi(\xi, \eta, \zeta, s)=0$ denote the second of these equations, the edge of regression is defined by (32) and

$$\text{(33)} \qquad \frac{\partial \phi}{\partial \xi}\left(\frac{\eta}{\rho}-1\right)-\frac{\partial \phi}{\partial \eta}\left(\frac{\xi}{\rho}+\frac{\zeta}{\tau}\right)+\frac{\partial \phi}{\partial \zeta}\frac{\eta}{\tau}+\frac{\partial \phi}{\partial s}=0.$$

For example, the family of osculating planes of a curve is defined with reference to the moving trihedral by $\zeta=0$. In this case the second of (32) is $\eta=0$, and (33) is $\frac{\xi}{\rho}+\frac{\zeta}{\tau}=0$. Hence the tangents are the characteristics, and the edge of regression is the curve; for, we have $\xi=\eta=\zeta=0$.

In like manner the family of normal planes is defined by $\xi=0$. Now the second of (32) is $\eta-\rho=0$; consequently the polar lines are the characteristics. Equation (33) reduces to $\zeta+\rho'\tau=0$; hence the locus of the centers of the osculating spheres is the edge of regression (cf. § 18). The envelope is called the *polar developable.*

The osculating spheres of a twisted curve constitute a family of surfaces which is readily studied by the foregoing methods. From (§ 18) it follows that the equation of these spheres is

$$\xi^2+\eta^2+\zeta^2-2\rho\eta+2\rho'\tau\zeta=0.$$

The second of equations (32) for this case is

$$\zeta\left[\frac{\rho}{\tau}+(\tau\rho')'\right]=0,$$

which, since spherical curves are not considered, reduces to $\zeta=0$. And equation (33) is $\eta=0$, so that the coördinates of the edge of regression are $\xi=\eta=\zeta=0$. Hence:

The osculating circles of a curve are the characteristics of its osculating spheres; and the curve itself is the edge of regression of the envelope of the spheres.

29. Envelope of spheres. Canal surfaces. We consider now any family of spheres of one parameter. Referred to the moving trihedral of the curve of centers, the equation of the spheres is

$$\xi^2+\eta^2+\zeta^2-r^2=0.$$

By means of (32) we find that a characteristic is the circle in which a sphere is cut by the plane

$$\xi+rr'=0.$$

The radius of this circle is equal to $r\sqrt{1-r'^2}$. Hence the characteristic is imaginary when $r'^2>1$, reduces to a point when $r=\pm s+\text{const.}$, and is real for $r'^2<1$.

By means of (33) we find that the coördinates of the edge of regression are given by

$$(34)\quad \xi=-rr',\quad \eta=[1-(rr')']\rho,\quad \zeta^2=r^2(1-r'^2)-\rho^2[1-(rr')']^2.$$

Hence the edge of regression consists of two parts with corresponding points symmetrically placed with respect to the osculating plane of the curve of centers C, unless

$$r^2(1-r'^2)-\rho^2[1-(rr')']^2=0.$$

When this condition is satisfied the edge is a single curve, and its points lie in the osculating planes of C. We have seen that this is the case with the osculating spheres of a curve. We shall show that when the above condition is satisfied the spheres osculate their edge of regression C_1.

We write the above equation in the form

$$(35)\qquad \rho[1-(rr')']=er\sqrt{1-r'^2},$$

where e is $+1$ or -1, so that ρ may be positive.

We have seen (§ 16) that the absolute and relative rates of change with s of the coördinates ξ, η, ζ of a point on C_1 are in the relations

$$\frac{\delta\xi}{\delta s}=\frac{d\xi}{ds}-\frac{\eta}{\rho}+1,\qquad \frac{\delta\eta}{\delta s}=\frac{d\eta}{ds}+\frac{\xi}{\rho}+\frac{\zeta}{\tau},\qquad \frac{\delta\zeta}{\delta s}=\frac{d\zeta}{ds}-\frac{\eta}{\tau}.$$

When the values (34) are substituted in the right-hand members of these equations, we obtain, in consequence of (35),

$$\frac{\delta\xi}{\delta s}=0,\qquad \frac{\delta\eta}{\delta s}=0,\qquad \frac{\delta\zeta}{\delta s}=-\frac{er\sqrt{1-r'^2}}{\tau}.$$

Hence the linear element δs_1 of C_1 is given by

$$\delta s_1 = \frac{er\sqrt{1-r'^2}}{\tau}\,ds,$$

and

$$(36)\qquad \frac{\delta\xi}{\delta s_1}=0,\qquad \frac{\delta\eta}{\delta s_1}=0,\qquad \frac{\delta\zeta}{\delta s_1}=-1.$$

Since these are the direction-cosines of the tangent to C_1, we see that this tangent is normal to the osculating plane to the curve of centers C. Moreover, these direction-cosines must satisfy (cf. I, 83) the equations

$$(37)\qquad \frac{\delta a}{\delta s}=\frac{da}{ds}-\frac{b}{\rho},\qquad \frac{\delta b}{\delta s}=\frac{db}{ds}+\frac{a}{\rho}+\frac{c}{\tau},\qquad \frac{\delta c}{\delta s}=\frac{dc}{ds}-\frac{b}{\tau}.$$

Hence we have

$$\frac{\delta^2\xi}{\delta s_1^2}=0,\qquad \frac{\delta^2\eta}{\delta s_1^2}=-\frac{1}{er\sqrt{1-r'^2}},\qquad \frac{\delta^2\zeta}{\delta s_1^2}=0,$$

from which it follows that the radius of curvature ρ_1 of C_1 is

$$(38)\qquad \rho_1 = ee'r\sqrt{1-r'^2},$$

where e' is $+1$ or -1, so that ρ_1 may be positive. Since, now, the direction-cosines of the principal normal have the values

$$(39)\qquad \rho_1\frac{\delta^2\xi}{\delta s_1^2}=0,\qquad \rho_1\frac{\delta^2\eta}{\delta s_1^2}=-e',\qquad \rho_1\frac{\delta^2\zeta}{\delta s_1^2}=0,$$

it follows that the principal normals to C and C_1 are parallel. Furthermore, since these quantities must satisfy equations (37), we have

$$\frac{\delta^3\xi}{\delta s_1^3}=\frac{\tau}{\rho\rho_1^2},\qquad \frac{\delta^3\eta}{\delta s_1^3}=\frac{e'}{\rho_1^2}\rho_1',\qquad \frac{\delta^3\zeta}{\delta s_1^3}=\frac{1}{\rho_1^2},$$

where ρ_1' denotes the derivative of ρ_1 with respect to s_1. By means of (I, 51) we find that the radius of torsion τ_1 of C_1 is given by $\tau\tau_1 = e'\rho\rho_1$.

From (38) we find $\rho_1' = \dfrac{rr'\tau}{\rho\rho_1}$, so that the radius R_1 of the osculating sphere of C_1 is given by $R_1^2=\rho_1^2+\rho_1'^2\tau_1^2=r^2$, and consequently the osculating spheres of C_1 are of the same radius as the given spheres.

The direction-cosines of the tangent, principal normal, and binormal to C_1 are found from (36) and (39) to be

$$\alpha_1 = \beta_1 = 0,\ \gamma_1 = -1;\quad l_1 = n_1 = 0,\ m_1 = -e';\quad \lambda_1 = -e',\ \mu_1 = \nu_1 = 0.$$

Hence the coördinates (I, 94) of the center of the osculating sphere of C_1 are reducible, in consequence of (34), to

$$\xi + l_1\rho_1 - \rho_1'\tau_1\lambda_1 = 0,\quad \eta + m_1\rho_1 - \rho_1'\tau_1\mu_1 = 0,\quad \zeta + n_1\rho_1 - \rho_1'\tau_1\nu_1 = 0.$$

Therefore we have the theorem:

When the edge of regression of a family of spheres of one parameter has only one branch, the spheres osculate the edge.

Since τ does not appear in equation (35), it follows that when r is given as a function of s, the intrinsic equations of the curve of centers are

$$\rho = \frac{er\sqrt{1-r'^2}}{1-(rr')'},\qquad \tau = f(s),$$

where the function $f(s)$ is arbitrary. Moreover, any curve will serve for the curve of centers of such an envelope of spheres. The determination of r requires the solution of equation (35) and consequently involves two arbitrary constants.

When all the spheres of a family have the same radius, the envelope is called a *canal surface*. From (34) it is seen that in this case a characteristic is a great circle. Moreover, equation (35) reduces to $\rho = r$. Hence a necessary and sufficient condition that the edge of regression of a canal surface consist of a single curve is that the curve of centers be of constant curvature and the radius of the sphere equal to the radius of first curvature of the curve.

GENERAL EXAMPLES

1. Let MN be a generator of the right conoid

$$x = u\cos v,\qquad y = u\sin v,\qquad z = 2k\operatorname{cosec} 2v,$$

M being the point in which it meets the z-axis. Show that the tangent plane at N meets the surface in a hyperbola which passes through M, and that as N moves along the generator the tangent at M to the hyperbola describes a plane.

2. A point moves on an ellipsoid $\frac{x^2}{a^2} + \frac{y^2}{b^2} + \frac{z^2}{c^2} = 1$, so that the direction of its motion always passes through the perpendicular from the center on the tangent plane at the point. Show that the path of the point is the curve in which the ellipsoid is cut by the surface $x^l y^m z^n = \text{const.}$, where $l : m : n = \frac{1}{b^2} - \frac{1}{c^2} : \frac{1}{c^2} - \frac{1}{a^2} : \frac{1}{a^2} - \frac{1}{b^2}$.

3. If each of the generators of a developable surface be revolved through the same angle about the tangent to an orthogonal trajectory of the generators at the point of intersection, the locus of these lines is a developable surface whose edge of regression is an evolute of the given trajectory.

4. Show that the edge of regression of the family of planes

$$(1-u^2)x + i(1+u^2)y + 2uz + f(u) = 0$$

is a minimal curve.

5. The developable surface which passes through the circles $x^2 + y^2 = a^2$, $z = 0$; $x^2 + z^2 = b^2$, $y = 0$ meets the plane $x = 0$ in an equilateral hyperbola.

6. Find the edge of regression of the developable surface which envelopes the surface $az = xy$ along the curve in which the latter is cut by the cylinder $x^2 = by$.

7. Find the envelope of the planes which pass through the center of an ellipsoid and cut it in sections of equal area.

8. The first and second curvatures of the edge of regression of the family of planes $\alpha x + \beta y + \gamma z = p$, where α, β, γ, p are functions of a single parameter u and $\alpha^2 + \beta^2 + \gamma^2 = 1$, are given by

$$\frac{1}{\rho} = \frac{\Delta^3}{D(\alpha'^2 + \beta'^2 + \gamma'^2)^{\frac{3}{2}}}, \qquad \frac{1}{\tau} = \frac{\Delta^2}{D},$$

where

$$\Delta = \begin{vmatrix} \alpha & \alpha' & \alpha'' \\ \beta & \beta' & \beta'' \\ \gamma & \gamma' & \gamma'' \end{vmatrix}, \qquad D = \begin{vmatrix} p & p' & p'' & p''' \\ \alpha & \alpha' & \alpha'' & \alpha''' \\ \beta & \beta' & \beta'' & \beta''' \\ \gamma & \gamma' & \gamma'' & \gamma''' \end{vmatrix}.$$

9. Derive the equations of the edge of regression of the rectifying developable by the method of § 28.

10. Derive the results of § 29 without the aid of the moving trihedral.

11. Find the envelope of the spheres whose diameters are the chords of a circle through a point of the latter.

12. Find the envelope and edge of regression of the spheres which pass through a fixed point and whose centers lie on a given curve.

13. Find the envelope and edge of regression of the spheres which have for diametral planes one family of circular sections of an ellipsoid.

14. Find the envelope and edge of regression of the family of ellipsoids $a^2\left(\frac{x^2}{\alpha^2} + \frac{y^2}{\beta^2}\right) + \frac{z^2}{a^2} = 1$, where a is the parameter.

15. Find the envelope of the family of spheres whose diameters are parallel chords of an ellipse.

16. Find the equations of the canal surface whose curve of centers is a circular helix and whose edge of regression has one branch. Determine the latter.

17. Find the envelope of the family of cones

$$(ax + x + y + z - 1)(ay + z) - ax(x + y + z - 1) = 0,$$

where a is the parameter.

CHAPTER III

LINEAR ELEMENT OF A SURFACE. DIFFERENTIAL PARAMETERS. CONFORMAL REPRESENTATION

30. Linear element. Upon a surface S, defined by equations in the parametric form

$$x = f_1(u, v), \qquad y = f_2(u, v), \qquad z = f_3(u, v), \tag{1}$$

we select any curve and write its equations $\phi(u, v) = 0$. From § 3 we have that the linear element of the curve is given by

$$ds^2 = dx^2 + dy^2 + dz^2, \tag{2}$$

where $dx = \frac{\partial x}{\partial u} du + \frac{\partial x}{\partial v} dv, \quad dy = \frac{\partial y}{\partial u} du + \frac{\partial y}{\partial v} dv, \quad dz = \frac{\partial z}{\partial u} du + \frac{\partial z}{\partial v} dv,$

the differentials du, dv satisfying the condition

$$\frac{\partial \phi}{\partial u} du + \frac{\partial \phi}{\partial v} dv = 0.$$

If we put

$$\left\{ \begin{aligned} E &= \left(\frac{\partial x}{\partial u}\right)^2 + \left(\frac{\partial y}{\partial u}\right)^2 + \left(\frac{\partial z}{\partial u}\right)^2, \\ F &= \frac{\partial x}{\partial u}\frac{\partial x}{\partial v} + \frac{\partial y}{\partial u}\frac{\partial y}{\partial v} + \frac{\partial z}{\partial u}\frac{\partial z}{\partial v}, \\ G &= \left(\frac{\partial x}{\partial v}\right)^2 + \left(\frac{\partial y}{\partial v}\right)^2 + \left(\frac{\partial z}{\partial v}\right)^2, \end{aligned} \right. \tag{3}$$

or, in abbreviated form,

$$E = \sum \left(\frac{\partial x}{\partial u}\right)^2, \qquad F = \sum \frac{\partial x}{\partial u}\frac{\partial x}{\partial v}, \qquad G = \sum \left(\frac{\partial x}{\partial v}\right)^2,$$

equation (2) becomes

$$ds^2 = E\,du^2 + 2F\,du\,dv + G\,dv^2. \tag{4}$$

The functions E, F, G thus defined were first used by Gauss.* When the surface is real, and likewise the curvilinear coördinates

* *Disquisitiones generales circa superficies curvas* (English translation by Morehead and Hiltebeitel), p. 18. Princeton, 1902. Unless otherwise stated, all references to Gauss are to this translation.

u, v, the functions $\sqrt{E}$, $\sqrt{G}$ are real. We shall understand also that the latter are positive. There is, however, an important exceptional case, namely when both E and G are zero (cf. § 35).

For any other curve equation (4) will have the same form, but the relation between du and dv will depend upon the curve. Consequently the value of ds, given by (4), is the element of arc of any curve upon the surface. It is called the *linear element of the surface* (cf. § 20). However, in order to avoid circumlocution, we shall frequently call the expression for ds^2 the linear element, that is, the right-hand member of equation (4), which is also called *the first fundamental quadratic form.* The coefficients of the latter, namely E, F, G, are called the *fundamental quantities of the first order.*

If, for the sake of brevity, we put

$$(5) \qquad A = \frac{\partial(y, z)}{\partial(u, v)} = \begin{vmatrix} \frac{\partial y}{\partial u} & \frac{\partial z}{\partial u} \\ \frac{\partial y}{\partial v} & \frac{\partial z}{\partial v} \end{vmatrix}, \quad B = \frac{\partial(z, x)}{\partial(u, v)}, \quad C = \frac{\partial(x, y)}{\partial(u, v)},$$

it follows from (3) and (5) that

$$(6) \qquad EG - F^2 = A^2 + B^2 + C^2.$$

Hence when the surface is real and likewise the parameters, the quantity $EG - F^2$ is different from zero unless A, B, and C are zero. But if A, B, and C are zero, it follows from (5) that u and v are not independent, and consequently equations (1) define a curve and not a surface. However, it may happen that for certain values of u and v all the quantities A, B, C vanish. The corresponding points are called *singular points* of the surface. These points may be isolated or constitute one or more curves upon the surface; such curves are called *singular lines.* In the following discussion only ordinary points will be considered.

From the preceding remarks it follows that for real surfaces, referred to real coördinate lines, the function H defined by

$$(7) \qquad H = \sqrt{EG - F^2}$$

is real, and it is positive by hypothesis.

31. Isotropic developable. The exceptional case, where the surface is imaginary and H is zero, is afforded by the tangent surface of a minimal curve. The equations of such a surface are (cf. § 22)

$$x = \int \frac{1-u^2}{2}\phi(u)\,du + \frac{1-u^2}{2}\phi(u)\,v,$$

$$y = i\int \frac{1+u^2}{2}\phi(u)\,du + \frac{i(1+u^2)}{2}\phi(u)\,v,$$

$$z = \int u\phi(u)\,du + u\phi(u)\,v,$$

where $\phi(u)$ is a function of u different from zero. It is readily found that $E = v^2\phi^2(u)$, $F = G = 0$, and consequently $EG - F^2 = 0$. This equation is likewise the sufficient condition that the surface be of the kind sought. For, when it is satisfied, the equation of the linear element can be written $ds^2 = (\sqrt{E}\,du + \sqrt{G}\,dv)^2$. If λ denote an integrating factor of $\sqrt{E}\,du + \sqrt{G}\,dv$, and a function u_1 be defined by the equation $\lambda(\sqrt{E}\,du + \sqrt{G}\,dv) = du_1$, the above equation becomes $ds^2 = \frac{1}{\lambda^2}du_1^2$. Hence, if we take for parametric curves $u_1 = \text{const.}$ and any other system for $v_1 = \text{const.}$, we have $F_1 = 0$, $G_1 = 0$. In other form these equations are

$$\frac{\partial x}{\partial u_1}\frac{\partial x}{\partial v_1} + \frac{\partial y}{\partial u_1}\frac{\partial y}{\partial v_1} + \frac{\partial z}{\partial u_1}\frac{\partial z}{\partial v_1} = 0, \qquad \left(\frac{\partial x}{\partial v_1}\right)^2 + \left(\frac{\partial y}{\partial v_1}\right)^2 + \left(\frac{\partial z}{\partial v_1}\right)^2 = 0.$$

In accordance with the last equation we put

$$\frac{\partial x}{\partial v_1} = k\frac{1-u_1^2}{2}, \qquad \frac{\partial y}{\partial v_1} = ik\frac{1+u_1^2}{2}, \qquad \frac{\partial z}{\partial v_1} = ku_1,$$

where k is undetermined.

By integration we have

$$\text{(i)} \qquad x = \frac{1-u_1^2}{2}\int k\,dv_1 + \lambda, \qquad y = i\frac{1+u_1^2}{2}\int k\,dv_1 + \mu, \qquad z = u_1\int k\,dv_1 + \nu,$$

λ, μ, ν being functions of u_1 alone. When these values are substituted in the first of the above equations of condition, we get

$$(1-u_1^2)\frac{\partial\lambda}{\partial u_1} + i(1+u_1^2)\frac{\partial\mu}{\partial u_1} + 2u_1\frac{\partial\nu}{\partial u_1} = 0,$$

to be satisfied by λ, μ, and ν.

The equation of the tangent plane to the surface (i) is reducible to

$$(1-u_1^2)(X-x) + i(1+u_1^2)(Y-y) + 2u_1(Z-z) = 0.$$

Hence the surface is developable. Since its edge of regression is a minimal curve (Ex. 4, p. 69), the theorem is proved. The surface is called an *isotropic developable.*

32. Transformation of coördinates. It is readily found that the functions E, F, G are unaltered in value by any change of the rectangular axes. But now we shall show that these functions change their values when there is a change of the curvilinear coördinates.

Let the transformation of coördinates be defined by the equations

$$u = u(u_1, v_1), \qquad v = v(u_1, v_1); \tag{8}$$

then we have

$$\frac{\partial x}{\partial u_1} = \frac{\partial x}{\partial u}\frac{\partial u}{\partial u_1} + \frac{\partial x}{\partial v}\frac{\partial v}{\partial u_1}, \qquad \frac{\partial x}{\partial v_1} = \frac{\partial x}{\partial u}\frac{\partial u}{\partial v_1} + \frac{\partial x}{\partial v}\frac{\partial v}{\partial v_1}.$$

If we put $E_1 = \sum\left(\frac{\partial x}{\partial u_1}\right)^2$, $F_1 = \sum \frac{\partial x}{\partial u_1}\frac{\partial x}{\partial v_1}$, $G_1 = \sum\left(\frac{\partial x}{\partial v_1}\right)^2$,

we find the relations

$$\begin{cases} E_1 = E\left(\frac{\partial u}{\partial u_1}\right)^2 + 2F\frac{\partial u}{\partial u_1}\frac{\partial v}{\partial u_1} + G\left(\frac{\partial v}{\partial u_1}\right)^2, \\ F_1 = E\frac{\partial u}{\partial u_1}\frac{\partial u}{\partial v_1} + F\left(\frac{\partial u}{\partial u_1}\frac{\partial v}{\partial v_1} + \frac{\partial u}{\partial v_1}\frac{\partial v}{\partial u_1}\right) + G\frac{\partial v}{\partial u_1}\frac{\partial v}{\partial v_1}, \\ G_1 = E\left(\frac{\partial u}{\partial v_1}\right)^2 + 2F\frac{\partial u}{\partial v_1}\frac{\partial v}{\partial v_1} + G\left(\frac{\partial v}{\partial v_1}\right)^2. \end{cases} \tag{9}$$

Hence the fundamental quantities of the first order assume new forms when there is a change of curvilinear coördinates.

From (8) we have, by differentiation,

$$du = \frac{\partial u}{\partial u_1}du_1 + \frac{\partial u}{\partial v_1}dv_1, \qquad dv = \frac{\partial v}{\partial u_1}du_1 + \frac{\partial v}{\partial v_1}dv_1.$$

Solving these equations for du_1, dv_1, we get

$$du_1 = \frac{1}{\delta}\left(\frac{\partial v}{\partial v_1}du - \frac{\partial u}{\partial v_1}dv\right), \qquad dv_1 = \frac{1}{\delta}\left(-\frac{\partial v}{\partial u_1}du + \frac{\partial u}{\partial u_1}dv\right),$$

where

$$\delta = \frac{\partial(u, v)}{\partial(u_1, v_1)}. \tag{10}$$

Hence we have

$$\begin{cases} \frac{\partial u_1}{\partial u} = \frac{1}{\delta}\frac{\partial v}{\partial v_1}, & \frac{\partial u_1}{\partial v} = -\frac{1}{\delta}\frac{\partial u}{\partial v_1}, \\ \frac{\partial v_1}{\partial u} = -\frac{1}{\delta}\frac{\partial v}{\partial u_1}, & \frac{\partial v_1}{\partial v} = \frac{1}{\delta}\frac{\partial u}{\partial u_1}, \end{cases} \tag{11}$$

so that

$$\frac{\partial(u_1, v_1)}{\partial(u, v)} = \frac{1}{\delta}. \tag{12}$$

From (9) we find the relation

$$E_1G_1 - F_1^2 = \delta^2(EG - F^2).$$

By means of this equation and the relations (11), we can transform equations (9) into the following:

$$
(13)\quad \begin{cases}
\dfrac{E_1}{E_1G_1-F_1^2}=\dfrac{E\left(\dfrac{\partial v_1}{\partial v}\right)^2-2F\dfrac{\partial v_1}{\partial u}\dfrac{\partial v_1}{\partial v}+G\left(\dfrac{\partial v_1}{\partial u}\right)^2}{EG-F^2},\\[2ex]
\dfrac{-F_1}{E_1G_1-F_1^2}=\dfrac{E\dfrac{\partial u_1}{\partial v}\dfrac{\partial v_1}{\partial v}-F\left(\dfrac{\partial u_1}{\partial u}\dfrac{\partial v_1}{\partial v}+\dfrac{\partial u_1}{\partial v}\dfrac{\partial v_1}{\partial u}\right)+G\dfrac{\partial u_1}{\partial u}\dfrac{\partial v_1}{\partial u}}{EG-F^2},\\[2ex]
\dfrac{G_1}{E_1G_1-F_1^2}=\dfrac{E\left(\dfrac{\partial u_1}{\partial v}\right)^2-2F\dfrac{\partial u_1}{\partial u}\dfrac{\partial u_1}{\partial v}+G\left(\dfrac{\partial u_1}{\partial u}\right)^2}{EG-F^2}.
\end{cases}
$$

33. Angles between curves. The element of area. Upon a parametric line $v=\text{const.}$ we take for positive sense the direction in which the parameter u increases, and likewise upon a curve $u=\text{const.}$ the direction in which v increases. If ds_v and ds_u denote the elements of arc of curves $v=\text{const.}$ and $u=\text{const.}$ respectively, we find, from (4),

$$
(14)\qquad ds_v=\sqrt{E}\,du,\qquad ds_u=\sqrt{G}\,dv.
$$

Hence, if $\alpha_v, \beta_v, \gamma_v$ and $\alpha_u, \beta_u, \gamma_u$ denote the direction-cosines of the tangents to these curves respectively, we have

$$
\alpha_v=\frac{1}{\sqrt{E}}\frac{\partial x}{\partial u},\quad \beta_v=\frac{1}{\sqrt{E}}\frac{\partial y}{\partial u},\quad \gamma_v=\frac{1}{\sqrt{E}}\frac{\partial z}{\partial u};
$$

$$
\alpha_u=\frac{1}{\sqrt{G}}\frac{\partial x}{\partial v},\quad \beta_u=\frac{1}{\sqrt{G}}\frac{\partial y}{\partial v},\quad \gamma_u=\frac{1}{\sqrt{G}}\frac{\partial z}{\partial v}.
$$

We have seen that through an ordinary point of a surface there passes one curve of parameter u and one of parameter v. If, as in fig. 11, ω denotes the angle, between 0° and 180°, formed by the positive directions of the tangents to these curves at the point, we have

$$
(15)\qquad \cos\omega=\alpha_u\alpha_v+\beta_u\beta_v+\gamma_u\gamma_v=\frac{F}{\sqrt{EG}},
$$

and

$$
(16)\qquad \sin\omega=\frac{\sqrt{EG-F^2}}{\sqrt{EG}}=\frac{H}{\sqrt{EG}}.
$$

When two families of curves upon a surface are such that through any point a curve of each family, and but one, passes, and when, moreover, the tangents at a point to the two curves through it are perpendicular, the curves are said to form an *orthogonal system.* From (15) we have the theorem:

A necessary and sufficient condition that the parametric lines upon a surface form an orthogonal system is that $F = 0$.

Consider the small quadrilateral (fig. 11) whose vertices are the points with the curvilinear coördinates (u, v), $(u + du, v)$, $(u, v + dv)$, $(u + du, v + dv)$. To within terms of higher order the opposite sides of the figure are equal. Consequently it is approximately a parallelogram whose sides are of length $\sqrt{E}\,du$ and $\sqrt{G}\,dv$ and the included angle is ω. The area of this parallelogram is called the *element of area of the surface.* Its expression is

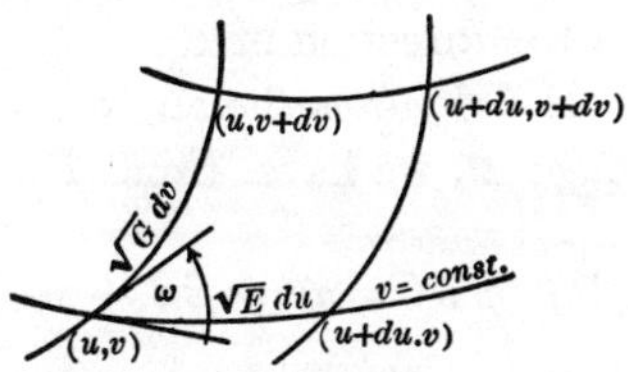

FIG. 11

$$(17) \quad d\Sigma = \sin\omega\sqrt{EG}\,dudv = H\,dudv.$$

If C is any curve on a surface, the direction-cosines α, β, γ of its tangent at a point have the form

$$\alpha = \frac{dx}{ds} = \left(\frac{\partial x}{\partial u}\frac{du}{ds} + \frac{\partial x}{\partial v}\frac{dv}{ds}\right), \quad \beta = \frac{dy}{ds} = \left(\frac{\partial y}{\partial u}\frac{du}{ds} + \frac{\partial y}{\partial v}\frac{dv}{ds}\right),$$

$$\gamma = \frac{dz}{ds} = \left(\frac{\partial z}{\partial u}\frac{du}{ds} + \frac{\partial z}{\partial v}\frac{dv}{ds}\right).$$

If we put $dv/du = \lambda$ and replace ds by the positive square root of the right-hand member of (4), the above expressions can be written

$$(18) \quad \begin{cases} \alpha = \dfrac{\dfrac{\partial x}{\partial u} + \lambda\dfrac{\partial x}{\partial v}}{\sqrt{E + 2F\lambda + G\lambda^2}}, \\[2ex] \beta = \dfrac{\dfrac{\partial y}{\partial u} + \lambda\dfrac{\partial y}{\partial v}}{\sqrt{E + 2F\lambda + G\lambda^2}}, \\[2ex] \gamma = \dfrac{\dfrac{\partial z}{\partial u} + \lambda\dfrac{\partial z}{\partial v}}{\sqrt{E + 2F\lambda + G\lambda^2}}. \end{cases}$$

From these results it is seen that the direction-cosines depend not upon the absolute values of du and dv, but upon their ratio λ. The value of λ is obtained by differentiation from the equation of C, namely

$$\phi(u, v) = 0. \tag{19}$$

Let C_1 be a second curve meeting C at a point M, and let the direction-cosines of the tangent to C_1 at M be α_1, β_1, γ_1. They are given by

$$\alpha_1 = \frac{\partial x}{\partial u}\frac{\delta u}{\delta s} + \frac{\partial x}{\partial v}\frac{\delta v}{\delta s}, \tag{20}$$

and similar expressions for β_1 and γ_1, where δ indicates variation in the direction of C_1.

If θ denotes the angle between the positive directions to C and C_1 at M, we have, from (18) and (20),

$$\cos\theta = \alpha\alpha_1 + \beta\beta_1 + \gamma\gamma_1 = \frac{E\,du\,\delta u + F(du\,\delta v + dv\,\delta u) + G\,dv\,\delta v}{ds\,\delta s}, \tag{21}$$

and

$$\sin\theta = \pm\sqrt{1 - \cos^2\theta} = \pm H\left(\frac{\delta u}{\delta s}\frac{dv}{ds} - \frac{\delta v}{\delta s}\frac{du}{ds}\right).$$

This ambiguity of sign is due to the fact that θ as defined is one of two angles which together are equal to 360°. We take the upper sign, thus determining θ. This gives

$$\sin\theta = H\left(\frac{\delta u}{\delta s}\frac{dv}{ds} - \frac{\delta v}{\delta s}\frac{du}{ds}\right). \tag{22}$$

The significance of the above choice will be pointed out shortly.

When in particular C_1 is the curve $v = \text{const.}$ through M, we have $\delta v = 0$ and $\delta s = \sqrt{E}\,\delta u$, so that

$$\cos\theta_0 = \frac{1}{\sqrt{E}}\left(E\frac{du}{ds} + F\frac{dv}{ds}\right), \qquad \sin\theta_0 = \frac{H}{\sqrt{E}}\frac{dv}{ds}. \tag{23}$$

From these equations we obtain

$$\tan\theta_0 = \frac{H\,dv}{E\,du + F\,dv}. \tag{24}$$

The angle ω between the positive half tangents to the parametric curves has been uniquely defined. Hence there is, in general, only one sense in which the tangent to the curve $v = \text{const.}$ can be brought into coincidence with the tangent to the curve $u = \text{const.}$ by a rotation of amount ω. We say that rotations in this direction are *positive*, in the opposite sense *negative*. From (23) it is seen that θ_0 is the angle described in the positive sense when the positive half tangent to the curve $v = \text{const.}$ is rotated into coincidence with the half tangent to C. And so in the general case θ, defined by (22), is the angle described in the positive rotation from the second curve to the first.

From equations (15), (16), and (23) we find

$$(25)\quad \cos(\omega - \theta_0) = \frac{1}{\sqrt{G}}\left(F\frac{du}{ds} + G\frac{dv}{ds}\right), \quad \sin(\omega - \theta_0) = \frac{H}{\sqrt{G}}\frac{du}{ds}.$$

These equations follow also directly from (20) and (21) by considering the curve $u = \text{const.}$ as the second line.

As an immediate consequence of equation (21) we have the theorem:

A necessary and sufficient condition that the tangents to two curves upon a surface at a point of meeting be perpendicular is

$$(26)\qquad E\,du\,\delta u + F(du\,\delta v + dv\,\delta u) + G\,dv\,\delta v = 0.$$

EXAMPLES

1. Show that when the equation of a surface is of the form $z = f(x, y)$, its linear element can be written

$$ds^2 = (1 + p^2)\,dx^2 + 2\,pq\,dx\,dy + (1 + q^2)\,dy^2,$$

where $p = \partial z/\partial x$, and $q = \partial z/\partial y$. Under what conditions do the lines $x = \text{const.}$, $y = \text{const.}$ form an orthogonal system?

2. Show that the parametric curves on the sphere

$$x = a\sin u\cos v, \qquad y = a\sin u\sin v, \qquad z = a\cos u$$

form an orthogonal system. Determine the two families of curves which meet the curves $v = \text{const.}$ under the angles $\pi/4$ and $3\,\pi/4$. Find the linear element of the surface when these new curves are parametric.

3. Find the equation of a curve on the paraboloid of revolution $x = u\cos v$, $y = u\sin v$, $z = u^2/2$, which meets the curves $v = \text{const.}$ under constant angle α and passes through two points (u_0, v_0), (u_1, v_1). Determine α as a function of u_0, v_0, u_1, v_1.

4. Find the differential equation of the curves upon the tangent surface of a curve which cut the generators under constant angle α.

5. Show that the equations of a curve which lies upon a right cone and cuts all the generators under the same angle are of the form $x = ce^{au}\cos u$, $y = ce^{au}\sin u$, $z = be^{au}$, where a, b, and c are constants. What is the projection of the curve upon a plane perpendicular to the axis of the cone? Find the radius of curvature of the curve.

6. Find the equations of the curves which bisect the angles between the parametric curves of the paraboloid in Ex. 3.

34. Families of curves. An equation of the form

$$\phi(u, v) = c, \tag{27}$$

where c is an arbitrary constant, defines an infinity of curves, or a *family of curves*, upon the surface. Through any point of the surface there passes a curve of the family. For, given the curvilinear coördinates of a point, when these values are substituted in (27) we obtain a value of c, say c_0; then evidently the curve $\phi = c_0$ passes through the point. We inquire whether this family of curves can be defined by another equation. Suppose it is possible, and let the equation be

$$\psi(u, v) = \kappa. \tag{28}$$

Since c and κ are constant along any curve and vary in passing from one curve to another, each is necessarily a function of the other. Hence ψ is a function of ϕ. Moreover, if ψ is any function of ϕ, equations (27) and (28) define the same family of curves.

From equation (24) it is seen that the direction, at any point, of the curve of the family through the point is determined by the value of dv/du. We obtain the latter from the equation

$$\frac{\partial \phi}{\partial u} du + \frac{\partial \phi}{\partial v} dv = 0, \tag{29}$$

which is derived from (27) by differentiation.

Let $\phi(u, v) = c$ be an integral of an ordinary differential equation of the first order and first degree, such as

$$M(u, v)\,du + N(u, v)\,dv = 0. \tag{30}$$

The curves defined by the former equation are called *integral curves* of equation (30). From the integral equation we get equation (29) by differentiation. It must be possible then to obtain equation (30) from the integral equation and (29). But c does not appear in (29), consequently the latter equation differs from (30) by a factor at most. Hence $M\frac{\partial \phi}{\partial v} - N\frac{\partial \phi}{\partial u} = 0$. Suppose, now, that we have another integral of (30), as $\psi(u, v) = e$. Then $M\frac{\partial \psi}{\partial v} - N\frac{\partial \psi}{\partial u} = 0$. The elimination of M and N from these equations gives $\frac{\partial(\phi, \psi)}{\partial(u, v)} = 0$; from which it follows that ψ is a function of ϕ. Moreover, ψ can by any function of ϕ. But we have seen that if ψ is a function of ϕ, the families of curves $\phi =$ const. and $\psi =$ const. are the same. Hence all integrals of equation (30) of the form $\phi = c$ or $\psi = e$ define the same family of curves. However, equation (30) may admit of an integral in which the constant of integration enters implicitly, as $F(u, v, c) = 0$. But if this be solved for c, we obtain one or more integrals of the form (27). Hence an equation of the form (30) defines one family of curves on a surface. Although the determination of the curves when thus defined requires the integration of the equation, the direction of any curve at a point is given directly by means of (24).

If at each point of intersection of a curve C_1 with the curves of a family the tangents to the two curves are perpendicular to one another, C_1 is called an *orthogonal trajectory of the curves.* Suppose that the family of curves is defined by equation (30). The relation between the ratios $\frac{dv}{du}$ and $\frac{\delta v}{\delta u}$, which determine the directions of the tangents to the two curves at the point of intersection, is given by equation (26). If we replace $\frac{\delta v}{\delta u}$ by $-\frac{M}{N}$, we obtain

$$(31) \qquad (EN - FM)\,du + (FN - GM)\,dv = 0.$$

But any integral curve of this equation is an orthogonal trajectory of the given curves. Hence a family of curves admits of a family of orthogonal trajectories. They are defined by equation (31), when the differential equation of the curves is in the form (30). But when the family is defined by a finite equation, such as (27), the equation of the orthogonal trajectories is

$$(32) \qquad \left(E\frac{\partial \phi}{\partial v} - F\frac{\partial \phi}{\partial u}\right)du + \left(F\frac{\partial \phi}{\partial v} - G\frac{\partial \phi}{\partial u}\right)dv = 0.$$

As an example, we consider the family of circles in the plane with centers on the x-axis whose equation is

$$x^2 + y^2 - 2ux = a^2, \tag{i}$$

where u is the parameter of the family and a is a constant. In order to find the orthogonal trajectories of these curves, we take the lines $x = \text{const.}$, $y = \text{const.}$ for parametric curves, in which case

$$E = G = 1, \qquad F = 0,$$

and write the equation (i) in the form (27), thus

$$x + \frac{1}{x}(y^2 - a^2) = 2u.$$

Now equation (32) is $2xy\,dx - (x^2 - y^2 + a^2)\,dy = 0$, of which the integral is

$$\frac{x^2 + a^2}{y} + y = 2v,$$

where v is the constant of integration. Hence the orthogonal trajectories are circles whose centers are on the y-axis.

An ordinary differential equation of the second degree, such as

$$R(u, v)\,du^2 + 2\,S(u, v)\,du\,dv + T(u, v)\,dv^2 = 0, \tag{33}$$

is equivalent to two equations of the first degree, which are found by solving this equation as a quadratic in dv. Hence equation (33) defines two families of curves upon the surface. We seek the condition that the curves of one family be the orthogonal trajectories of the other, or, in other words, the condition that (33) be the equation of an orthogonal system, as previously defined. If k_1 and k_2 denote the two values of $\frac{dv}{du}$ obtained from (33), we have

$$\kappa_1 + \kappa_2 = -\frac{2S}{T}, \qquad \kappa_1\kappa_2 = \frac{R}{T}.$$

From (26) it follows that the condition that the two directions at a point corresponding to κ_1 and κ_2 be perpendicular is

$$E + F(\kappa_1 + \kappa_2) + G\kappa_1\kappa_2 = 0.$$

If the above values are substituted in this equation, we have the condition sought; it is

$$ET + GR - 2FS = 0. \tag{34}$$

35. Minimal curves on a surface. An equation of the form (33) is obtained by equating to zero the first fundamental form of a surface. This gives

$$E\,du^2 + 2\,F\,du\,dv + G\,dv^2 = 0,$$

and it defines the double family of imaginary curves of length zero which lie on the surface. In this case equation (34) reduces to $EG - F^2 = 0$; hence the minimal lines on a surface form an orthogonal system only when the surface is an isotropic developable (§ 31).

An important example of these lines is furnished by the system on the sphere. If we take a sphere of unit radius and center at the origin, its equation, $x^2 + y^2 + z^2 = 1$, can be written in either of the forms

$$\frac{x + iy}{1 - z} = \frac{1 + z}{x - iy} = u,$$

$$\frac{x - iy}{1 - z} = \frac{1 + z}{x + iy} = v,$$

where u and v denote the respective ratios, and evidently are conjugate imaginaries. If these four equations are solved for x, y, z, we find

$$(35) \qquad x = \frac{u + v}{uv + 1}, \qquad y = \frac{i(v - u)}{uv + 1}, \qquad z = \frac{uv - 1}{uv + 1}.$$

From these expressions we find that the linear element, in terms of the parameters u and v, is given by

$$(36) \qquad ds^2 = \frac{4\,du\,dv}{(1 + uv)^2}.$$

Hence the curves $u = \text{const.}$ and $v = \text{const.}$ are the lines of length zero.

Eliminating u from the first two and the last two of equations (35), we get

$$(37) \qquad \begin{cases} i(v^2 + 1)x + (1 - v^2)y - 2\,iv = 0, \\ i(v^2 + 1)z + 2\,vy + i(1 - v^2) = 0. \end{cases}$$

Hence all the points of a curve $v = \text{const.}$ lie on the line

$$i(v^2 + 1)X + (1 - v^2)Y - 2\,iv = 0,$$

$$i(v^2 + 1)Z + 2\,vY + i(1 - v^2) = 0,$$

where X, Y, Z denote current coördinates. In consequence of (35), these equations can be written

$$\frac{X-x_0}{v^2-1}=\frac{Y-y_0}{i(v^2+1)}=\frac{Z-z_0}{-2v},$$

where x_0, y_0, z_0 are the coördinates of a particular point. In like manner the curves $u = \text{const.}$ are the minimal lines

$$\frac{X-x_0}{1-u^2}=\frac{Y-y_0}{i(1+u^2)}=\frac{Z-z_0}{2u}.$$

EXAMPLES

1. Show that the most general orthogonal system of circles in the plane is that of the example in § 34.

2. Show that on the right conoid $x = u\cos v$, $y = u\sin v$, $z = av$, the curves $du^2 - (u^2 + a^2)\,dv^2 = 0$ form an orthogonal system.

3. When the coefficients of the linear elements of two surfaces,

$$ds_1^2 = E_1du^2 + 2F_1dudv + G_1dv^2, \quad ds_2^2 = E_2du^2 + 2F_2dudv + G_2dv^2,$$

are not proportional, and points with the same curvilinear coördinates on each of the surfaces are said to correspond, there is a unique orthogonal system on one surface corresponding to an orthogonal system on the other; its equation is

$$(F_1E_2 - F_2E_1)du^2 + (E_2G_1 - E_1G_2)\,dudv + (G_1F_2 - G_2F_1)\,dv^2 = 0.$$

4. If θ_1 and θ_2 are solutions of the equation

$$\frac{\partial^2\theta}{\partial\alpha\,\partial\beta} - \frac{1}{2}\frac{\partial\log\lambda}{\partial\alpha}\frac{\partial\theta}{\partial\beta} - \lambda\theta = 0,$$

where λ is any function of α and β, the equations

$$x + iy = \int \theta_1^2 d\alpha + \frac{1}{\lambda}\left(\frac{\partial\theta_1}{\partial\beta}\right)^2 d\beta,$$

$$x - iy = \int \theta_2^2 d\alpha + \frac{1}{\lambda}\left(\frac{\partial\theta_2}{\partial\beta}\right)^2 d\beta,$$

$$z = i\int \theta_1\theta_2 d\alpha + \frac{1}{\lambda}\frac{\partial\theta_1}{\partial\beta}\frac{\partial\theta_2}{\partial\beta}\,d\beta,$$

define a surface referred to its minimal lines.

36. Variation of a function. Let S be a surface referred to any system of coördinates u, v, and let $\phi(u, v)$ be a function of u and v. When the values of the coördinates of a point M of the surface are substituted in ϕ, we obtain a number c; and consequently the curve

$$\phi(u, v) = c \tag{38}$$

passes through M. In a displacement from M along this curve the value of ϕ remains the same, but in any other direction it changes and the rate of change is given by

$$\frac{d\phi}{ds}=\frac{\frac{\partial\phi}{\partial u}+\frac{\partial\phi}{\partial v}k}{\sqrt{E+2Fk+Gk^2}},$$

where $k=dv/du$ determines the direction. As thus written it is understood that the denominator of the right-hand member is positive.

For the present we consider the absolute value of $\frac{d\phi}{ds}$, and write

$$A=\left|\frac{d\phi}{ds}\right|=e\frac{\frac{\partial\phi}{\partial u}+\frac{\partial\phi}{\partial v}k}{\sqrt{E+2Fk+Gk^2}}, \tag{39}$$

where e is ± 1 according as the sign of the numerator is positive or negative. The minimum value of A is zero and corresponds to the direction along the curve (38). In order to find the maximum value we equate to zero the derivative of A with respect to k. This gives

$$\left(E\frac{\partial\phi}{\partial v}-F\frac{\partial\phi}{\partial u}\right)+\left(F\frac{\partial\phi}{\partial v}-G\frac{\partial\phi}{\partial u}\right)k=0.$$

From (32) it follows that this value of k determines the direction at right angles to the tangent to $\phi=c$ at the point. By substituting this value of k in (39) we get the maximum value of A. Hence:

The differential quotient $\frac{d\phi}{ds}$ *of a function* $\phi(u, v)$ *at a point on a surface varies in value with the direction from the point. It equals zero in the direction tangent to the curve* $\phi=c$, *and attains its greatest absolute value in the direction normal to this curve, this value being*

$$\left|\frac{d\phi}{ds}\right|=\frac{\sqrt{E\left(\frac{\partial\phi}{\partial v}\right)^2-2F\frac{\partial\phi}{\partial u}\frac{\partial\phi}{\partial v}+G\left(\frac{\partial\phi}{\partial u}\right)^2}}{\sqrt{EG-F^2}}. \tag{40}$$

A means of representing graphically the magnitude of the differential quotient A for any direction is given by the following theorem:

If in the tangent plane to a surface at a point M *the positive half tangents at* M, *corresponding to all values of* k, *positive and negative,* *be drawn, and on them the corresponding lengths A be laid off from M, the locus of the extremities of these lengths is a circle tangent to the curve $\phi =$ const.*

The proof of this theorem is simplified if we effect a transformation of curvilinear coördinates. Thus we take for the new coördinate lines the curves $\phi=$ const. and their orthogonal trajectories. We let the former be denoted by $u_1=$ const. and the latter by $v_1=$ const., and indicate by subscript 1 functions in terms of these parameters. Now $F_1=0$, so that

$$A=\frac{du_1}{ds}=\frac{1}{\sqrt{E_1+G_1k_1^2}},$$

where k_1 denotes the value of dv_1/du_1 which determines a given direction, and the maximum length is $(E_1)^{-\frac{1}{2}}$. From (23) we have

$$\cos\theta_0=\frac{\sqrt{E_1}}{\sqrt{E_1+G_1k_1^2}},\qquad \sin\theta_0=\frac{\sqrt{G_1}k_1}{\sqrt{E_1+G_1k_1^2}},$$

where θ_0 is the angle which the given direction makes with the tangent to the curve $v_1=$ const. Hence if we regard the tangents at M to the curves $v_1=$ const. and $u_1=$ const. as axes of coördinates in the tangent plane, the coördinates of the end of a segment of length A are

$$\frac{\sqrt{E_1}}{E_1+G_1k_1^2},\qquad \frac{\sqrt{G_1}k_1}{E_1+G_1k_1^2}.$$

The distance from this point to the mid-point of the maximum segment, measured along the tangent to $v_1=$ const., is readily found to be $\dfrac{1}{2\sqrt{E_1}}$, which proves the theorem.

37. Differential parameters of the first order. If we put

$$(41)\qquad \Delta_1\phi=\frac{E\left(\dfrac{\partial\phi}{\partial v}\right)^2-2F\dfrac{\partial\phi}{\partial u}\dfrac{\partial\phi}{\partial v}+G\left(\dfrac{\partial\phi}{\partial u}\right)^2}{EG-F^2},$$

equation (40) can be written

$$\left(\frac{d\phi}{ds}\right)^2=\Delta_1\phi,$$

where now the differential quotient corresponds to the direction normal to the curve $\phi=$ const. The left-hand member of this equation is evidently independent of the nature of the parameters u and v to which the surface is referred. Consequently the same is true of the right-hand member. Hence $\Delta_1\phi$ is unchanged in value when there is any change of parameters whatever. The full significance of this result is as follows. Given a new set of parameters defined by $u = f_1(u_1, v_1)$, $v = f_2(u_1, v_1)$; let $\phi_1(u_1, v_1)$ denote the result of substituting these expressions for u and v in $\phi(u, v)$, and write the linear element thus:

$$ds^2 = E_1 du_1^2 + 2 F_1 du_1 dv_1 + G_1 dv_1^2.$$

The invariance of $\Delta_1\phi$ under this transformation is expressed by the identical equation

$$\frac{E\left(\frac{\partial\phi}{\partial v}\right)^2 - 2F\frac{\partial\phi}{\partial u}\frac{\partial\phi}{\partial v} + G\left(\frac{\partial\phi}{\partial u}\right)^2}{EG - F^2} = \frac{E_1\left(\frac{\partial\phi_1}{\partial v_1}\right)^2 - 2F_1\frac{\partial\phi_1}{\partial u_1}\frac{\partial\phi_1}{\partial v_1} + G_1\left(\frac{\partial\phi_1}{\partial u_1}\right)^2}{E_1G_1 - F_1^2}.$$

We leave it to the reader to verify this directly with the aid of equations (9). The invariant $\Delta_1\phi$ is called the *differential parameter of the first order;* this name and the notation are due to Lamé.*

Consider for the moment the partial differential equation

$$\Delta_1\phi = 0 \tag{42}$$

and a solution $\phi = \text{const.}$ From the latter we get, by differentiation,

$$\frac{\partial\phi}{\partial u}du + \frac{\partial\phi}{\partial v}dv = 0.$$

If we replace $\frac{\partial\phi}{\partial u}$ and $\frac{\partial\phi}{\partial v}$ in (42) by dv and $-du$, which are evidently proportional to them, we obtain

$$E du^2 + 2 F du dv + G dv^2 = 0.$$

Hence the integral curves of equation (42) are lines of length zero, and conversely if $\phi = \text{const.}$ is a line of length zero, the function ϕ is a solution of equation (42).

Another particular case is that in which $\Delta_1\phi$ is a function of ϕ, say

$$\Delta_1\phi = F(\phi). \tag{43}$$

* *Leçons sur les coordonnées curvilignes et leurs diverses applications*, p. 5. Paris, 1859.

From (41) it is seen that when we put

$$\theta = \int \frac{d\phi}{\sqrt{F(\phi)}}$$

equation (43) becomes

$$\Delta_1 \theta = 1. \tag{44}$$

As defined, θ is a function of ϕ; hence the family of curves $\theta = \text{const.}$ is the same as the family $\phi = \text{const.}$ Suppose we have such a family, and we take the curves $\theta = \text{const.}$ for the curves $u = \text{const.}$ and their orthogonal trajectories for $v = \text{const.}$, thus effecting a change of parameters. Since $\Delta_1 u = 1$, it follows from (41) that $E = 1$, and consequently the linear element is

$$ds^2 = du^2 + G\,dv^2. \tag{45}$$

Since now the linear element of a curve $v = \text{const.}$ is du, the length of the curve between its points of intersection with two curves $u = u_0$ and $u = u_1$ is $u_1 - u_0$. Moreover, this length is the same for the segment of every curve $v = \text{const.}$ between these two curves. For this reason the latter curves are said to be *parallel.* Conversely, in order that the curves $u = \text{const.}$ of an orthogonal system be parallel, it is necessary that the linear element of the curves $v = \text{const.}$ be independent of v. Hence E must be a function of u alone, which, by a transformation of coördinates, can be made equal to unity. Hence we have the theorem:

A necessary and sufficient condition that the curves of a family $\phi = \text{const.}$ be parallel is that $\Delta_1 \phi$ be a function of ϕ.

Let $\phi = \text{const.}$ and $\psi = \text{const.}$ be the equations of two curves upon a surface, through a point M, and let θ denote the angle between the tangents at M. If we put

$$\Delta_1(\phi, \psi) = \frac{E\dfrac{\partial \phi}{\partial v}\dfrac{\partial \psi}{\partial v} - F\left(\dfrac{\partial \phi}{\partial v}\dfrac{\partial \psi}{\partial u} + \dfrac{\partial \phi}{\partial u}\dfrac{\partial \psi}{\partial v}\right) + G\dfrac{\partial \phi}{\partial u}\dfrac{\partial \psi}{\partial u}}{EG - F^2}, \tag{46}$$

the expression (21) for $\cos\theta$ can be written

$$\cos\theta = \frac{\Delta_1(\phi, \psi)}{\sqrt{\Delta_1 \phi} \cdot \sqrt{\Delta_1 \psi}}. \tag{47}$$

Since $\cos\theta$ is an invariant for transformations of coördinates, it follows from this equation that $\Delta_1(\phi, \psi)$ also is an invariant. It is called the *mixed differential parameter of the first order.* An immediate consequence of (47) is that

$$\Delta_1(\phi, \psi) = 0$$

is the condition of orthogonality of the curves $\phi = \text{const.}$ and $\psi = \text{const.}$

Now equation (22) can be written

$$\sin\theta = \frac{1}{H\sqrt{\Delta_1\phi}\sqrt{\Delta_1\psi}}\left(\frac{\partial\phi}{\partial u}\frac{\partial\psi}{\partial v} - \frac{\partial\phi}{\partial v}\frac{\partial\psi}{\partial u}\right),$$

which by means of the function $\Theta(u, v)$, defined thus by Darboux,*

$$\Theta(\phi, \psi) = \frac{1}{H}\left(\frac{\partial\phi}{\partial u}\frac{\partial\psi}{\partial v} - \frac{\partial\phi}{\partial v}\frac{\partial\psi}{\partial u}\right), \tag{48}$$

can be written in the abbreviated form

$$\sin\theta = \frac{\Theta(\phi, \psi)}{\sqrt{\Delta_1\phi}\cdot\sqrt{\Delta_1\psi}}. \tag{49}$$

Since all the functions in this identity except $\Theta(\phi, \psi)$ are known to be invariants, we have a proof that it also is an invariant. It is a mixed differential parameter of the first order. From (47) and (49) it follows that

$$\Delta_1^2(\phi, \psi) + \Theta^2(\phi, \psi) = \Delta_1\phi\cdot\Delta_1\psi; \tag{50}$$

consequently the three invariants defined thus far are not independent of one another.

From (41) and (46) it follows that

$$\Delta_1 u = \frac{G}{H^2}, \quad \Delta_1(u, v) = \frac{-F}{H^2}, \quad \Delta_1 v = \frac{E}{H^2},$$

and from these we find

$$\Theta^2(u, v) = \Delta_1 u\cdot\Delta_1 v - \Delta_1^2(u, v) = \frac{1}{H^2}. \tag{51}$$

Consequently

$$E = \frac{\Delta_1 v}{\Theta^2(u, v)}, \quad F = \frac{-\Delta_1(u, v)}{\Theta^2(u, v)}, \quad G = \frac{\Delta_1 u}{\Theta^2(u, v)}. \tag{52}$$

Equations (13) can be written in similar form.

* *Leçons*, Vol. III, p. 197.

Another result of these equations is the following. If the parameters of the surface are changed in accordance with the equations

$$u_1 = u_1(u, v), \qquad v_1 = v_1(u, v),$$

and the resulting linear element is written,

$$ds^2 = E_1 du_1^2 + 2 F_1 du_1 dv_1 + G_1 dv_1^2,$$

the value of E_1 is given by

$$E_1 = \frac{\Delta_1 v_1}{\Theta^2(u_1, v_1)} = \frac{E\left(\frac{\partial v_1}{\partial v}\right)^2 - 2F\frac{\partial v_1}{\partial u}\frac{\partial v_1}{\partial v} + G\left(\frac{\partial v_1}{\partial u}\right)^2}{\left[\frac{\partial (u_1, v_1)}{\partial (u, v)}\right]^2};$$

and F_1 and G_1 are found in like manner. In consequence of (51) these equations are equivalent to (13), which were found by direct calculation.

38. Differential parameters of the second order. Thus far we have considered differential invariants of the first order only. We introduce now one of the second order, discovered by Beltrami.* To this end we study the integral

$$\Pi = \iint \Delta_1(\phi, \psi)\, d\Sigma = \iint \Delta_1(\phi, \psi)\, H\, du\, dv,$$

for an ordinary portion of the surface bounded by a closed curve C (cf. § 33). For convenience we put

$$(53) \qquad M = \frac{G\frac{\partial \psi}{\partial u} - F\frac{\partial \psi}{\partial v}}{H}, \qquad N = \frac{E\frac{\partial \psi}{\partial v} - F\frac{\partial \psi}{\partial u}}{H},$$

so that, in consequence of (46), we have

$$\Pi = \iint \left(M\frac{\partial \phi}{\partial u} + N\frac{\partial \phi}{\partial v}\right) du\, dv.$$

This may be written

$$\Pi = \iint \left[\frac{\partial}{\partial u}(M\phi) + \frac{\partial}{\partial v}(N\phi)\right] du\, dv - \iint \phi \left(\frac{\partial M}{\partial u} + \frac{\partial N}{\partial v}\right) du\, dv.$$

If we apply Green's theorem to the first integral, this equation reduces to

$$(54) \qquad \Pi = \int_c \phi\,(M\, dv - N\, du) - \iint \phi \left(\frac{\partial M}{\partial u} + \frac{\partial N}{\partial v}\right) du\, dv,$$

* Ricerche di analisi applicata alla geometria, *Giornale di matematiche*, Vol. II (1864), p. 365.

where the first integral is curvilinear and is taken about C in the customary manner. Evidently du and dv refer to a displacement along C. If we indicate by δ variations in directions normal to C and directed toward the exterior of the contour, then from (23) and (25) it follows that

$$\text{(55)} \qquad \frac{E\,du + F\,dv}{ds} = \frac{-H\,\delta v}{\delta s}, \qquad \frac{F\,du + G\,dv}{ds} = \frac{H\,\delta u}{\delta s}.$$

Hence $$M\,dv - N\,du = \left(\frac{\partial\psi}{\partial u}\frac{\delta u}{\delta s} + \frac{\partial\psi}{\partial v}\frac{\delta v}{\delta s}\right)ds = \frac{\delta\psi}{\delta s}\,ds,$$

so that $$\Pi = \iint \Delta_1(\phi, \psi)\,d\Sigma = \int_c \phi\,\frac{\delta\psi}{\delta s}\,ds - \iint \phi\,\frac{1}{H}\left(\frac{\partial M}{\partial u} + \frac{\partial N}{\partial v}\right)d\Sigma.$$

All of the terms in this equation, with the exception of $\frac{1}{H}\left(\frac{\partial M}{\partial u} + \frac{\partial N}{\partial v}\right)$, are independent of the choice of parameters. Hence the latter is an invariant. It is called *the differential parameter of the second order* and is denoted by $\Delta_2\psi$. In consequence of (53) we have

$$\text{(56)} \qquad \Delta_2\psi = \frac{1}{H}\left\{\frac{\partial}{\partial u}\left(\frac{G\dfrac{\partial\psi}{\partial u} - F\dfrac{\partial\psi}{\partial v}}{H}\right) + \frac{\partial}{\partial v}\left(\frac{E\dfrac{\partial\psi}{\partial v} - F\dfrac{\partial\psi}{\partial u}}{H}\right)\right\}.$$

In the foregoing discussion it has been assumed that only real quantities appear. But all these results can be obtained directly from algebraic considerations of quadratic differential forms* without any hypothesis regarding the character of the variables; hence the differential parameters can be used for any kind of curvilinear coördinates.

In addition to $\Delta_2\phi$ there are other differential invariants of the second order, such as

$$\Delta_1\Delta_1\phi, \qquad \Delta_1(\phi, \Delta_1\phi), \qquad \Theta(\phi, \Delta_1\phi).$$

And $$\Delta_1\Delta_1(\phi, \psi), \qquad \Delta_1(\Delta_1\phi, \Delta_1\psi), \qquad \Theta(\Delta_1\phi, \Delta_1\psi)$$

are mixed invariants of the second order. In like manner we can find a group of invariants of the third order; for instance,

$$\Delta_1\Delta_1\Delta_1\phi, \quad \Delta_1\Delta_1(\phi, \Delta_1\phi), \quad \Delta_1\Delta_2\phi, \quad \Delta_2\Delta_1\phi, \cdots.$$

* Cf. Bianchi, *Lezioni di geometria differenziale*, Vol. I, chap. ii. Pisa, 1902.

These invariants and others, which can be obtained by an evident extension of this method, involve functions ϕ, ψ, $\cdots$, E, F, G, and their derivatives.

Conversely, we shall show * that every invariant of the form

$$I = f\left(E, F, G, \frac{\partial E}{\partial u}, \cdots, \frac{\partial G}{\partial v}, \cdots, \phi, \frac{\partial \phi}{\partial u}, \cdots, \psi, \frac{\partial \psi}{\partial u}, \cdots\right),$$

where ϕ, ψ, $\cdots$ are independent functions, is expressible by means of the symbols Δ and Θ. Already we have seen that E, F, and G can be expressed in terms of $\Delta_1 u$, $\Delta_1 v$, and $\Delta_1(u, v)$. Moreover, from (48) it follows that

$$\frac{\partial \lambda}{\partial u} = H\Theta(\lambda, v), \qquad \frac{\partial \lambda}{\partial v} = H\Theta(u, \lambda),$$

when λ is any function whatever. Hence all the terms in I can be expressed in terms of the symbols Δ and Θ, applied to

$$u, \quad v, \quad \phi, \quad \psi, \quad \cdots.$$

Since u and v do not appear explicitly in I, we can effect a change of parameters, replacing u and v by ϕ and ψ respectively, and consequently we express I in terms of ϕ, ψ, $\cdots$, and the differential invariants obtained by applying the operators Δ and Θ to these functions. In case ϕ is the only function appearing in I, we can take for ψ, in the change of parameters, any invariant of ϕ, such as $\Delta_1\phi$ or $\Delta_2\phi$, so long as it is not a function of ϕ, E, F, or G.

EXAMPLES

1. When the linear element of a surface is in the form

$$ds^2 = \lambda(du^2 + dv^2),$$

where λ is a function of u and v, both u and v are solutions of the equation $\Delta_2\theta = 0$, the differential parameter being formed with respect to the right-hand member.

2. Show that on the surface

$$x = u\cos v, \qquad y = u\sin v, \qquad z = av + \phi(u),$$

the curves $u =$ const. are parallel.

3. When the linear element is in the form

$$ds^2 = \cos^2\alpha\, du^2 + \sin^2\alpha\, dv^2,$$

where α is a function of u and v, both u and v are solutions of the equation

$$\Delta_1(\theta, \Delta_1\theta) = 2\,\Delta_2\theta\,(\Delta_1\theta - 1).$$

* Cf. Beltrami, *l.c.*, p. 357.

4. If the curves $\phi=$ const., $\psi=$ const. form an orthogonal system on a surface, the projection on the x-axis of any displacement on the surface is given by

$$dx=\frac{dx}{ds}\frac{d\psi}{\sqrt{\Delta\psi}}+\frac{dx}{d\sigma}\frac{d\phi}{\sqrt{\Delta\phi}},$$

where ds and $d\sigma$ are the elements of length of the curves $\phi=$ const., $\psi=$ const. respectively.

5. If f and ϕ are any functions of u and v, then

$$\Delta_1 f=\left(\frac{\partial f}{\partial u}\right)^2\Delta_1 u+2\frac{\partial f}{\partial u}\frac{\partial f}{\partial v}\Delta_1(u,v)+\left(\frac{\partial f}{\partial v}\right)^2\Delta_1 v,$$

$$\Delta_1(f,\phi)=\frac{\partial f}{\partial u}\frac{\partial \phi}{\partial u}\Delta_1 u+\left(\frac{\partial f}{\partial u}\frac{\partial \phi}{\partial v}+\frac{\partial f}{\partial v}\frac{\partial \phi}{\partial u}\right)\Delta_1(u,v)+\frac{\partial f}{\partial v}\frac{\partial \phi}{\partial v}\Delta_1 v,$$

$$\Delta_2 f=\frac{\partial f}{\partial u}\Delta_2 u+\frac{\partial f}{\partial v}\Delta_2 v+\frac{\partial^2 f}{\partial u^2}\Delta_1 u+2\frac{\partial^2 f}{\partial u\,\partial v}\Delta_1(u,v)+\frac{\partial^2 f}{\partial v^2}\Delta_1 v.$$

39. Symmetric coördinates. We have seen that through every point of a surface there pass two minimal curves which lie entirely on the surface, and that these curves are defined by the differential equation

$$E\,du^2+2\,F\,dudv+G\,dv^2=0.$$

If the finite equations of these curves be written

$$\alpha(u,v)=\text{const.},\qquad \beta(u,v)=\text{const.},$$

it follows from (42) that

(57) $$\Delta_1(\alpha)=0,\qquad \Delta_1(\beta)=0.$$

Since for any parameters

(58) $$E=\frac{\Delta_1 v}{\Theta^2(u,v)},\qquad F=-\frac{\Delta_1(u,v)}{\Theta^2(u,v)},\qquad G=\frac{\Delta_1 u}{\Theta^2(u,v)},$$

when the curves $\alpha=$ const., $\beta=$ const., are taken as parametric, the corresponding coefficients E and G are zero, and consequently the linear element of the surface has the form

(59) $$ds^2=\lambda\,d\alpha\,d\beta,$$

where, in general, λ is a function of α and β. Conversely, as follows from (58), when the linear element has the form (59) equations (57) are satisfied and the parametric curves are minimal. Hence the only transformations of coördinates which preserve this form of the linear element are those which leave the minimal lines parametric, that is

(60) $$\left\{\begin{array}{l}\alpha_1=F(\alpha),\quad \beta_1=F_1(\beta),\\ \text{or}\\ \alpha_1=F(\beta),\quad \beta_1=F_1(\alpha),\end{array}\right.$$

where F and F_1 are arbitrary functions. Whenever the linear element has the form (59), we say that the parameters are *symmetric*. The above results are given by the theorem:

When α and β are symmetric coördinates of a surface, any two arbitrary functions of α and β respectively are symmetric coördinates, and they are the only ones.

The general linear element of a surface can be written as the product of two factors, namely

$$ds^2 = \left(\sqrt{E}\,du + \frac{F+iH}{\sqrt{E}}\,dv\right)\left(\sqrt{E}\,du + \frac{F-iH}{\sqrt{E}}\,dv\right). \tag{61}$$

If t and t_1 denote integrating factors of the respective terms of the right-hand member of this equation, a pair of symmetric coördinates is given by the quadratures

$$\begin{cases} t\left(\sqrt{E}\,du + \dfrac{F+iH}{\sqrt{E}}\,dv\right) = d\alpha, \\ t_1\left(\sqrt{E}\,du + \dfrac{F-iH}{\sqrt{E}}\,dv\right) = d\beta. \end{cases} \tag{62}$$

When these values are substituted in (61), and the result is compared with (59), it is seen that $\lambda = \dfrac{1}{tt_1}$.

The first of equations (62) can be replaced by

$$t\sqrt{E} = \frac{\partial\alpha}{\partial u}, \qquad t\frac{F+iH}{\sqrt{E}} = \frac{\partial\alpha}{\partial v}.$$

Eliminating t from these equations, we have

$$\frac{E\dfrac{\partial\alpha}{\partial v} - F\dfrac{\partial\alpha}{\partial u}}{H} = i\frac{\partial\alpha}{\partial u}. \tag{63}$$

If this equation be multiplied by $\dfrac{F-iH}{E}$, the result can be reduced to

$$\frac{F\dfrac{\partial\alpha}{\partial v} - G\dfrac{\partial\alpha}{\partial u}}{H} = i\frac{\partial\alpha}{\partial v}. \tag{64}$$

From these equations it follows that

$$\frac{\partial}{\partial v}\left(\frac{E\frac{\partial \alpha}{\partial v}-F\frac{\partial \alpha}{\partial u}}{H}\right)=\frac{\partial}{\partial u}\left(\frac{F\frac{\partial \alpha}{\partial v}-G\frac{\partial \alpha}{\partial u}}{H}\right),$$

or, by (56),

$$\Delta_2 \alpha = 0. \tag{65}$$

It is readily found that β also satisfies this condition.

40. Isothermic and isometric parameters. When the surface is real, and the coördinates also, the factors in (61) are conjugate imaginary. Hence the conjugate imaginary of t can be taken for t_1. In this case α and β are conjugate imaginary also. In what follows we assume that this choice has been made, and write

$$\alpha = \phi + i\psi, \qquad \beta = \phi - i\psi. \tag{66}$$

If these values be substituted in (59), we get

$$ds^2 = \lambda(d\phi^2 + d\psi^2). \tag{67}$$

At once we see that the curves $\phi = \text{const.}$ and $\psi = \text{const.}$ form an orthogonal system. Moreover, the elements of arc of these lines are $\sqrt{\lambda}d\psi$ and $\sqrt{\lambda}d\phi$ respectively. Consequently when the increments $d\phi$ and $d\psi$ are taken equal, the four points (ϕ, ψ), $(\phi + d\phi, \psi)$, $(\phi, \psi + d\psi)$, $(\phi + d\phi, \psi + d\psi)$ are the vertices of a small square. Hence the curves $\phi = \text{const.}$ and $\psi = \text{const.}$ divide the surface into a network of small squares. On this account these curves are called *isometric curves*, and ϕ and ψ *isometric parameters*. These lines are of importance in the theory of heat, and are termed *isothermal* or *isothermic*, which names are used in this connection as synonymous with *isometric*.

Whenever the linear element can be put in the symmetric form, equations similar to (66) give at once a set of isometric parameters. And conversely, the knowledge of a set of isometric parameters leads at once to a set of symmetric parameters. But we have seen that when one system of symmetric parameters is known, all the others are given by equations of the form (60). Hence we have the theorem:

Given any pair of real isometric parameters ϕ, ψ for a surface; every other pair ϕ_1, ψ_1 is given by equations of the form

$$\phi_1 + i\psi_1 = F(\phi \pm i\psi), \qquad \phi_1 - i\psi_1 = F_0(\phi \mp i\psi),$$

where F and F_0 are any functions conjugate imaginary to one another.

Consider, for instance, the case

$$\phi_1 + i\psi_1 = F(\phi + i\psi). \tag{68}$$

From the Cauchy-Riemann differential equations

$$\frac{\partial \phi_1}{\partial \phi} = \frac{\partial \psi_1}{\partial \psi}, \qquad \frac{\partial \phi_1}{\partial \psi} = -\frac{\partial \psi_1}{\partial \phi}, \tag{69}$$

it follows that ϕ_1 and ψ_1 are functions of both ϕ and ψ. Hence the curves $\phi_1 = \text{const.}$, $\psi_1 = \text{const.}$ are different from the system $\phi = \text{const.}$, $\psi = \text{const.}$ Similar results hold when $+i$ is replaced by $-i$ in the argument of the right-hand member of (68). Hence

There is a double infinity of isometric systems of lines upon a surface; when one system is known all the others can be found directly.

If the value (66) for α be substituted in the first of equations (57), the resulting equation is reducible to

$$\Delta_1\phi - \Delta_1\psi + 2i\Delta_1(\phi, \psi) = 0.$$

Since ϕ and ψ are real, this equation is equivalent to

$$\Delta_1\phi = \Delta_1\psi, \qquad \Delta_1(\phi, \psi) = 0. \tag{70}$$

From (58) it is seen that these equations are the condition that $E = G$, $F = 0$, when ϕ and ψ are the parameters. Hence equations (70) are the necessary and sufficient conditions that ϕ and ψ be isometric parameters.

Again, when α in (65) is replaced by $\phi + i\psi$, and all the functions are real, we have

$$\Delta_2\phi = 0, \qquad \Delta_2\psi = 0. \tag{71}$$

Conversely, when we have a function ϕ satisfying the first of these equations, the expression

$$\frac{F\dfrac{\partial \phi}{\partial u} - E\dfrac{\partial \phi}{\partial v}}{H}\,du + \frac{G\dfrac{\partial \phi}{\partial u} - F\dfrac{\partial \phi}{\partial v}}{H}\,dv$$

is an exact differential. Call it $d\psi$; then

$$\frac{F\dfrac{\partial \phi}{\partial u} - E\dfrac{\partial \phi}{\partial v}}{H} = \frac{\partial \psi}{\partial u}, \qquad \frac{G\dfrac{\partial \phi}{\partial u} - F\dfrac{\partial \phi}{\partial v}}{H} = \frac{\partial \psi}{\partial v}. \tag{72}$$

If these equations be solved for $\frac{\partial \phi}{\partial u}$, $\frac{\partial \phi}{\partial v}$, we get

$$\text{(73)} \qquad \frac{E\frac{\partial \psi}{\partial v} - F\frac{\partial \psi}{\partial u}}{H} = \frac{\partial \phi}{\partial u}, \qquad \frac{F\frac{\partial \psi}{\partial v} - G\frac{\partial \psi}{\partial u}}{H} = \frac{\partial \phi}{\partial v}.$$

When we express the condition $\frac{\partial}{\partial v}\left(\frac{\partial \phi}{\partial u}\right) = \frac{\partial}{\partial u}\left(\frac{\partial \phi}{\partial v}\right)$, we find that $\Delta_2 \psi = 0$. Moreover, these two functions ϕ and ψ satisfy (70), in consequence of (72) and (73), and therefore they are isometric parameters. Hence:

A necessary and sufficient condition that ϕ be the isometric parameter of one family of an isometric system on a surface is that $\Delta_2 \phi = 0$; the isometric parameter of the other family is given by a quadrature.

Incidentally we remark that if u and v are a pair of isometric parameters, equations (72) and (73) reduce to (69).

41. Isothermic orthogonal systems. If the linear element of a surface is given in the form (67) and the parameters are changed in accordance with the equations

$$\phi = f_1(u), \qquad \psi = f_2(v),$$

the linear element becomes

$$ds^2 = \lambda(f_1'^2 du^2 + f_2'^2 dv^2),$$

where the accents indicate differentiation. However, this transformation of parameters has not changed the coördinate lines; the coefficients are now no longer equal, but in the relation

$$\text{(74)} \qquad \frac{E}{G} = \frac{U}{V},$$

where U and V denote functions of u and v respectively.

Conversely, when this relation is satisfied the linear element may be written

$$ds^2 = \lambda(U\,du^2 + V\,dv^2),$$

and by the transformation of coördinates,

$$\text{(75)} \qquad \phi = \int \sqrt{U}\,du, \qquad \psi = \int \sqrt{V}\,dv,$$

it is brought to the form (67), whatever be U and V; and the coördinate lines are unaltered. Hence:

A necessary and sufficient condition that an orthogonal system of parametric lines on a surface form an isothermic system is that the coefficients of the corresponding linear element satisfy a relation of the form (74).

We seek now the necessary and sufficient condition which a function $\omega(u, v)$ must satisfy in order that the curves $\omega=$ const. and their orthogonal trajectories form an isothermic system. Either ω, or a function of it, is the isothermic parameter of the curves $\omega=$ const. We denote this parameter by ϕ; then $\phi=f(\omega)$. Since ϕ must be a solution of equations (71), we have, on substitution,

$$(76) \qquad \Delta_2\omega \cdot f'(\omega)+\Delta_1\omega \cdot f''(\omega)=0,$$

where the primes indicate differentiation with respect to ω. If this equation is written in the form

$$\frac{\Delta_2\omega}{\Delta_1\omega}=-\frac{f''(\omega)}{f'(\omega)},$$

we see that the ratio of the two differential parameters is a function of ω. Conversely, if this ratio is a function of ω, the function $f(\omega)$, obtained by two quadratures from

$$(77) \qquad f'(\omega)=e^{-\int\frac{\Delta_2\omega}{\Delta_1\omega}d\omega},$$

will satisfy equations (71). Hence:

A necessary and sufficient condition that a family of curves $\omega=$ const. and their orthogonal trajectories form an isothermic system is that the ratio of $\Delta_2\omega$ and $\Delta_1\omega$ be a function of ω.

Suppose we have such a function ω; then the orthogonal trajectories of the curves $\omega=$ const. can be found by quadrature; for, the differential equation of these trajectories is

$$(78) \qquad \left(E\frac{\partial\omega}{\partial v}-F\frac{\partial\omega}{\partial u}\right)du+\left(F\frac{\partial\omega}{\partial v}-G\frac{\partial\omega}{\partial u}\right)dv=0.$$

If equation (76) be written in the form

$$\frac{\partial}{\partial v}\left[f'(\omega)\,\frac{E\dfrac{\partial\omega}{\partial v}-F\dfrac{\partial\omega}{\partial u}}{H}\right]+\frac{\partial}{\partial u}\left[f'(\omega)\,\frac{G\dfrac{\partial\omega}{\partial u}-F\dfrac{\partial\omega}{\partial v}}{H}\right]=0,$$

it is seen that an integrating factor of equation (78) is $f'(\omega)/H$, where $f'(\omega)$ is given by (77). Hence $f(\omega)$ and the function ψ, obtained by the quadrature

$$(79)\qquad \frac{\partial\psi}{\partial u} = -f'(\omega)\frac{E\frac{\partial\omega}{\partial v} - F\frac{\partial\omega}{\partial u}}{H}, \qquad \frac{\partial\psi}{\partial v} = f'(\omega)\frac{G\frac{\partial\omega}{\partial u} - F\frac{\partial\omega}{\partial v}}{H},$$

are a pair of isometric parameters. From these equations and (77) it follows that

$$\Delta_1\psi = e^{-2\int\frac{\Delta_2\omega}{\Delta_1\omega}d\omega}\,\Delta_1\omega,$$

and consequently, by means of (52), the linear element can be given the form

$$(80)\qquad ds^2 = \frac{1}{\Delta_1\omega}\left(d\omega^2 + e^{2\int\frac{\Delta_2\omega}{\Delta_1\omega}d\omega}\,d\psi^2\right).$$

The linear element of the plane referred to rectangular axes is $ds^2 = dx^2 + dy^2$. Consequently x and y are isothermic parameters, and we have the theorem:

The plane curves whose equations are obtained by equating to constants the real and imaginary parts of any function of $x + iy$ or $x - iy$ form an isothermal orthogonal system; and every such system can be obtained in this way.

For example, consider
$$\phi + i\psi = \frac{c^2}{x - iy},$$
where c is any constant. From this it follows that

$$\phi = \frac{c^2x}{x^2 + y^2}, \qquad \psi = \frac{c^2y}{x^2 + y^2}.$$

Hence the circles $\phi = \text{const.}$, $\psi = \text{const.}$ form an isothermal orthogonal system, and ϕ and ψ are isothermic parameters.

The above system of circles is a particular case of the system considered in § 34. We inquire whether the latter also form an isothermal system. If we put

$$\omega = x + \frac{1}{x}(y^2 - a^2),$$

we find that
$$\Delta_1\omega = \frac{1}{x^2}(\omega^2 + 4a^2), \qquad \Delta_2\omega = \frac{2\omega}{x^2}.$$

Hence the ratio of $\Delta_1\omega$ and $\Delta_2\omega$ is a function of ω, and consequently the system of circles is isothermal. From (77) it follows that the isothermic parameter of the first family is $\phi = \frac{1}{2a}\tan^{-1}\frac{\omega}{2a}$, and the parameter of the orthogonal family is

$$\psi = \frac{1}{2a}\tanh^{-1}\frac{\varpi}{2a}, \qquad \varpi = y + \frac{x^2 + a^2}{y}.$$

EXAMPLES

1. Show that the meridians and parallels on a sphere form an isothermal orthogonal system, and determine the isothermic parameters.

2. Show that a system of confocal ellipses and hyperbolas form an isothermal orthogonal system in the plane.

3. Show that the surface

$$\frac{x}{a}=\sqrt{\frac{(a^2-u)(a^2-v)}{(a^2-b^2)(a^2-c^2)}},\quad \frac{y}{b}=\sqrt{\frac{(b^2-u)(b^2-v)}{(b^2-a^2)(b^2-c^2)}},\quad \frac{z}{c}=\sqrt{\frac{(c^2-u)(c^2-v)}{(c^2-a^2)(c^2-b^2)}}$$

is an ellipsoid, and that the parametric curves form an isothermal orthogonal system.

4. Find the curves which bisect the angles between the parametric curves on the surface

$$\frac{x}{a}=\frac{u+v}{2},\quad \frac{y}{b}=\frac{u-v}{2},\quad z=\frac{uv}{2},$$

and show that they form an isothermal orthogonal system.

5. Determine $\phi(v)$ so that on the right conoid $x=u\cos v$, $y=u\sin v$, $z=\phi(v)$ the parametric curves form an isothermal orthogonal system, and show that the curves which bisect the angles between the parametric curves form a system of the same kind.

6. Express the results of Ex. 4, page 82, in terms of the parameters ϕ and ψ defined by (66).

42. Conformal representation. When a one-to-one correspondence of any kind is established between the points of two surfaces, either surface may be said to be *represented* on the other. Thus, if we roll out a cylindrical surface upon a plane and say that the points of the surface correspond to the respective points of the plane into which they are developed, we have a representation of the surface upon the plane. Furthermore, as there is no stretching or folding of the surface in this development of it upon the plane, lengths of lines and the magnitude of angles are unaltered. It is evidently impossible to make such a representation of every surface upon a plane, and, in general, two surfaces of this kind do not admit of such a representation upon one another. However, it is possible, as we shall see, to represent one surface upon another in such a way that the angles between corresponding lines on the surfaces are equal. In this case we say that one surface has *conformal representation* on the other.

In order to obtain the condition to be satisfied for a conformal representation of two surfaces S and S', we imagine that they are referred to a corresponding system of real lines in terms of the same parameters u, v, and that corresponding points have the same curvilinear coördinates. We write their linear elements in the respective forms

$$ds^2 = E\,du^2 + 2\,F\,du\,dv + G\,dv^2, \quad ds'^2 = E'\,du^2 + 2\,F'\,du\,dv + G'\,dv^2.$$

Since the angles ω and ω' between the coördinate lines at corresponding points must be equal, it is necessary that

$$\text{(81)} \qquad \frac{F}{\sqrt{EG}} = \frac{F'}{\sqrt{E'G'}}.$$

If θ_0 and θ_0' denote the angles which a curve on S and the corresponding curve on S' respectively make with the curves $v = \text{const.}$ at points of the former curves, we have, from (23) and (25),

$$\sin\theta_0 = \frac{H}{\sqrt{E}}\frac{dv}{ds}, \qquad \sin(\omega - \theta_0) = \frac{H}{\sqrt{G}}\frac{du}{ds},$$

$$\sin\theta_0' = \frac{H'}{\sqrt{E'}}\frac{dv}{ds'}, \qquad \sin(\omega' - \theta_0') = \frac{H'}{\sqrt{G'}}\frac{du}{ds'}.$$

By hypothesis $\omega' = \pm\,\omega$ and $\theta_0' = \pm\,\theta_0$, according as the angles have the same or opposite sense. Hence we have

$$\text{(82)} \qquad \frac{H'}{\sqrt{E'}}\frac{dv}{ds'} = \pm\frac{H}{\sqrt{E}}\frac{dv}{ds}, \qquad \frac{H'}{\sqrt{G'}}\frac{du}{ds'} = \pm\frac{H}{\sqrt{G}}\frac{du}{ds},$$

according to the sense of the angles. From these equations we find

$$\frac{\sqrt{E}}{\sqrt{G}} = \frac{\sqrt{E'}}{\sqrt{G'}},$$

which, in combination with (81), may be written

$$\text{(83)} \qquad \frac{E'}{E} = \frac{F'}{F} = \frac{G'}{G} = t^2,$$

where t^2 denotes the factor of proportionality, a function of u and v in general. From (83) it follows at once that

$$\text{(84)} \qquad ds'^2 = t^2\,ds^2.$$

And so when the proportion (83) is satisfied, the equations (81) and (82) follow. Hence we have the theorem:

A necessary and sufficient condition that the representation of two surfaces referred to a corresponding system of lines be conformal is that the first fundamental coefficients of the two surfaces be proportional, the factor of proportionality being a function of the parameters; the representation is direct or inverse according as the relative positions of the positive half tangents to the parametric curves on the two surfaces are the same or symmetric.

Later we shall find means of obtaining conformal representations.

From (84) it follows that small arcs measured from corresponding points on S and S' along corresponding curves are in the same ratio, the factor of proportionality being in general a function of the position of the point. Conversely, when the ratio is the same for all curves at a point, there is a relation such as (84), with t a function of u and v at most. And since it holds for all directions, we must have the proportion (83). On this account we may say that two surfaces are represented conformally upon one another when in the neighborhood of each pair of homologous points corresponding small lengths are proportional.

43. Isometric representation. Applicable surfaces. When in particular the factor t is equal to unity, corresponding small lengths are equal as well as angles. In this case the representation is said to be *isometric*, and the two surfaces are said to be *applicable*. The significance of the latter term is that the portion of one surface in the neighborhood of every point can be so bent as to be made to coincide with the corresponding portion of the other surface without stretching or duplication. It is evident that such an application of one surface upon another necessitates a continuous array of surfaces applicable to both S and S_1. This process of transformation is called *deformation*, and S_1 is called a *deform* of S and vice versa. An example of this is afforded by the rolling of a cylinder on a plane.

Although a conformal representation can be established between any two surfaces, it is not true, as we shall see later, that any two surfaces admit of an isometric representation upon one another. From time to time we shall meet with examples of applicable surfaces, and in a later chapter we shall discuss at length problems which arise concerning the applicability of surfaces. However, we consider here an example afforded by the tangent surface of a twisted curve.

We recall that if x, y, z are the coördinates of a point on the curve, expressed in terms of the arc, the equations of the surface are of the form $\xi = x + x't, \quad \eta = y + y't, \quad \zeta = z + z't,$

and the linear element of the surface is

$$d\sigma^2 = \left(1 + \frac{t^2}{\rho^2}\right) ds^2 + 2\, ds dt + dt^2,$$

where ρ denotes the radius of curvature of the curve.

Since this expression does not involve the radius of torsion, it follows that the tangent surfaces to all curves which have the same intrinsic equation $\rho = f(s)$ are applicable in such a way that points on the curves determined by the same value of s correspond. As there is a plane curve with this equation, the surface is applicable to the plane in such a way that points of the surface correspond to points of the plane on the convex side of the plane curve.

The tangents to a curve are the characteristics of the osculating planes as the point of osculation moves along the curve, and consequently they are the axes of rotation of the osculating plane as it moves enveloping the surface. Instead of rolling the plane over the tangent surface, we may roll the surface over the plane and bring all of its points into coincidence with the plane. It is in this sense that the surface is developable upon a plane, and for this reason it is called a developable surface (cf. § 27). Later it will be shown that every surface applicable to the plane is the tangent surface of a curve (§ 64).

44. Conformal representation of a surface upon itself. We return to the consideration of conformal representation, and remark that another consequence of equations (83) is that the minimal curves correspond upon S and S'. Conversely, when two surfaces are referred to a corresponding system of lines, if the minimal lines on the two surfaces correspond, equations (83) must hold. Hence:

A necessary and sufficient condition that the representation of two surfaces upon one another be conformal is that the minimal lines correspond.

If the minimal lines upon the two surfaces are known and taken as parametric, the linear elements are of the form

$$(85) \qquad ds^2 = \lambda\, d\alpha d\beta, \qquad ds'^2 = \lambda_1 d\alpha_1 d\beta_1.$$

Hence a conformal representation is defined in the most general way by the equations

(86) $$\alpha_1 = F(\alpha), \qquad \beta_1 = F_1(\beta),$$

or

(87) $$\alpha_1 = F(\beta), \qquad \beta_1 = F_1(\alpha),$$

where F and F_1 are arbitrary functions which must be conjugate imaginary when the surfaces are real.

Instead of interpreting (85) as the linear elements of two surfaces referred to their minimal lines, we can look upon them as the linear element of the same surface in terms of two sets of parameters referring to the minimal lines. From this point of view equations (86) and (87) define the most general conformal representation of a surface upon itself. If we limit our consideration to real surfaces and put, as before,

$$\alpha = \phi + i\psi, \quad \beta = \phi - i\psi, \quad \alpha_1 = \phi_1 + i\psi_1, \quad \beta_1 = \phi_1 - i\psi_1,$$

the functions ϕ, ψ and ϕ_1, ψ_1 are pairs of isothermic parameters. Now equations (86), (87) may be written

(88) $$\phi_1 + i\psi_1 = F(\phi \pm i\psi).$$

Consequently we have the theorem:

When a pair of isothermic parameters ϕ, ψ of a surface are known and the surface is referred to the lines $\phi = const.$, $\psi = const.$, the most general conformal representation of the surface upon itself is obtained by making a point (ϕ, ψ) correspond to the point (ϕ_1, ψ_1), into which it can be transformed in accordance with equation (88).

As a corollary of this theorem, we have:

When a pair of isothermic parameters is known for each of two surfaces, all the conformal representations of one surface upon the other can be found directly.

Consider two pairs of isothermic parameters ϕ, ψ and ϕ_1, ψ_1 for a surface S, and suppose their relation is

(89) $$\phi_1 + i\psi_1 = F(\phi + i\psi).$$

If two curves C and C_1 are in correspondence in this representation, their parametric equations must be the same functional relation between the parameters, namely,

$$f(\phi, \psi) = 0, \qquad f(\phi_1, \psi_1) = 0.$$

Denote by θ and θ_1 the angles which C and C_1 make with the curves $\psi = \text{const.}$ and $\psi_1 = \text{const.}$ respectively. If we write the linear element of S in the two forms

$$ds^2 = \lambda(d\phi^2 + d\psi^2), \qquad ds_1^2 = \lambda_1(d\phi_1^2 + d\psi_1^2),$$

it follows from (23) that

$$\cos\theta = \frac{d\phi}{\sqrt{d\phi^2 + d\psi^2}}, \qquad \sin\theta = \frac{d\psi}{\sqrt{d\phi^2 + d\psi^2}},$$

$$\cos\theta_1 = \frac{d\phi_1}{\sqrt{d\phi_1^2 + d\psi_1^2}}, \qquad \sin\theta_1 = \frac{d\psi_1}{\sqrt{d\phi_1^2 + d\psi_1^2}}.$$

From these expressions we derive the following:

$$e^{2i\theta} = \frac{d\phi + i\,d\psi}{d\phi - i\,d\psi}, \qquad e^{2i\theta_1} = \frac{d\phi_1 + i\,d\psi_1}{d\phi_1 - i\,d\psi_1},$$

so that in consequence of (89) we have

$$e^{2i(\theta_1 - \theta)} = \frac{F'(\phi + i\psi)}{F_0'(\phi - i\psi)}, \tag{90}$$

where F_0 is the function conjugate to F, and the accents indicate differentiation with respect to the argument. If Γ and Γ_1 are another pair of corresponding curves, and their angles are denoted by $\bar{\theta}$ and $\bar{\theta}_1$, it follows from (90) that

$$e^{2i(\bar{\theta}_1 - \bar{\theta})} = e^{2i(\theta_1 - \theta)},$$

or

$$\bar{\theta}_1 - \theta_1 = \bar{\theta} - \theta.$$

For, the right-hand member of (90) is merely a function of the position of the point and is independent of directions. Hence in any conformal representation defined by an equation of the form (89) the angles between corresponding curves have the same sense.

When, now, the correspondence satisfies the equation

$$\phi_1 + i\psi_1 = F(\phi - i\psi),$$

the equation analogous to (90) is

$$e^{2i(\theta_1 + \theta)} = \frac{F'(\phi - i\psi)}{F_0'(\phi + i\psi)}.$$

Hence

$$\theta_1 - \bar{\theta}_1 = \bar{\theta} - \theta;$$

consequently the corresponding angles are equal in the inverse sense.

45. Conformal representation of the plane. For the plane the preceding theorem may be stated thus:

The most general real conformal representation of the plane upon itself is obtained by making a point (x, y) correspond to the point (x_1, y_1), where $x_1 + iy_1$ is any function of $x + iy$ or $x - iy$.

We recall the example of § 41, namely

$$\text{(i)} \qquad x_1 + iy_1 = \frac{c^2}{x - iy},$$

where c is a real constant. This equation is equivalent to

$$\text{(ii)} \qquad x = \frac{c^2 x_1}{x_1^2 + y_1^2}, \qquad y = \frac{c^2 y_1}{x_1^2 + y_1^2}, \qquad (x^2 + y^2)(x_1^2 + y_1^2) = c^4,$$

and also to

$$\text{(iii)} \qquad x_1 = \frac{c^2 x}{x^2 + y^2}, \qquad y_1 = \frac{c^2 y}{x^2 + y^2}.$$

Hence the parallels $x = \text{const.}$ and $y = \text{const.}$, in the xy-plane, are represented in the x_1y_1-plane by circles which pass through the origin and have their centers on the respective axes. Conversely, these circles in the xy-plane correspond to the parallels in the x_1y_1-plane.

If we put

$$x^2 + y^2 = r^2, \qquad x_1^2 + y_1^2 = r_1^2,$$

equations (ii) and (iii) may be written

$$\text{(iv)} \qquad \frac{x}{r} = \frac{x_1}{r_1}, \qquad \frac{y}{r} = \frac{y_1}{r_1}, \qquad rr_1 = c^2.$$

Hence corresponding points are on the same line through the origin, and their distances from it are such that $rr_1 = c^2$. On this account equations (iv) are said to define an *inversion with respect to the circle* $x^2 + y^2 = c^2$, or, since $r_1 = c^2/r$, *a transformation by reciprocal radii vectores.*

From § 44 it follows that corresponding angles are equal in the inverse sense.

For the case

$$\text{(v)} \qquad x_1 + iy_1 = \frac{c^2}{x + iy}$$

the equations analogous to (iv) are

$$\frac{x}{r} = \frac{x_1}{r_1}, \qquad \frac{y}{r} = -\frac{y_1}{r_1}.$$

Hence the point $P_1(x_1, y_1)$ corresponding to $P(x, y)$ lies on the line which is the reflection in the x-axis of the line OP, and at the distance $OP_1 = c^2/r$. Evidently this transformation is the combination of an inversion and the transformation $x_1 = x$, $y_1 = -y$.

One finds that the transformations (i) and (v) have the following properties:

Every straight line is transformed into a circle which passes through the origin; and conversely.

Every circle which does not pass through the origin is transformed into a circle.

We propose now the problem of finding the most general conformal transformation of the plane into itself, which changes circles *not passing through the origin* into circles. In solving it we refer the plane to symmetric parameters α, β, where

$$\alpha = x + iy, \qquad \beta = x - iy.$$

The equation of any circle which does not pass through the origin is of the form

$$c\alpha\beta + a\alpha + b\beta + d = 0, \tag{91}$$

where a, b, c, d are constants; when the circle is real a and b must be conjugate imaginaries and c real. Equation (91) defines β as a function of α. If we differentiate the equation three times with respect to α, and eliminate the constants from the resulting equations, we find

$$3\,\beta''^2 - 2\,\beta'\beta''' = 0, \tag{92}$$

where the accent indicates differentiation with respect to α. Moreover, as equation (91) contains three independent constants, it is the general integral of (92).

We know that the most general conformal representation of the plane upon itself is given by

$$\alpha_1 = A(\alpha), \qquad \beta_1 = B(\beta), \tag{93}$$

or

$$\alpha_1 = B(\beta), \qquad \beta_1 = A(\alpha). \tag{94}$$

Our problem reduces, therefore, to the determination of functions A and B, such that the equation

$$3\,\beta_1''^2 - 2\,\beta_1'\beta_1''' = 0, \tag{95}$$

where the accent indicates differentiation with respect to α_1, can be transformed by (93) or (94) into (92).

We consider first equations (93), which we write

$$\alpha = A_1(\alpha_1), \qquad \beta_1 = B(\beta).$$

Now

$$\beta_1' = \frac{dB}{d\beta}\cdot\frac{d\beta}{d\alpha}\cdot\frac{d\alpha}{d\alpha_1} = B'\beta' A_1'.$$

In like manner we find β_1'' and β_1'''. When their values are substituted in (95) we get, since A_1' and B' are different from zero,

$$3\,\beta''^2 - 2\,\beta'\beta''' + \frac{1}{B'^2}(3\,B''^2 - 2\,B'B''')\beta'^4 + \frac{1}{A_1'^4}(3\,A_1''^2 - 2\,A_1'A_1''')\beta'^2 = 0.$$

Since equation (95) must be directly transformable into (92), it follows that

$$(96)\qquad 3\,B''^2-2\,B'B'''=0,\qquad 3\,A_1''^2-2\,A_1'A_1'''=0.$$

As these equations are of the form (92), their general integrals are similar to (91). Hence the most general forms of (93) for our problem are

$$(97)\qquad \alpha_1=\frac{a_1\alpha+a_2}{a_3\alpha+a_4},\qquad \beta_1=\frac{b_1\beta+b_2}{b_3\beta+b_4}.$$

Moreover, when these values are substituted in an equation in α_1, β_1 of the form (91), the resulting equation in α and β is of this form.

Equation (91) may likewise be looked upon as defining α in terms of β, so that α, as a function of β, satisfies an equation of the form (92); similarly for α_1 as a function of β_1. Hence if we had used (94), we should have been brought to results analogous to (97); and therefore the most general forms of (94) for our problem are

$$(98)\qquad \alpha_1=\frac{b_1\beta+b_2}{b_3\beta+b_4},\qquad \beta_1=\frac{a_1\alpha+a_2}{a_3\alpha+a_4}.$$

Hence:

When a plane is defined in symmetric parameters α, β, the most general conformal representation of the plane upon itself, for which circles correspond to circles or straight lines, is given by (97) *or* (98).*

EXAMPLES

1. Deduce the equations which define the most general conformal representation of a surface with the linear element $ds^2 = du^2 + (a^2 - u^2)\,dv^2$ upon itself.

2. Show that the surfaces

$$x = u\cos v,\qquad y = u\sin v,\qquad z = av,$$

$$x = u\cos v,\qquad y = u\sin v,\qquad z = a\cosh^{-1}\frac{u}{a},$$

are applicable. Find the curve in which a plane through the z-axis cuts the latter surface, and deduce the equations of the conformal representation of these surfaces on the plane.

3. When the representation is defined by (97), what are the coördinates of the center and radius of the circle in the α_1-plane which corresponds to the circle of center (c, d) and radius r in the α-plane?

* The transformations (97) and (98) play an important rôle in the theory of functions. For a more detailed study of them the reader is referred to the treatises of Picard, Darboux, and Forsyth.

4. Show that in the conformal representation (97) there are, in general, two distinct points, each of which corresponds to itself; also that if γ and δ are the values of α at these points, then

$$\frac{\alpha_1-\gamma}{\alpha_1-\delta}=\frac{\alpha-\gamma}{\alpha-\delta}K, \quad \text{where } K=\frac{a_1+a_4-\sqrt{(a_1-a_4)^2+4\,a_2a_3}}{a_1+a_4+\sqrt{(a_1-a_4)^2+4\,a_2a_3}}.$$

5. Find the condition that the origin be the only point which corresponds to itself, and show that if the quantities a_1, a_2, a_3, a_4 are real, a circle in the α-plane through the origin O corresponds to a circle in the α_1-plane through O and touching the other circle; also that a circle touching the x-axis at O corresponds to itself.

6. The equation $2\,\alpha_1=(a-b)\,\alpha+\dfrac{a+b}{\alpha}$, where a and b are constants, defines a conformal representation of the plane upon itself, such that circles about the origin and straight lines through the latter in the α-plane correspond to confocal ellipses and hyperbolas in the α_1-plane.

7. In the conformal representation $\alpha_1=\log\alpha$ to lines parallel to the x- and y-axes in the α_1-plane there correspond lines through the origin and circles about it in the α-plane, and to any orthogonal system of straight lines in the α_1-plane an orthogonal system of logarithmic spirals in the α-plane.

46. Surfaces of revolution. By definition a *surface of revolution* is the surface generated by a plane curve when the plane of the curve is made to rotate about a line in the plane. The various positions of the curve are called the *meridians* of the surface, and the circles described by each point of the curve in the revolution are called the *parallels*. We take the axis of rotation for the z-axis, and for x-axis and y-axis any two lines perpendicular to one another, and to the z-axis, and meeting it in the same point. For any position of the plane the equation of the curve may be written $z=\phi(r)$, where r denotes the distance of a point of the curve from the z-axis. We let v denote the angle which the plane, in any of its positions, makes with the xz-plane. Hence the equations of the surface are

$$x=r\cos v, \qquad y=r\sin v, \qquad z=\phi(r). \tag{99}$$

The linear element is

$$ds^2=[1+\phi'^2(r)]\,dr^2+r^2dv^2. \tag{100}$$

If we put

$$u=\int\frac{1}{r}\sqrt{1+\phi'^2}\,dr, \tag{101}$$

the linear element is transformed into

$$ds^2=\lambda\,(du^2+dv^2), \tag{102}$$

where λ is a function of u, which shows that the meridians and parallels form an isothermal system. As this change of parameters does not change the parametric lines, the equations

$$x = u, \qquad y = v,$$

define a conformal representation of the surface of revolution upon the plane in which the meridians and parallels correspond to the straight lines $x =$ const. and $y =$ const. respectively.

By definition a *loxodromic curve* on a surface of revolution is a curve which cuts the meridians under constant angle. Evidently it is represented on the plane by a straight line. Hence loxodromic curves on a surface of revolution (99) are given by

$$a\int \frac{1}{r}\sqrt{1+\phi'^2}\,dr + bv + c = 0,$$

where a, b, c are constants.

Incidentally we have the theorem:

When the linear element of a surface is reducible to the form

$$ds^2 = \lambda(du^2 + dv^2),$$

where λ is a function of u or v alone, the surface is applicable to a surface of revolution.

For, suppose that λ is a function of u alone. Put $r = \sqrt{\lambda}$ and solve this equation for u as a function of r. If the resulting expression be substituted in (101), we find, by a quadrature, the function $\phi(r)$, for which equations (99) define the surface of revolution with the given linear element.

When, in particular, the surface of revolution is the unit sphere, with center at the origin, we have

$$r = \sin u, \qquad z = \sqrt{1 - r^2} = \cos u,$$

where u is the angle which the radius vector of the point makes with the positive z-axis. Now

$$u_1 = \int \frac{1}{r}\sqrt{1+\phi'^2}\,dr = \log\tan\frac{u}{2}.$$

Hence the equations of correspondence are

$$x = \log\tan\frac{u}{2}, \qquad y = v.$$

This representation is called a *Mercator chart* of the sphere upon the plane. It is used in making maps of the earth for mariners. A path represented by a straight line on the chart cuts the meridians at constant angle.

47. Conformal representations of the sphere. We have found (§ 35) that when the unit sphere, with center at the origin, is referred to minimal lines, its equations are

$$(103) \qquad x=\frac{\alpha+\beta}{\alpha\beta+1}, \qquad y=i\frac{(\beta-\alpha)}{\alpha\beta+1}, \qquad z=\frac{\alpha\beta-1}{\alpha\beta+1},$$

where α and β are conjugate imaginary. Hence the parametric equation of any real circle on the sphere is of the form

$$c\alpha\beta+a\alpha+b\beta+d=0,$$

where a and b are conjugate imaginary and c and d are real. From this it follows that the problem of finding any conformal representation of the sphere upon the plane with circles of the former in correspondence with circles or straight lines of the latter, is the same problem analytically as the determination of this kind of representation of the plane upon itself. Hence, from the results of § 45, it follows that

All conformal representations of the sphere (103) *upon a plane, with circles of the former corresponding to circles or straight lines of the latter, are defined by*

$$(104) \qquad x_1 \pm iy_1=\frac{a_1\alpha+a_2}{a_3\alpha+a_4}, \qquad x_1 \mp iy_1=\frac{b_1\beta+b_2}{b_3\beta+b_4}.^*$$

We wish to consider in particular the case in which the sphere is represented on the xy-plane in such a way that the great circle determined by this plane corresponds with itself point for point.

From (103) we have that the equations of this circle are

$$\alpha\beta=1, \qquad x=\frac{\alpha^2+1}{2\alpha}, \qquad y=i\frac{1-\alpha^2}{2\alpha}.$$

* The representation with the lower signs is the combination of the one with the upper sign and the transformation $\alpha_1=\beta$, $\beta_1=\alpha$, which from (103) is seen to transform a figure on the sphere into the figure symmetrical with respect to the xz-plane.

When these values are substituted in (104) it is found that we must have

$$a_1 = a_4, \qquad b_1 = b_4, \qquad a_2 = a_3 = b_2 = b_3 = 0,$$

so that the particular form of (104)* is equivalent to

$$x_1 = \frac{1}{2}(\alpha + \beta), \qquad y_1 = \frac{i}{2}(\beta - \alpha).$$

From these equations and (103) we find that the equations of the straight lines joining corresponding points on the sphere and plane are reducible to

$$\frac{X}{\alpha + \beta} = \frac{Y}{i(\beta - \alpha)} = \frac{1 - Z}{2}.$$

For all values of α and β these lines pass through the point $(0, 0, 1)$. Hence a point of the plane corresponding to a given point P upon the sphere is the point of intersection with the plane of the line joining P with the pole $(0, 0, 1)$. This form of representation is called the *stereographic projection* of the sphere upon the plane. It is evident that a line in the plane corresponds to a circle on the sphere; this circle is determined by the plane of the pole and the given line.

We will close this chapter with a few remarks about the conformal representation of the sphere upon itself. From the foregoing results we know that every such representation of the sphere (103) is given by equations of similar form in α_1, β_1, where the latter are given by (86) or (87), and that for conformal representations with circles in correspondence α_1 and β_1 have the values (97) or (98).

We consider in particular the case

$$\alpha_1 = \frac{a_1\alpha + a_2}{a_3\alpha + a_4}, \qquad \beta_1 = \frac{a_4\beta - a_3}{-a_2\beta + a_1}. \tag{105}$$

The expressions of the linear elements of the sphere are found to be reducible to

$$ds^2 = \frac{4\,d\alpha d\beta}{(1 + \alpha\beta)^2}, \qquad ds_1^2 = \frac{4\,d\alpha_1 d\beta_1}{(1 + \alpha_1\beta_1)^2} = \frac{4\,d\alpha d\beta}{(1 + \alpha\beta)^2}.$$

* Here we have used the upper signs in (104).

Hence equations (105) define an isometric representation of the sphere upon itself. Since angles are preserved in the same sense by (105), this representation may be looked upon as determining a motion of configuration upon the sphere into new positions upon it. The stationary points in the general motion, if there are any, correspond to values of α and β, which are roots of the respective equations

$$(106) \quad a_3 t^2 + (a_4 - a_1) t - a_2 = 0, \qquad a_2 u^2 + (a_4 - a_1) u - a_3 = 0.$$

If t_1 and t_2 are the roots of the former, those of the latter are $-1/t_1$ and $-1/t_2$. Hence there are four points stationary in the motion; their curvilinear coördinates are

$$\left(t_1, -\frac{1}{t_1}\right), \quad \left(t_2, -\frac{1}{t_2}\right), \quad \left(t_1, -\frac{1}{t_2}\right), \quad \left(t_2, -\frac{1}{t_1}\right).$$

From (103) it is seen that the first two are at infinity, and the last two determine points on the sphere, so that the motion is a rotation about these points. If the z-axis is taken for the axis of rotation, we have from (103) that the roots of (106) must be ∞ and 0; hence $a_2 = a_3 = 0$, so that (105) becomes

$$\alpha_1 = \frac{a_1}{a_4}\alpha, \qquad \beta_1 = \frac{a_4}{a_1}\beta.$$

If the rotation is real, these equations must be of the form

$$\alpha_1 = e^{i\omega}\alpha, \qquad \beta_1 = e^{-i\omega}\beta,$$

where ω is the angle of rotation.

EXAMPLES

1. Find the equations of the surface of revolution with the linear element $ds^2 = du^2 + (a^2 - u^2)\,dv^2$.

2. Find the loxodromic curves on the surface

$$x = u\cos v, \quad y = u\sin v, \quad z = a\cosh^{-1}\frac{u}{a},$$

and find the equations of the surface when referred to an orthogonal system of these curves.

3. Find the general equations of the conformal representation of the oblate spheroid upon the plane.

4. Show that for the surface generated by the revolution of the evolute of the catenary about the base of the latter the linear element is reducible to $ds^2 = du^2 + u\,dv^2$.

5. A great circle on the unit sphere cuts the meridian $v = 0$ in latitude α under angle a. Find the equation of its stereographic projection.

6. Determine the stereographic projection of the curve $x = a \sin u \cos u$, $y = a \cos^2 u$, $z = a \sin u$ from the pole $(0, a, 0)$.

GENERAL EXAMPLES

1. When there is a one-to-one point correspondence between two surfaces, the cross-ratio of four tangents to one surface at a point is equal to the cross-ratio of the corresponding tangents to the other.

2. Given the paraboloid

$$x = 2\,au\cos v, \quad y = 2\,bu\sin v, \quad z = 2\,u^2(a\cos^2 v + b\sin^2 v),$$

where a and b are constants. Determine the equation of the curves on the surface, such that the tangent planes along a curve make a constant angle with the xy-plane. Show that the generators of the developable Σ, enveloped by these planes, make a constant angle with the z-axis, and express the coördinates of the edge of regression in terms of v.

3. Find the orthogonal trajectories of the generators of the surface Σ in Ex. 2. Show that they are plane curves and that their projections on the xy-plane are involutes of the projection of the edge of regression.

4. Let C be a curve on a cone of revolution which cuts the generators under constant angle, and C_1 the locus of the centers of curvature of C. Show that C_1 lies upon a cone whose elements it cuts under constant angle.

5. When the polar developable of a curve is developed upon a plane, the curve degenerates into a point.

6. When the rectifying developable of a curve is developed upon a plane, the curve becomes a straight line.

7. Determine $\phi(v)$ so that the right conoid,

$$x = u\cos v, \quad y = u\sin v, \quad z = \phi(v),$$

shall be applicable to a surface of revolution.

8. Determine the equations of a conformal representation of the plane upon itself for which the parallels to the axes in the α_1-plane correspond to lines through a point (a, b) and circles concentric about it in the α-plane.

9. The equation $\alpha_1 = c\sin\alpha$, where c is a constant, defines a conformal representation of the plane upon itself such that the lines parallel to the axes in the α-plane correspond to confocal ellipses and hyperbolas in the α_1-plane.

10. In the conformal representation of the plane upon itself, given by $\alpha_1 = \alpha^2$, to lines parallel to the axes in the α_1-plane there correspond equilateral hyperbolas in the α-plane, and to the pencil of rays through a point in the α_1-plane and the circles concentric about it there corresponds a system of equilateral hyperbolas through the corresponding point in the α-plane and a family of confocal Cassini ovals.

11. When the sides of a triangle upon a surface of revolution are loxodromic curves, the sum of the three angles is equal to two right angles.

12. The only conformal perspective representation of a sphere upon a plane is given by (104).

13. Show that equations (105) and the equations obtained from (105) by the interchange of α and β define the most general isometric representation of the sphere upon itself.

14. Let each of two surfaces S, S_1 be defined in terms of parameters u, v, and let points on each with the same values of the parameters correspond. If $H = H_1$, where the latter is the function for S_1 analogous to H for S, corresponding elements of area are equal and the representation is said to be *equivalent.** If $H \neq H_1$ and the parameters of S are changed in accordance with the equations $u' = \phi(u, v)$, $v' = \psi(u, v)$, the condition that the equations $u' = u$, $v' = v$ define an equivalent representation of S and S_1 is
$$\frac{\partial \phi}{\partial u}\frac{\partial \psi}{\partial v} - \frac{\partial \phi}{\partial v}\frac{\partial \psi}{\partial u} = \frac{H(u, v)}{H_1(\phi, \psi)}.$$

15. Under what conditions do the equations
$$x' = a_1 x + a_2 y + a_3, \qquad y' = b_1 x + b_2 y + b_3$$
define an equivalent representation of the plane upon itself ?

16. Show that the equations
$$x = \int r\sqrt{1 + \phi'^2}\, dr, \qquad y = v,$$
determine an equivalent representation of the surface of revolution (99) upon the plane.

17. Given a sphere and circumscribed circular cylinder. If the points at which a perpendicular to the axis of the latter meets the two surfaces correspond, the representation is equivalent.

18. Find an equivalent representation of the sphere upon the plane such that the parallel circles correspond to lines parallel to the y-axis and the meridians to ellipses for which the extremities of one of the principal axes are $(a, 0)$, $(-a, 0)$.

* German writers call it "flächentreu."

CHAPTER IV

GEOMETRY OF A SURFACE IN THE NEIGHBORHOOD OF A POINT

48. Fundamental coefficients of the second order. In this chapter we study the form of a surface in the neighborhood of a point M of it, and the character of the curves which lie upon the surface and pass through the point. We recall that the tangents at M to all these curves lie in a plane, — *the tangent plane* to the surface at the point.

The equation of the tangent plane at $M(x, y, z)$, namely (II, 11), may be written

$$(1) \qquad (\xi - x)X + (\eta - y)Y + (\zeta - z)Z = 0,$$

where we have put

$$(2) \qquad X = \frac{1}{H}\begin{vmatrix} \frac{\partial y}{\partial u} & \frac{\partial z}{\partial u} \\ \frac{\partial y}{\partial v} & \frac{\partial z}{\partial v} \end{vmatrix}, \quad Y = \frac{1}{H}\begin{vmatrix} \frac{\partial z}{\partial u} & \frac{\partial x}{\partial u} \\ \frac{\partial z}{\partial v} & \frac{\partial x}{\partial v} \end{vmatrix}, \quad Z = \frac{1}{H}\begin{vmatrix} \frac{\partial x}{\partial u} & \frac{\partial y}{\partial u} \\ \frac{\partial x}{\partial v} & \frac{\partial y}{\partial v} \end{vmatrix}.$$

We define the positive direction of the normal (§ 25) to be that for which the functions X, Y, Z are the direction-cosines. From this definition it follows that the tangents to the curves $v = \text{const.}$ and $u = \text{const.}$ at a point and the normal at the point have the same mutual orientation as the x-, y-, and z-axes.

From (2) follow the identities

$$(3) \qquad \sum X\frac{\partial x}{\partial u} = 0, \qquad \sum X\frac{\partial x}{\partial v} = 0,$$

which express the fact that the normal is perpendicular to the tangents to the coördinate curves. In consequence of these identities the expression for the distance p from a point M' $(u + du, v + dv)$ to the tangent plane at M is of the second order in du and dv. It may be written

$$(4) \qquad p = \Sigma X\, dx = \tfrac{1}{2}(D\, du^2 + 2D'\, du dv + D''\, dv^2) + e,$$

where e denotes the aggregate of terms of the third and higher orders in du and dv, and the functions D, D', D'' are defined by

$$D=\sum X\frac{\partial^2 x}{\partial u^2},\quad D'=\sum X\frac{\partial^2 x}{\partial u\,\partial v},\quad D''=\sum X\frac{\partial^2 x}{\partial v^2}. \tag{5}$$

If equations (3) be differentiated with respect to u and v respectively, we get

$$\begin{cases}\sum X\dfrac{\partial^2 x}{\partial u^2}+\sum\dfrac{\partial X}{\partial u}\dfrac{\partial x}{\partial u}=0, & \sum X\dfrac{\partial^2 x}{\partial u\,\partial v}+\sum\dfrac{\partial X}{\partial v}\dfrac{\partial x}{\partial u}=0,\\ \sum X\dfrac{\partial^2 x}{\partial u\,\partial v}+\sum\dfrac{\partial X}{\partial u}\dfrac{\partial x}{\partial v}=0, & \sum X\dfrac{\partial^2 x}{\partial v^2}+\sum\dfrac{\partial X}{\partial v}\dfrac{\partial x}{\partial v}=0.\end{cases} \tag{6}$$

And so equations (5) may be written

$$\begin{cases}D=\sum X\dfrac{\partial^2 x}{\partial u^2}=-\sum\dfrac{\partial X}{\partial u}\dfrac{\partial x}{\partial u},\\ D'=\sum X\dfrac{\partial^2 x}{\partial u\,\partial v}=-\sum\dfrac{\partial X}{\partial u}\dfrac{\partial x}{\partial v}=-\sum\dfrac{\partial X}{\partial v}\dfrac{\partial x}{\partial u},\\ D''=\sum X\dfrac{\partial^2 x}{\partial v^2}=-\sum\dfrac{\partial X}{\partial v}\dfrac{\partial x}{\partial v}.\end{cases} \tag{7}$$

The quadratic differential form

$$\Phi=D\,du^2+2\,D'\,du\,dv+D''\,dv^2 \tag{8}$$

is called the *second fundamental form* of the surface, and the functions D, D', D'' the *fundamental coefficients of the second order*. We leave it to the reader to show that these coefficients, like those of the first order, are invariant for any displacement of the surface in space.

Later we shall have occasion to use two sets of formulas which will now be derived.

From the equations of definition,

$$E=\sum\left(\frac{\partial x}{\partial u}\right)^2,\quad F=\sum\frac{\partial x}{\partial u}\frac{\partial x}{\partial v},\quad G=\sum\left(\frac{\partial x}{\partial v}\right)^2, \tag{9}$$

we get, by differentiation and simple reduction, the following:

$$\begin{cases}\sum\dfrac{\partial x}{\partial u}\dfrac{\partial^2 x}{\partial u^2}=\dfrac{1}{2}\dfrac{\partial E}{\partial u}, & \sum\dfrac{\partial x}{\partial v}\dfrac{\partial^2 x}{\partial u^2}=\dfrac{\partial F}{\partial u}-\dfrac{1}{2}\dfrac{\partial E}{\partial v},\\ \sum\dfrac{\partial x}{\partial u}\dfrac{\partial^2 x}{\partial u\,\partial v}=\dfrac{1}{2}\dfrac{\partial E}{\partial v}, & \sum\dfrac{\partial x}{\partial v}\dfrac{\partial^2 x}{\partial u\,\partial v}=\dfrac{1}{2}\dfrac{\partial G}{\partial u},\\ \sum\dfrac{\partial x}{\partial u}\dfrac{\partial^2 x}{\partial v^2}=\dfrac{\partial F}{\partial v}-\dfrac{1}{2}\dfrac{\partial G}{\partial u}, & \sum\dfrac{\partial x}{\partial v}\dfrac{\partial^2 x}{\partial v^2}=\dfrac{1}{2}\dfrac{\partial G}{\partial v}.\end{cases} \tag{10}$$

Again, if the expressions (9) be substituted in the left-hand members of the following equations, the reduced results may be written by means of (2) in the form indicated:

$$(11)\qquad \begin{cases} E\dfrac{\partial x}{\partial v}-F\dfrac{\partial x}{\partial u}=H\left(Y\dfrac{\partial z}{\partial u}-Z\dfrac{\partial y}{\partial u}\right), \\ F\dfrac{\partial x}{\partial v}-G\dfrac{\partial x}{\partial u}=H\left(Y\dfrac{\partial z}{\partial v}-Z\dfrac{\partial y}{\partial v}\right). \end{cases}$$

Similar identities can be found by permuting the letters x, y, z; X, Y, Z.

From the fundamental relation

$$X^2+Y^2+Z^2=1$$

we obtain, by differentiation with respect to u and v respectively, the identities

$$(12)\qquad \sum X\frac{\partial X}{\partial u}=0, \qquad \sum X\frac{\partial X}{\partial v}=0.$$

These equations and (7) constitute a system of three equations linear in $\frac{\partial X}{\partial u}$, $\frac{\partial Y}{\partial u}$, $\frac{\partial Z}{\partial u}$, and a system linear in $\frac{\partial X}{\partial v}$, $\frac{\partial Y}{\partial v}$, $\frac{\partial Z}{\partial v}$. Solving for $\frac{\partial X}{\partial u}$ and for $\frac{\partial X}{\partial v}$, we find, by means of (11),

$$(13)\qquad \begin{cases} \dfrac{\partial X}{\partial u}=\dfrac{FD'-GD}{H^2}\dfrac{\partial x}{\partial u}+\dfrac{FD-ED'}{H^2}\dfrac{\partial x}{\partial v}, \\ \dfrac{\partial X}{\partial v}=\dfrac{FD''-GD'}{H^2}\dfrac{\partial x}{\partial u}+\dfrac{FD'-ED''}{H^2}\dfrac{\partial x}{\partial v}. \end{cases}$$

The expressions for $\frac{\partial Y}{\partial u}, \cdots, \frac{\partial Z}{\partial v}$ are obtained by replacing x by y and z respectively.

By means of these equations we shall prove that a real surface whose first and second fundamental coefficients are in proportion, thus

$$(14)\qquad \frac{D}{E}=\frac{D'}{F}=\frac{D''}{G}=-\lambda,$$

where λ denotes the factor of proportionality, is a sphere or a plane. We assume that the minimal lines are parametric. In consequence we have

$$E=G=D=D''=0,$$

so that equations (13) become

$$(15)\qquad \frac{\partial X}{\partial u}=\lambda\frac{\partial x}{\partial u}, \qquad \frac{\partial X}{\partial v}=\lambda\frac{\partial x}{\partial v}.$$

The function λ must satisfy the condition

$$\frac{\partial}{\partial v}\left(\lambda\frac{\partial x}{\partial u}\right)=\frac{\partial}{\partial u}\left(\lambda\frac{\partial x}{\partial v}\right),$$

which reduces to $\frac{\partial\lambda}{\partial v}\frac{\partial x}{\partial u}-\frac{\partial\lambda}{\partial u}\frac{\partial x}{\partial v}=0$. Moreover, we have two other equations of condition, obtained from the above by replacing x by y and z respectively. Since the proportion

$$\frac{\partial x}{\partial u}:\frac{\partial y}{\partial u}:\frac{\partial z}{\partial u}=\frac{\partial x}{\partial v}:\frac{\partial y}{\partial v}:\frac{\partial z}{\partial v}$$

is not possible for a real surface, we must have $\frac{\partial\lambda}{\partial u}=\frac{\partial\lambda}{\partial v}=0$; that is, λ is a constant. When λ is zero the functions X, Y, Z given by (15) are constant, and consequently the surface is a plane. When λ is any other constant, we get, by integration from (15),

$$X=\lambda x+a,\qquad Y=\lambda y+b,\qquad Z=\lambda z+c,$$

where a, b, c are constants. From these equations we obtain $(\lambda x+a)^2+(\lambda y+b)^2+(\lambda z+c)^2=1$. Since this is the general equation of a sphere, it follows that the above condition is necessary as well as sufficient.

49. Radius of normal curvature. Consider on a surface S any curve C through a point M. The direction of its tangent, MT, is determined by a value of dv/du. Let ϖ denote the angle which the positive direction of the normal to the surface makes with the positive direction of the principal normal to C at M, angles being measured toward the positive binormal. If we use the notation of the first chapter, and take the arc of C for its parameter, we have

$$\cos\varpi=Xl+Ym+Zn=\rho\left(X\frac{d^2x}{ds^2}+Y\frac{d^2y}{ds^2}+Z\frac{d^2z}{ds^2}\right).$$

In terms of $\frac{du}{ds}$ and $\frac{dv}{ds}$ the derivatives in the parenthesis have the forms

$$\frac{d^2x}{ds^2}=\frac{\partial^2x}{\partial u^2}\left(\frac{du}{ds}\right)^2+2\frac{\partial^2x}{\partial u\partial v}\frac{du}{ds}\frac{dv}{ds}+\frac{\partial^2x}{\partial v^2}\left(\frac{dv}{ds}\right)^2,$$

so that the above equation is equivalent to

$$\text{(16)}\qquad \frac{\cos\varpi}{\rho}=\frac{D\,du^2+2D'du\,dv+D''dv^2}{E\,du^2+2F\,du\,dv+G\,dv^2}.$$

As the right-hand member of this equation depends only upon the curvilinear coördinates of the point and the direction of MT, it is the same for all curves with this tangent at M. Since ρ is positive, the angle ϖ cannot be greater than a right angle for one curve tangent to MT, if it is less than a right angle for any other curve tangent to MT; and vice versa. We consider in particular the curve in which the surface is cut by the plane determined by MT and the normal to the surface at M. We call it the *normal section* tangent to MT, and let ρ_n denote its radius. Since the right-hand member of equation (16) is the same for C and the normal section tangent to it, we have

$$\frac{\cos \bar{\omega}}{\rho} = \frac{e}{\rho_n}, \tag{17}$$

where e is $+1$ or -1, according as $\bar{\omega}$ is less or greater than a right angle; for ρ and ρ_n are positive. Equation (17) gives the following theorem of Meusnier:

The center of curvature of any curve upon a surface is the projection upon its osculating plane of the center of curvature of the normal section tangent to the curve at the point.

In order to avoid the ambiguous sign in (17), we introduce a new function R which is equal to ρ_n when $0 \leqq \bar{\omega} \leqq \pi/2$, and to $-\rho_n$ when $\pi/2 \leqq \bar{\omega} \leqq \pi$, and call it the *radius of normal curvature* of the surface for the given direction MT. As thus defined, R is given by

$$\frac{1}{R} = \frac{D\,du^2 + 2\,D'\,du\,dv + D''\,dv^2}{E\,du^2 + 2\,F\,du\,dv + G\,dv^2}. \tag{18}$$

Now we may state Meusnier's theorem as follows:

If a segment, equal to twice the radius of normal curvature for a given direction at a point on a surface, be laid off from the point on the normal to the surface, and a sphere be described with the segment for diameter, the circle in which the sphere is met by the osculating plane of a curve with the given direction at the point is the circle of curvature of the curve.

50. Principal radii of normal curvature. If we put $t = \dfrac{dv}{du}$, equation (18) becomes

$$\frac{1}{R} = \frac{D + 2\,D't + D''t^2}{E + 2\,Ft + Gt^2}. \tag{19}$$

When the proportion (14) is satisfied, R is the same for all values of t, being ∞ for the plane, and the constant $-1/\lambda$ for the sphere. For any other surface R varies continuously with t. And so we seek the values of t for which R is a maximum or minimum. To this end we differentiate the above expression with respect to t and put the result equal to zero. This gives

$$(20)\quad (D'+D''t)(E+2Ft+Gt^2)-(F+Gt)(D+2D't+D''t^2)=0,$$

or

$$(21)\quad (FD''-GD')t^2+(ED''-GD)t+(ED'-FD)=0.$$

Without any loss of generality we can assume that the parametric curves are such that $E\neq 0$, so that we have the identity

$$(22)\quad \begin{aligned}(ED''-GD)^2-4(FD''-D'G)(ED'-FD)\\ =4\frac{H^2}{E^2}(ED'-FD)^2+\left[ED''-GD-\frac{2F}{E}(ED'-FD)\right]^2.\end{aligned}$$

When the surface is real, and the parameters also, the right-hand member of this equation is positive. Since the left-hand member is the discriminant of equation (21), the latter has two real and distinct roots.* When the test (III, 34) is applied to equation (21), it is found that the two directions at a point determined by the roots of (21) are perpendicular. Hence:

At every ordinary point of a surface there is a direction for which the radius of normal curvature is a maximum and a direction for which it is a minimum, and they are at right angles to one another.

These limiting values of R are called the *principal radii of normal curvature* at the point. They are equal to each other for the plane and the sphere, and these are the only real surfaces with this property.

From (20) and (19) we have

$$\frac{D'+D''t}{F+Gt}=\frac{D+D't}{E+Ft}=\frac{1}{R}.$$

Hence the following relations hold between the principal radii and the corresponding values of t:

$$(23)\quad \begin{cases}E+Ft-R(D+D't)=0,\\ F+Gt-R(D'+D''t)=0.\end{cases}$$

* In order that the two roots be equal, the discriminant must vanish. This is impossible for real surfaces other than spheres and planes, as seen from (22). For an imaginary surface of this kind referred to its lines of length zero, we have from (21) that D or D'' is zero, since $F\neq 0$. The vanishing of the discriminant is also the necessary and sufficient condition that the numerator and denominator in (19) have a common factor.

When t is eliminated from these equations, we get the equation

$$(24)\quad (DD''-D'^2)R^2-(ED''+GD-2FD')R+(EG-F^2)=0,$$

whose roots are the principal radii. If these roots be denoted by ρ_1 and ρ_2, we have

$$(25)\quad \begin{cases} \dfrac{1}{\rho_1}+\dfrac{1}{\rho_2}=\dfrac{ED''+GD-2FD'}{H^2}, \\ \dfrac{1}{\rho_1\rho_2}=\dfrac{DD''-D'^2}{H^2}. \end{cases}$$

Although equations (14) hold at all points of a sphere and a plane, and for no other surface, it may happen that for certain particular points of a surface they are satisfied. At such points R, as given by (19), is the same for all directions, and the equation (21) vanishes identically. When points of this kind exist they are called *umbilical points* of the surface.

EXAMPLES

1. When the equation of the surface is $z=f(x, y)$, show that

$$X, Y, Z=\frac{-p, -q, 1}{\sqrt{1+p^2+q^2}},$$

$$D, D', D''=\frac{r, s, t}{\sqrt{1+p^2+q^2}},$$

where
$$p=\frac{\partial z}{\partial x},\quad q=\frac{\partial z}{\partial y},\quad r=\frac{\partial^2 z}{\partial x^2},\quad s=\frac{\partial^2 z}{\partial x\,\partial y},\quad t=\frac{\partial^2 z}{\partial y^2}.$$

2. Show that the normals to the right conoid

$$x=u\cos v,\quad y=u\sin v,\quad z=\phi(v)$$

along a generator form a hyperbolic paraboloid.

3. Show that the principal radii of normal curvature of a right conoid at a point differ in sign.

4. Find the expression for the radius of normal curvature of a surface of revolution at a point in the direction of the loxodromic curve through it, which makes the angle α with the meridians.

5. Show that the meridians and parallels on a surface of revolution, $x=u\cos v$, $y=u\sin v$, $z=\phi(u)$, are the directions in which the radius of normal curvature is maximum and minimum; that the principal radii are given by

$$\frac{1}{\rho_1}=\frac{\phi''(u)}{(1+\phi'^2)^{\frac{3}{2}}},\qquad \frac{1}{\rho_2}=\frac{\phi'(u)}{u(1+\phi'^2)^{\frac{1}{2}}};$$

and that ρ_2 is the segment of the normal between the point of the surface and the intersection of the normal with the z-axis.

6. Show that $\Delta_1 x=1-X^2$ and $\Delta_1(x, y)=-XY$, where the differential parameters are formed with respect to the linear element of the surface.

51. Lines of curvature. Equations of Rodrigues. We have seen that the curves defined by equation (21), written

$$(26) \quad (ED'-FD)\,du^2+(ED''-GD)\,du\,dv+(FD''-GD')\,dv^2=0,$$

form an orthogonal system. As defined, the two curves of the system through a point on the surface determine the directions at the point for which the radii of normal curvature have their maximum and minimum values. These curves are called the *lines of curvature*, and their tangents at a point the *principal directions* for the point. They possess another geometric property which we shall now find.

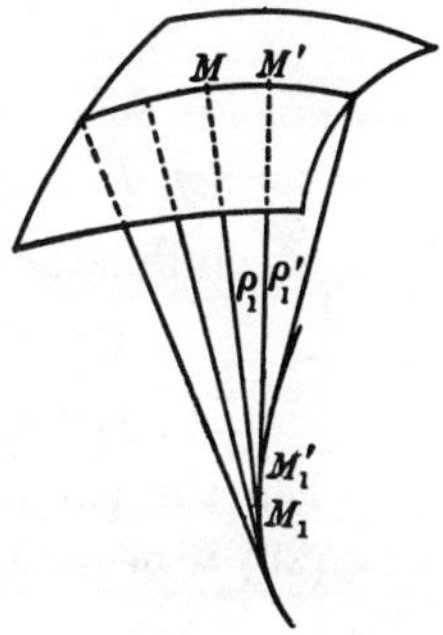

FIG. 12

The normals to a surface along a curve form a ruled surface. In order that the surface be developable, the normals must be tangent to a curve (§ 27), as in fig. 12. If the coördinates of a point M_1 on the normal at a point M be denoted by x_1, y_1, z_1, we have

$$x_1 = x + rX, \qquad y_1 = y + rY, \qquad z_1 = z + rZ,$$

where r denotes the length MM_1. If M_1 be a point of the edge of regression, we must have

$$\frac{dx + r\,dX + X\,dr}{X} = \frac{dy + r\,dY + Y\,dr}{Y} = \frac{dz + r\,dZ + Z\,dr}{Z}.$$

Multiplying the numerators and the denominators of the respective members by X, Y, Z, and combining, we find that the common ratio is dr. Hence the above equations reduce to

$$dx + r\,dX = 0, \qquad dy + r\,dY = 0, \qquad dz + r\,dZ = 0;$$

or, when the parametric coördinates are used,

$$(27) \quad \begin{cases} \dfrac{\partial x}{\partial u}du + \dfrac{\partial x}{\partial v}dv + r\left(\dfrac{\partial X}{\partial u}du + \dfrac{\partial X}{\partial v}dv\right) = 0, \\ \dfrac{\partial y}{\partial u}du + \dfrac{\partial y}{\partial v}dv + r\left(\dfrac{\partial Y}{\partial u}du + \dfrac{\partial Y}{\partial v}dv\right) = 0, \\ \dfrac{\partial z}{\partial u}du + \dfrac{\partial z}{\partial v}dv + r\left(\dfrac{\partial Z}{\partial u}du + \dfrac{\partial Z}{\partial v}dv\right) = 0. \end{cases}$$

If these equations be multiplied by $\frac{\partial x}{\partial u}, \frac{\partial y}{\partial u}, \frac{\partial z}{\partial u}$ respectively and added, and by $\frac{\partial x}{\partial v}, \frac{\partial y}{\partial v}, \frac{\partial z}{\partial v}$ respectively and added, we get

$$E\,du + F\,dv - r(D\,du + D'\,dv) = 0,$$
$$F\,du + G\,dv - r(D'\,du + D''\,dv) = 0.$$

But these equations are the same as (23). Hence:

The normals to a surface along a curve of it form a ruled surface which is a developable only when the curve is a line of curvature; in this case the points of the edge of regression are the centers of normal curvature of the surface in the direction of the curve.

The coördinates of the *principal centers of curvature* are

$$(28) \qquad \begin{cases} x_1 = x + \rho_1 X, & y_1 = y + \rho_1 Y, & z_1 = z + \rho_1 Z; \\ x_2 = x + \rho_2 X, & y_2 = y + \rho_2 Y, & z_2 = z + \rho_2 Z. \end{cases}$$

When the parametric curves are the lines of curvature, equation (26) is necessarily of the form

$$(29) \qquad \lambda\, du dv = 0,$$

and consequently we must have $ED' - FD = 0$, $FD'' - GD' = 0$. Since $ED'' - GD \neq 0$, these equations are equivalent to

$$(30) \qquad F = 0, \qquad D' = 0.$$

Conversely, when these conditions are satisfied equation (26) reduces to the form (29). Hence:

A necessary and sufficient condition that the lines of curvature be parametric is that F and D′ be zero.

Let the lines of curvature be parametric, and let ρ_1 and ρ_2 denote the principal radii of normal curvature of the surface in the directions of the lines of curvature $v = \text{const.}$ and $u = \text{const.}$ respectively. From (19) we find

$$(31) \qquad \frac{1}{\rho_1} = \frac{D}{E}, \qquad \frac{1}{\rho_2} = \frac{D''}{G},$$

and equations (13) become

$$(32) \qquad \begin{cases} \dfrac{\partial x}{\partial u} = -\rho_1 \dfrac{\partial X}{\partial u}, & \dfrac{\partial y}{\partial u} = -\rho_1 \dfrac{\partial Y}{\partial u}, & \dfrac{\partial z}{\partial u} = -\rho_1 \dfrac{\partial Z}{\partial u}, \\ \dfrac{\partial x}{\partial v} = -\rho_2 \dfrac{\partial X}{\partial v}, & \dfrac{\partial y}{\partial v} = -\rho_2 \dfrac{\partial Y}{\partial v}, & \dfrac{\partial z}{\partial v} = -\rho_2 \dfrac{\partial Z}{\partial v}. \end{cases}$$

These equations are called the *equations of Rodrigues*.

52. Total and mean curvature. Of fundamental importance in the discussion of the nature of a surface in the neighborhood of a point are the product and the sum of the principal curvatures at the point. They are called the *total curvature* * of the surface at the point and the *mean curvature* respectively. If they be denoted by K and K_m, we have, from (25),

$$(33) \qquad \begin{cases} K = \dfrac{1}{\rho_1 \rho_2} = \dfrac{DD'' - D'^2}{H^2}, \\ K_m = \dfrac{1}{\rho_1} + \dfrac{1}{\rho_2} = \dfrac{ED'' + GD - 2FD'}{H^2}. \end{cases}$$

When K is positive at a point M, the two principal radii have the same sign, and consequently the two centers of principal curvature lie on the same side of the tangent plane. As all the centers of curvature of other normal sections lie between these two, the portion of the surface in the neighborhood of M lies entirely on one side of the tangent plane. This can be seen also in another way. Since H^2 is positive, we must have $DD'' - D'^2 > 0$. Hence the distance from a near-by point to the tangent plane at M, since it is proportional to the fundamental form Φ (§ 48), does not change sign as dv/du is varied.

When K is negative at M, the principal radii differ in sign, and consequently part of the surface lies on one side of the tangent plane and part on the other. In particular there are two directions, given by

$$D\,du^2 + 2D'\,dudv + D''\,dv^2 = 0,$$

for which the normal curvature is zero. In these directions the distances of the near-by points of the surface from the tangent plane, as given by (4), are quantities of the third order at least. Hence these lines are the tangents at M to the curve in which the tangent plane at M meets the surface.

At the points for which K is zero, one of the principal radii is infinite. At these points Φ has the form $(\sqrt{D}\,du + \sqrt{D''}\,dv)^2$ and vanishes in the direction $\sqrt{D}\,du + \sqrt{D''}\,dv = 0$. But as dv/du passes through the value given by this equation, Φ does not change sign. Hence the surface lies on one side of the tangent plane and is tangent to it along the above direction.

* The total curvature is sometimes called the *Gaussian curvature*, after the celebrated geometer who suggested it as a suitable measure of the curvature at a point. Cf. Gauss, p. 15.

An anchor ring, or tore, is a surface with points of all three kinds. Such a surface may be generated by the rotation of a circle of radius a about an axis in the plane of the circle and at a distance b ($> a$) from the center of the circle. The points at the distance b from the axis lie in two circles, and the tangent plane to the tore at a point of either of the circles is tangent all along the circle. Hence the surface has zero curvature at all points of these circles. At every point whose distance from the axis is greater than b the surface lies on one side of the tangent plane, whereas, when the distance is less than b, the tangent plane cuts the surface.

There are surfaces for which K is positive at every point, as, for example, the ellipsoid and the elliptic paraboloid. Moreover, for the hyperboloid of one sheet and the hyperbolic paraboloid the curvature is negative at every point. Surfaces of the former type are called *surfaces of positive curvature*, of the latter type *surfaces of negative curvature.*

Later (§ 64) we shall prove that when K is zero at all points of a surface the latter is developable, and conversely.

53. Equation of Euler. Dupin indicatrix. When the lines of curvature are parametric, equation (18) can be written, in consequence of (III, 23) and (31), in the form

$$\frac{1}{R}=\frac{\cos^2\theta}{\rho_1}+\frac{\sin^2\theta}{\rho_2}, \tag{34}$$

where θ is the angle between the directions whose radii of normal curvature are R and ρ_1. Equation (34) is called the *equation of Euler.*

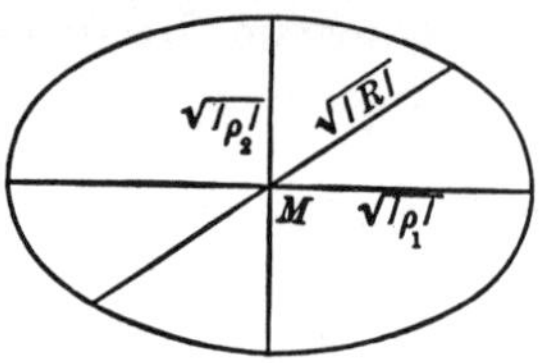

Fig. 13

When the total curvature K at a point is positive, ρ_1 and ρ_2 for the point have the same sign, and R has this sign for all directions. If the tangents to the lines of curvature at the point M be taken for coördinate axes, with respect to which ξ and η are coördinates, and segments of length $\pm\sqrt{|R|}$ be laid off from M in the two directions corresponding to R, the locus of the end points of these segments is the ellipse (fig. 13) whose equation is

$$\frac{\xi^2}{|\rho_1|}+\frac{\eta^2}{|\rho_2|}=1.$$

This ellipse is called the *Dupin indicatrix* for the point. When, in particular, ρ_1 and ρ_2 are equal, the indicatrix is a circle. Hence the Dupin indicatrix at an umbilical point is a circle (§ 50). For this reason such a point is sometimes called a *circular point.*

When K is negative ρ_1 and ρ_2 differ in sign, and consequently certain values of R are positive and the others are negative. In the directions for which R is positive we lay off the segments $\pm\sqrt{R}$, and in the other directions $\pm\sqrt{-R}$. The locus of the end points of these segments consists of the conjugate hyperbolas (fig. 14) whose equations are

$$\frac{\xi^2}{\rho_1}+\frac{\eta^2}{\rho_2}=1, \quad \frac{\xi^2}{\rho_1}+\frac{\eta^2}{\rho_2}=-1.$$

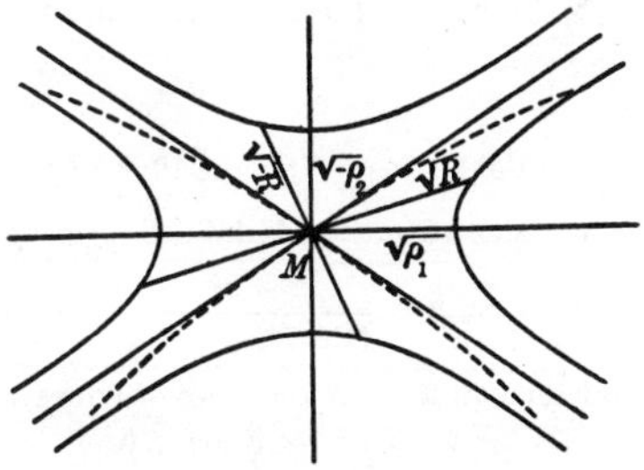

FIG. 14

We remark that R is infinite for the directions given by

$$(35) \qquad \tan^2\theta=-\frac{\rho_2}{\rho_1},$$

or, in other words, in the directions of the asymptotes to the hyperbolas. The above locus is the *Dupin indicatrix* for the point.

Finally, when $K=0$ the equation of the indicatrix is of one of the forms

$$\xi^2=|\rho_1|, \qquad \eta^2=|\rho_2|,$$

that is, a pair of parallel straight lines. In view of the foregoing results, a point of a surface is called *elliptic*, *hyperbolic*, or *parabolic*, according as the total curvature at the point is positive, negative, or zero.

In consequence of (4) the expression for the distance p upon the tangent plane to a surface at a point M from a near-by point P of the surface is given by

$$\frac{E\,du^2}{\rho_1}+\frac{G\,dv^2}{\rho_2}=2p,$$

to within terms of higher order. But $\sqrt{E}\,du$ and $\sqrt{G}\,dv$ are the distances, to within terms of higher order, of P from the normal planes to the surface at M in the directions of the lines of curvature. Hence the plane parallel to the tangent plane and at a distance p from it cuts the surface in the curve

$$\frac{\xi^2}{\rho_1}+\frac{\eta^2}{\rho_2}=2p, \qquad \zeta=p.$$

Evidently this is a conic similar to the Dupin indicatrix at an elliptic or parabolic point, and to a part of the indicatrix at a hyperbolic point.

EXAMPLES

1. Show that the meridians and parallels of a surface of revolution are its lines of curvature, and determine the character of the developable surfaces formed by the normals to the surface along these lines.

2. Show that the parametric lines on the surface

$$x = \frac{a}{2}(u + v), \qquad y = \frac{b}{2}(u - v), \qquad z = \frac{uv}{2},$$

are straight lines. Find the lines of curvature.

3. When a surface is defined by $z = f(x, y)$, the expressions for the curvatures are

$$K = \frac{rt - s^2}{(1 + p^2 + q^2)^2}, \qquad K_m = \frac{(1 + p^2)t + (1 + q^2)r - 2pqs}{(1 + p^2 + q^2)^{\frac{3}{2}}},$$

and the equation of the lines of curvature is

$$[(1 + p^2)s - pqr]\,dx^2 + [(1 + p^2)t - (1 + q^2)r]\,dxdy + [pqt - (1 + q^2)s]\,dy^2 = 0.$$

4. The principal radii of the surface $y \cos\frac{z}{a} - x \sin\frac{z}{a} = 0$ at a point (x, y, z) are equal to $\pm\frac{x^2 + y^2 + a^2}{a}$. Find the lines of curvature.

5. Derive the equations of the tore, defined in § 52, and prove therefrom the results stated.

6. The sum of the normal curvatures in two orthogonal directions is constant.

7. The Euler equation can be written

$$R = \frac{2\rho_1\rho_2}{\rho_1 + \rho_2 - (\rho_1 - \rho_2)\cos 2\theta}.$$

54. Conjugate directions at a point. Conjugate systems. Two curves on a surface through a point M are said to have *conjugate directions* when their tangents at M coincide with conjugate diameters of the Dupin indicatrix for the point. These tangents are also parallel to conjugate diameters of the conic in which the surface is cut by a plane parallel to the tangent plane to M and very near it. Let P denote a point of this conic and N the point in which its plane a cuts the normal at M. The tangent plane to the surface at P meets the plane a in the tangent line at P to the conic. Moreover, this tangent line is parallel to the diameter conjugate to NP. Hence as P approaches M this tangent line approaches the diameter of the Dupin indicatrix, which is conjugate to the diameter in the direction MP. Hence we have (cf. § 27):

The characteristic of the tangent plane to a surface, as the point of contact moves along a curve, is the tangent to the surface in the direction conjugate to the curve.

By means of this theorem we derive the analytical condition for conjugate directions.

If the equation of the tangent plane is

$$(\xi - x)X + (\eta - y)Y + (\zeta - z)Z = 0,$$

ξ, η, ζ being current coördinates, the characteristic is defined by this equation, and

$$(\xi - x)\frac{dX}{ds} + (\eta - y)\frac{dY}{ds} + (\zeta - z)\frac{dZ}{ds} = 0,$$

where s is the arc of the curve along which the point of contact moves. If δx, δy, δz denote increments of x, y, z in the direction conjugate to the curve, we have, from the above equations,

$$\delta x dX + \delta y dY + \delta z dZ = 0.$$

If increments of u and v in the conjugate direction be denoted by δu and δv, this equation may be written

$$D\,du\delta u + D'(du\delta v + dv\delta u) + D''dv\delta v = 0. \tag{36}$$

The directions conjugate to any curve of the family

$$\phi(u, v) = \text{const.} \tag{37}$$

are given by

$$\left(D\frac{\partial\phi}{\partial v} - D'\frac{\partial\phi}{\partial u}\right)\delta u + \left(D'\frac{\partial\phi}{\partial v} - D''\frac{\partial\phi}{\partial u}\right)\delta v = 0. \tag{38}$$

As this is a differential equation of the first order and first degree, it defines a one-parameter family of curves. These curves and the curves $\phi = \text{const.}$ are said to form a *conjugate system*. Moreover, any two families of curves are said to form a conjugate system when the tangents to a curve of each family at their point of intersection have conjugate directions.

From (36) it follows that the curves conjugate to the curves $v = \text{const.}$ are defined by $D\,\delta u + D'\delta v = 0$. Consequently, in order that they be the curves $u = \text{const.}$, we must have D' equal to zero. As the converse also is true, we have:

A necessary and sufficient condition that the parametric curves form a conjugate system is that D' be zero.

We have seen (§ 51) that the lines of curvature are characterized by the property that, when they are parametric, the coefficients F and D' are zero. Hence:

The lines of curvature form a conjugate system and the only orthogonal conjugate system.

If the lines of curvature are parametric, and the angles which a pair of conjugate directions make with the tangent to the curve $v = \text{const.}$ are denoted by θ and θ', we have

$$\tan\theta = \sqrt{\frac{G}{E}}\frac{dv}{du}, \qquad \tan\theta' = \sqrt{\frac{G}{E}}\frac{\delta v}{\delta u},$$

so that equation (36) may be put in the form

$$\tan\theta\tan\theta' = -\frac{\rho_2}{\rho_1}, \tag{39}$$

which is the well-known equation of conjugate directions of a conic.

55. Asymptotic lines. Characteristic lines. When θ' is equal to θ, equation (39) reduces to (35). Hence the *asymptotic directions* are self-conjugate. If in equation (36) we put $\delta v/\delta u = dv/du$, we obtain

$$D\,du^2 + 2\,D'\,du dv + D''\,dv^2 = 0, \tag{40}$$

which determines, consequently, the asymptotic directions at each point of the surface. This equation defines a double family of curves upon the surface, two of which pass through each point and admit as tangents the asymptotic directions at the point. They are called the *asymptotic lines* of the surface.

The asymptotic lines are imaginary on surfaces of positive curvature, real on surfaces of negative curvature, and consist of a single real family on a surface of zero curvature.

Recalling the results of § 52, we say that the tangent plane to a surface at a point cuts the surface in asymptotic lines in the neighborhood of the point. As an immediate consequence, we have that the generators of a ruled surface form one family of asymptotic lines.

Since an asymptotic line is self-conjugate, the characteristics of the tangent plane as the point of contact moves along an asymptotic line are the tangents to the latter. Hence the osculating plane of an asymptotic line at a point is the tangent plane to the surface at the point, and consequently the asymptotic line is the edge of regression of the developable circumscribing the surface along the asymptotic line. This follows also from equation (16).

From (40) we have the theorem:

A necessary and sufficient condition that the asymptotic lines upon a surface be parametric is that

$$D = D'' = 0.$$

If these equations hold, and, furthermore, the parametric curves are orthogonal, it is seen from (33) that the mean curvature is zero, and conversely. Hence:

A necessary and sufficient condition that the asymptotic lines form an orthogonal system is that the mean curvature of the surface be zero.

A surface whose mean curvature is zero at every point is called a *minimal surface*. At each of its points the Dupin indicatrix consists of two conjugate equilateral hyperbolas.

By means of (39) we find that the angle between conjugate directions is given by

$$\tan(\theta' - \theta) = \frac{\rho_2 \cot\theta + \rho_1 \tan\theta}{\rho_2 - \rho_1}.$$

If we consider only real lines, this angle can be zero only for surfaces of negative curvature, in which case the directions are asymptotic. It is natural, therefore, to seek the conjugate directions upon a surface of positive curvature for which the included angle is a minimum. To this end we differentiate the right-hand member of the above equation with respect to θ and equate the result to zero. The result is reducible to

$$\tan\theta = \pm\sqrt{\frac{\rho_2}{\rho_1}}. \tag{41}$$

Then from (39) we have

$$\tan\theta' = \mp\sqrt{\frac{\rho_2}{\rho_1}}.$$

From these equations it follows that $\theta' = -\theta$, and

$$\tan(\theta' - \theta) = \pm\frac{2\sqrt{\rho_1\rho_2}}{\rho_2 - \rho_1}.$$

Conversely, when $\theta' = -\theta$ equation (39) becomes (41). Hence:

Upon a surface of positive curvature there is a unique conjugate system for which the angle between the directions at any point is the minimum angle between conjugate directions at the point; it is the only conjugate system whose directions are symmetric with respect to the directions of the lines of curvature.

These lines are called the *characteristic lines*. It is of interest to note that equations (35) and (41) are similar, and that the real asymptotic directions upon a surface of negative curvature are symmetric with respect to the directions of the lines of curvature.

As just seen, if θ is the angle which one characteristic line makes with the line of curvature $v = \text{const.}$ at a point, the other characteristic line makes the angle $-\theta$. Hence the radii of normal curvature for these directions are equal, and consequently a necessary and sufficient condition that the characteristic curves of a surface be parametric is

$$\frac{D}{E} = \frac{D''}{G}, \qquad D' = 0. \tag{42}$$

56. Corresponding systems on two surfaces. By reasoning similar to that of § 34 we establish the theorem:

A necessary and sufficient condition that the curves defined by $R\,du^2 + 2\,S\,du dv + T\,dv^2 = 0$ *form a conjugate system upon a surface is*

$$RD'' + TD - 2\,SD' = 0.$$

From this we have at once:

If the second quadratic forms of two surfaces S *and* S_1 *are* $D\,du^2 + 2\,D'du dv + D''dv^2$ *and* $D_1 du^2 + 2\,D_1' du dv + D_1'' dv^2$, *and if a point on one surface is said to correspond to the point on the other with the same values of* u *and* v, *the equation*

$$\begin{vmatrix} du^2 & D_1'' & D'' \\ -du dv & D_1' & D' \\ dv^2 & D_1 & D \end{vmatrix} = 0 \tag{43}$$

defines a system of curves which is conjugate for both surfaces.

By the methods of § 50 we prove that these curves are real when either or both of the surfaces S, S_1 is of positive curvature. If the curvature of S is negative and it is referred to its asymptotic lines, the above equation reduces to

$$D_1 du^2 - D_1'' dv^2 = 0.$$

Hence the system is real when D_1 and D_1'' have the same sign, that is, when the curvature of S_1 is positive.

Another consequence of the above theorem is:

A necessary and sufficient condition that asymptotic lines on one of two surfaces S, S_1 correspond to a conjugate system on the other is

$$DD_1''+D''D_1-2D'D_1'=0. \tag{44}$$

EXAMPLES

1. Find the curves on the general surface of revolution which are conjugate to the loxodromic curves which cut the meridians under the angle α.

2. Find the curves on the general right conoid, Ex. 1, p. 56, which are conjugate to the orthogonal trajectories of the generators.

3. When the equations of a surface are of the form

$$x = U_1, \qquad y = V_1, \qquad z = U_2 + V_2,$$

where U_1 and U_2 are functions of u alone, and V_1 and V_2 of v alone, the parametric curves are plane and form a conjugate system.

4. Prove that the sum of normal radii at a point in conjugate directions is constant.

5. When a surface of revolution is referred to its meridians and parallels, the asymptotic lines can be found by quadratures.

6. Find the asymptotic lines on the surface

$$x = a(1+\cos u)\cot v, \qquad y = a(1+\cos u), \qquad z = \frac{a\cos u}{\sin v}.$$

7. Determine the asymptotic lines upon the surface $z = y\sin x$ and their orthogonal trajectories. Show that the x-axis belongs to one of the latter families.

8. Find the asymptotic lines on the surface $2y^3 - 2xyz + z^2 = 0$, and determine their projections on the xy-plane.

9. Prove that the product of the normal radii in conjugate directions is a maximum for characteristic lines and a minimum for lines of curvature.

10. When the parametric lines are any whatever, the equation of characteristic lines is

$$[D(GD-ED'')-2D'(FD-ED')]\,du^2+2[D'(GD+ED'')-2FDD'']\,du\,dv$$
$$+[2D'(GD'-FD'')-D''(GD-ED'')]\,dv^2=0.$$

57. Geodesic curvature. Geodesics. Consider a curve C upon a surface and the tangent plane to the surface at a point M of C. Project orthogonally upon this tangent plane the portion of the curve in the neighborhood of M, and let C' denote this projection. The curve C' is a normal section of the projecting cylinder, and C is a curve upon the latter, tangent to C' at M. Hence the theorem of Meusnier can be applied to these two curves. If $1/\rho_g$ denotes the curvature of C' and ψ the angle between the principal normal to C and the positive direction of the normal to the cylinder at M, we have

$$\frac{1}{\rho_g} = \frac{\cos\psi}{\rho}. \tag{45}$$

In order to connect this result with others, it is necessary to define the positive direction of the normal to the cylinder. This normal lies in the tangent plane to the surface. We make the convention that the positive directions of the tangent to the curve, the normal to the cylinder, and the normal to the surface shall have the same mutual orientations as the positive x-, y-, and z-axes. From this choice of direction it follows that if, as usual, the direction-cosines of the tangent to the curve be dx/ds, dy/ds, dz/ds, then those of the normal to the cylinder are

$$Y\frac{dz}{ds} - Z\frac{dy}{ds}, \qquad Z\frac{dx}{ds} - X\frac{dz}{ds}, \qquad X\frac{dy}{ds} - Y\frac{dx}{ds}. \tag{46}$$

The curvature of C' is called the *geodesic curvature* of C, and ρ_g the *radius of geodesic curvature*. And the center of curvature of C' is called the *center of geodesic curvature* of C.

From its definition the geodesic curvature is positive or negative according as the osculating plane of C lies on one side or the other of the normal plane to the surface through the tangent to C. From (45) it follows that the center of first curvature of C is the projection upon its osculating plane of the center of geodesic curvature. Moreover, the former is also the projection of the center of curvature of the normal section tangent to C (§ 49). Hence the plane through a point M of C, normal to the line joining the centers of normal and geodesic curvature at M, is the osculating plane of C for this point, and its intersection with the join is the center of first curvature.

By definition (§ 49) ϖ denotes the angle which the positive direction of the normal to the surface makes with the positive direction of the principal normal to C, angles being measured toward the binormal. Hence equation (45) can be written

$$\frac{1}{\rho_g} = \frac{\sin\varpi}{\rho}. \tag{47}$$

These various quantities are represented in fig. 15, for which the tangent to the curve is normal to the plane of the paper, and is directed toward the reader. The directed lines MP, MB, MK, MN represent respectively the positive directions of the principal normal and binormal of the curve and the normals to the projecting cylinder and to the surface.

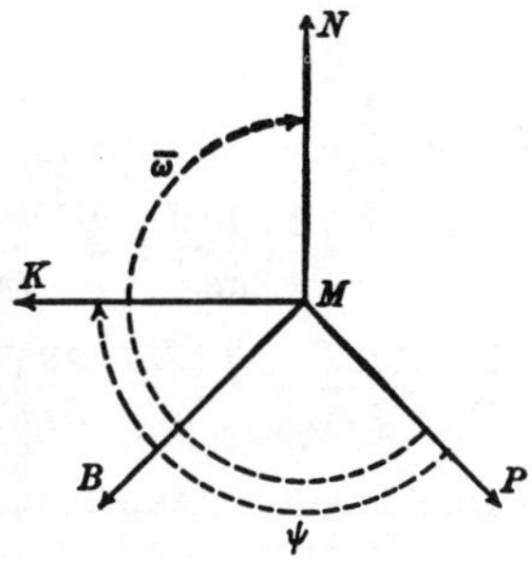

FIG. 15

A curve whose principal normal at every point coincides with the normal to the surface upon which it lies, is called a *geodesic*. From (45) it follows that a geodesic may also be defined as a curve whose geodesic curvature is zero at every point. For example, the meridians of a surface of revolution are geodesics, as follows from the results in § 46. A twisted curve is a geodesic on its rectifying developable, and when a straight line lies on a surface, it is a geodesic for the surface. Later we shall make an extensive study of geodesics, but now we desire to find an expression for the geodesic curvature in terms of the fundamental quantities of the surface and the equation of the curve.

58. Fundamental formulas. The direction-cosines of the principal normal are (§ 8)

$$\rho\frac{d^2x}{ds^2}, \quad \rho\frac{d^2y}{ds^2}, \quad \rho\frac{d^2z}{ds^2}.$$

Consequently, by means of (46), equation (45) may be put in the form

$$\frac{1}{\rho_g}=\sum\left(Y\frac{dz}{ds}-Z\frac{dy}{ds}\right)\frac{d^2x}{ds^2}. \tag{48}$$

Expressed as functions of u and v, the quantities $\frac{dx}{ds}, \frac{d^2x}{ds^2}$ are of the form

$$\frac{dx}{ds}=\frac{\partial x}{\partial u}\frac{du}{ds}+\frac{\partial x}{\partial v}\frac{dv}{ds},$$

$$\frac{d^2x}{ds^2}=\frac{\partial^2 x}{\partial u^2}\left(\frac{du}{ds}\right)^2+2\frac{\partial^2 x}{\partial u\partial v}\frac{du}{ds}\frac{dv}{ds}+\frac{\partial^2 x}{\partial v^2}\left(\frac{dv}{ds}\right)^2+\frac{\partial x}{\partial u}\frac{d^2u}{ds^2}+\frac{\partial x}{\partial v}\frac{d^2v}{ds^2}.$$

When these expressions are substituted in (48), and in the reduction we make use of (10) and (11), we obtain

$$\frac{1}{\rho_g} = \frac{1}{H} \begin{vmatrix} E\dfrac{du}{ds} + F\dfrac{dv}{ds}, & L \\ F\dfrac{du}{ds} + G\dfrac{dv}{ds}, & M \end{vmatrix}, \tag{49}$$

where L and M have the significance

$$\begin{cases} L = \dfrac{1}{2}\dfrac{\partial E}{\partial u}\left(\dfrac{du}{ds}\right)^2 + \dfrac{\partial E}{\partial v}\dfrac{du}{ds}\dfrac{dv}{ds} + \left(\dfrac{\partial F}{\partial v} - \dfrac{1}{2}\dfrac{\partial G}{\partial u}\right)\left(\dfrac{dv}{ds}\right)^2 + E\dfrac{d^2u}{ds^2} + F\dfrac{d^2v}{ds^2}, \\ M = \left(\dfrac{\partial F}{\partial u} - \dfrac{1}{2}\dfrac{\partial E}{\partial v}\right)\left(\dfrac{du}{ds}\right)^2 + \dfrac{\partial G}{\partial u}\dfrac{du}{ds}\dfrac{dv}{ds} + \dfrac{1}{2}\dfrac{\partial G}{\partial v}\left(\dfrac{dv}{ds}\right)^2 + F\dfrac{d^2u}{ds^2} + G\dfrac{d^2v}{ds^2}. \end{cases}$$

From this it is seen that the geodesic curvature of a curve depends upon E, F, G, and is entirely independent of D, D', D''.

Suppose that the parametric lines form an orthogonal system, and that the radius of geodesic curvature of a curve $v = \text{const.}$ be denoted by ρ_{gu}. In this case $F = 0$, $ds = \sqrt{E}\,du$. Hence the above equation reduces to

$$\frac{1}{\rho_{gu}} = -\frac{1}{\sqrt{EG}}\frac{\partial \sqrt{E}}{\partial v}. \tag{50}$$

In like manner we find that the geodesic curvature of a curve $u = \text{const.}$ is given by

$$\frac{1}{\rho_{gv}} = \frac{1}{\sqrt{EG}}\frac{\partial \sqrt{G}}{\partial u}. \tag{51}$$

As an immediate consequence of these equations we have the theorem:

When the parametric lines upon a surface form an orthogonal system, a necessary and sufficient condition that the curves $v = \text{const.}$ or $u = \text{const.}$ be geodesics is that E be a function of u alone or G of v alone respectively.

It will now be shown that ρ_{gu} is expressible as a function of differential parameters of v formed with respect to the linear element (III, 4).

From the definition of these parameters (§§ 37, 38) it follows that when $F = 0$

$$\Delta_1 v = \frac{1}{G}, \qquad \Delta_1(v, \sqrt{G}) = \frac{1}{G}\frac{\partial \sqrt{G}}{\partial v}, \qquad \Delta_2 v = \frac{1}{\sqrt{EG}}\frac{\partial}{\partial v}\sqrt{\frac{E}{G}}.$$

Hence, by substitution in (50), we obtain

$$\frac{1}{\rho_{gu}} = -\left[\frac{\Delta_2 v}{\sqrt{\Delta_1 v}} + \Delta_1\left(v, \frac{1}{\sqrt{\Delta_1 v}}\right)\right]. \tag{52}$$

In like manner, we find

$$\frac{1}{\rho_{gv}} = \left[\frac{\Delta_2 u}{\sqrt{\Delta_1 u}} + \Delta_1\left(u, \frac{1}{\sqrt{\Delta_1 u}}\right)\right]. \tag{53}$$

Thus we have shown that the geodesic curvature of a parametric line is a differential parameter of the curvilinear coördinate of the line. Since this curvature is a geometrical property of a line, it is necessarily independent of the choice of parameters, and thus is an invariant. This was evident *a priori*, but we have just shown that it is an invariant of the differential parameter type.

From the definition of the positive direction of the normal to a surface (§ 48), and the normal to the cylinder of projection, it follows that the latter for a curve $v = \text{const.}$ is the direction in which v increases, whereas, for a curve $u = \text{const.}$, it is the direction in which u decreases. Hence, if the latter curves be defined by $-u = \text{const.}$, equations (52) and (53) have the same sign.

If, now, we imagine the surface referred to another parametric system, for which the linear element is

$$ds^2 = E\,du^2 + 2\,F\,du\,dv + G\,dv^2, \tag{54}$$

the curve whose geodesic curvature is given by (50) will be defined by an equation such as $\phi(u, v) = \text{const.}$ And if the sign of ϕ be such that ϕ is increasing in the direction of the normal of its projecting cylinder, its geodesic curvature will be given by

$$\frac{1}{\rho_g} = -\left[\frac{\Delta_2 \phi}{\sqrt{\Delta_1 \phi}} + \Delta_1\left(\phi, \frac{1}{\sqrt{\Delta_1 \phi}}\right)\right], \tag{55}$$

where the differential parameters are formed with respect to (54).

If two surfaces are applicable, and points on each with the same curvilinear coördinates correspond, the geodesic curvature of the curve $\phi = \text{const.}$ on each at corresponding points will be the same in consequence of (55). Hence:

Upon two applicable surfaces the geodesic curvature of corresponding curves, at corresponding points, is the same.

When the second member of equation (55) is developed by (III, 46, 56), we have

$$\frac{1}{\rho_g}=-\frac{1}{H}\left\{\frac{1}{\sqrt{\Delta_1\phi}}\left[\frac{\partial}{\partial u}\left(\frac{G\frac{\partial\phi}{\partial u}-F\frac{\partial\phi}{\partial v}}{H}\right)+\frac{\partial}{\partial v}\left(\frac{E\frac{\partial\phi}{\partial v}-F\frac{\partial\phi}{\partial u}}{H}\right)\right]\right.$$

$$\left.+\frac{G\frac{\partial\phi}{\partial u}-F\frac{\partial\phi}{\partial v}}{H}\frac{\partial}{\partial u}\frac{1}{\sqrt{\Delta_1\phi}}+\frac{E\frac{\partial\phi}{\partial v}-F\frac{\partial\phi}{\partial u}}{H}\frac{\partial}{\partial v}\frac{1}{\sqrt{\Delta_1\phi}}\right\}$$

$$=-\frac{1}{H}\left\{\frac{\partial}{\partial u}\left(\frac{G\frac{\partial\phi}{\partial u}-F\frac{\partial\phi}{\partial v}}{H\sqrt{\Delta_1\phi}}\right)+\frac{\partial}{\partial v}\left(\frac{E\frac{\partial\phi}{\partial v}-F\frac{\partial\phi}{\partial u}}{H\sqrt{\Delta_1\phi}}\right)\right\}.$$

Hence we have the formula of Bonnet*:

$$\frac{1}{\rho_g}=\frac{1}{H}\left\{\frac{\partial}{\partial u}\left(\frac{F\frac{\partial\phi}{\partial v}-G\frac{\partial\phi}{\partial u}}{\left[E\left(\frac{\partial\phi}{\partial v}\right)^2-2F\frac{\partial\phi}{\partial v}\frac{\partial\phi}{\partial u}+G\left(\frac{\partial\phi}{\partial u}\right)^2\right]^{\frac{1}{2}}}\right)\right. \tag{56}$$

$$\left.+\frac{\partial}{\partial v}\left(\frac{F\frac{\partial\phi}{\partial u}-E\frac{\partial\phi}{\partial v}}{\left[E\left(\frac{\partial\phi}{\partial v}\right)^2-2F\frac{\partial\phi}{\partial v}\frac{\partial\phi}{\partial u}+G\left(\frac{\partial\phi}{\partial u}\right)^2\right]^{\frac{1}{2}}}\right)\right\}.$$

In particular, the geodesic curvature of the parametric curves, when the latter do not form an orthogonal system, is given by

$$\left\{\begin{aligned}\frac{1}{\rho_{gu}}&=\frac{1}{H}\left(\frac{\partial}{\partial u}\frac{F}{\sqrt{E}}-\frac{\partial}{\partial v}\sqrt{E}\right),\\ \frac{1}{\rho_{gv}}&=-\frac{1}{H}\left(\frac{\partial}{\partial v}\frac{F}{\sqrt{G}}-\frac{\partial}{\partial u}\sqrt{G}\right).\end{aligned}\right. \tag{57}$$

The geodesic curvature of a curve of the family, defined by the differential equation

$$M\,du+N\,dv=0,$$

has the value

$$\frac{1}{\rho_g}=\frac{1}{H}\left\{\frac{\partial}{\partial u}\left(\frac{FN-GM}{\sqrt{EN^2-2FMN+GM^2}}\right)\right. \tag{58}$$

$$\left.+\frac{\partial}{\partial v}\left(\frac{FM-EN}{\sqrt{EN^2-2FMN+GM^2}}\right)\right\}.$$

* **Mémoire sur la théorie générale des surfaces,** *Journal de l'École Polytechnique,* **Cahier 32 (1848), p. 1.**

In illustration of the preceding results, we establish the theorem:

When the curves of an orthogonal system have constant geodesic curvature, the system is isothermal.

When the surface is referred to these lines, and the linear element is written $ds^2 = Edu^2 + Gdv^2$, the condition that the geodesic curvature of these curves be constant is, by (50) and (51),

$$\text{(i)} \qquad \frac{1}{\sqrt{EG}}\frac{\partial\sqrt{G}}{\partial u} = U_1, \qquad \frac{1}{\sqrt{EG}}\frac{\partial\sqrt{E}}{\partial v} = V_1,$$

where U_1 and V_1 are functions of u and v respectively. If these equations are differentiated with respect to v and u respectively, we get

$$\frac{\partial^2\log\sqrt{G}}{\partial u\,\partial v} - \frac{\partial\log\sqrt{E}}{\partial v}\frac{\partial\log\sqrt{G}}{\partial u} = 0, \qquad \frac{\partial^2\log\sqrt{E}}{\partial u\,\partial v} - \frac{\partial\log\sqrt{E}}{\partial v}\frac{\partial\log\sqrt{G}}{\partial u} = 0.$$

Subtracting, we obtain
$$\frac{\partial^2}{\partial u\,\partial v}\log\frac{E}{G} = 0.$$

Hence E/G is equal to the ratio of a function of u and a function of v, and the system is isothermic. In terms of isothermic parameters, equations (i) are of the form

$$\frac{1}{\lambda^2}\frac{\partial\lambda}{\partial u} = U', \qquad \frac{1}{\lambda^2}\frac{\partial\lambda}{\partial v} = V',$$

and the linear element is

$$\text{(ii)} \qquad ds^2 = \frac{du^2 + dv^2}{(U+V)^2}.$$

It is evident that the meridians and parallels on a surface of revolution form such a system. The same is true likewise of an orthogonal system of small circles on a sphere.

59. Geodesic torsion. We have just seen that when a curve is defined by a finite equation or a differential equation, its geodesic curvature can be found directly. The same is true of the normal curvature of the surface in the direction of the curve by (18). Then from (16) and (47) follow the expressions for ρ and $\bar{\omega}$. In order to define the curve it remains for us to obtain an expression for the torsion.

From the definition of $\bar{\omega}$ it follows that

$$\text{(59)} \qquad \sin\bar{\omega} = X\lambda + Y\mu + Z\nu,$$

where λ, μ, ν are the direction-cosines of the binormal. If this equation is differentiated with respect to the arc of the curve, and the Frenet formulas (I, 50) are used in the reduction, we get

$$\text{(60)} \qquad \cos\bar{\omega}\left(\frac{d\bar{\omega}}{ds} - \frac{1}{\tau}\right) = \sum\lambda\frac{dX}{ds}.$$

From (I, 37, 41) we have

$$\lambda = \rho\left(\frac{dy}{ds}\frac{d^2z}{ds^2} - \frac{dz}{ds}\frac{d^2y}{ds^2}\right), \qquad \mu = \rho\left(\frac{dz}{ds}\frac{d^2x}{ds^2} - \frac{dx}{ds}\frac{d^2z}{ds^2}\right),$$

$$\nu = \rho\left(\frac{dx}{ds}\frac{d^2y}{ds^2} - \frac{dy}{ds}\frac{d^2x}{ds^2}\right),$$

and

$$\rho\sum X\frac{d^2x}{ds^2} = \sum Xl = \cos\overline{\omega}.$$

Moreover, from (13), we obtain the identity

$$\frac{dX}{ds}\frac{dy}{ds} - \frac{dY}{ds}\frac{dx}{ds} = \frac{Z}{H}\left[(ED' - FD)\left(\frac{du}{ds}\right)^2 + (ED'' - GD)\frac{du}{ds}\frac{dv}{ds} + (FD'' - GD')\left(\frac{dv}{ds}\right)^2\right].$$

Consequently equation (60) is equivalent to

$$\cos\overline{\omega}\left(\frac{1}{\tau} - \frac{d\overline{\omega}}{ds} - \frac{1}{T}\right) = 0, \tag{61}$$

where $1/T$ has the value

$$\frac{1}{T} = \frac{(FD - ED')\,du^2 + (GD - ED'')\,dudv + (GD' - FD'')\,dv^2}{H(E\,du^2 + 2\,F\,dudv + G\,dv^2)}. \tag{62}$$

When $\cos\overline{\omega}$ is different from zero, that is, when the curve is not an asymptotic line, equation (61) becomes

$$\frac{1}{\tau} - \frac{d\overline{\omega}}{ds} = \frac{1}{T}. \tag{63}$$

As the expression for T involves only the fundamental coefficients and dv/du, we have the following theorem of Bonnet:

The function $\frac{1}{\tau} - \frac{d\overline{\omega}}{ds}$ *is the same for all curves which have the same tangent at a common point.*

Among these curves there is one geodesic, and only one, for it will be shown later (§ 85) that one geodesic and only one passes through a given point and has a given direction at the point. At every point of this geodesic $\overline{\omega}$ is equal to 0° or 180°, and consequently $\tau = T$. Hence the value of T for a given point and direction is that of the radius of torsion of the geodesic with this direction. The function T is therefore called the *radius of geodesic torsion* of the curve. From (63) it is seen that T is the radius of torsion of any curve whose osculating plane makes a constant angle with the tangent plane.*

When the numerator of the right-hand member of equation (62) is equated to zero, we have the differential equation of lines of curvature. Hence:

A necessary and sufficient condition that the geodesic torsion of a curve be zero at a point is that the curve be tangent to a line of curvature at the point.

The geodesic torsion of the parametric lines is given by

$$\frac{1}{T_u}=\frac{FD-ED'}{EH}, \qquad \frac{1}{T_v}=\frac{GD'-FD''}{GH}. \tag{64}$$

When these lines form an orthogonal system T_u and T_v differ only in sign. Consequently the geodesic torsion at the point of meeting of two curves cutting orthogonally is the same to within the sign.

Thus far in the consideration of equation (61) we have excluded the case of asymptotic lines. In considering them now, we assume that they are parametric. The direction-cosines of the tangent and binormal to a curve $v=\text{const.}$ in this case are

$$\alpha=\frac{1}{\sqrt{E}}\frac{\partial x}{\partial u}, \qquad \beta=\frac{1}{\sqrt{E}}\frac{\partial y}{\partial u}, \qquad \gamma=\frac{1}{\sqrt{E}}\frac{\partial z}{\partial u};$$

$$\lambda=\epsilon X, \qquad \mu=\epsilon Y, \qquad \nu=\epsilon Z,$$

where ϵ is $+1$ or -1. Consequently the direction-cosines of the principal normal have the values

$$l=\frac{\epsilon}{\sqrt{E}}\left(Y\frac{\partial z}{\partial u}-Z\frac{\partial y}{\partial u}\right),$$

and similar expressions for m and n.

When in the Frenet formulas

$$\frac{d\lambda}{ds}=\frac{l}{\tau}, \qquad \frac{d\mu}{ds}=\frac{m}{\tau}, \qquad \frac{d\nu}{ds}=\frac{n}{\tau},$$

we substitute the above values, and in the reduction make use of (11) and (13), we get

$$\frac{1}{\tau}=-\frac{D'}{H}=-\sqrt{-K}. \tag{65}$$

* Thus far exception must be made of asymptotic lines, but later this restriction will be removed.

In like manner, the torsion of the asymptotic lines $u = \text{const.}$ is found to be $\sqrt{-K}$. But from (64) we find that the geodesic torsion in the direction of the asymptotic lines is $\mp\sqrt{-K}$. Hence equation (63) is true for the asymptotic lines as well as for all other curves on the surface.

Incidentally we have established the following theorem of Enneper:

The square of the torsion of a real asymptotic line at a point is equal to the absolute value of the total curvature of the surface at the point; the radii of torsion of the asymptotic lines through a point differ only in sign.

The following theorem of Joachimsthal is an immediate consequence of (63):

When two surfaces meet under a constant angle, the line of intersection is a line of curvature of both or neither; and conversely, when the curve of intersection of two surfaces is a line of curvature of both they meet under constant angle.

For, if ϖ_1, ϖ_2 denote the values of ϖ for the two surfaces, and T_1, T_2 the values of T, we have, by subtracting the two equations of the form (63), that $T_1 = T_2$, which proves the first part of the theorem. Conversely, if $1/T_1 = 1/T_2 = 0$, we have $\frac{d}{ds}(\omega_1 - \omega_2) = 0$, and consequently the surfaces meet under constant angle.

EXAMPLES

1. Show that the radius of geodesic curvature of a parallel on a surface of revolution is the same at all points of the parallel, and determine its geometrical significance.

2. Find the geodesic curvature of the parametric lines on the surface

$$x = \frac{a}{2}(u + v), \qquad y = \frac{b}{2}(u - v), \qquad z = \frac{uv}{2}.$$

3. Given a family of loxodromic curves upon a surface of revolution which cut the meridians under the same angle α; show that the geodesic curvature of all these curves is the same at their points of intersection with a parallel.

4. Straight lines on a surface are the only asymptotic lines which are geodesics.

5. Show that the geodesic torsion of a curve is given by

$$\frac{1}{T} = \frac{1}{2}\left(\frac{1}{\rho_1} - \frac{1}{\rho_2}\right)\sin 2\theta,$$

where θ denotes the angle which the direction of the curve at a point makes with the line of curvature $v = \text{const.}$ through the point.

6. Every geodesic line of curvature is a plane curve.

7. Every plane geodesic line is a line of curvature.

8. When a surface is cut by a plane or a sphere under constant angle, it is a line of curvature on the surface, and conversely.

9. If the curves of one family of an isothermal orthogonal system have constant geodesic curvature, the curves of the other family have the same property.

60. Spherical representation. In the discussion of certain properties of a surface S it is of advantage to make a representation of S upon the unit sphere * by drawing radii of the sphere parallel to the positive directions of the normals to S, and taking the extremities of the radii as *spherical images* of the corresponding points on S. As a point M moves along a curve on S, its image m describes a curve on the sphere. If we limit our consideration to a portion of the surface in which no two normals are parallel, the portions of the surface and sphere will be in a one-to-one correspondence. This map of the surface upon the sphere is called the *spherical representation* of the surface, or the *Gaussian representation.* It was first employed by Gauss in his treatment of the curvature of surfaces.†

The coördinates of m are the direction-cosines of the normal to the surface, namely X, Y, Z, so that if we put

$$\mathcal{E}=\sum\left(\frac{\partial X}{\partial u}\right)^2, \qquad \mathcal{F}=\sum\frac{\partial X}{\partial u}\frac{\partial X}{\partial v}, \qquad \mathcal{G}=\sum\left(\frac{\partial X}{\partial v}\right)^2, \tag{66}$$

the square of the linear element of the spherical representation is

$$d\sigma^2=\mathcal{E}du^2+2\mathcal{F}dudv+\mathcal{G}dv^2. \tag{67}$$

In § 48 we established the following equations:

$$\left\{\begin{aligned}\frac{\partial X}{\partial u}&=\frac{FD'-GD}{H^2}\frac{\partial x}{\partial u}+\frac{FD-ED'}{H^2}\frac{\partial x}{\partial v},\\ \frac{\partial X}{\partial v}&=\frac{FD''-GD'}{H^2}\frac{\partial x}{\partial u}+\frac{FD'-ED''}{H^2}\frac{\partial x}{\partial v}.\end{aligned}\right. \tag{68}$$

By means of these relations and similar ones in Y and Z, the functions $\mathcal{E}$, $\mathcal{F}$, $\mathcal{G}$ may be given the forms

$$\left\{\begin{aligned}\mathcal{E}&=\frac{1}{H^2}[GD^2-2FDD'+ED'^2],\\ \mathcal{F}&=\frac{1}{H^2}[GDD'-F(DD''+D'^2)+ED'D''],\\ \mathcal{G}&=\frac{1}{H^2}[GD'^2-2FD'D''+ED''^2],\end{aligned}\right. \tag{69}$$

or, in terms of the total and mean curvatures (§ 52),

$$\mathcal{E}=K_mD-KE, \qquad \mathcal{F}=K_mD'-KF, \qquad \mathcal{G}=K_mD''-KG. \tag{70}$$

* The sphere of unit radius and center at the origin of coördinates. † *L.c.*, p. 9.

In consequence of these relations the linear element (67) may be given the form

$$(71)\quad d\sigma^2 = K_m(D\,du^2 + 2\,D'du\,dv + D''dv^2) - K(E\,du^2 + 2\,F\,du\,dv + G\,dv^2),$$

and, by (18),

$$(72)\qquad d\sigma^2 = \left(\frac{K_m}{R} - K\right)ds^2.$$

From (70) we have also

$$(73)\qquad \mathcal{H} = \sqrt{\mathcal{E}\mathcal{G} - \mathcal{F}^2} = \epsilon K H,$$

where ϵ is ± 1, according as K is positive or negative.

Equations (69) are linear in E, F, G. Solving for the latter, we have

$$(74)\qquad \begin{cases} E = \dfrac{1}{\mathcal{H}^2}[\mathcal{G}D^2 - 2\,\mathcal{F}DD' + \mathcal{E}D'^2], \\[2ex] F = \dfrac{1}{\mathcal{H}^2}[\mathcal{G}DD' - \mathcal{F}(DD'' + D'^2) + \mathcal{E}D'D''], \\[2ex] G = \dfrac{1}{\mathcal{H}^2}[\mathcal{G}D'^2 - 2\,\mathcal{F}D'D'' + \mathcal{E}D''^2]. \end{cases}$$

In seeking the differential equation of the lines of curvature from the definition that the normals to the surface along such a curve form a developable surface, we found (§ 51) that for a displacement in the direction of a line of curvature we have

$$\frac{\partial x}{\partial u}du + \frac{\partial x}{\partial v}dv + r\left(\frac{\partial X}{\partial u}du + \frac{\partial X}{\partial v}dv\right) = 0,$$

and similar equations in y and z, where r denotes the radius of principal curvature for the direction. If these equations be multiplied respectively by $\frac{\partial X}{\partial u}, \frac{\partial Y}{\partial u}, \frac{\partial Z}{\partial u}$ and added, and likewise by $\frac{\partial X}{\partial v}$, $\frac{\partial Y}{\partial v}, \frac{\partial Z}{\partial v}$ and added, the resulting equations may be written

$$D\,du + D'\,dv - r(\mathcal{E}\,du + \mathcal{F}\,dv) = 0,$$

$$D'\,du + D''\,dv - r(\mathcal{F}\,du + \mathcal{G}\,dv) = 0.$$

Eliminating r, we have as the equation of the lines of curvature

$$(75)\quad (D\mathcal{F} - D'\mathcal{E})\,du^2 + (D\mathcal{G} - D''\mathcal{E})\,du\,dv + (D'\mathcal{G} - D''\mathcal{F})\,dv^2 = 0.$$

Again, the elimination of du and dv gives the equation of the principal radii in the form

$$(76)\quad (\mathcal{E}\mathcal{G}-\mathcal{F}^2)r^2-(\mathcal{E}D''+\mathcal{G}D-2\mathcal{F}D')r+(DD''-D'^2)=0,$$

so that

$$(77)\quad \begin{cases} \rho_1+\rho_2=\dfrac{\mathcal{E}D''+\mathcal{G}D-2\mathcal{F}D'}{\mathcal{H}^2}, \\ \rho_1\rho_2=\dfrac{DD''-D'^2}{\mathcal{H}^2}. \end{cases}$$

These results enable us to write equations (74) thus:

$$(78)\quad \begin{cases} E=(\rho_1+\rho_2)D-\rho_1\rho_2\mathcal{E}, \\ F=(\rho_1+\rho_2)D'-\rho_1\rho_2\mathcal{F}, \\ G=(\rho_1+\rho_2)D''-\rho_1\rho_2\mathcal{G}. \end{cases}$$

61. Relations between a surface and its spherical representation. Since the radius of normal curvature R is a function of the direction except when the surface is a sphere, we obtain from (72) the following theorem:

A necessary and sufficient condition that the spherical representation of a surface be conformal is that the surface be minimal or a sphere.

As a consequence of this theorem we have that every orthogonal system on a minimal surface is represented on the sphere by an orthogonal system. From (70) it is seen that if a surface is not minimal, the parametric systems on both the surface and the sphere can be orthogonal only when D' is zero, that is, when the lines of curvature are parametric. Hence we have:

The lines of curvature of a surface are represented on the sphere by an orthogonal system; this is a characteristic property of lines of curvature, unless the surface be minimal.

This theorem follows also as a direct consequence of the theorem:

A necessary and sufficient condition that the tangents to a curve upon a surface and to its image at corresponding points be parallel is that the curve be a line of curvature.

In order to prove this theorem we assume that the curve is parametric, $v=$ const. Then the condition of parallelism is

$$\frac{\partial X}{\partial u}:\frac{\partial Y}{\partial u}:\frac{\partial Z}{\partial u}=\frac{\partial x}{\partial u}:\frac{\partial y}{\partial u}:\frac{\partial z}{\partial u}.$$

From (68) it follows that in this case $(FD - ED')$ must be zero. But the latter is the condition that the curves $v =$ const. be lines of curvature (§ 51). Moreover, from (32) it follows that the positive half-tangents to a line of curvature and its spherical representation have the same or contrary sense according as the corresponding radius of normal curvature is negative or positive.

In consequence of (7) the equation (40) of the asymptotic directions may be written

$$dxdX + dydY + dzdZ = 0.$$

And so we have the theorem:

The tangents to an asymptotic line and to its spherical representation at corresponding points are perpendicular to one another; this property is characteristic of asymptotic lines.

It is evident that the direction-cosines of the normal to the sphere are equal to X, Y, Z, to within sign at most. Let them be denoted by $\mathcal{X}$, $\mathcal{Y}$, $\mathcal{Z}$; then

$$\mathcal{X} = \frac{1}{\mathcal{H}}\left(\frac{\partial Y}{\partial u}\frac{\partial Z}{\partial v} - \frac{\partial Z}{\partial u}\frac{\partial Y}{\partial v}\right). \tag{79}$$

When expressions similar to (68) are substituted for the quantities in the parentheses, the latter expression is reducible to KHX. Hence, in consequence of (73), we have

$$\mathcal{X} = \epsilon X, \qquad \mathcal{Y} = \epsilon Y, \qquad \mathcal{Z} = \epsilon Z, \tag{80}$$

where $\epsilon = \pm 1$ according as the curvature of the surface is positive or negative.

From the above it follows that according as a point of a surface is elliptic or hyperbolic the positive sides of the tangent planes at corresponding points of the surface and the sphere are the same or different. Suppose, for the moment, that the lines of curvature are parametric. From our convention about the positive direction of the normal to a surface, and the above results, it follows that both the tangents to the parametric curves through a point M have the same sense as the corresponding tangents to the sphere, or both have the opposite sense, when M is an elliptic point; but that one tangent has the same sense as the corresponding tangent to the sphere, and the other the opposite sense, when the point is hyperbolic. Hence, when a point describes a closed curve on a surface its image describes a closed curve on the sphere in the same or opposite sense according as the surface has positive or negative curvature. We say that the areas inclosed by these curves have the same or opposite signs in these respective cases.

Suppose now that we consider a small parallelogram on the surface, whose vertices are the points (u, v), $(u + du, v)$, $(u, v + dv)$, and $(u + du, v + dv)$. The vertices of the corresponding parallelogram on the sphere have the same curvilinear coördinates, and the areas are $H\,du\,dv$ and $\epsilon \mathcal{H}\,du\,dv$, ϵ being ± 1 according as the surface has positive or negative curvature in the neighborhood of the point (u, v). The limiting value of the ratio of the spherical and the surface areas as the vertices of the latter approach the point (u, v) is a measure of the curvature of the surface similar to that of a plane curve. In consequence of (73) this limiting value is the Gaussian curvature K. Since any closed area may be looked upon as made up of such small parallelograms, we have the following theorem of Gauss:

The limit of the ratio of the area of a closed portion of a surface to the area of the spherical image of it, as the former converges to a point, is equal in value to the product of the principal radii at the point.

Since the normals to a developable surface along a generator are parallel, there can be no closed area for which there are not two normals which are parallel. Hence spherical representation, as defined in § 60, applies only to nondevelopable surfaces, but so far as the preceding theorem goes, it is not necessary to make this exception; for the total curvature of a developable surface is zero (§ 64), and the area of the spherical image of any closed area on such a surface is zero.

The fact that the Gaussian curvature is zero at all points of a developable surface, whereas such a surface is surely curved, makes this measure not altogether satisfactory, and so others have been suggested. Thus, Sophie Germain * advocated the mean curvature, and Casorati † has put forward the expression $\frac{1}{2}\left(\frac{1}{\rho_1^2} + \frac{1}{\rho_2^2}\right)$. But according to the first, the curvature of a minimal surface is zero, and according to the second, a minimal surface has the same curvature as a sphere. Hence the Gaussian curvature continues to be the one most frequently used, which may be due largely to an important property of it to be discussed later (§ 64).

* *Crelle*, Vol. VII (1831), p. 1. † *Acta Mathematica*, Vol. XIV (1890), p. 95.

62. Helicoids. We apply the preceding results in a study of an important class of surfaces called the *helicoids*. A helicoid is generated by a curve, plane or twisted, which is rotated about a fixed line as axis, and at the same time translated in the direction of the axis with a velocity which is in constant ratio with the velocity of rotation. A section of the surface by a plane through the axis is called a *meridian*. All the meridians are equal plane curves, and the surface can be generated by a meridian moving with the same velocities as the given curve. The particular motion described is called helicoidal motion, and so we may say that any helicoid can be generated by a plane curve with helicoidal motion.

In order to determine the equations of a helicoid in parametric form, we take the axis of rotation for the z-axis, and let u denote the distance of a point of the surface from the axis, and v the angle made by the plane through the point and the axis with the xz-plane in the positive direction of rotation. If the equation of the generating curve in any position of its plane is $z = \phi(u)$, the equations of the surface are

$$(81) \qquad x = u \cos v, \qquad y = u \sin v, \qquad z = \phi(u) + av,$$

where a denotes the constant ratio of the velocities; it is called the *parameter* of the helicoidal motion. When, in particular, a is zero, these equations define any surface of revolution. Moreover, when $\phi(u)$ is a constant, the curves $v = \text{const.}$ are straight lines perpendicular to the axis, and so the surface is a right conoid. It is called the *right helicoid*.

By calculation we obtain from (81)

$$(82) \qquad E = 1 + \phi'^2, \qquad F = a\phi', \qquad G = u^2 + a^2,$$

where the accent indicates differentiation with respect to u. From the method of generation it follows that the curves $v = \text{const.}$ are meridians, and $u = \text{const.}$ are helices on the helicoids, and circles on surfaces of revolution. From (82) it is seen that these curves form an orthogonal system only on surfaces of revolution and on the right helicoid. Moreover, from (57) it is found that the geodesic curvature of the meridians is zero only when a is zero or ϕ' is a constant. In the latter case the meridian is a straight line perpendicular to the axis or oblique, according as ϕ' is zero or not.

Hence the meridians of surfaces of revolution and of the ruled helicoids are geodesics.

The orthogonal trajectories of the helices upon a helicoid are determined by the equation (cf. III, 31)

$$a\phi' du + (u^2 + a^2)\, dv = 0.$$

Hence, if we put $$v_1 = \int \frac{a\phi'}{u^2 + a^2}\, du + v,$$

the curves $v_1 = \text{const.}$ are the orthogonal trajectories, and their equations in finite form are found by a quadrature. In terms of the parameters u and v_1 the linear element is

$$ds^2 = \left(1 + \frac{u^2\phi'^2}{u^2 + a^2}\right) du^2 + (u^2 + a^2)\, dv_1^2. \tag{83}$$

As an immediate consequence of this result we have that the helices and their orthogonal trajectories on any helicoid form an isothermal system.

From (83) and (§ 46) we have the theorem of Bour:

Every helicoid is applicable to some surface of revolution, and helices on the former correspond to the parallels on the latter.

We derive also the following expressions:

$$X, Y, Z = \frac{a \sin v - u\phi' \cos v, \quad -(a \cos v + u\phi' \sin v), \quad u}{\sqrt{u^2(1 + \phi'^2) + a^2}}, \tag{84}$$

and

$$D, D', D'' = \frac{u\phi'', \quad -a, \quad u^2\phi'}{\sqrt{u^2(1 + \phi'^2) + a^2}}. \tag{85}$$

From (84) it follows that a meridian is a normal section of a surface of revolution at all its points, and consequently is a line of curvature (Ex. 7, p. 140). This is evident also from the equation of the lines of curvature of a helicoid, namely

$$\begin{aligned} a[1 + \phi'^2 + u\phi'\phi'']\, du^2 + [(u^2 + a^2)u\phi'' - (1 + \phi'^2)u^2\phi']\, du\, dv \\ - a[u^2\phi'^2 + u^2 + a^2]\, dv^2 = 0. \end{aligned} \tag{86}$$

Moreover, the meridians are lines of curvature of those helicoids, for which ϕ satisfies the condition

$$1 + \phi'^2 + u\phi'\phi'' = 0.$$

By integration this gives

$$\phi = \sqrt{c^2 - u^2} - c \log \frac{c + \sqrt{c^2 - u^2}}{u}.$$

When the surface is the right helicoid the expressions for D and D'' vanish. Hence the meridians and helices are the asymptotic lines. Moreover, these lines form an orthogonal system, so that the surface is a minimal surface (§ 55). Since the tangent planes to a surface along an asymptotic line are its osculating planes, if the surface is a ruled minimal surface, the generators are the principal normals of all the curved asymptotic lines. But a circular helix is the only Bertrand curve whose principal normals are the principal normals of an infinity of curves (§ 19). Hence we have the theorem of Catalan:

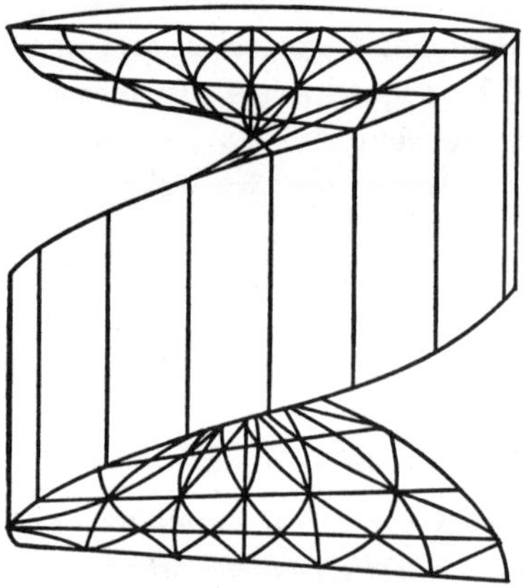

FIG. 16

The right helicoid is the only real minimal ruled surface.

In fig. 16 are represented the asymptotic lines and lines of curvature of a right helicoid.

For any other helicoid the equation of the asymptotic lines is

$$u\phi'' du^2 - 2\, a\, du dv + u^2\phi' dv^2 = 0. \tag{87}$$

As the coefficients in (86) and (87) are functions of u alone, we have the theorem:

When a helicoid is referred to its meridians and helices, the asymptotic lines and the lines of curvature can be found by quadratures.

EXAMPLES

1. Show that the spherical representation of the lines of curvature of a surface of revolution is isothermal.

2. The osculating planes of a line of curvature and of its spherical representation at corresponding points are parallel.

3. The angles between the asymptotic directions at a point on a surface and between their spherical representation are equal or supplementary, according as the surface has positive or negative curvature at the point.

4. Show that the helicoidal surface

$$x = u\cos v, \qquad y = u\sin v, \qquad z = bv + b\int\left[\frac{(u^2+b^2)}{(u^2-b^2)}\right]^{\frac{1}{2}}\frac{du}{u}$$

is minimal.

5. The total curvature of a helicoid is constant along a helix.

6. The orthogonal trajectories of the helices upon a helicoid are geodesics.

7. If the fundamental functions E, F, G of a surface are functions of a single parameter u, the surface is applicable to a surface of revolution.

8. Find the equations of the helicoid generated by a circle of constant radius whose plane passes through the axis and the lines of curvature on the surface; also find the equations of the surface in terms of parameters referring to the meridians and their orthogonal trajectories.

GENERAL EXAMPLES

1. If a pencil of planes be drawn through a tangent MT to a surface, and if lengths be laid off on the normals at M to the sections of the surface by these planes equal to the curvature of the sections, the locus of the end points is a straight line normal to the plane determined by MT and the normal to the surface at M.

2. If P is a point of a developable surface, P_0 the point where the generator through P touches the edge of regression, t the length P_0P, ρ and τ the radii of curvature and torsion of the edge of regression, then the principal radii of the surface are given by

$$\frac{1}{\rho_1} = 0, \qquad \frac{1}{\rho_2} = \frac{\rho}{\tau t}.$$

3. For the surface of revolution of a parabola about its directrix, the principal radii are in constant ratio.

4. The equations $x = a\cos u$, $y = a\sin u$, $z \pm uv$ define a family of circular helices which pass through the point $A\,(a, 0, 0)$ of the cylinder; each helix has an involute whose points are at the distance c from A (cf. I, 106). Find the surface which is the locus of these involutes; show that the tangents to the helices are normal to this surface; find also the lines of curvature upon the latter.

5. The surfaces defined by the equations (cf. § 25)

$$1 + p^2 + q^2 = q^2 f(y), \qquad x + pz = \phi(p)$$

have a system of lines of curvature in planes parallel to the xz-plane and to the y-axis respectively.

6. The equations

$$y - \alpha x = 0, \qquad x^2 + y^2 + z^2 - 2\beta x - a^2 = 0,$$

where α and β are parameters, define all the circles through the points $(0, 0, a)$, $(0, 0, -a)$. Show that the circles determined by a relation $\beta = f(\alpha)$ are the characteristics of a family of spheres, except when $f(\alpha)$ is a linear function; also that the circles are lines of curvature on the envelope of these spheres.

7. If one of the lines of curvature of a developable surface lies upon a sphere, the other nonrectilinear lines of curvature lie on concentric spheres.

8. If P is a point on a surface, P_0 the center of normal curvature of the line bisecting the angle between the lines of curvature, and P_1, P_2 the centers of normal curvature in two directions equally inclined to the first, then the four points P, P_1, P_0, P_2 form a harmonic range.

9. If $R_1, R_2, R_3, \cdots, R_m$ denote the radii of normal curvature of m sections of a surface which make equal angles $2\pi/m$ with one another, and $m > 2$, then

$$\frac{1}{m}\left(\frac{1}{R_1}+\frac{1}{R_2}+\cdots+\frac{1}{R_m}\right)=\frac{1}{2}\left(\frac{1}{\rho_1}+\frac{1}{\rho_2}\right).$$

10. If the Dupin indicatrix at a point P of a surface is an ellipse, and through either one of the asymptotes of its focal hyperbola two planes be drawn perpendicular to one another, their intersections with the tangent plane are conjugate directions on the surface.

11. All curves tangent to an asymptotic line at a point M, and whose osculating planes are not tangent to the surface at M, have M for a point of inflection.

12. The normal curvature of an orthogonal trajectory of an asymptotic line is equal to the mean curvature of the surface at the point of intersection.

13. The surface of revolution whose equations are

$$x = u\cos v, \qquad y = u\sin v, \qquad z = a\log\left(u+\sqrt{u^2-a^2}\right)$$

is generated by the rotation of a catenary about its axis; it is called the *catenoid*. Show that it is the only minimal surface of revolution.

14. When the osculating plane of a line of curvature makes a constant angle with the tangent plane to the surface, the line of curvature is plane.

15. A plane line of curvature is represented on the unit sphere by a circle.

16. The cylinder whose right section is the curve defined by the intrinsic equation $\rho = a - s^2/b$, where a and b are positive constants, has the characteristic property that upon it lie curves of curvature $\sqrt{\dfrac{a+b}{a^2b}}$, whose geodesic curvature is $1/\sqrt{ab}$.

17. When a surface is referred to an orthogonal system of lines, and the radii of geodesic curvature of the curves $v =$ const. and $u =$ const. are ρ_{gu}, ρ_{gv} respectively, the geodesic curvature of the curve which makes an angle θ_0 with the lines $v =$ const. is given by

$$\frac{1}{\rho_g}=\frac{d\theta_0}{ds}+\frac{\cos\theta_0}{\rho_{gu}}+\frac{\sin\theta_0}{\rho_{gv}}.$$

18. When a surface is referred to an orthogonal system of lines, and ρ_g, s denote the radius of geodesic curvature and the arc for one system of isogonal trajectories of the parametric lines, and ρ_g', s' the similar functions for the orthogonal trajectories of the former, then whatever be the direction of the first curves the quantity $\dfrac{d}{ds}\dfrac{1}{\rho_g}+\dfrac{d}{ds'}\dfrac{1}{\rho_g'}$ is constant at a point.

19. If ρ and ρ' denote the radii of first curvature of a line of curvature and its spherical representation, and also ρ_g and ρ_g' the radii of geodesic curvature of these curves, then

$$\frac{ds}{\rho}=\frac{d\sigma}{\rho'}, \qquad \frac{ds}{\rho_g}=\frac{d\sigma}{\rho_g'},$$

where ds and $d\sigma$ are the linear elements of the curves.

20. When a surface is referred to its lines of curvature, and θ_0, θ_0' denote the angles which a curve on the surface and its spherical representation make with the curves $v =$ const., the radii of geodesic curvature of these curves, denoted by ρ_g and ρ_g' respectively, are in the relation

$$d\theta_0 - \frac{ds}{\rho_g} = d\theta_0' - \frac{d\sigma}{\rho_g'}.$$

21. When the curve

$$x = f(u) \cos u, \qquad y = f(u) \sin u, \qquad z = -\frac{1}{a}\int f^2(u)\, du$$

is subjected to a helicoidal motion of parameter a about the z-axis, the various positions of this curve are orthogonal trajectories of the helices, and also geodesics on the surface.

22. When a curve is subjected to a continuous rotation about an axis, and at the same time to a homothetic transformation with respect to a point of the axis, such that the tangent to the locus described by a point of the curve makes a constant angle with the axis, the locus of the resulting curves is called a *spiral surface.* Show that if the z-axis be taken for the axis of rotation and the origin for the center of the transformation, the equations of the surface are of the form

$$x = f(u)\, e^{hv} \cos(u + v), \qquad y = f(u)\, e^{hv} \sin(u + v), \qquad z = \phi(u)\, e^{hv},$$

where h is a constant.

23. A spiral surface can be generated in the following manner: Let C be a curve, l any line, and P a point on the latter; if each point M on C describes an isogonal trajectory of the generators on the circular cone with vertex P and axis l in such a way that the perpendicular upon l, from the moving point M, revolves about l with constant velocity, the locus of these curves is a spiral surface (cf. Ex. 5, § 33).

24. Show that the orthogonal trajectories of the curves $u =$ const., in Ex. 22, can be found by quadratures, and that the linear element can be put in the form

$$ds^2 = e^{2\beta}(d\alpha^2 + A^2 d\beta^2),$$

where A is a function of α alone.

25. Show that the lines of curvature, minimal lines, and asymptotic lines upon a spiral surface can be found by quadrature.

CHAPTER V

FUNDAMENTAL EQUATIONS. THE MOVING TRIHEDRAL

63. Christoffel symbols. In this chapter we derive the necessary and sufficient equations of condition to be satisfied by six functions, $E, F, G; D, D', D''$, in order that they may be the fundamental quantities for a surface.

For the sake of brevity we make use of two sets of symbols, suggested by Christoffel,* which represent certain functions of the coefficients of a quadratic differential form and their derivatives of the first order. If the differential form is

$$a_{11}du_1^2 + 2\,a_{12}du_1du_2 + a_{22}du_2^2,$$

the first set of symbols is defined by

$$\begin{bmatrix} ik \\ l \end{bmatrix} = \frac{1}{2}\left(\frac{\partial a_{il}}{\partial u_k} + \frac{\partial a_{kl}}{\partial u_i} - \frac{\partial a_{ik}}{\partial u_l}\right),$$

where each of the subscripts i, k, l has one of the values 1 and 2.†
From this definition it follows that

$$\begin{bmatrix} ik \\ l \end{bmatrix} = \begin{bmatrix} ki \\ l \end{bmatrix}.$$

When these symbols are used in connection with the first fundamental quadratic form of a surface $ds^2 = E\,du^2 + 2\,F\,dudv + G\,dv^2$, they are found to have the following significance:

$$(1)\qquad \begin{cases} \begin{bmatrix} 11 \\ 1 \end{bmatrix} = \dfrac{1}{2}\dfrac{\partial E}{\partial u}, & \begin{bmatrix} 11 \\ 2 \end{bmatrix} = \dfrac{\partial F}{\partial u} - \dfrac{1}{2}\dfrac{\partial E}{\partial v}, \\[2ex] \begin{bmatrix} 12 \\ 1 \end{bmatrix} = \dfrac{1}{2}\dfrac{\partial E}{\partial v}, & \begin{bmatrix} 12 \\ 2 \end{bmatrix} = \dfrac{1}{2}\dfrac{\partial G}{\partial u}, \\[2ex] \begin{bmatrix} 22 \\ 1 \end{bmatrix} = \dfrac{\partial F}{\partial v} - \dfrac{1}{2}\dfrac{\partial G}{\partial u}, & \begin{bmatrix} 22 \\ 2 \end{bmatrix} = \dfrac{1}{2}\dfrac{\partial G}{\partial v}. \end{cases}$$

* *Crelle*, Vol. LXX, pp. 241–245.

† This equation defines these symbols for a quadratic form of any number of variables $u_1, \cdots, u_n$. In this case i, k, l take the values $1, \cdots, n$.

The second set of symbols is defined by the equation

$$\begin{Bmatrix} ik \\ \nu \end{Bmatrix} = A_{\nu 1}\begin{bmatrix} ik \\ 1 \end{bmatrix} + A_{\nu 2}\begin{bmatrix} ik \\ 2 \end{bmatrix},$$

where $A_{\nu l}$ denotes the algebraic complement of $a_{\nu l}$ in the discriminant $a_{11}a_{22} - a_{12}^2$ divided by the discriminant itself. With reference to the first fundamental quadratic form these symbols mean

$$A_{11} = \frac{G}{H^2}, \qquad A_{12} = \frac{-F}{H^2}, \qquad A_{22} = \frac{E}{H^2},$$

and

$$(2)\begin{cases} \begin{Bmatrix} 11 \\ 1 \end{Bmatrix} = \dfrac{G\dfrac{\partial E}{\partial u} + F\dfrac{\partial E}{\partial v} - 2F\dfrac{\partial F}{\partial u}}{2H^2}, & \begin{Bmatrix} 11 \\ 2 \end{Bmatrix} = \dfrac{-F\dfrac{\partial E}{\partial u} - E\dfrac{\partial E}{\partial v} + 2E\dfrac{\partial F}{\partial u}}{2H^2}, \\ \begin{Bmatrix} 12 \\ 1 \end{Bmatrix} = \dfrac{G\dfrac{\partial E}{\partial v} - F\dfrac{\partial G}{\partial u}}{2H^2}, & \begin{Bmatrix} 12 \\ 2 \end{Bmatrix} = \dfrac{E\dfrac{\partial G}{\partial u} - F\dfrac{\partial E}{\partial v}}{2H^2}, \\ \begin{Bmatrix} 22 \\ 1 \end{Bmatrix} = \dfrac{-F\dfrac{\partial G}{\partial v} - G\dfrac{\partial G}{\partial u} + 2G\dfrac{\partial F}{\partial v}}{2H^2}, & \begin{Bmatrix} 22 \\ 2 \end{Bmatrix} = \dfrac{E\dfrac{\partial G}{\partial v} + F\dfrac{\partial G}{\partial u} - 2F\dfrac{\partial F}{\partial v}}{2H^2}. \end{cases}$$

From these equations we derive the following identities:

$$(3) \qquad \frac{\partial \log H}{\partial u} = \begin{Bmatrix} 11 \\ 1 \end{Bmatrix} + \begin{Bmatrix} 12 \\ 2 \end{Bmatrix}, \qquad \frac{\partial \log H}{\partial v} = \begin{Bmatrix} 12 \\ 1 \end{Bmatrix} + \begin{Bmatrix} 22 \\ 2 \end{Bmatrix}.$$

With the aid of these identities we derive from (III, 15, 16) the expressions

$$(4)\ \frac{\partial \omega}{\partial u} = -H\left(\frac{1}{E}\begin{Bmatrix} 11 \\ 2 \end{Bmatrix} + \frac{1}{G}\begin{Bmatrix} 12 \\ 1 \end{Bmatrix}\right), \quad \frac{\partial \omega}{\partial v} = -H\left(\frac{1}{E}\begin{Bmatrix} 12 \\ 2 \end{Bmatrix} + \frac{1}{G}\begin{Bmatrix} 22 \\ 1 \end{Bmatrix}\right).$$

From the above definition of the symbols $\begin{Bmatrix} ik \\ \nu \end{Bmatrix}$ we obtain the following important relation:

$$\begin{bmatrix} ik \\ l \end{bmatrix} = a_{l1}\begin{Bmatrix} ik \\ 1 \end{Bmatrix} + a_{l2}\begin{Bmatrix} ik \\ 2 \end{Bmatrix}.$$

64. The equations of Gauss and of Codazzi. The first two of equations (IV, 10) and the equation

$$(5) \qquad X\frac{\partial^2 x}{\partial u^2} + Y\frac{\partial^2 y}{\partial u^2} + Z\frac{\partial^2 z}{\partial u^2} = D$$

form a consistent set of equations linear in $\frac{\partial^2 x}{\partial u^2}, \frac{\partial^2 y}{\partial u^2}, \frac{\partial^2 z}{\partial u^2}$, and the determinant is equal to H. Solving for $\frac{\partial^2 x}{\partial u^2}$, we get

$$\frac{\partial^2 x}{\partial u^2} = \begin{Bmatrix} 11 \\ 1 \end{Bmatrix} \frac{\partial x}{\partial u} + \begin{Bmatrix} 11 \\ 2 \end{Bmatrix} \frac{\partial x}{\partial v} + DX;$$

similar equations hold for y and z. Proceeding in like manner with the other equations (IV, 10) and

$$(6) \qquad \sum X \frac{\partial^2 x}{\partial u \, \partial v} = D', \qquad \sum X \frac{\partial^2 x}{\partial v^2} = D'',$$

we get the following *equations of Gauss*:

$$(7) \qquad \begin{cases} \dfrac{\partial^2 x}{\partial u^2} = \begin{Bmatrix} 11 \\ 1 \end{Bmatrix} \dfrac{\partial x}{\partial u} + \begin{Bmatrix} 11 \\ 2 \end{Bmatrix} \dfrac{\partial x}{\partial v} + DX, \\[2ex] \dfrac{\partial^2 x}{\partial u \, \partial v} = \begin{Bmatrix} 12 \\ 1 \end{Bmatrix} \dfrac{\partial x}{\partial u} + \begin{Bmatrix} 12 \\ 2 \end{Bmatrix} \dfrac{\partial x}{\partial v} + D'X, \\[2ex] \dfrac{\partial^2 x}{\partial v^2} = \begin{Bmatrix} 22 \\ 1 \end{Bmatrix} \dfrac{\partial x}{\partial u} + \begin{Bmatrix} 22 \\ 2 \end{Bmatrix} \dfrac{\partial x}{\partial v} + D''X. \end{cases}$$

For convenience of reference we recall from § 48 the equations

$$(8) \qquad \begin{cases} \dfrac{\partial X}{\partial u} = \dfrac{FD' - GD}{H^2} \dfrac{\partial x}{\partial u} + \dfrac{FD - ED'}{H^2} \dfrac{\partial x}{\partial v}, \\[2ex] \dfrac{\partial X}{\partial v} = \dfrac{FD'' - GD'}{H^2} \dfrac{\partial x}{\partial u} + \dfrac{FD' - ED''}{H^2} \dfrac{\partial x}{\partial v}. \end{cases}$$

The conditions of integrability of the Gauss equations (7) are

$$\frac{\partial}{\partial v}\left(\frac{\partial^2 x}{\partial u^2}\right) = \frac{\partial}{\partial u}\left(\frac{\partial^2 x}{\partial u \, \partial v}\right), \qquad \frac{\partial}{\partial v}\left(\frac{\partial^2 x}{\partial u \, \partial v}\right) = \frac{\partial}{\partial u}\left(\frac{\partial^2 x}{\partial v^2}\right).$$

By means of (7) and (8) these equations are reducible to the forms

$$(9) \qquad a_1 \frac{\partial x}{\partial u} + b_1 \frac{\partial x}{\partial v} + c_1 X = 0, \qquad a_2 \frac{\partial x}{\partial u} + b_2 \frac{\partial x}{\partial v} + c_2 X = 0,$$

where $a_1, a_2, \cdots, c_2$ are determinate functions of E, F, G; D, D', D'' and their derivatives. Since equations similar to (9) hold for y and z, we must have

$$(10) \qquad a_1 = 0, \quad a_2 = 0, \quad b_1 = 0, \quad b_2 = 0, \quad c_1 = 0, \quad c_2 = 0.$$

When the expressions for a_1, a_2, b_1, and b_2 are calculated, it is found that the first four equations are equivalent to the following:

$$(11)\quad \begin{cases} \dfrac{DD''-D'^2}{H^2}F = \dfrac{\partial}{\partial u}\begin{Bmatrix}12\\1\end{Bmatrix} - \dfrac{\partial}{\partial v}\begin{Bmatrix}11\\1\end{Bmatrix} + \begin{Bmatrix}12\\2\end{Bmatrix}\begin{Bmatrix}12\\1\end{Bmatrix} - \begin{Bmatrix}11\\2\end{Bmatrix}\begin{Bmatrix}22\\1\end{Bmatrix}, \\ \dfrac{DD''-D'^2}{H^2}E = \dfrac{\partial}{\partial v}\begin{Bmatrix}11\\2\end{Bmatrix} - \dfrac{\partial}{\partial u}\begin{Bmatrix}12\\2\end{Bmatrix} + \begin{Bmatrix}11\\1\end{Bmatrix}\begin{Bmatrix}12\\2\end{Bmatrix} + \begin{Bmatrix}11\\2\end{Bmatrix}\begin{Bmatrix}22\\2\end{Bmatrix} \\ \qquad\qquad - \begin{Bmatrix}12\\1\end{Bmatrix}\begin{Bmatrix}11\\2\end{Bmatrix} - \begin{Bmatrix}12\\2\end{Bmatrix}^2, \\ \dfrac{DD''-D'^2}{H^2}G = \dfrac{\partial}{\partial u}\begin{Bmatrix}22\\1\end{Bmatrix} - \dfrac{\partial}{\partial v}\begin{Bmatrix}12\\1\end{Bmatrix} + \begin{Bmatrix}22\\2\end{Bmatrix}\begin{Bmatrix}12\\1\end{Bmatrix} + \begin{Bmatrix}22\\1\end{Bmatrix}\begin{Bmatrix}11\\1\end{Bmatrix} \\ \qquad\qquad - \begin{Bmatrix}12\\2\end{Bmatrix}\begin{Bmatrix}22\\1\end{Bmatrix} - \begin{Bmatrix}12\\1\end{Bmatrix}^2, \\ \dfrac{DD''-D'^2}{H^2}F = \dfrac{\partial}{\partial v}\begin{Bmatrix}12\\2\end{Bmatrix} - \dfrac{\partial}{\partial u}\begin{Bmatrix}22\\2\end{Bmatrix} + \begin{Bmatrix}12\\1\end{Bmatrix}\begin{Bmatrix}12\\2\end{Bmatrix} - \begin{Bmatrix}22\\1\end{Bmatrix}\begin{Bmatrix}11\\2\end{Bmatrix}. \end{cases}$$

When the expressions for the Christoffel symbols are substituted in these equations the latter reduce to the single equation

$$(12)\qquad \frac{DD''-D'^2}{H^2} = \frac{1}{2H}\left\{\frac{\partial}{\partial u}\left[\frac{F}{EH}\frac{\partial E}{\partial v} - \frac{1}{H}\frac{\partial G}{\partial u}\right] + \frac{\partial}{\partial v}\left[\frac{2}{H}\frac{\partial F}{\partial u} - \frac{1}{H}\frac{\partial E}{\partial v} - \frac{F}{EH}\frac{\partial E}{\partial u}\right]\right\}.$$

This equation was discovered by Gauss, and is called the *Gauss equation* of condition upon the fundamental functions. The left-hand member of the equation is the expression for the total curvature of the surface. Hence we have the celebrated theorem of Gauss *:

The expression for the total curvature of a surface is a function of the fundamental coefficients of the first order and of their derivatives of the first and second orders.

When the expressions for c_1 and c_2 are calculated, we find that the last two of equations (10) are

$$(13)\quad \begin{cases} \dfrac{\partial D}{\partial v} - \dfrac{\partial D'}{\partial u} - \begin{Bmatrix}12\\1\end{Bmatrix}D + \left(\begin{Bmatrix}11\\1\end{Bmatrix} - \begin{Bmatrix}12\\2\end{Bmatrix}\right)D' + \begin{Bmatrix}11\\2\end{Bmatrix}D'' = 0, \\ \dfrac{\partial D''}{\partial u} - \dfrac{\partial D'}{\partial v} + \begin{Bmatrix}22\\1\end{Bmatrix}D + \left(\begin{Bmatrix}22\\2\end{Bmatrix} - \begin{Bmatrix}12\\1\end{Bmatrix}\right)D' - \begin{Bmatrix}12\\2\end{Bmatrix}D'' = 0. \end{cases}$$

* *L.c.*, p. 20.

These are the *Codazzi equations*, so called because they are equivalent to the equations found by Codazzi *; however, it should be mentioned that Mainardi was brought to similar results somewhat earlier.† It is sometimes convenient to have these equations written in the form

$$(13') \quad \left\{ \begin{aligned} &\frac{\partial}{\partial v}\frac{D}{H} - \frac{\partial}{\partial u}\frac{D'}{H} + \begin{Bmatrix} 22 \\ 2 \end{Bmatrix}\frac{D}{H} - 2\begin{Bmatrix} 12 \\ 2 \end{Bmatrix}\frac{D'}{H} + \begin{Bmatrix} 11 \\ 2 \end{Bmatrix}\frac{D''}{H} = 0, \\ &\frac{\partial}{\partial u}\frac{D''}{H} - \frac{\partial}{\partial v}\frac{D'}{H} + \begin{Bmatrix} 22 \\ 1 \end{Bmatrix}\frac{D}{H} - 2\begin{Bmatrix} 12 \\ 1 \end{Bmatrix}\frac{D'}{H} + \begin{Bmatrix} 11 \\ 1 \end{Bmatrix}\frac{D''}{H} = 0, \end{aligned} \right.$$

which reduce readily to (13) by means of (3).

With the aid of equations (7) we find that the conditions of integrability of equations (8) and similar ones in Y and Z reduce to (13).

From the preceding theorem and the definition of applicable surfaces (§ 43) follows the theorem:

Two applicable surfaces have the same total curvature at corresponding points.

As a consequence we have:

Every surface applicable to a plane is the tangent surface of a twisted curve.

For, when a surface is applicable to a plane its linear element is reducible to $ds^2 = du^2 + dv^2$, and consequently its total curvature is zero at every point by (12). From (IV, 73) it follows that

$$\mathcal{H}^2 = \left[\frac{\partial(Y, Z)}{\partial(u, v)}\right]^2 + \left[\frac{\partial(Z, X)}{\partial(u, v)}\right]^2 + \left[\frac{\partial(X, Y)}{\partial(u, v)}\right]^2 = 0.$$

Hence X, Y, Z are functions of a single parameter, and therefore the surface is the tangent surface of a twisted curve (cf. § 27).

Incidentally we have proved the theorem:

When K is zero at all points of a surface the latter is developable, and conversely.

* Sulle coordinate curvilinee d'una superficie e dello spazio, *Annali*, Ser. 3, Vol. II (1868), p. 269.

† *Giornale dell' Istituto Lombardo*, Vol. IX, p. 395.

65. Fundamental theorem. When the lines of curvature are parametric, the Gauss and Codazzi equations (12), (13) reduce to

$$(14)\qquad \begin{cases} \dfrac{DD''}{\sqrt{EG}}+\dfrac{\partial}{\partial u}\left(\dfrac{1}{\sqrt{E}}\dfrac{\partial\sqrt{G}}{\partial u}\right)+\dfrac{\partial}{\partial v}\left(\dfrac{1}{\sqrt{G}}\dfrac{\partial\sqrt{E}}{\partial v}\right)=0, \\ \dfrac{\partial}{\partial v}\left(\dfrac{D}{\sqrt{E}}\right)-\dfrac{D''}{G}\dfrac{\partial\sqrt{E}}{\partial v}=0, \\ \dfrac{\partial}{\partial u}\left(\dfrac{D''}{\sqrt{G}}\right)-\dfrac{D}{E}\dfrac{\partial\sqrt{G}}{\partial u}=0. \end{cases}$$

The direction-cosines of the tangents to the parametric curves, $v=\text{const.}$ and $u=\text{const.}$, have the respective values

$$(15)\qquad \begin{cases} X_1,\quad Y_1,\quad Z_1=\dfrac{1}{\sqrt{E}}\dfrac{\partial x}{\partial u},\quad \dfrac{1}{\sqrt{E}}\dfrac{\partial y}{\partial u},\quad \dfrac{1}{\sqrt{E}}\dfrac{\partial z}{\partial u}, \\ X_2,\quad Y_2,\quad Z_2=\dfrac{1}{\sqrt{G}}\dfrac{\partial x}{\partial v},\quad \dfrac{1}{\sqrt{G}}\dfrac{\partial y}{\partial v},\quad \dfrac{1}{\sqrt{G}}\dfrac{\partial z}{\partial v}. \end{cases}$$

By means of equations (7) and (8) we find

$$(16)\qquad \begin{cases} \dfrac{\partial X_1}{\partial u}=-\dfrac{1}{\sqrt{G}}\dfrac{\partial\sqrt{E}}{\partial v}X_2+\dfrac{D}{\sqrt{E}}X, & \dfrac{\partial X_1}{\partial v}=\dfrac{1}{\sqrt{E}}\dfrac{\partial\sqrt{G}}{\partial u}X_2, \\ \dfrac{\partial X_2}{\partial u}=\dfrac{1}{\sqrt{G}}\dfrac{\partial\sqrt{E}}{\partial v}X_1, & \dfrac{\partial X_2}{\partial v}=-\dfrac{1}{\sqrt{E}}\dfrac{\partial\sqrt{G}}{\partial u}X_1+\dfrac{D''}{\sqrt{G}}X, \\ \dfrac{\partial X}{\partial u}=-\dfrac{D}{\sqrt{E}}X_1, & \dfrac{\partial X}{\partial v}=-\dfrac{D''}{\sqrt{G}}X_2, \end{cases}$$

and similar equations obtained by replacing X_1, X_2, X by Y_1, Y_2, Y respectively, and by Z_1, Z_2, Z. From (15) we have

$$(17)\qquad \begin{cases} x=\displaystyle\int\sqrt{E}X_1du+\sqrt{G}X_2dv, \\ y=\displaystyle\int\sqrt{E}Y_1du+\sqrt{G}Y_2dv, \\ z=\displaystyle\int\sqrt{E}Z_1du+\sqrt{G}Z_2dv. \end{cases}$$

We proceed to the proof of the converse theorem:

Given four functions, E, G, D, D'', satisfying equations (14); *there exists a surface for which E, 0, G; D, 0, D'' are the fundamental quantities of the first and second order respectively.*

In the first place we remark that all the conditions of integrability of the equations (16) are satisfied in consequence of (14). Hence these equations admit sets of particular solutions whose values for the initial values of u and v are arbitrary. From the form of equations (16) it follows (cf. § 13) that, if two such sets of particular solutions be denoted by X_1, X_2, X and Y_1, Y_2, Y, then

$$(18)\qquad \begin{cases} X_1^2 + X_2^2 + X^2 = \text{const.}, \\ Y_1^2 + Y_2^2 + Y^2 = \text{const.}, \\ X_1 Y_1 + X_2 Y_2 + XY = \text{const.} \end{cases}$$

From the theory of differential equations we know that there exist three particular sets of solutions X_1, X_2, X; Y_1, Y_2, Y; Z_1, Z_2, Z, which for the initial values of u and v have the values 1, 0, 0; 0, 1, 0; 0, 0, 1. In this case equations (18) become

$$(19)\qquad \begin{cases} X_1^2 + X_2^2 + X^2 = 1, \\ Y_1^2 + Y_2^2 + Y^2 = 1, \\ X_1 Y_1 + X_2 Y_2 + XY = 0, \end{cases}$$

which are true for all values of u and v. In like manner we have

$$(19')\qquad \begin{cases} Z_1^2 + Z_2^2 + Z^2 = 1, \\ X_1 Z_1 + X_2 Z_2 + XZ = 0, \\ Y_1 Z_1 + Y_2 Z_2 + YZ = 0. \end{cases}$$

From (16) it follows that the expressions in the right-hand members of (17) are exact differentials, and that the surface defined by these equations has, for its linear element and its second quadratic form, the expressions

$$(20)\qquad E\,du^2 + G\,dv^2, \qquad D\,du^2 + D''dv^2$$

respectively.

Suppose, now, that we had a second system of three sets of solutions of equations (16) satisfying the conditions (19), (19′). By a motion in space we could make these X's, Y's, and Z's equal to the corresponding ones of the first system for the initial values of u and v. But then, because of the relations similar to (18), they would be equal for all values of u and v, as shown in § 13. Hence, to within a motion in space, a surface is determined by two quadratic forms (20). As in § 13, it can be shown that the solution of equations (16) reduces to the integration of an equation of Riccati.

Later* we shall find that the direction-cosines of any two perpendicular lines in the tangent plane to a surface, and of the normal to the surface, satisfy a system of equations similar in form to (16). Moreover, these equations possess the property that sets of solutions satisfy the conditions (18) when the parametric lines are any whatever. Hence the choice of lines of curvature as parametric lines simplifies the preceding equations, but the result is a general one. Consequently we have the following fundamental theorem:

When the coefficients of two quadratic forms,

$$E du^2 + 2F du dv + G dv^2, \qquad D du^2 + 2D' du dv + D'' dv^2,$$

satisfy the equations of Gauss and Codazzi, there exists a surface, unique to within its position in space, for which these forms are respectively the first and second fundamental quadratic forms; and the determination of the surface requires the integration of a Riccati equation and quadratures.

From (III, 3), (5) and (6), it follows that if E, F, G; D, D', D'' are the fundamental functions for a surface of coördinates (x, y, z), the surface symmetric with respect to the origin, that is, the surface with the coördinates $(-x, -y, -z)$, has the fundamental functions E, F, G; $-D$, $-D'$, $-D''$. Moreover, in consequence of the above theorem, two surfaces whose fundamental quantities bear such a relation can be moved in space so that they will be symmetric with respect to a point. Two surfaces of this kind will be treated as the same surface.

EXAMPLES

1. When the lines of curvature of a surface form an isothermal system, the surface is said to be *isothermic*. Show that surfaces of revolution are isothermic.

2. Show that the hyperbolic paraboloid

$$x = \frac{a}{2}(u + v), \qquad y = \frac{b}{2}(u - v), \qquad z = \frac{uv}{2}$$

is isothermic.

3. When a surface is isothermic, and the linear element, expressed in terms of parameters referring to the lines of curvature, is $ds^2 = \lambda^2(du^2 + dv^2)$, the equations of Codazzi and Gauss are reducible to

$$\frac{\partial}{\partial u}\log\lambda = \frac{\rho_1}{\rho_2}\frac{1}{\rho_1 - \rho_2}\frac{\partial \rho_2}{\partial u}, \qquad \frac{\partial}{\partial v}\log\lambda = \frac{\rho_2}{\rho_1}\frac{1}{\rho_2 - \rho_1}\frac{\partial \rho_1}{\partial v},$$

$$\left(\frac{\partial^2}{\partial u^2} + \frac{\partial^2}{\partial v^2}\right)\log\lambda + \frac{\lambda^2}{\rho_1\rho_2} = 0.$$

4. Find the form of equations (11), (13) when the surface is defined in terms of symmetric coördinates (cf. § 39).

* Cf. § 69. Consult also Scheffers, Vol. II, pp. 310 *et seq.*; Bianchi, Vol. I, pp. 122–124.

5. Show that K is equal to zero for the tangent surface of a twisted curve, taking the linear element of the latter in the form (105), § 20.

6. Show that the total curvature of the surface of revolution of the tractrix about its axis is negative and constant.

7. Establish the following formulas, in which the differential parameters are formed with respect to the form $Edu^2 + 2\,Fdudv + Gdv^2$:

$$\Delta_1 X = \frac{X_1^2}{\rho_1^2} + \frac{X_2^2}{\rho_2^2}, \qquad \Delta_1(Y, Z) = \frac{Y_1 Z_1}{\rho_1^2} + \frac{Y_2 Z_2}{\rho_2^2},$$

$$\Theta(y, z) = X, \qquad \Theta(Y, Z) = \frac{X}{\rho_1 \rho_2},$$

where the quantities have the same significance as in § 65.

8. Deduce the identity $\qquad \Delta_2 x = \left(\frac{1}{\rho_1} + \frac{1}{\rho_2}\right) X,$

and show therefrom that the curves in which a minimal surface is cut by a family of parallel planes and the orthogonal trajectories of these curves form an isothermal system.

66. Fundamental equations in another form. We have seen in § 61 that if X, Y, Z denote the direction-cosines of the normal to a surface, the direction-cosines of the normal to the spherical representation of the surface are ϵX, ϵY, ϵZ, where ϵ is ± 1 according as the curvature of the surface is positive or negative. If, then, the second fundamental quantities for the sphere be denoted by $\mathcal{D}$, $\mathcal{D}'$, $\mathcal{D}''$, we have

$$\mathcal{D} = -\epsilon \mathcal{E}, \qquad \mathcal{D}' = -\epsilon \mathcal{F}, \qquad \mathcal{D}'' = -\epsilon \mathcal{G}, \tag{21}$$

so that for the sphere equations (7) become

$$\left\{\begin{aligned} \frac{\partial^2 X}{\partial u^2} &= \begin{Bmatrix} 11 \\ 1 \end{Bmatrix}' \frac{\partial X}{\partial u} + \begin{Bmatrix} 11 \\ 2 \end{Bmatrix}' \frac{\partial X}{\partial v} - \mathcal{E} X, \\ \frac{\partial^2 X}{\partial u \partial v} &= \begin{Bmatrix} 12 \\ 1 \end{Bmatrix}' \frac{\partial X}{\partial u} + \begin{Bmatrix} 12 \\ 2 \end{Bmatrix}' \frac{\partial X}{\partial v} - \mathcal{F} X, \\ \frac{\partial^2 X}{\partial v^2} &= \begin{Bmatrix} 22 \\ 1 \end{Bmatrix}' \frac{\partial X}{\partial u} + \begin{Bmatrix} 22 \\ 2 \end{Bmatrix}' \frac{\partial X}{\partial v} - \mathcal{G} X, \end{aligned}\right. \tag{22}$$

where the Christoffel symbols $\begin{Bmatrix} rs \\ t \end{Bmatrix}'$ are formed with respect to the linear element of the spherical representation, namely

$$d\sigma^2 = \mathcal{E}\, du^2 + 2\,\mathcal{F}\, dudv + \mathcal{G}\, dv^2. \tag{23}$$

The conditions of integrability of equations (22) are reducible by means of the latter to

$$A_1 \frac{\partial X}{\partial u} + B_1 \frac{\partial X}{\partial v} = 0, \qquad A_2 \frac{\partial X}{\partial u} + B_2 \frac{\partial X}{\partial v} = 0,$$

where A_1, A_2, B_1, B_2 are the functions obtained from the quantities a_1, a_2, b_1, b_2 respectively of § 64 by replacing $\frac{DD''-D'^2}{H^2}$, E, F, G by 1, $\mathcal{E}$, $\mathcal{F}$, $\mathcal{G}$ respectively. Since the above equations must be satisfied by Y and Z, the quantities A_1, A_2, B_1, B_2 must be zero. This gives the single equation of condition

$$(24)\quad \frac{1}{2\mathcal{H}}\left[\frac{\partial}{\partial u}\left(\frac{\mathcal{F}}{\mathcal{E}\mathcal{H}}\frac{\partial \mathcal{E}}{\partial v}-\frac{1}{\mathcal{H}}\frac{\partial \mathcal{G}}{\partial u}\right)+\frac{\partial}{\partial v}\left(\frac{2}{\mathcal{H}}\frac{\partial \mathcal{F}}{\partial u}-\frac{1}{\mathcal{H}}\frac{\partial \mathcal{E}}{\partial v}-\frac{\mathcal{F}}{\mathcal{E}\mathcal{H}}\frac{\partial \mathcal{E}}{\partial u}\right)\right]=1.$$

Moreover, the Codazzi equations (13′) become, in consequence of (21),

$$(25)\quad \begin{cases} \dfrac{\partial}{\partial v}\left(\dfrac{\mathcal{E}}{\mathcal{H}}\right)-\dfrac{\partial}{\partial u}\left(\dfrac{\mathcal{F}}{\mathcal{H}}\right)+\begin{Bmatrix}22\\2\end{Bmatrix}'\dfrac{\mathcal{E}}{\mathcal{H}}-2\begin{Bmatrix}12\\2\end{Bmatrix}'\dfrac{\mathcal{F}}{\mathcal{H}}+\begin{Bmatrix}11\\2\end{Bmatrix}'\dfrac{\mathcal{G}}{\mathcal{H}}=0, \\ \dfrac{\partial}{\partial u}\left(\dfrac{\mathcal{G}}{\mathcal{H}}\right)-\dfrac{\partial}{\partial v}\left(\dfrac{\mathcal{F}}{\mathcal{H}}\right)+\begin{Bmatrix}22\\1\end{Bmatrix}'\dfrac{\mathcal{E}}{\mathcal{H}}-2\begin{Bmatrix}12\\1\end{Bmatrix}'\dfrac{\mathcal{F}}{\mathcal{H}}+\begin{Bmatrix}11\\1\end{Bmatrix}'\dfrac{\mathcal{G}}{\mathcal{H}}=0, \end{cases}$$

which vanish identically.

If equations (IV, 13) be solved for $\frac{\partial x}{\partial u}$ and $\frac{\partial x}{\partial v}$, we get

$$(26)\quad \begin{cases} \dfrac{\partial x}{\partial u}=\dfrac{\mathcal{F}D'-\mathcal{G}D}{\mathcal{H}^2}\dfrac{\partial X}{\partial u}+\dfrac{\mathcal{F}D-\mathcal{E}D'}{\mathcal{H}^2}\dfrac{\partial X}{\partial v}, \\ \dfrac{\partial x}{\partial v}=\dfrac{\mathcal{F}D''-\mathcal{G}D'}{\mathcal{H}^2}\dfrac{\partial X}{\partial u}+\dfrac{\mathcal{F}D'-\mathcal{E}D''}{\mathcal{H}^2}\dfrac{\partial X}{\partial v}. \end{cases}$$

By means of equations (22) the condition of integrability of these equations, namely

$$\frac{\partial}{\partial v}\left(\frac{\partial x}{\partial u}\right)=\frac{\partial}{\partial u}\left(\frac{\partial x}{\partial v}\right)$$

and similar conditions in y and z, reduce to

$$(27)\quad \begin{cases} \dfrac{\partial D}{\partial v}-\dfrac{\partial D'}{\partial u}-\begin{Bmatrix}12\\1\end{Bmatrix}'D+\left[\begin{Bmatrix}11\\1\end{Bmatrix}'-\begin{Bmatrix}12\\2\end{Bmatrix}'\right]D'+\begin{Bmatrix}11\\2\end{Bmatrix}'D''=0, \\ \dfrac{\partial D''}{\partial u}-\dfrac{\partial D'}{\partial v}+\begin{Bmatrix}22\\1\end{Bmatrix}'D+\left[\begin{Bmatrix}22\\2\end{Bmatrix}'-\begin{Bmatrix}12\\1\end{Bmatrix}'\right]D'-\begin{Bmatrix}12\\2\end{Bmatrix}'D''=0. \end{cases}$$

Hence two quadratic forms

$$\mathcal{E}\,du^2+2\mathcal{F}\,dudv+\mathcal{G}\,dv^2, \qquad D\,du^2+2D'\,dudv+D''dv^2,$$

whose coefficients satisfy the conditions (24), (27), may be taken as the linear element of the spherical representation of a surface and as the second quadratic form of the latter. When X, Y, Z are known, the cartesian coördinates of the surface can be found by quadratures (26); however, the determination of the former requires the solution of a Riccati equation.

If the equations

$$D = -\sum \frac{\partial x}{\partial u}\frac{\partial X}{\partial u}, \quad D' = -\sum \frac{\partial x}{\partial u}\frac{\partial X}{\partial v} = -\sum \frac{\partial x}{\partial v}\frac{\partial X}{\partial u}, \quad D'' = -\sum \frac{\partial x}{\partial v}\frac{\partial X}{\partial v},$$

be differentiated with respect to u and v, the resulting equations may be reduced by means of (7) and (22) to the form:*

$$(28)\quad \left\{\begin{aligned}
\frac{\partial D}{\partial u} &= \begin{Bmatrix}11\\1\end{Bmatrix} D + \begin{Bmatrix}11\\2\end{Bmatrix} D' + \begin{Bmatrix}11\\1\end{Bmatrix}' D + \begin{Bmatrix}11\\2\end{Bmatrix}' D',\\
\frac{\partial D}{\partial v} &= \begin{Bmatrix}12\\1\end{Bmatrix} D + \begin{Bmatrix}12\\2\end{Bmatrix} D' + \begin{Bmatrix}12\\1\end{Bmatrix}' D + \begin{Bmatrix}12\\2\end{Bmatrix}' D',\\
\frac{\partial D'}{\partial u} &= \begin{Bmatrix}11\\1\end{Bmatrix} D' + \begin{Bmatrix}11\\2\end{Bmatrix} D'' + \begin{Bmatrix}12\\1\end{Bmatrix}' D + \begin{Bmatrix}12\\2\end{Bmatrix}' D',\\
\frac{\partial D'}{\partial v} &= \begin{Bmatrix}12\\1\end{Bmatrix} D' + \begin{Bmatrix}12\\2\end{Bmatrix} D'' + \begin{Bmatrix}22\\1\end{Bmatrix}' D + \begin{Bmatrix}22\\2\end{Bmatrix}' D',\\
\frac{\partial D'}{\partial u} &= \begin{Bmatrix}12\\1\end{Bmatrix} D + \begin{Bmatrix}12\\2\end{Bmatrix} D' + \begin{Bmatrix}11\\1\end{Bmatrix}' D' + \begin{Bmatrix}11\\2\end{Bmatrix}' D'',\\
\frac{\partial D'}{\partial v} &= \begin{Bmatrix}22\\1\end{Bmatrix} D + \begin{Bmatrix}22\\2\end{Bmatrix} D' + \begin{Bmatrix}12\\1\end{Bmatrix}' D' + \begin{Bmatrix}12\\2\end{Bmatrix}' D'',\\
\frac{\partial D''}{\partial u} &= \begin{Bmatrix}12\\1\end{Bmatrix} D' + \begin{Bmatrix}12\\2\end{Bmatrix} D'' + \begin{Bmatrix}12\\1\end{Bmatrix}' D' + \begin{Bmatrix}12\\2\end{Bmatrix}' D'',\\
\frac{\partial D''}{\partial v} &= \begin{Bmatrix}22\\1\end{Bmatrix} D' + \begin{Bmatrix}22\\2\end{Bmatrix} D'' + \begin{Bmatrix}22\\1\end{Bmatrix}' D' + \begin{Bmatrix}22\\2\end{Bmatrix}' D''.
\end{aligned}\right.$$

67. Tangential coördinates. Mean evolute. A surface may be looked upon not only as the locus of a point whose position depends upon two parameters, but also as the envelope of its tangent planes. This family of planes depends upon one or two parameters according as the surface is developable or not. We considered the former case in § 27, and now take up the latter.

If W denotes the algebraic distance from the origin to the tangent plane to a surface S at the point $M(x, y, z)$, then

$$(29)\qquad W = xX + yY + zZ.$$

If this equation is differentiated with respect to u and v, the resulting equations are reducible, in consequence of (IV, 3), to

$$(30)\qquad \frac{\partial W}{\partial u} = \sum x\frac{\partial X}{\partial u}, \qquad \frac{\partial W}{\partial v} = \sum x\frac{\partial X}{\partial v}.$$

* Cf. Bianchi, Vol. I, p. 157.

The three equations (29), (30) are linear in x, y, z, and in consequence of (IV, 79, 80) their determinant is equal to $\epsilon/\mathcal{H}$. Hence we have

$$x = \frac{\epsilon}{\mathcal{H}} \begin{vmatrix} W & Y & Z \\ \frac{\partial W}{\partial u} & \frac{\partial Y}{\partial u} & \frac{\partial Z}{\partial u} \\ \frac{\partial W}{\partial v} & \frac{\partial Y}{\partial v} & \frac{\partial Z}{\partial v} \end{vmatrix},$$

and similar expressions for y and z. From (IV, 11) we deduce the identities

$$(31)\quad \begin{cases} Y\dfrac{\partial Z}{\partial u} - Z\dfrac{\partial Y}{\partial u} = \dfrac{\epsilon}{\mathcal{H}}\left(-\mathcal{F}\dfrac{\partial X}{\partial u} + \mathcal{E}\dfrac{\partial X}{\partial v}\right), \\ Y\dfrac{\partial Z}{\partial v} - Z\dfrac{\partial Y}{\partial v} = \dfrac{\epsilon}{\mathcal{H}}\left(-\mathcal{G}\dfrac{\partial X}{\partial u} + \mathcal{F}\dfrac{\partial X}{\partial v}\right). \end{cases}$$

By means of these equations the above expression for x is reducible to

$$x = WX + \frac{1}{\mathcal{H}^2}\left[\mathcal{G}\frac{\partial W}{\partial u}\frac{\partial X}{\partial u} - \mathcal{F}\left(\frac{\partial W}{\partial u}\frac{\partial X}{\partial v} + \frac{\partial W}{\partial v}\frac{\partial X}{\partial u}\right) + \mathcal{E}\frac{\partial W}{\partial v}\frac{\partial X}{\partial v}\right].$$

Hence we have

$$(32)\quad x = WX + \Delta_1'(W, X), \quad y = WY + \Delta_1'(W, Y), \quad z = WZ + \Delta_1'(W, Z),$$

the differential parameters being formed with respect to (23).

Conversely, if we have four functions X, Y, Z, W of u and v, such that the first three satisfy the identity

$$(33)\quad X^2 + Y^2 + Z^2 = 1,$$

equations (32) define the surface for which X, Y, Z are the direction-cosines of the tangent plane, and W is the distance of the latter from the origin. For, from (33), we have

$$\sum X\frac{\partial X}{\partial u} = 0, \qquad \sum X\frac{\partial X}{\partial v} = 0,$$

in consequence of which and formulas (22) we find from (32) that

$$\sum X\frac{\partial x}{\partial u} = 0, \qquad \sum X\frac{\partial x}{\partial v} = 0.$$

Moreover, equation (29) also follows from (32). Hence a surface is completely defined by the functions X, Y, Z, W, which are called the *tangential coördinates* of the surface.*

* Cf. Weingarten, *Festschrift der Technischen Hochschule zu Berlin* (1884); Bianchi, Vol. I, pp. 172–174; Darboux, Vol. I, pp. 234–248.

When equations (30) are differentiated, we obtain

$$\frac{\partial^2 W}{\partial u^2}=-D+\sum x\frac{\partial^2 X}{\partial u^2},\qquad \frac{\partial^2 W}{\partial u\,\partial v}=-D'+\sum x\frac{\partial^2 X}{\partial u\,\partial v},$$
$$\frac{\partial^2 W}{\partial v^2}=-D''+\sum x\frac{\partial^2 X}{\partial v^2}.$$

By means of (22), (29), and (30) these equations are reducible to

$$\text{(34)}\qquad \left\{\begin{aligned} D&=-\left[\frac{\partial^2 W}{\partial u^2}-\begin{Bmatrix}11\\1\end{Bmatrix}'\frac{\partial W}{\partial u}-\begin{Bmatrix}11\\2\end{Bmatrix}'\frac{\partial W}{\partial v}+\mathcal{E}W\right],\\ D'&=-\left[\frac{\partial^2 W}{\partial u\,\partial v}-\begin{Bmatrix}12\\1\end{Bmatrix}'\frac{\partial W}{\partial u}-\begin{Bmatrix}12\\2\end{Bmatrix}'\frac{\partial W}{\partial v}+\mathcal{F}W\right],\\ D''&=-\left[\frac{\partial^2 W}{\partial v^2}-\begin{Bmatrix}22\\1\end{Bmatrix}'\frac{\partial W}{\partial u}-\begin{Bmatrix}22\\2\end{Bmatrix}'\frac{\partial W}{\partial v}+\mathcal{G}W\right].\end{aligned}\right.$$

When these expressions for D, $D',$ D'' are substituted in the expression (IV, 77) for $\rho_1+\rho_2$, the latter becomes

$$\begin{aligned}\rho_1+\rho_2=&-\frac{1}{\mathcal{H}^2}\left[\mathcal{E}\frac{\partial^2 W}{\partial v^2}-2\mathcal{F}\frac{\partial^2 W}{\partial u\,\partial v}+\mathcal{G}\frac{\partial^2 W}{\partial u^2}\right]-2W\\ &+\frac{1}{\mathcal{H}^2}\left[\begin{Bmatrix}22\\1\end{Bmatrix}'\mathcal{E}-2\begin{Bmatrix}12\\1\end{Bmatrix}'\mathcal{F}+\begin{Bmatrix}11\\1\end{Bmatrix}'\mathcal{G}\right]\frac{\partial W}{\partial u}\\ &+\frac{1}{\mathcal{H}^2}\left[\begin{Bmatrix}22\\2\end{Bmatrix}'\mathcal{E}-2\begin{Bmatrix}12\\2\end{Bmatrix}'\mathcal{F}+\begin{Bmatrix}11\\2\end{Bmatrix}'\mathcal{G}\right]\frac{\partial W}{\partial v}.\end{aligned}$$

By means of (25) this equation can be written in the form

$$\text{(35)}\qquad \rho_1+\rho_2=-(\Delta_2' W+2W),$$

where the differential parameter is formed with respect to the linear element (23) of the sphere.

Moreover, if $\Delta_{22}'\theta$ denotes the following expression,

$$\text{(36)}\qquad \begin{aligned}\Delta_{22}'\theta=\frac{1}{\mathcal{E}\mathcal{G}-\mathcal{F}^2}\Bigg[&\left(\frac{\partial^2\theta}{\partial u^2}-\begin{Bmatrix}11\\1\end{Bmatrix}'\frac{\partial\theta}{\partial u}-\begin{Bmatrix}11\\2\end{Bmatrix}'\frac{\partial\theta}{\partial v}\right)\\ &\left(\frac{\partial^2\theta}{\partial v^2}-\begin{Bmatrix}22\\1\end{Bmatrix}'\frac{\partial\theta}{\partial u}-\begin{Bmatrix}22\\2\end{Bmatrix}'\frac{\partial\theta}{\partial v}\right)\\ &-\left(\frac{\partial^2\theta}{\partial u\,\partial v}-\begin{Bmatrix}12\\1\end{Bmatrix}'\frac{\partial\theta}{\partial u}-\begin{Bmatrix}12\\2\end{Bmatrix}'\frac{\partial\theta}{\partial v}\right)^2\Bigg],\end{aligned}$$

it follows from (34) that

$$\text{(37)}\qquad \rho_1\rho_2=\frac{DD''-D'^2}{\mathcal{E}\mathcal{G}-\mathcal{F}^2}=\Delta_{22}'W+W\Delta_2'W+W^2.$$

In passing we shall prove that $\Delta_{22}\theta$ is a differential parameter by showing that it is expressible in the form

$$\Delta_{22}\theta = \frac{2\Delta_2\theta\Delta_1(\theta, \Delta_1\theta) - \Delta_1\Delta_1\theta}{4\Delta_1\theta}. \tag{38}$$

Without loss of generality we take

$$E\,du^2 + G\,dv^2 \tag{39}$$

as the quadratic form, with respect to which these differential parameters are formed. Then

$$\Delta_1 u = \frac{1}{E}, \quad \Delta_2 u = \frac{1}{2E}\left(\frac{1}{G}\frac{\partial G}{\partial u} - \frac{1}{E}\frac{\partial E}{\partial u}\right), \quad \Delta_1(u, \Delta_1 u) = -\frac{1}{E^3}\frac{\partial E}{\partial u}$$

$$\Delta_1\Delta_1 u = \frac{E\left(\frac{\partial E}{\partial v}\right)^2 + G\left(\frac{\partial E}{\partial u}\right)^2}{E^5 G}, \quad \Delta_{22} u = -\frac{1}{4E^3G}\left[\frac{\partial E}{\partial u}\frac{\partial G}{\partial u} + \left(\frac{\partial E}{\partial v}\right)^2\right].$$

By substitution we find

$$\Delta_{22} u = \frac{2\Delta_2 u\,\Delta_1(u, \Delta_1 u) - \Delta_1\Delta_1 u}{4\Delta_1 u}.$$

Since the terms in the right-hand member are differential parameters, their values are independent of the choice of parameters u and v, in terms of which (39) is expressed. Hence equation (38) is an identity.

The coördinates x_0, y_0, z_0 of the point on the normal to a surface halfway between the centers of principal curvature have the expressions

$$x_0 = x + \tfrac{1}{2}(\rho_1 + \rho_2)X, \quad y_0 = y + \tfrac{1}{2}(\rho_1 + \rho_2)Y, \quad z_0 = z + \tfrac{1}{2}(\rho_1 + \rho_2)Z.$$

The surface S_0 enveloped by the plane through this point, which is parallel to the tangent plane to the given surface, is called the *mean evolute* of the latter.

If W_0 denotes the distance from the origin to this plane, we have

$$W_0 = \Sigma X x_0 = W + \tfrac{1}{2}(\rho_1 + \rho_2). \tag{40}$$

By means of (35) this may be written

$$W_0 = -\tfrac{1}{2}\Delta_2' W. \tag{41}$$

EXAMPLES

1. Derive the equations of the lines of curvature and the expressions for the principal radii in terms of W, when the parametric lines on the sphere are

(i) meridians and parallels;
(ii) the imaginary generators.

Show that in the latter case the curves corresponding to the generators lie symmetrically with respect to the lines of curvature.

2. Let W_1 and W_2 denote the distances from the origin to the planes through the normal to a surface and the tangents to the lines of curvature $v=$ const., $u=$ const. respectively, so that we have

$$W_1 = xX_1 + yY_1 + zZ_1, \qquad W_2 = xX_2 + yY_2 + zZ_2.$$

Show that

$$\Delta_1 W = \frac{W_1^2}{\rho_1^2} + \frac{W_2^2}{\rho_2^2}, \qquad \Delta_1(W, x) = -\frac{X_1 W_1}{\rho_1} - \frac{X_2 W_2}{\rho_2},$$

$$\Delta_1(W, X) = \frac{X_1 W_1}{\rho_1^2} + \frac{X_2 W_2}{\rho_2^2}, \qquad \Theta(W, x) = -\frac{X_2 W_1}{\rho_1} + \frac{X_1 W_2}{\rho_2},$$

the differential parameters being formed with respect to $E du^2 + 2 F du dv + G dv^2$.

3. If $2q = x^2 + y^2 + z^2$, then we have

$$\Delta_1 q = 2q - W^2, \quad \Delta_2 q = 2 + W\left(\frac{1}{\rho_1} + \frac{1}{\rho_2}\right), \quad \Delta_1(q, W) = -\left(\frac{W_1^2}{\rho_1} + \frac{W_2^2}{\rho_2}\right).$$

4. Show that when the lines of curvature are parametric

$$-\frac{1}{\rho_1}\frac{\partial q}{\partial u} = \frac{\partial W}{\partial u}, \qquad -\frac{1}{\rho_2}\frac{\partial q}{\partial v} = \frac{\partial W}{\partial v}.$$

5. The determination of surfaces whose mean evolute is a point is the same problem as finding isothermal systems of lines on the sphere.

68. The moving trihedral. The fundamental equations of condition may be given another form, in which they are frequently used by French writers. In deriving them we refer the surface to a moving set of rectangular axes called the trihedral T. Its vertex M is a point of the surface, the xy-plane is tangent to the surface at M, and the positive z-axis coincides with the positive direction of the normal to the surface at M. The position of the x- and y-axes is determined by the angle U which the tangent to the curve $v=$ const. through M makes with the x-axis, U being a given function of u and v.

In Chapter I we considered another moving trihedral, consisting of the tangent, principal normal, and binormal of a twisted curve. Let us associate such a trihedral with the curve $v=$ const. through

M and call it the trihedral t_u. We have found (§ 16) that the variations of the direction-cosines a', b', c' of a line L, fixed in space, with reference to t_u, as its vertex moves along the curve which we call C_u, are given by

$$\frac{da'}{ds_u}=\frac{b'}{\rho_u},\quad \frac{db'}{ds_u}=-\left(\frac{a'}{\rho_u}+\frac{c'}{\tau_u}\right),\quad \frac{dc'}{ds_u}=\frac{b'}{\tau_u}, \tag{42}$$

where ρ_u, τ_u denote the radii of first and second curvature of C_u, and ds_u its linear element; evidently the latter may be replaced by $\sqrt{E}\,du$.

The direction-cosines of L with respect to the trihedral T have the values

$$\left\{\begin{aligned} a &= a'\cos U-(b'\sin\varpi_u - c'\cos\varpi_u)\sin U,\\ b &= a'\sin U+(b'\sin\varpi_u - c'\cos\varpi_u)\cos U,\\ c &= b'\cos\varpi_u + c'\sin\varpi_u,\end{aligned}\right. \tag{43}$$

where ϖ_u is the angle which the positive direction of the z-axis makes with the positive direction of the principal normal to C_u at M, the angle being measured toward the positive direction of the binormal of C_u. From equations (42) and (43) we obtain the following:

$$\frac{\partial a}{\partial u}=br-cq,\quad \frac{\partial b}{\partial u}=cp-ar,\quad \frac{\partial c}{\partial u}=aq-bp, \tag{44}$$

where p, q, r have the following significance:

$$\left\{\begin{aligned} p &= \sqrt{E}\left[\cos U\left(\frac{d\varpi_u}{ds_u}-\frac{1}{\tau_u}\right)+\sin U\frac{\cos\varpi_u}{\rho_u}\right],\\ q &= \sqrt{E}\left[\sin U\left(\frac{d\varpi_u}{ds_u}-\frac{1}{\tau_u}\right)-\cos U\frac{\cos\varpi_u}{\rho_u}\right],\\ r &= \sqrt{E}\left(\frac{\sin\varpi_u}{\rho_u}-\frac{dU}{ds_u}\right).\end{aligned}\right. \tag{45}$$

If, in like manner, we consider the trihedral t_v of the curve $u=$ const. through M, denoted by C_v, we obtain the equations

$$\frac{\partial a}{\partial v}=br_1-cq_1,\quad \frac{\partial b}{\partial v}=cp_1-ar_1,\quad \frac{\partial c}{\partial v}=aq_1-bp_1,$$

where p_1, q_1, r_1 can be obtained from (45) by replacing $\sqrt{E}$, U, s_u, ϖ_u, ρ_u, τ_u by $\sqrt{G}$, V, s_v, ϖ_v, ρ_v, τ_v. As V denotes the angle which the tangent to the curve C_v at M makes with the x-axis, we have

$$V-U=\omega. \tag{46}$$

If the vertex M moves along a curve other than a parametric line, that is, along a curve determined by a value of dv/du, the variations of a, b, c are evidently given by

$$\frac{\partial a}{\partial u}\frac{du}{ds}+\frac{\partial a}{\partial v}\frac{dv}{ds}, \qquad \frac{\partial b}{\partial u}\frac{du}{ds}+\frac{\partial b}{\partial v}\frac{dv}{ds}, \qquad \frac{\partial c}{\partial u}\frac{du}{ds}+\frac{\partial c}{\partial v}\frac{dv}{ds},$$

in which the differential quotients have the above values.

69. Fundamental equations of condition. Suppose that we associate with the trihedral T a second trihedral T_0 whose vertex O is fixed in space, about which it revolves in such a manner that its edges are always parallel to the corresponding edges of T, as the vertex of the latter moves over the surface in a given manner. The position of T_0 is completely determined by the nine direction-cosines of its edges with three mutually perpendicular lines L_1, L_2, L_3 through O. Call these direction-cosines a_1, b_1, c_1; a_2, b_2, c_2; a_3, b_3, c_3. These functions must satisfy the equations

$$(47) \qquad \begin{cases} \dfrac{\partial a}{\partial u}=br-cq, & \dfrac{\partial b}{\partial u}=cp-ar, & \dfrac{\partial c}{\partial u}=aq-bp, \\[2ex] \dfrac{\partial a}{\partial v}=br_1-cq_1, & \dfrac{\partial b}{\partial v}=cp_1-ar_1, & \dfrac{\partial c}{\partial v}=aq_1-bp_1. \end{cases}$$

If we equate the two values of $\dfrac{\partial^2 a}{\partial u\,\partial v}$, obtained from the first two of these equations, and in the reduction of the resulting equation make use of (47), we find

$$b\left(\frac{\partial r}{\partial v}-\frac{\partial r_1}{\partial u}-pq_1+qp_1\right)=c\left(\frac{\partial q}{\partial v}-\frac{\partial q_1}{\partial u}-rp_1+pr_1\right).$$

Since this equation must be true when b and c have the values b_1, c_1; b_2, c_2; b_3, c_3, the expressions in parenthesis must be equal to zero. Proceeding in the same manner with $\dfrac{\partial^2 b}{\partial u\,\partial v}$ and $\dfrac{\partial^2 c}{\partial u\,\partial v}$, we obtain the following fundamental equations *:

$$(48) \qquad \begin{cases} \dfrac{\partial p}{\partial v}-\dfrac{\partial p_1}{\partial u}=qr_1-rq_1, \\[2ex] \dfrac{\partial q}{\partial v}-\dfrac{\partial q_1}{\partial u}=rp_1-pr_1, \\[2ex] \dfrac{\partial r}{\partial v}-\dfrac{\partial r_1}{\partial u}=pq_1-qp_1. \end{cases}$$

* These equations were first obtained by Combescure, *Annales de l'École Normale*, Ser. 1, Vol. IV (1867), p. 108; cf. also Darboux, Vol. I, p. 48.

These necessary conditions upon the six functions $p, \cdots, r_1$, in order that the nine functions $a_1, \cdots, c_3$ may determine the position of the trihedral T_0, are also sufficient conditions. The proof of this is similar to that given in § 65.

Equations (47) have been obtained by Darboux * from a study of the motion of the trihedral T_0. He has called $p, q, \cdots, r_1$ the *rotations*.

We return to the consideration of the moving trihedrals T and t_u. Let (x, y, z) and (x', y', z') denote the coördinates of a point P with respect to T and t_u respectively. Between these coördinates the following relations hold:

$$(49) \qquad \begin{cases} x = x' \cos U - (y' \sin \overline{\omega}_u - z' \cos \overline{\omega}_u) \sin U, \\ y = x' \sin U + (y' \sin \overline{\omega}_u - z' \cos \overline{\omega}_u) \cos U, \\ z = y' \cos \overline{\omega}_u + z' \sin \overline{\omega}_u. \end{cases}$$

If in a displacement of P absolute increments with respect to the trihedral t_u at M be indicated by δ, and increments relative to these moving axes by d, we have, from § 16,

$$(50) \quad \frac{\delta x'}{ds_u} = \frac{dx'}{ds_u} - \frac{y'}{\rho_u} + 1, \qquad \frac{\delta y'}{ds_u} = \frac{dy'}{ds_u} + \frac{x'}{\rho_u} + \frac{z'}{\tau_u}, \qquad \frac{\delta z'}{ds_u} = \frac{dz'}{ds_u} - \frac{y'}{\tau_u}.$$

From (49), (50), and (45) we obtain the following †:

$$\frac{\delta x}{\partial u} = \frac{dx}{\partial u} + \sqrt{E} \cos U - ry + qz,$$

$$\frac{\delta y}{\partial u} = \frac{dy}{\partial u} + \sqrt{E} \sin U - pz + rx,$$

$$\frac{\delta z}{\partial u} = \frac{dz}{\partial u} - qx + py.$$

Equations similar to these follow also from the consideration of the trihedral t_v. Hence, when the trihedral T moves over the surface with its vertex M describing a curve determined by a value of dv/du, the increments of the coördinates of a point $P(x, y, z)$, in the directions of the axes of the trihedral, in the

* *L.c.*, Vol. I, chaps. i and v.

† In deriving these equations we have made use of the fact that equations (49) define a transformation of coördinates, and consequently hold when the coördinates are replaced by the projections of an absolute displacement of P.

absolute displacement of P, which may also be moving relative to these axes, have the values *

$$(51)\quad \begin{cases} \delta x = dx + \xi\, du + \xi_1 dv + (q\, du + q_1 dv) z - (r\, du + r_1 dv) y, \\ \delta y = dy + \eta\, du + \eta_1 dv + (r\, du + r_1 dv) x - (p\, du + p_1 dv) z, \\ \delta z = dz \qquad\qquad\qquad\qquad + (p\, du + p_1 dv) y - (q\, du + q_1 dv) x, \end{cases}$$

where we have put

$$(52)\quad \begin{cases} \xi = \sqrt{E} \cos U, \quad \eta = \sqrt{E} \sin U, \\ \xi_1 = \sqrt{G} \cos V = \sqrt{G} \cos(\omega + U), \quad \eta_1 = \sqrt{G} \sin V = \sqrt{G} \sin(\omega + U). \end{cases}$$

The coördinates of M are $(0, 0, 0)$, so that the increments of its displacements are

$$(53)\qquad \delta x = \xi\, du + \xi_1 dv, \qquad \delta y = \eta\, du + \eta_1 dv, \qquad \delta z = 0.$$

If (x_1, y_1, z_1) denote the coördinates of M with respect to the fixed axes formed by the lines L_1, L_2, L_3 previously defined, it follows that

$$(54)\qquad \frac{\partial x_1}{\partial u} = \xi a_1 + \eta b_1, \qquad \frac{\partial x_1}{\partial v} = \xi_1 a_1 + \eta_1 b_1,$$

and similar expressions for y_1 and z_1, where a_1, b_1, c_1; a_2, b_2, c_2; a_3, b_3, c_3 are the direction-cosines of the fixed axes with reference to the moving axes. Since the latter satisfy equations (47), the conditions that the two values of $\frac{\partial^2 x_1}{\partial u\, \partial v}$ obtained from (54) be equal, and similarly for $\frac{\partial^2 y_1}{\partial u\, \partial v}$ and $\frac{\partial^2 z_1}{\partial u\, \partial v}$, are

$$(55)\qquad \begin{cases} \dfrac{\partial \xi}{\partial v} - \dfrac{\partial \xi_1}{\partial u} = \eta r_1 - \eta_1 r, \\[2ex] \dfrac{\partial \eta}{\partial v} - \dfrac{\partial \eta_1}{\partial u} = \xi_1 r - \xi r_1, \\[2ex] \eta_1 p - \eta p_1 + \xi q_1 - \xi_1 q = 0. \end{cases}$$

When we have ten functions $\xi, \xi_1, \eta, \eta_1, p, p_1, q, q_1, r, r_1$, satisfying these conditions and (48), the functions $a_1, \cdots, c_3$ can be found by the solution of a Riccati equation, and x_1, y_1, z_1 by quadratures. Hence equations (48) and (55) are sufficient as well as necessary, and consequently are equivalent to the Gauss and Codazzi equations.

* Cf. Darboux, Vol. II, p. 348.

70. Linear element. Lines of curvature. From (53) we see that the linear element of the surface is

(56) $$ds^2 = (\xi\, du + \xi_1 dv)^2 + (\eta\, du + \eta_1 dv)^2.$$

Hence a necessary and sufficient condition that the parametric lines be orthogonal is

(57) $$\xi\xi_1 + \eta\eta_1 = 0.$$

For a sphere of radius c the coördinates of the center are $(0, 0, -c)$, it being assumed that the positive normal is directed outwards. As this is a fixed point, it follows from equations (51) that whatever be the value of dv/du we must have

$$\xi\, du + \xi_1 dv - (q\, du + q_1 dv)\, c = 0,$$
$$\eta\, du + \eta_1 dv + (p\, du + p_1 dv)\, c = 0,$$

and consequently

(58) $$\frac{\xi}{q} = \frac{\eta}{-p} = \frac{\xi_1}{q_1} = \frac{\eta_1}{-p_1} = c.$$

Conversely, when these equations are satisfied, the point $(0, 0, -c)$ is fixed in space, and therefore the surface is a sphere. Moreover, suppose that we have a proportion such as (58), where the factor of proportionality is not necessarily constant. For the moment call it t. When the values from (58) are substituted in (55) and reduction is made in accordance with (48) we get

$$\eta \frac{\partial t}{\partial v} - \eta_1 \frac{\partial t}{\partial u} = 0, \qquad \xi \frac{\partial t}{\partial v} - \xi_1 \frac{\partial t}{\partial u} = 0.$$

Hence t is constant unless $\xi\eta_1 - \xi_1\eta$ is zero, which, from (56) and § 31, is seen to be possible only in case the surface is isotropic developable.

By definition (§ 51) a line of curvature is a curve along which the normals to the surface form a developable surface. When the vertex is displaced along one of these lines, a point $(0, 0, \rho)$ must move in such a way that δx and δy are zero. Hence we must have

$$\xi\, du + \xi_1 dv + (q\, du + q_1 dv)\, \rho = 0,$$
$$\eta\, du + \eta_1 dv - (p\, du + p_1 dv)\, \rho = 0.$$

Eliminating ρ and dv/du respectively, we obtain the equation of the lines of curvature,

(59) $$(\xi\, du + \xi_1 dv)(p\, du + p_1 dv) + (\eta\, du + \eta_1 dv)(q\, du + q_1 dv) = 0,$$

and the equation of the principal radii,

(60) $$\rho^2 (pq_1 - qp_1) + \rho(q\eta_1 - q_1\eta + p\xi_1 - p_1\xi) + (\xi\eta_1 - \eta\xi_1) = 0.$$

From (59) it follows that a necessary and sufficient condition that the parametric lines be the lines of curvature is

(61) $$\xi p + \eta q = 0, \qquad \xi_1 p_1 + \eta_1 q_1 = 0.$$

We may replace these equations by

$$p = \lambda\eta, \qquad q = -\lambda\xi, \qquad p_1 = \lambda_1\eta_1, \qquad q_1 = -\lambda_1\xi_1,$$

thus introducing two auxiliary functions λ and λ_1. When these values are substituted in the third of (55), we have

$$(\lambda - \lambda_1)(\xi\xi_1 + \eta\eta_1) = 0.$$

If λ and λ_1 are equal, the above equations are of the form (58), which were seen to be characteristic of the sphere and the isotropic developable. Hence the second factor is zero, so that equations (61) may be replaced by

$$(62) \qquad \xi\xi_1 + \eta\eta_1 = 0, \qquad pp_1 + qq_1 = 0,$$

of which a particular case is

$$(63) \qquad \xi_1 = \eta = 0, \qquad p = q_1 = 0.$$

From (52) it follows that in the latter case the x- and y-axes are tangent to the curves $v = \text{const.}$ and $u = \text{const.}$ We shall consider this case later.

From (60) and (52) we find that the expression for the total curvature of the surface is

$$K = \frac{1}{\rho_1\rho_2} = \frac{pq_1 - p_1q}{\xi\eta_1 - \xi_1\eta} = \frac{pq_1 - p_1q}{\sqrt{EG}\sin\omega},$$

where ω denotes the angle between the parametric curves. Hence the third of equations (48) may be written

$$(64) \qquad \frac{\sqrt{EG}\sin\omega}{\rho_1\rho_2} = \frac{H}{\rho_1\rho_2} = \frac{\partial r}{\partial v} - \frac{\partial r_1}{\partial u}.$$

71. Conjugate directions and asymptotic directions. Spherical representation. We have found (§ 54) that the direction in the tangent plane conjugate to a given direction is the characteristic of this plane as it envelopes the surface in the given direction. Hence, from the point of view of the moving trihedral, the direction conjugate to a displacement, determined by a value of dv/du, is the line in the xy-plane which passes through the origin, and which does not experience an absolute displacement in the direction of the z-axis. From the third of equations (51) it is seen that the equation of this line is

$$(65) \qquad (p\,du + p_1dv)\,y - (q\,du + q_1dv)\,x = 0.$$

If the increments of u and v, corresponding to a displacement in the direction of this line, be indicated by d_1u and d_1v, the quantities x and y are proportional to $(\xi\, d_1u + \xi_1 d_1v)$ and $(\eta\, d_1u + \eta_1 d_1v)$. When x and y in (65) are replaced by these values, the resulting equation may be reduced to

$$(66)\quad (p\eta - q\xi)\, dud_1u + (p\eta_1 - q\xi_1)\, dud_1v + (p_1\eta - q_1\xi)\, d_1udv + (p_1\eta_1 - q_1\xi_1)\, d_1vdv = 0.$$

In consequence of (55) the coefficients of dud_1v and d_1udv are equal, so that the equation is symmetrical with respect to the two sets of differentials, thus establishing the fact that the relation between a line and its conjugate is reciprocal.

In order that the parametric lines be conjugate, equation (66) must be satisfied by $du = 0$ and $d_1v = 0$. Hence we must have

$$(67)\quad p\eta_1 - q\xi_1 = 0, \qquad p_1\eta - q_1\xi = 0.$$

It should be noticed that equations (61) are a consequence of the first of (62) and (67). Hence we have the result that the lines of curvature form the only orthogonal conjugate system.

From (66) it follows that the asymptotic directions are given by

$$(68)\quad (p\eta - q\xi)\, du^2 + (p\eta_1 - q\xi_1 + p_1\eta - q_1\xi)\, dudv + (p_1\eta_1 - q_1\xi_1)\, dv^2 = 0.$$

The spherical representation of a surface is traced out by the point m, whose coördinates are (0, 0, 1) with respect to the trihedral T_0 of fixed vertex. From (51) we find that the projections of a displacement of m, corresponding to a displacement along the surface, are

$$(69)\quad \delta X = q\, du + q_1 dv, \qquad \delta Y = -(p\, du + p_1 dv), \qquad \delta Z = 0.$$

Hence the linear element of the spherical representation is

$$(70)\quad d\sigma^2 = (q\, du + q_1 dv)^2 + (p\, du + p_1 dv)^2.$$

The line defined by (65) is evidently perpendicular to the direction of the displacement of m, as given by (69). Hence the tangent to the spherical representation of a curve upon a surface is perpendicular to the direction conjugate to the curve at the corresponding point. Therefore the tangents to a line of curvature and its representation are parallel, whereas an asymptotic direction and its representation are perpendicular (§ 61).

72. Fundamental relations and formulas. From equations (53) and (69) we have, for the point M on the surface,

$$(71)\qquad \begin{cases} \dfrac{\delta x}{\partial u}=\xi, & \dfrac{\delta y}{\partial u}=\eta, & \dfrac{\delta z}{\partial u}=0, \\ \dfrac{\delta x}{\partial v}=\xi_1, & \dfrac{\delta y}{\partial v}=\eta_1, & \dfrac{\delta z}{\partial v}=0; \end{cases}$$

and

$$(72)\qquad \begin{cases} \dfrac{\delta X}{\partial u}=q, & \dfrac{\delta Y}{\partial u}=-p, & \dfrac{\delta Z}{\partial u}=0, \\ \dfrac{\delta X}{\partial v}=q_1, & \dfrac{\delta Y}{\partial v}=-p_1, & \dfrac{\delta Z}{\partial v}=0. \end{cases}$$

Consequently the following relations hold between the fundamental coefficients, the rotations, and the translations:

$$(73)\qquad \begin{cases} E=\xi^2+\eta^2, & F=\xi\xi_1+\eta\eta_1, & G=\xi_1^2+\eta_1^2, \\ D=p\eta-q\xi, & D'=p_1\eta-q_1\xi=p\eta_1-q\xi_1, & D''=p_1\eta_1-q_1\xi_1, \\ \mathcal{E}=p^2+q^2, & \mathcal{F}=pp_1+qq_1, & \mathcal{G}=p_1^2+q_1^2. \end{cases}$$

When, in particular, the parametric system on a surface is orthogonal, and the x- and y- axes of the trihedral are tangent to the curves $v=$ const. and $u=$ const. through the vertex, equations (52) are

$$(74)\qquad \xi=\sqrt{E},\qquad \eta=\xi_1=0,\qquad \eta_1=\sqrt{G},$$

and equations (55) reduce to

$$(75)\qquad r=-\frac{1}{\sqrt{G}}\frac{\partial\sqrt{E}}{\partial v},\qquad r_1=\frac{1}{\sqrt{E}}\frac{\partial\sqrt{G}}{\partial u},\qquad p\sqrt{G}+q_1\sqrt{E}=0.$$

Moreover, equations (45) and the similar ones for p_1, q_1, r_1 become

$$(76)\qquad \begin{cases} p=\sqrt{E}\left(\dfrac{d\varpi_u}{ds_u}-\dfrac{1}{\tau_u}\right), & q=-\sqrt{E}\dfrac{\cos\varpi_u}{\rho_u}, & r=\sqrt{E}\dfrac{\sin\varpi_u}{\rho_u}, \\ p_1=\sqrt{G}\dfrac{\cos\varpi_v}{\rho_v}, & q_1=\sqrt{G}\left(\dfrac{d\varpi_v}{ds_v}-\dfrac{1}{\tau_v}\right), & r_1=\sqrt{G}\dfrac{\sin\varpi_v}{\rho_v}. \end{cases}$$

The first two of equations (75) lead, by means of (76), to

$$(77)\qquad \frac{\sin\varpi_u}{\rho_u}=-\frac{1}{\sqrt{EG}}\frac{\partial\sqrt{E}}{\partial v},\qquad \frac{\sin\varpi_v}{\rho_v}=\frac{1}{\sqrt{EG}}\frac{\partial\sqrt{G}}{\partial u},$$

which follow also from § 58.

The third of equations (75) establishes the fact, previously remarked in § 59, that the geodesic torsion in two orthogonal directions differs only in sign.

The variations of the direction-cosines X_1, Y_1, Z_1 of the tangent to the curve $v = \text{const.}$ are represented by the motion of the point (1, 0, 0) of the trihedral T_0 with fixed vertex. From (51) we have

$$
(78) \qquad \begin{cases} \dfrac{\delta X_1}{\partial u} = 0, & \dfrac{\delta Y_1}{\partial u} = r, & \dfrac{\delta Z_1}{\partial u} = -q, \\[2ex] \dfrac{\delta X_1}{\partial v} = 0, & \dfrac{\delta Y_1}{\partial v} = r_1, & \dfrac{\delta Z_1}{\partial v} = -q_1. \end{cases}
$$

From these equations we see that as a point describes a curve $v = \text{const.}$, namely C_u, the tangent to this curve undergoes an infinitesimal rotation consisting of two components, one in amount $r\,du$ about the normal to the surface and the other, $-q\,du$, about the line in the tangent plane perpendicular to the tangent to C_u. Consequently, by their definition, the geodesic and normal curvature of C_u are $r/\sqrt{E}$ and $-q/\sqrt{E}$ respectively. Moreover, it is seen from (72) that as a point describes C_u the normal to the surface undergoes a rotation consisting of the components $q\,du$ about the line in the tangent plane perpendicular to the tangent, and $-p\,du$ about the tangent. Hence, if C_u were a geodesic, the torsion would be $p/\sqrt{E}$ to within the sign at least. Thus by geometrical considerations we have obtained the fundamental relations (76).

We suppose now that the parametric system is any whatever. From the definition of the differential parameters (§ 37) it follows that

$$E = H^2 \Delta_1 v, \qquad G = H^2 \Delta_1 u.$$

Consequently if P, Q, R denote functions similar to p, q, r, for a general curve

$$\phi(u, v) = \text{const.},$$

which passes through M and whose tangent makes the angle Φ with the moving x-axis, we have, from (45),

$$
(79) \qquad \begin{cases} P = H_1 \sqrt{\Delta_1 \phi} \left[\cos \Phi \left(\dfrac{d\varpi}{ds} - \dfrac{1}{\tau} \right) + \sin \Phi \dfrac{\cos \varpi}{\rho} \right], \\[2ex] Q = H_1 \sqrt{\Delta_1 \phi} \left[\sin \Phi \left(\dfrac{d\varpi}{ds} - \dfrac{1}{\tau} \right) - \cos \Phi \dfrac{\cos \varpi}{\rho} \right], \\[2ex] R = H_1 \sqrt{\Delta_1 \phi} \left[\dfrac{\sin \varpi}{\rho} - \dfrac{d\Phi}{ds} \right], \end{cases}
$$

where by (III, 51) $H_1^{-2} = \Delta_1 \phi \, \Delta_1 \psi - \Delta_1^2(\phi, \psi)$ and $\psi = \text{const.}$ defines any other family of curves.

Moreover, equations analogous to (44) are

$$\frac{da}{ds} = \frac{bR - cQ}{H_1 \sqrt{\Delta_1 \phi}}, \qquad \frac{db}{ds} = \frac{cP - aR}{H_1 \sqrt{\Delta_1 \phi}}, \qquad \frac{dc}{ds} = \frac{aQ - bP}{H_1 \sqrt{\Delta_1 \phi}}.$$

If now in
$$\frac{da}{ds} = \frac{\partial a}{\partial u}\frac{du}{ds} + \frac{\partial a}{\partial v}\frac{dv}{ds}$$
we replace the expressions for $\frac{\partial a}{\partial u}$ and $\frac{\partial a}{\partial v}$ from (47), and similarly for db/ds and dc/ds, we obtain
$$Pds = H_1\sqrt{\Delta_1\phi}\,(p\,du + p_1 dv), \qquad Qds = H_1\sqrt{\Delta_1\phi}\,(q\,du + q_1 dv),$$
$$Rds = H_1\sqrt{\Delta_1\phi}\,(r\,du + r_1 dv).$$

From these equations and (79) we derive the following fundamental formulas:

$$(80)\quad \begin{cases} \left(\dfrac{d\varpi}{ds} - \dfrac{1}{\tau}\right)ds = \cos\Phi\,(p\,du + p_1 dv) + \sin\Phi\,(q\,du + q_1 dv), \\ \dfrac{\cos\varpi}{\rho}\,ds = \sin\Phi\,(p\,du + p_1 dv) - \cos\Phi\,(q\,du + q_1 dv), \\ \dfrac{\sin\varpi}{\rho} = \dfrac{d\Phi}{ds} + r\dfrac{du}{ds} + r_1\dfrac{dv}{ds}. \end{cases}$$

By means of the last of equations (80) we shall express the geodesic curvature of a curve in terms of the functions E, F, G, of their derivatives, and of the angle θ which the curve makes with the curve $v = \text{const.}$ If we take the x-axis of the trihedral tangent to the curve $v = \text{const.}$, we obtain from the last of (80), in consequence of (45),
$$\frac{1}{\rho_g} = \frac{d\theta}{ds} + \frac{\sqrt{E}}{\rho_{gu}}\frac{du}{ds} + \left(\frac{\sqrt{G}}{\rho_{gv}} - \frac{\partial\omega}{\partial v}\right)\frac{dv}{ds}.$$
From (III, 15, 16) we obtain
$$\frac{\partial\omega}{\partial v} = \frac{1}{H}\left[\frac{F}{2EG}\left(G\frac{\partial E}{\partial v} + E\frac{\partial G}{\partial v}\right) - \frac{\partial F}{\partial v}\right].$$
When this value and the expressions for ρ_{gu} and ρ_{gv} (IV, 57) are substituted in the above equation, we have the formula desired:
$$(81)\quad \frac{1}{\rho_g} = \frac{d\theta}{ds} + \frac{1}{H}\left(\frac{\partial F}{\partial u} - \frac{F}{2E}\frac{\partial E}{\partial u} - \frac{1}{2}\frac{\partial E}{\partial v}\right)\frac{du}{ds} + \frac{1}{2H}\left(\frac{\partial G}{\partial u} - \frac{F}{E}\frac{\partial E}{\partial v}\right)\frac{dv}{ds}.$$

EXAMPLES

1. A necessary and sufficient condition that the origin of the trihedral T be the only point in the moving xy-plane which generates a surface to which this plane is tangent, is that the surface be nondevelopable.

2. Determine ρ so that the point of coördinates $(\rho, 0, 0)$ with respect to T shall describe a surface to which the x-axis of T is normal; examine the case when the lines of curvature are parametric and the x-axis is tangent to the curve $v = \text{const.}$

3. When the parametric curves are minimal lines for both the surface and the sphere, it is necessary that

$$\eta = i\xi, \qquad \eta_1 = -i\xi_1, \qquad q = -ip, \qquad q_1 = ip_1,$$

or

$$\eta = -i\xi, \qquad \eta_1 = i\xi_1, \qquad q = ip, \qquad q_1 = -ip_1;$$

in this case the parametric curves on the surface form a conjugate system, and the surface is minimal (cf. § 55).

4. When the asymptotic lines on a surface form an orthogonal system, we must have

$$p\xi_1 + q\eta_1 = 0, \qquad p_1\xi + q_1\eta = 0,$$

in which case the surface is minimal.

5. When the lines of curvature are parametric, and the x-axis of T is tangent to the curve $v = \text{const.}$, equations (80) reduce to

$$\frac{1}{\tau} - \frac{d\overline{\omega}}{ds} = \left(\frac{1}{\rho_1} - \frac{1}{\rho_2}\right)\sin\Phi\cos\Phi, \qquad \frac{\cos\overline{\omega}}{\rho} = \frac{\cos^2\Phi}{\rho_1} + \frac{\sin^2\Phi}{\rho_2},$$

$$\frac{\sin\overline{\omega}}{\rho} = \frac{d\Phi}{ds} + \frac{1}{\rho_2 - \rho_1}\left(\frac{q}{p_1}\frac{\partial\rho_1}{\partial v}\frac{du}{ds} + \frac{p_1}{q}\frac{\partial\rho_2}{\partial u}\frac{dv}{ds}\right).$$

6. When the second equation in Ex. 5 is differentiated with respect to s, the resulting equation is reducible to

$$\frac{\cos\overline{\omega}}{\rho^2}\frac{d\rho}{ds} + \frac{\sin\overline{\omega}}{\rho}\left(3\frac{d\overline{\omega}}{ds} - \frac{2}{\tau}\right) = q^2\frac{\partial\rho_1}{\partial u}\left(\frac{du}{ds}\right)^3 + 3q^2\frac{\partial\rho_1}{\partial v}\left(\frac{du}{ds}\right)^2\frac{dv}{ds}$$
$$+ 3p_1^2\frac{\partial\rho_2}{\partial u}\frac{du}{ds}\left(\frac{dv}{ds}\right)^2 + p_1^2\frac{\partial\rho_2}{\partial v}\left(\frac{dv}{ds}\right)^3.$$

7. On a surface a given curve makes the angle Φ with the x-axis of a trihedral T; the point P_0 of coördinates $\cos\Phi, \sin\Phi, 0$ with reference to the parallel trihedral T_0 with fixed vertex, describes the spherical indicatrix of the tangent to the curve; the direction-cosines of the tangent to this curve are

$$-\sin\Phi\sin\overline{\omega}, \qquad \cos\Phi\sin\overline{\omega}, \qquad \cos\overline{\omega},$$

where $\overline{\omega}$ has the significance indicated in § 49, and the linear element is ds/ρ; derive therefrom by means of (51) the second and third of formulas (80).

8. The point B, whose coördinates with reference to T_0 of Ex. 7 are

$$\sin\Phi\cos\overline{\omega}, \qquad -\cos\Phi\cos\overline{\omega}, \qquad \sin\overline{\omega},$$

describes the spherical indicatrix of the binormal to the given curve on the surface, and its linear element is ds/τ; derive therefrom the first of formulas (80).

73. Parallel surfaces. We inquire under what conditions the normals to a surface are normal to a second surface. In order that this be possible, there must exist a function t such that the point of coördinates $(0, 0, t)$, with reference to the trihedral T, describes a surface to which the moving z-axis is constantly normal. Hence we must have $\delta z = 0$, and consequently, by equations (51), t must be a constant, which may have any value whatever. We have, therefore, the theorem:

If segments of constant length be laid off upon the normals to a surface, these segments being measured from the surface, the locus of their other end points is a surface with the same normals as the given surface.

These two surfaces are said to be *parallel*. Evidently there is an infinity of surfaces parallel to a given surface, and all of them have the same spherical representation.

Consider the surface for which t has the value a, and call it $\bar{S}$. From (51) it follows that the projections on the axes of T of a displacement on $\bar{S}$ have the values

$$\left\{\begin{aligned} \delta\bar{x} &= \xi\,du + \xi_1 dv + (q\,du + q_1 dv)\,a, \\ \delta\bar{y} &= \eta\,du + \eta_1 dv - (p\,du + p_1 dv)\,a. \end{aligned}\right. \tag{82}$$

Comparing these results with (53), we see that the displacements on the two surfaces corresponding to the same value of dv/du are parallel only in case equation (59) is satisfied, that is, when the point describes a line of curvature on S. But from a characteristic property of lines of curvature (§ 51) it follows that the lines of curvature on the two surfaces correspond. Hence we have the theorem:

The tangents to corresponding lines of curvature of two parallel surfaces at corresponding points are parallel.

From (82) and (73) we have the following expressions for the first fundamental quantities of $\bar{S}$:

$$\left\{\begin{aligned} \bar{E} &= E - 2\,aD + a^2\mathcal{E}, \\ \bar{F} &= F - 2\,aD' + a^2\mathcal{F}, \\ \bar{G} &= G - 2\,aD'' + a^2\mathcal{G}, \end{aligned}\right. \tag{83}$$

or, in consequence of (IV, 78),

$$\left\{\begin{aligned} \bar{E} &= E\left(1 - \frac{a^2}{\rho_1\rho_2}\right) + Da\left[a\left(\frac{1}{\rho_1} + \frac{1}{\rho_2}\right) - 2\right], \\ \bar{F} &= F\left(1 - \frac{a^2}{\rho_1\rho_2}\right) + D'a\left[a\left(\frac{1}{\rho_1} + \frac{1}{\rho_2}\right) - 2\right], \\ \bar{G} &= G\left(1 - \frac{a^2}{\rho_1\rho_2}\right) + D''a\left[a\left(\frac{1}{\rho_1} + \frac{1}{\rho_2}\right) - 2\right]. \end{aligned}\right. \tag{84}$$

The moving trihedral for $\bar{S}$ can be taken parallel to T for S, and thus the rotations are the same for both trihedrals; and from (82) it follows that the translations have the values

$$\bar{\xi}=\xi+aq, \qquad \bar{\xi}_1=\xi_1+aq_1, \qquad \bar{\eta}=\eta-ap, \qquad \bar{\eta}_1=\eta_1-ap_1.$$

On substituting in the equations for $\bar{S}$ analogous to (59), (60), (66), we obtain the fundamental equations for $\bar{S}$ in terms of the functions for S. Also from (73) we have the following expressions for the second fundamental coefficients for $\bar{S}$:

$$\bar{D}=D-a\mathcal{E}, \qquad \bar{D}'=D'-a\mathcal{F}, \qquad \bar{D}''=D''-a\mathcal{G}. \tag{85}$$

Since the centers of principal curvature of a surface and its parallel at corresponding points are the same, it follows that

$$\bar{\rho}_1=\rho_1-a, \qquad \bar{\rho}_2=\rho_2-a. \tag{86}$$

Suppose that we have a surface whose total curvature is constant and equal to $1/c^2$. Evidently a sphere of radius c is of this kind, but later (Chapter VIII) it will be shown that there is a large group of surfaces with this property. We call them *spherical surfaces*.

From (86) we have $\quad (\bar{\rho}_1+a)(\bar{\rho}_2+a)=c^2,$

so that if we take $a=\pm c$, we obtain

$$\frac{1}{\bar{\rho}_1}+\frac{1}{\bar{\rho}_2}=\pm\frac{1}{c}\cdot$$

Hence we have the theorem of Bonnet:*

With every surface of constant total curvature $1/c^2$ there are associated two surfaces of mean curvature $\pm 1/c$; they are parallel to the former and at the distances $\mp c$ from it.

And conversely,

With every surface whose mean curvature is constant and different from zero there are associated two parallel surfaces, one of which has constant total curvature and the other constant mean curvature.

74. Surfaces of center. As a point M moves over a surface S the corresponding centers of principal curvature M_1 and M_2 describe two surfaces S_1 and S_2, which are called the *surfaces of center* of S. Let C_1 and C_2 be the lines of curvature of S through M, and D_1 and D_2 the developable surfaces formed by the normals to S along C_1

* *Nouvelles annales de mathématiques*, Ser. 1, Vol. XII (1853), p. 433.

and C_2 respectively. The edge of regression of D_1, denoted by Γ_1, is a curve on S_1 (see fig. 17), and consequently S_1 is the locus of one set of evolutes of the curves C_1 on S. Similarly S_2 is the locus of a set of evolutes of the curves C_2 on S. For this reason S_1 and S_2 are said to constitute the *evolute* of S, and S is their *involute*. Evidently any surface parallel to S is also an involute of S_1 and S_2.

The line M_1M_2, as a generator of D_1, is tangent to Γ_1 at M_1, and, as a generator of D_2, it is tangent to Γ_2 at M_2. Hence it is a common tangent of the surfaces S_1 and S_2. From this it follows that the developable surface D_1 meets S_1 along Γ_1 and envelopes S_2 along a curve Γ_2'. Its generators are consequently tangent to the curves conjugate to Γ_2' (§ 54). In particular, the generator M_1M_2 is tangent to Γ_2, and therefore the directions of Γ_2 and Γ_2' at M_2 are conjugate. Similar results follow from the consideration of D_2. Hence:

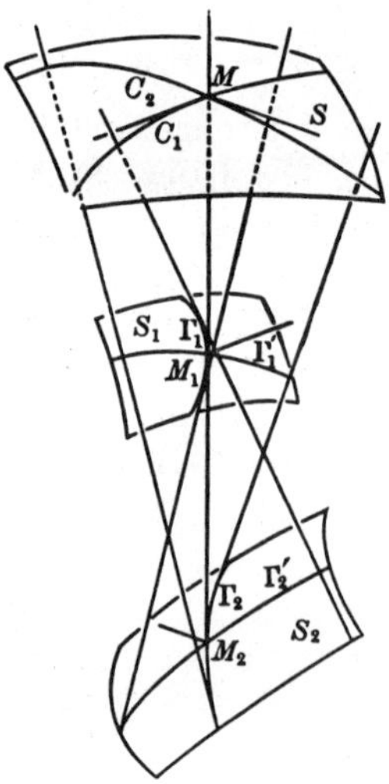

FIG. 17

On the surfaces of center of a surface S the curves corresponding to the lines of curvature of S form a conjugate system.

Since the developable D_1 envelopes S_2, the tangent plane to S_2 at M_2 is the tangent plane to D_1 at this point. But the tangent plane at M_2 is tangent to D_1 all along M_1M_2 (§ 25), and consequently it is determined by M_1M_2 and the tangent to C_1 at M. Hence the normal to S_2 at M_2 is parallel to the tangent to C_2 at M. In like manner, the normal to S_1 at M_1 is parallel to the tangent to C_1 at M. Thus, through each normal to S we have two perpendicular planes, of which one is tangent to one surface of center and the other to the second surface. But each of these planes is at the same time tangent to one of the developables, and is the osculating plane of its edge of regression. Hence, at every point of one of these curves, the osculating plane is perpendicular to the tangent plane to the sheet of the evolute upon which it lies, and so we have the theorem:

The edges of regression of the developable surfaces formed by the normals to a surface along the lines of curvature of one family are geodesics on the surface of center which is the locus of these edges; and the developable surfaces formed by the normals along the lines of curvature in the other family envelope this surface of center along the curves conjugate to these geodesics.

In the following sections we shall obtain, in an analytical manner, the results just deduced geometrically.

75. Fundamental quantities for surfaces of center. As the trihedral T moves over the surface S the point $(0, 0, \rho_1)$ describes the surface of center S_1. Let the lines of curvature on S be parametric, and the x-axis of T be tangent, to the curve $v=$ const. Now

$$\eta=\xi_1=p=q_1=0, \qquad q=-\frac{\sqrt{E}}{\rho_1}, \qquad p_1=\frac{\sqrt{G}}{\rho_2}, \tag{87}$$

so that the first two of equations (48) may be put in the form

$$\begin{cases} \dfrac{\sqrt{E}}{\sqrt{G}}\left(\dfrac{1}{\rho_1}-\dfrac{1}{\rho_2}\right)r_1=\left(\dfrac{1}{\rho_1}-\dfrac{1}{\rho_2}\right)\dfrac{1}{\sqrt{G}}\dfrac{\partial\sqrt{G}}{\partial u}=\dfrac{\partial}{\partial u}\left(\dfrac{1}{\rho_2}\right), \\ \dfrac{\sqrt{G}}{\sqrt{E}}\left(\dfrac{1}{\rho_1}-\dfrac{1}{\rho_2}\right)r=\left(\dfrac{1}{\rho_2}-\dfrac{1}{\rho_1}\right)\dfrac{1}{\sqrt{E}}\dfrac{\partial\sqrt{E}}{\partial v}=\dfrac{\partial}{\partial v}\left(\dfrac{1}{\rho_1}\right). \end{cases} \tag{88}$$

The projections on the moving axes of the absolute displacement of M_1 corresponding to a displacement of M on S are found from (51) to be

$$\delta x_1=0, \quad \delta y_1=(\eta_1-\rho_1 p_1)\,dv=\sqrt{G}\left(1-\frac{\rho_1}{\rho_2}\right)dv, \quad \delta z_1=d\rho_1. \tag{89}$$

Hence the linear element of S_1 is

$$ds_1^2=d\rho_1^2+G\left(1-\frac{\rho_1}{\rho_2}\right)^2 dv^2; \tag{90}$$

consequently the curves $\rho_1=$ const. on S_1 are the orthogonal trajectories of the curves $v=$ const., which are the edges of regression, Γ_1, of the developables of the normals to S along the lines of curvature $v=$ const.

Let us consider the moving trihedral T_1 for S_1 formed by the tangents to the curves $v=$ const. and $\rho_1=$ const. at M_1 and the normal at this point. From (89) it follows that the first tangent has the same direction and sense as the normal to S, and that the second tangent has the same direction as the tangent to $u=$ const. on S, the sense being the same or different according as $(1-\rho_1/\rho_2)$ is positive or negative. And the normal to S_1 has the same direction as the tangent to $v = \text{const.}$ on S, and the contrary or same sense accordingly.

If then we indicate with an accent quantities referring to the moving trihedral T_1, we have

$$(91)\qquad \begin{cases} x' = z - \rho_1, & y' = \epsilon y, & z' = -\epsilon x, \\ a' = c, & b' = \epsilon b, & c' = -\epsilon a, \end{cases}$$

where ϵ is ± 1 according as $(1 - \rho_1/\rho_2)$ is positive or negative. From (89) it follows that

$$(92)\qquad \begin{cases} \xi' = \dfrac{\partial \rho_1}{\partial u}, \quad \xi_1' = \dfrac{\partial \rho_1}{\partial v}, \quad \eta' = 0, \quad \eta_1' = \epsilon\sqrt{G}\left(1 - \dfrac{\rho_1}{\rho_2}\right), \\ \zeta' = \zeta_1' = 0. \end{cases}$$

When the values (91) are substituted in equations for T_1 similar to equations (47), we find

$$(93)\qquad \begin{cases} p' = r, & q' = \epsilon q, & r' = -\epsilon p = 0, \\ p_1' = r_1, & q_1' = \epsilon q_1 = 0, & r_1' = -\epsilon p_1. \end{cases}$$

Since r' is zero, it follows from (76) that the curves $v = \text{const.}$ are geodesics, as found geometrically.

The various fundamental equations for S_1 may now be obtained by substituting these values in the corresponding equations of the preceding sections. Thus, from (73) we have

$$(94)\qquad E_1 = \left(\frac{\partial \rho_1}{\partial u}\right)^2, \qquad F_1 = \frac{\partial \rho_1}{\partial u}\frac{\partial \rho_1}{\partial v}, \qquad G_1 = \left(\frac{\partial \rho_1}{\partial v}\right)^2 + G\left(1 - \frac{\rho_1}{\rho_2}\right)^2,$$

which follow likewise from (90); and also

$$(95)\qquad D_1 = \frac{\epsilon\sqrt{E}}{\rho_1}\frac{\partial \rho_1}{\partial u}, \qquad D_1' = 0, \qquad D_1'' = -\frac{\epsilon\rho_1}{\rho_2^2}\frac{G}{\sqrt{E}}\frac{\partial \rho_2}{\partial u}.$$

Hence the parametric curves on S_1 form a conjugate system (cf. § 54).

The equation of the lines of curvature may be written

$$(96)\quad r\frac{\partial \rho_1}{\partial u}du^2 + \left[r_1\frac{\partial \rho_1}{\partial u} + r\frac{\partial \rho_1}{\partial v} + qp_1(\rho_2 - \rho_1)\right]du\,dv + r_1\frac{\partial \rho_1}{\partial v}dv^2 = 0,$$

and the equation of the asymptotic directions is

$$(97)\qquad \frac{E}{\rho_1^2}\frac{\partial \rho_1}{\partial u}du^2 - \frac{G}{\rho_2^2}\frac{\partial \rho_2}{\partial u}dv^2 = 0.$$

The expression for K_1, the total curvature of S_1, is

$$K_1 = -\frac{1}{(\rho_1-\rho_2)^2}\frac{\frac{\partial\rho_2}{\partial u}}{\frac{\partial\rho_1}{\partial u}}. \tag{98}$$

From (80) and (93) it follows that the geodesic curvature at M_1 of the curve on S_1 which makes the angle Φ_1 with the curve $v = \text{const.}$ through M_1 is given by

$$\frac{1}{\rho_g} = \frac{d\Phi_1}{ds_1} - \epsilon p_1 \frac{dv}{ds_1}.$$

Hence the radius of geodesic curvature of a curve $\rho_1 = \text{const.}$, that is, a curve for which Φ_1 is a right angle, has, in consequence of (87), the value $\rho_1 - \rho_2$. In accordance with § 57 the center of geodesic curvature is found by measuring off the distance $\rho_1 - \rho_2$, in the negative direction, on the z-axis of the trihedral T. Consequently M_2 is this center of curvature. Hence we have the following theorem of Beltrami:

The centers of geodesic curvature of the curves $\rho_1 = \text{const.}$ on S_1 and of $\rho_2 = \text{const.}$ on S_2 are the corresponding points on S_2 and S_1 respectively.

For the sheet S_2 of the evolute we find the following results:

$$ds_2^2 = E\left(1 - \frac{\rho_2}{\rho_1}\right)^2 du^2 + d\rho_2^2; \tag{90'}$$

the equation of the lines of curvature is

$$r\frac{\partial\rho_2}{\partial u}du^2 + \left[r\frac{\partial\rho_2}{\partial v} + r_1\frac{\partial\rho_2}{\partial u} + p_1 q(\rho_2 - \rho_1)\right]du\,dv + r_1\frac{\partial\rho_2}{\partial v}dv^2 = 0; \tag{96'}$$

the equation of the asymptotic lines is

$$\frac{E}{\rho_1^2}\frac{\partial\rho_1}{\partial v}du^2 - \frac{G}{\rho_2^2}\frac{\partial\rho_2}{\partial v}dv^2 = 0; \tag{97'}$$

the expression for the total curvature is

$$K_2 = -\frac{1}{(\rho_1-\rho_2)^2}\frac{\frac{\partial\rho_1}{\partial v}}{\frac{\partial\rho_2}{\partial v}}. \tag{98'}$$

In consequence of these results we are led to the following theorems of Ribaucour,* the proof of which we leave to the reader:

A necessary and sufficient condition that the lines of curvature upon S_1 *and* S_2 *correspond is that* $\rho_1 - \rho_2 = c$ *(a constant); then* $K_1 = K_2 = -1/c^2$*, and the asymptotic lines upon* S_1 *and* S_2 *correspond.*

A necessary and sufficient condition that the asymptotic lines on S_1 *and* S_2 *correspond is that there exist a functional relation between* ρ_1 *and* ρ_2.

76. Surfaces complementary to a given surface. We have just seen that the normals to a surface are tangent to a family of geodesics on each surface of centers. Now we prove the converse:

The tangents to a family of geodesics on a surface S_1 *are normal to an infinity of parallel surfaces.*

Let the geodesics and their orthogonal trajectories be taken for the curves $v = \text{const.}$ and $u = \text{const.}$ respectively, and the parameters chosen so that the linear element has the form

$$ds_1^2 = du^2 + G_1 dv^2.$$

We refer the surface to the trihedral formed by the tangents to the parametric curves and the normal, the x-axis being tangent to the curve $v = \text{const.}$ Upon the latter we lay off from the point M_1 of the surface a length λ, and let P denote the other extremity. As M_1 moves over the surface the projections of the corresponding displacements of P have the values

$$d\lambda + du, \quad \left(\sqrt{G_1} + \lambda \frac{\partial \sqrt{G_1}}{\partial u}\right) dv, \quad -\lambda(q\,du + q_1 dv). \tag{99}$$

In order that the locus of P be normal to the lines M_1P, we must have $d\lambda + du = 0$, and consequently

$$\lambda = -u + c,$$

where c denotes the constant of integration whose value determines a particular one of the family of parallel surfaces. If the direction-cosines of M_1P with reference to fixed axes be X_1, Y_1, Z_1, the coördinates of the surface S, for which $c = 0$, are given by

$$x = x_1 - uX_1, \qquad y = y_1 - uY_1, \qquad z = z_1 - uZ_1,$$

where x_1, y_1, z_1 are the coördinates of M_1.

* *Comptes Rendus*, Vol. LXXIV (1872), p. 1399.

The surface S_1 is one of the surfaces of center of S. In order to find the other, S_2, we must determine λ so that the locus of P is tangent at P to the xz-plane of the moving trihedral. The condition for this is

$$\frac{\partial\sqrt{G_1}}{\partial u}\lambda+\sqrt{G_1}=0.$$

Hence S_2 is given by

$$x_2=x_1-\frac{\sqrt{G_1}}{\frac{\partial\sqrt{G_1}}{\partial u}}X_1,\quad y_2=y_1-\frac{\sqrt{G_1}}{\frac{\partial\sqrt{G_1}}{\partial u}}Y_1,\quad z_2=z_1-\frac{\sqrt{G_1}}{\frac{\partial\sqrt{G_1}}{\partial u}}Z_1,$$

and the principal radii of S are expressed by

$$(100)\qquad \rho_1=u,\qquad \rho_2=u-\frac{\sqrt{G_1}}{\frac{\partial\sqrt{G_1}}{\partial u}}.$$

Bianchi * calls S_2 the *surface complementary* to S_1 for the given geodesic system.

Beltrami has suggested the following geometrical proof of the above theorem. Of the involutes of the geodesics $v=\text{const.}$ we consider the single infinity which meet S_1 in one of the orthogonal trajectories $u=u_0$. We shall prove that the locus of these curves is a surface S, normal to the tangents to the geodesics. Consider the tangents to the geodesics at the points of meeting of the latter with a second orthogonal trajectory $u=u_1$. The segments of these tangents between the points of contact M and the points P of meeting with S are equal to one another, because they are equal to the length of the geodesics between the curves $u=u_0$ and $u=u_1$. Hence, as M moves along an orthogonal trajectory $u=u_1$ of the lines MP, P describes a second orthogonal trajectory of the latter. Moreover, as M moves along a geodesic, P describes an involute which is necessarily orthogonal to MP. Since two directions on S are perpendicular to MP, the latter is normal to S.

EXAMPLES

1. Obtain the results of § 73 concerning parallel surfaces without making use of the moving trihedral.

2. Show that the surfaces parallel to a surface of revolution are surfaces of revolution.

* Vol. I, p. 293.

3. Determine the conjugate systems upon a surface such that the corresponding curves on a parallel surface form a conjugate system.

4. Determine the character of a surface S such that its asymptotic lines correspond to conjugate lines upon a parallel surface, and find the latter surface.

5. Show that when the parametric curves are the lines of curvature of a surface, the characteristics of the yz-plane and xz-plane respectively of the moving trihedral whose x-axis is tangent to the curve $v = \text{const.}$ at the point are given by

$$(r\,du + r_1\,dv)\,y - q\,(z - \rho_1)\,du = 0,$$
$$(r\,du + r_1\,dv)\,x - p_1\,(z - \rho_2)\,dv = 0;$$

and show that these equations give the directions on the surfaces S_1 and S_2 which are conjugate to the direction determined by dv/du.

6. Show that for a canal surface (§ 29) one surface of centers is the curve of centers of the spheres and the other is the polar developable of this curve.

7. The surfaces of center of a helicoid are helicoids of the same axis and parameter as the given surface.

GENERAL EXAMPLES

1. If t is an integrating factor of $\sqrt{E}\,du + \dfrac{F + iH}{\sqrt{E}}\,dv$, and t_0 the conjugate imaginary function, then $\Delta_2 \log \sqrt{tt_0}$ is equal to the total curvature of the quadratic form $E\,du^2 + 2\,F\,dudv + G\,dv^2$, all the functions in the latter being real.

2. Show that the sphere is the only real surface such that its first and second fundamental quadratic forms can be the second and first forms respectively of another surface.

3. Show that there exists a surface referred to its lines of curvature with the linear element $ds^2 = e^{au}(du^2 + dv^2)$, where α is a constant, and that the surface is developable.

4. When a minimal surface is referred to its minimal lines

$$D = \phi\,(u), \qquad D' = 0, \qquad D'' = \psi\,(v);$$

hence the lines of curvature and asymptotic lines can be found by quadratures.

5. Establish the following identities in which the differential parameters are formed with respect to the linear element:

$$\Delta_1(x, X) = -\frac{X_1^2}{\rho_1} - \frac{X_2^2}{\rho_2}, \qquad \Theta\,(x, X) = X_1 X_2\left(\frac{1}{\rho_1} - \frac{1}{\rho_2}\right),$$
$$\Delta_1(x, Y) = -\frac{X_1 Y_1}{\rho_1} - \frac{X_2 Y_2}{\rho_2}, \qquad \Theta\,(x, Y) = \frac{X_2 Y_1}{\rho_1} - \frac{X_1 Y_2}{\rho_2}.$$

6. Prove that (cf. Ex. 2, p. 166)

$$\Delta_2 X = -\frac{X_1}{\sqrt{E}}\frac{\partial}{\partial u}\left(\frac{1}{\rho_1} + \frac{1}{\rho_2}\right) - \frac{X_2}{\sqrt{G}}\frac{\partial}{\partial v}\left(\frac{1}{\rho_1} + \frac{1}{\rho_2}\right) - X\left(\frac{1}{\rho_1^2} + \frac{1}{\rho_2^2}\right),$$
$$\Delta_2 W = -\left(\frac{1}{\rho_1} + \frac{1}{\rho_2}\right) - W\left(\frac{1}{\rho_1^2} + \frac{1}{\rho_2^2}\right) - \frac{W_1}{\sqrt{E}}\frac{\partial}{\partial u}\left(\frac{1}{\rho_1} + \frac{1}{\rho_2}\right) - \frac{W_2}{\sqrt{G}}\frac{\partial}{\partial v}\left(\frac{1}{\rho_1} + \frac{1}{\rho_2}\right).$$

7. Show that $$x^2 + y^2 + z^2 = W'^2 + \Delta_1' W,$$
the differential parameter being formed with respect to (23).

8. A necessary and sufficient condition that all the curves of an orthogonal system on a surface be geodesics is that the surface be developable.

9. If the geodesic curvature of the curves of an orthogonal system is constant (different from zero) all over the surface, the latter is a surface of constant negative curvature.

10. When the linear element of a surface is in the form
$$ds^2 = du^2 + 2\cos\omega\, du dv + dv^2,$$
the parametric curves are said to form an *equidistantial* system. Show that in this case the coördinates of the surface are integrals of the system
$$\frac{\dfrac{\partial^2 x}{\partial u\,\partial v}}{\dfrac{\partial y}{\partial u}\dfrac{\partial z}{\partial v} - \dfrac{\partial z}{\partial u}\dfrac{\partial y}{\partial v}} = \frac{\dfrac{\partial^2 y}{\partial u\,\partial v}}{\dfrac{\partial z}{\partial u}\dfrac{\partial x}{\partial v} - \dfrac{\partial z}{\partial v}\dfrac{\partial x}{\partial u}} = \frac{\dfrac{\partial^2 z}{\partial u\,\partial v}}{\dfrac{\partial x}{\partial u}\dfrac{\partial y}{\partial v} - \dfrac{\partial x}{\partial v}\dfrac{\partial y}{\partial u}}.$$

11. If the curves $v =$ const., $u =$ const. form an equidistantial system, the tangents to the curves $v =$ const. are orthogonal to the lines joining the centers of geodesic curvature of the curves $u =$ const. and of their orthogonal trajectories.

12. Of all the displacements of a trihedral T corresponding to a small displacement of its vertex M over the surface there are two which reduce to rotations; they occur when M describes either of the lines of curvature through the point, and the axes of rotation are situated in the planes perpendicular to the lines of curvature, each axis passing through one of the centers of principal curvature.

13. When a surface is referred to its lines of curvature, the curves defined by
$$q^2 \frac{\partial \rho_1}{\partial u} du^3 + 3\, q^2 \frac{\partial \rho_1}{\partial v} du^2 dv + 3\, p_1^2 \frac{\partial \rho_2}{\partial u} du dv^2 + p_1^2 \frac{\partial \rho_2}{\partial v} dv^3 = 0$$
possess the property that the normal sections in these directions at a point are straight lines, or are superosculated by their circles of curvature (cf. Ex. 9, p. 21; Ex. 6, p. 177). These curves are called the *superosculating lines* of the surface.

14. Show that the superosculating lines on a surface and on a parallel surface correspond.

15. Show that the Gauss equation (64) can be put in the following form due to Liouville:
$$\frac{\sqrt{EG}\sin\omega}{\rho_1\rho_2} = \frac{\partial}{\partial v}\left(\frac{\sqrt{E}}{\rho_{gu}}\right) - \frac{\partial}{\partial u}\left(\frac{\sqrt{G}}{\rho_{gv}}\right) + \frac{\partial^2 \omega}{\partial u\,\partial v},$$
where ρ_{gu} and ρ_{gv} denote the radii of geodesic curvature of the curves $v =$ const. and $u =$ const. respectively.

16. When the parametric curves form an orthogonal system, the equation of Ex. 15 may be written
$$\frac{1}{\rho_1\rho_2} = \frac{1}{\sqrt{G}}\frac{\partial}{\partial v}\left(\frac{1}{\rho_{gu}}\right) - \frac{1}{\sqrt{E}}\frac{\partial}{\partial u}\left(\frac{1}{\rho_{gv}}\right) - \frac{1}{\rho_{gu}^2} - \frac{1}{\rho_{gv}^2}.$$

17. Determine the surfaces which are such that one of them and a parallel divide harmonically the segment between the centers of principal curvature.

18. Determine the surfaces which are such that one of them and a parallel admit of an equivalent representation (cf. Ex. 14, p. 113) with lines of curvature corresponding.

19. Derive the following properties of the surface

$$x = \frac{a^2 - b^2}{ab}\,\frac{uv}{u+v}, \quad y = \frac{\sqrt{a^2 - b^2}}{b}\,\frac{v\sqrt{b^2 - u^2}}{u+v}, \quad z = \frac{\sqrt{a^2 - b^2}}{a}\,\frac{u\sqrt{v^2 - a^2}}{u+v}:$$

(i) the parametric lines are plane lines of curvature;
(ii) the principal radii of curvature are $\rho_1 = v$, $\rho_2 = -u$;
(iii) the surface is algebraic of the fourth order;
(iv) the surfaces of center are focal conics.

20. Given a curve C upon a surface S and the ruled surface formed by the tangents to S which are perpendicular to C at its points M; the point of each generator MN at which the tangent plane to the ruled surface is perpendicular to the tangent plane at M to S is the center of geodesic curvature of C at M; when the ruled surface is developable, this center of geodesic curvature is the point of contact of MN with the edge of regression.

21. If two surfaces have the same spherical representation of their lines of curvature, the locus of the point dividing the join of corresponding points in constant ratio is a surface with the same representation.

22. The locus of the centers of geodesic curvature of a line of curvature is an evolute of the latter.

23. Show that when $E, F, G;\ D, D', D''$ of a surface are functions of a single parameter, the surface is a helicoid, or a surface of revolution.

CHAPTER VI

SYSTEMS OF CURVES. GEODESICS

77. Asymptotic lines. We have said that the asymptotic lines on a surface are the double family of curves whose tangents at any point are determined in direction by the differential equation

$$D\,du^2 + 2\,D'du\,dv + D''dv^2 = 0.$$

These directions are imaginary and distinct at an elliptic point, real and distinct at a hyperbolic point, and real and coincident at a parabolic point. If we exclude the latter points from our discussion, the asymptotic lines may be taken for parametric curves. A necessary and sufficient condition that they be parametric is (§ 55)

$$D = D'' = 0. \tag{1}$$

Then from (IV, 25) we have

$$K = -\frac{D'^2}{H^2} = -\frac{1}{\rho^2}, \tag{2}$$

where ρ as thus defined is called the *radius of total curvature*. The Codazzi equations (V, 13′) may be written

$$\frac{\partial \log \rho}{\partial u} = 2\begin{Bmatrix}12\\2\end{Bmatrix}, \qquad \frac{\partial \log \rho}{\partial v} = 2\begin{Bmatrix}12\\1\end{Bmatrix}, \tag{3}$$

of which the condition of integrability is

$$\frac{\partial}{\partial u}\begin{Bmatrix}12\\1\end{Bmatrix} = \frac{\partial}{\partial v}\begin{Bmatrix}12\\2\end{Bmatrix}. \tag{4}$$

In consequence of (V, 3) this is equivalent to

$$\frac{\partial}{\partial u}\begin{Bmatrix}22\\2\end{Bmatrix} = \frac{\partial}{\partial v}\begin{Bmatrix}11\\1\end{Bmatrix}. \tag{5}$$

In § 64 we saw that K is a function of E, F, G and their derivatives. Hence equations (3) are two conditions upon the coefficients of a quadratic form

$$E\,du^2 + 2\,F\,du\,dv + G\,dv^2; \tag{6}$$

that it may be the linear element of a surface referred to its asymptotic lines. When these conditions are satisfied the function D' is given by (2) to within sign. Hence, if we make no distinction between a surface and its symmetric with respect to a point, from § 65 follows the theorem:

A necessary and sufficient condition that a quadratic form (6) *be the linear element of a surface referred to its asymptotic lines is that its coefficients satisfy equations* (3); *when they are satisfied, the surface is unique.*

For example, suppose that the total curvature of the surface is the same at every point, thus

$$K = -\frac{1}{a^2},$$

where a is a constant. In this case equations (3) are

$$G\frac{\partial E}{\partial v} - F\frac{\partial G}{\partial u} = 0, \qquad F\frac{\partial E}{\partial v} - E\frac{\partial G}{\partial u} = 0,$$

which, since $H^2 \neq 0$, are equivalent to

$$\frac{\partial E}{\partial v} = 0, \qquad \frac{\partial G}{\partial u} = 0.$$

Hence E is a function of u alone, and G a function of v alone. By a suitable choice of the parameters these two functions may be given the value a^2, so that the linear element of the surface can be written

$$ds^2 = a^2(du^2 + 2\cos\omega\, du dv + dv^2), \tag{7}$$

where ω denotes the angle between the asymptotic lines. Thus far the Codazzi equations are satisfied and only the Gauss equation (V, 12) remains to be considered. When the above values are substituted, this becomes

$$\frac{\partial^2\omega}{\partial u\, \partial v} = \sin\omega. \tag{8}$$

Hence to every solution of this equation there corresponds a surface of constant curvature $-\frac{1}{a^2}$ whose linear element is given by (7).

The equation of the lines of curvature is $du^2 - dv^2 = 0$, so that if we put $u + v = 2u_1$, $u - v = 2v_1$, the quantities u_1 and v_1 are parameters of the lines of curvature, and in terms of these the equation of the asymptotic lines is $du_1^2 - dv_1^2 = 0$. Hence, when either the asymptotic lines or the lines of curvature are known upon a surface of constant curvature, the other system can be found by quadratures.

When the asymptotic lines are parametric, the Gauss equations (V, 7) may be written

$$\begin{cases} \dfrac{\partial^2\theta}{\partial u^2} + a\dfrac{\partial\theta}{\partial u} + b\dfrac{\partial\theta}{\partial v} = 0, \\[2ex] \dfrac{\partial^2\theta}{\partial v^2} + a_1\dfrac{\partial\theta}{\partial u} + b_1\dfrac{\partial\theta}{\partial v} = 0, \end{cases} \tag{9}$$

where a, b, a_1, b_1 are determinate functions of u and v, and in consequence of (5)

$$\frac{\partial a}{\partial v}=\frac{\partial b_1}{\partial u}. \tag{10}$$

Conversely, if two such equations admit three real linearly independent integrals $f_1(u, v)$, $f_2(u, v)$, $f_3(u, v)$, the equations

$$x=f_1(u, v), \qquad y=f_2(u, v), \qquad z=f_3(u, v)$$

define a surface on which the parametric curves are the asymptotic lines. For, by the elimination of a, b, a_1, b_1 from the six equations obtained by replacing θ in (9) by x, y, z we get

$$\begin{vmatrix} \frac{\partial^2 x}{\partial u^2} & \frac{\partial^2 y}{\partial u^2} & \frac{\partial^2 z}{\partial u^2} \\ \frac{\partial x}{\partial u} & \frac{\partial y}{\partial u} & \frac{\partial z}{\partial u} \\ \frac{\partial x}{\partial v} & \frac{\partial y}{\partial v} & \frac{\partial z}{\partial v} \end{vmatrix}=0, \qquad \begin{vmatrix} \frac{\partial^2 x}{\partial v^2} & \frac{\partial^2 y}{\partial v^2} & \frac{\partial^2 z}{\partial v^2} \\ \frac{\partial x}{\partial u} & \frac{\partial y}{\partial u} & \frac{\partial z}{\partial u} \\ \frac{\partial x}{\partial v} & \frac{\partial y}{\partial v} & \frac{\partial z}{\partial v} \end{vmatrix}=0,$$

which are equivalent to (1), in consequence of (IV, 2, 5).*

As an example, consider the equations

$$\frac{\partial^2\theta}{\partial u^2}=0, \qquad \frac{\partial^2\theta}{\partial v^2}=0,$$

of which the general integral is $auv+bu+cv+d$, where a, b, c, d are constants. By choosing the fixed axes suitably, the most general form of the equations of the surface may be put in the form

$$x=b_1u+c_1v, \qquad y=b_2u+c_2v, \qquad z=a_3uv+b_3u+c_3v.$$

From these equations it is seen that all the asymptotic lines are straight lines, so that the surface is a quadric. Moreover, by the elimination of u and v from these equations we have an equation of the form $z=ax^2+2hxy+by^2+cx+dy$. Hence the surface is a paraboloid.

78. Spherical representation of asymptotic lines. From (IV, 77) we have that the total curvature of a surface, referred to its asymptotic lines, may be expressed in the form

$$K=-\frac{\mathcal{H}^2}{D'^2}, \tag{11}$$

where $\mathcal{H}^2=\mathcal{E}\mathcal{G}-\mathcal{F}^2$, the linear element of the spherical represen tation being $\quad d\sigma^2=\mathcal{E}du^2+2\mathcal{F}dudv+\mathcal{G}dv^2.$

* Darboux, Vol. I, p. 138. It should be noticed that the above result shows that the condition that equations (9) admit three independent integrals carries with it not only (10) but all other conditions of integrability.

From this result and (2) it follows that *

$$(12) \qquad \rho = \frac{D'}{\mathcal{H}}.$$

Hence the fundamental relations (IV, 74) reduce to

$$(13) \qquad E = \rho^2 \mathcal{E}, \qquad F = -\rho^2 \mathcal{F}, \qquad G = \rho^2 \mathcal{G},$$

and equations (V, 26) may be written

$$(14) \qquad \frac{\partial x}{\partial u} = \frac{\rho}{\mathcal{H}}\left(\mathcal{F}\frac{\partial X}{\partial u} - \mathcal{E}\frac{\partial X}{\partial v}\right), \qquad \frac{\partial x}{\partial v} = \frac{\rho}{\mathcal{H}}\left(-\mathcal{G}\frac{\partial X}{\partial u} + \mathcal{F}\frac{\partial X}{\partial v}\right).$$

Moreover, the Codazzi equations (V, 27) are reducible to

$$(15) \qquad \frac{\partial \log \rho}{\partial u} = -2\begin{Bmatrix}12\\2\end{Bmatrix}', \qquad \frac{\partial \log \rho}{\partial v} = -2\begin{Bmatrix}12\\1\end{Bmatrix}'.$$

Consider now the converse problem:

To determine the condition to be satisfied by a parametric system of lines on the sphere in order that they may serve as the spherical representation of the asymptotic lines on a surface.

First of all, equations (15) must satisfy the condition of integrability. Then ρ is obtainable by a quadrature. The corresponding values of x, y, z found from equations (14) and from similar ones are the coördinates of a surface upon which the asymptotic lines are parametric. For, it follows from (14) that

$$(16) \qquad \sum \frac{\partial x}{\partial u}\frac{\partial X}{\partial u} = 0, \qquad \sum \frac{\partial x}{\partial v}\frac{\partial X}{\partial v} = 0.$$

Furthermore, ρ is determined to within a constant factor; consequently the same is true of x, y, z; therefore the surface is unique to within homothetic transformations. Hence we have the following theorem of Dini:

A necessary and sufficient condition that a double family of curves upon the sphere be the spherical representation of the asymptotic lines upon a surface is that $\mathcal{E}$, $\mathcal{F}$, $\mathcal{G}$ satisfy the equation

$$(17) \qquad \frac{\partial}{\partial u}\begin{Bmatrix}12\\1\end{Bmatrix}' = \frac{\partial}{\partial v}\begin{Bmatrix}12\\2\end{Bmatrix}';$$

the corresponding surfaces are homothetic transforms of one another, and their Cartesian coördinates are found by quadratures.

* The choice $\rho = -D'/\mathcal{H}$ gives the surface symmetric to the one corresponding to (12), as is seen from (14), and hence may be neglected.

When equations (1) obtain, the fundamental equations (V, 28) lead to the identities

$$
(18)\qquad \begin{cases} \begin{Bmatrix}11\\1\end{Bmatrix}=\begin{Bmatrix}11\\1\end{Bmatrix}'-2\begin{Bmatrix}12\\2\end{Bmatrix}', & \begin{Bmatrix}22\\2\end{Bmatrix}=\begin{Bmatrix}22\\2\end{Bmatrix}'-2\begin{Bmatrix}12\\1\end{Bmatrix}', \\ \begin{Bmatrix}12\\1\end{Bmatrix}=-\begin{Bmatrix}12\\1\end{Bmatrix}', & \begin{Bmatrix}12\\2\end{Bmatrix}=-\begin{Bmatrix}12\\2\end{Bmatrix}', \\ \begin{Bmatrix}11\\2\end{Bmatrix}=-\begin{Bmatrix}11\\2\end{Bmatrix}', & \begin{Bmatrix}22\\1\end{Bmatrix}=-\begin{Bmatrix}22\\1\end{Bmatrix}'. \end{cases}
$$

The third and fourth of these equations are consequences also of (3) and (15).

79. Formulas of Lelieuvre. Tangential equations. In consequence of (V, 31) equations (14) may be put in the form

$$
(19)\qquad \frac{\partial x}{\partial u}=\epsilon\rho\left(Z\frac{\partial Y}{\partial u}-Y\frac{\partial Z}{\partial u}\right),\qquad \frac{\partial x}{\partial v}=-\epsilon\rho\left(Z\frac{\partial Y}{\partial v}-Y\frac{\partial Z}{\partial v}\right),
$$

where ϵ is ± 1 according as the curvature of the surface is positive or negative. Hence, if we put

$$
(20)\qquad \nu_1=\sqrt{-\epsilon\rho}X,\qquad \nu_2=\sqrt{-\epsilon\rho}Y,\qquad \nu_3=\sqrt{-\epsilon\rho}Z,
$$

we have the following formulas due to Lelieuvre: *

$$
(21)\qquad \begin{cases} \dfrac{\partial x}{\partial u}=\begin{vmatrix}\nu_2 & \nu_3\\ \dfrac{\partial\nu_2}{\partial u} & \dfrac{\partial\nu_3}{\partial u}\end{vmatrix}, & \dfrac{\partial x}{\partial v}=-\begin{vmatrix}\nu_2 & \nu_3\\ \dfrac{\partial\nu_2}{\partial v} & \dfrac{\partial\nu_3}{\partial v}\end{vmatrix}, \\ \dfrac{\partial y}{\partial u}=\begin{vmatrix}\nu_3 & \nu_1\\ \dfrac{\partial\nu_3}{\partial u} & \dfrac{\partial\nu_1}{\partial u}\end{vmatrix}, & \dfrac{\partial y}{\partial v}=-\begin{vmatrix}\nu_3 & \nu_1\\ \dfrac{\partial\nu_3}{\partial v} & \dfrac{\partial\nu_1}{\partial v}\end{vmatrix}, \\ \dfrac{\partial z}{\partial u}=\begin{vmatrix}\nu_1 & \nu_2\\ \dfrac{\partial\nu_1}{\partial u} & \dfrac{\partial\nu_2}{\partial u}\end{vmatrix}, & \dfrac{\partial z}{\partial v}=-\begin{vmatrix}\nu_1 & \nu_2\\ \dfrac{\partial\nu_1}{\partial v} & \dfrac{\partial\nu_2}{\partial v}\end{vmatrix}. \end{cases}
$$

The conditions of integrability of these equations are

$$
\frac{\dfrac{\partial^2\nu_1}{\partial u\,\partial v}}{\nu_1}=\frac{\dfrac{\partial^2\nu_2}{\partial u\,\partial v}}{\nu_2}=\frac{\dfrac{\partial^2\nu_3}{\partial u\,\partial v}}{\nu_3}.
$$

* *Bulletin des Sciences Mathématiques*, Vol. XII (1888), p. 126.

By means of (V, 22) and (15) we find from (20) that the common ratio of these equations is $\frac{1}{\sqrt{\rho}}\frac{\partial^2\sqrt{\rho}}{\partial u\,\partial v}-\mathcal{F}$. Consequently ν_1, ν_2, ν_3 are solutions of the equation

$$\frac{\partial^2\theta}{\partial u\,\partial v}=\left(\frac{1}{\sqrt{\rho}}\frac{\partial^2\sqrt{\rho}}{\partial u\,\partial v}-\mathcal{F}\right)\theta.$$

Conversely, we have the theorem:

Given three particular integrals ν_1, ν_2, ν_3 of an equation of the form

$$\frac{\partial^2\theta}{\partial u\,\partial v}=\lambda\theta, \tag{22}$$

where λ is any function whatever of u and v; the surface, whose coördinates are given by the corresponding quadratures (21), *has the parametric curves for asymptotic lines, and the total curvature of the surface is measured by*

$$K=-\frac{1}{\rho^2}=-\frac{1}{(\nu_1^2+\nu_2^2+\nu_3^2)^2}. \tag{23}$$

For, from (21), it is readily seen that ν_1, ν_2, ν_3 are proportional to the direction-cosines of the normal to the surface. And if these direction-cosines be given by (20), we are brought to (19), from which we see that the conditions (16) are satisfied.

Take, for example, the simplest case $\frac{\partial^2\theta}{\partial u\,\partial v}=0$, and three solutions

$$\nu_i=\phi_i(u)+\psi_i(v). \qquad (i=1, 2, 3)$$

The coördinates of the surface are

$$x=\psi_2\phi_3-\psi_3\phi_2+\int(\phi_2\phi_3'-\phi_2'\phi_3)\,du-\int(\psi_2\psi_3'-\psi_3\psi_2')\,dv,$$

and similar expressions for y and z. When, in particular, we take

$$\phi_i(u)=a_iu+b_i, \qquad \psi_i(v)=\alpha_iv+\beta_i,$$

the expressions for x, y, z are of the form $auv+bu+cv+d$, and consequently the surface is a paraboloid.

From equations (V, 22, 34) it follows that when the asymptotic lines are parametric, the tangential coördinates X, Y, Z, W are solutions of the equations

$$\left\{\begin{aligned}\frac{\partial^2\theta}{\partial u^2}&=\begin{Bmatrix}11\\1\end{Bmatrix}'\frac{\partial\theta}{\partial u}+\begin{Bmatrix}11\\2\end{Bmatrix}'\frac{\partial\theta}{\partial v}-\mathcal{E}\theta,\\ \frac{\partial^2\theta}{\partial v^2}&=\begin{Bmatrix}22\\1\end{Bmatrix}'\frac{\partial\theta}{\partial u}+\begin{Bmatrix}22\\2\end{Bmatrix}'\frac{\partial\theta}{\partial v}-\mathcal{G}\theta.\end{aligned}\right. \tag{24}$$

EXAMPLES

1. Upon a nondevelopable surface straight lines are the only plane asymptotic lines.

2. The asymptotic lines on a minimal surface form an orthogonal isothermal system, and their spherical images also form such a system.

3. Show that of all the surfaces with the linear element $ds^2 = du^2 + (u^2 + a^2)\,dv^2$, one has the parametric curves for asymptotic lines and another for lines of curvature. Determine these two surfaces.

4. The normals to a ruled surface along a generator are parallel to a plane. Prove conversely, by means of the formulas of Lelieuvre, that if the normals to a surface along the asymptotic lines in one system are parallel to a plane, which differs with the curve, the surface is ruled.

5. If we take $\nu_1 = u$, $\nu_2 = v$, $\nu_3 = \phi(v)$, the formulas of Lelieuvre define the most general right conoid.

6. If the asymptotic lines in one system on a surface be represented on the sphere by great circles, the surface is ruled.

80. Conjugate systems of parametric lines. Inversions. It is our purpose now to consider the case where the parametric lines of a surface form a conjugate system. As thus defined, the characteristics of the tangent plane, as it envelops the surface along a curve $v =$ const., are the tangents to the curves $u =$ const. at their points of intersection with the former curve; and similarly for a plane enveloping along a curve $u =$ const.

The analytical condition that the parametric lines form a conjugate system is (§ 54)

$$D' = 0. \tag{25}$$

It follows immediately from equations (V, 7) that x, y, z are solutions of an equation of the type

$$\frac{\partial^2\theta}{\partial u\,\partial v} + a\frac{\partial\theta}{\partial u} + b\frac{\partial\theta}{\partial v} = 0, \tag{26}$$

where a and b are functions of u and v, or constants. By a method similar to that of § 77 we prove the converse theorem:

If $f_1(u, v)$, $f_2(u, v)$, $f_3(u, v)$ be three linearly independent real solutions of an equation of the type (26), *the equations*

$$x = f_1(u, v), \qquad y = f_2(u, v), \qquad z = f_3(u, v) \tag{27}$$

*define a surface upon which the parametric curves form a conjugate system.**

* Cf. Darboux, Vol. I, p. 122.

We have seen that the lines of curvature form the only orthogonal conjugate system. Hence, in order that the parametric lines on the surface (27) be lines of curvature, we must have

$$F = \frac{\partial x}{\partial u}\frac{\partial x}{\partial v} + \frac{\partial y}{\partial u}\frac{\partial y}{\partial v} + \frac{\partial z}{\partial u}\frac{\partial z}{\partial v} = 0.$$

But this is equivalent to the condition that $x^2 + y^2 + z^2$ also be a solution of equation (26), as is seen by substitution. Hence we have the theorem of Darboux *:

If x, y, z, $x^2 + y^2 + z^2$ are particular solutions of an equation of the form (26), the first three serve for the rectangular coördinates of a surface, upon which the parametric lines are the lines of curvature.

Darboux † has applied this result to the proof of the following theorem:

When a surface is transformed by an inversion into a second surface, the lines of curvature of the former become lines of curvature of the latter.

By definition an inversion, or a transformation by reciprocal radii, is given by

$$(28) \qquad x_1 = \frac{c^2 x}{x^2 + y^2 + z^2}, \qquad y_1 = \frac{c^2 y}{x^2 + y^2 + z^2}, \qquad z_1 = \frac{c^2 z}{x^2 + y^2 + z^2},$$

where c denotes a constant. From these equations we find that

$$(29) \qquad (x^2 + y^2 + z^2)(x_1^2 + y_1^2 + z_1^2) = c^4,$$

and by solving for x, y, z,

$$(30) \qquad x = \frac{c^2 x_1}{x_1^2 + y_1^2 + z_1^2}, \qquad y = \frac{c^2 y_1}{x_1^2 + y_1^2 + z_1^2}, \qquad z = \frac{c^2 z_1}{x_1^2 + y_1^2 + z_1^2}.$$

If, now, the substitution $\theta = \dfrac{\sigma}{x_1^2 + y_1^2 + z_1^2}$

be effected upon equation (26), the resulting equation in σ will admit, in consequence of (29) and (30), the solutions x_1, y_1, z_1, c^4, and therefore is of the form

$$(31) \qquad \frac{\partial^2 \sigma}{\partial u \, \partial v} + \alpha \frac{\partial \sigma}{\partial u} + \beta \frac{\partial \sigma}{\partial v} = 0.$$

* Vol. I, p. 136. † Vol. I, p. 207.

Moreover, equation (26) admits unity for a particular solution, and consequently $x_1^2 + y_1^2 + z_1^2$ is a solution of (31), which proves the theorem.

As an example, we consider a cone of revolution. Its lines of curvature are the elements of the cone and the circular sections. When a transformation by reciprocal radii, whose pole is any point, is applied to the cone, the transform S has two families of circles for its lines of curvature, in consequence of the above theorem and the fact that circles and straight lines, not through the pole, are transformed into circles. Moreover, the cone is the envelope of a family of spheres whose centers lie on its axis, and also of the one-parameter family of tangent planes; the latter pass through the vertex. Since tangency is preserved in this transformation, the surface S is in two ways the envelope of a family of spheres: all the spheres of one family pass through a point, and the centers of the spheres of the other family lie in the plane determined by the axis of the cone and the pole.

81. Surfaces of translation. The simplest form of equation (26) is

$$\frac{\partial^2\theta}{\partial u\,\partial v} = 0,$$

in which case equations (27) are of the type

$$(32) \qquad x = U_1 + V_1, \qquad y = U_2 + V_2, \qquad z = U_3 + V_3,$$

where U_1, U_2, U_3 are any functions whatever of u alone, and V_1, V_2, V_3 any functions of v alone. This surface may be generated by effecting upon the curve

$$x_1 = U_1, \qquad y_1 = U_2, \qquad z_1 = U_3$$

a translation in which each of its points describes a curve congruent with the curve

$$x_2 = V_1, \qquad y_2 = V_2, \qquad z_2 = V_3.$$

In like manner it may be generated by a translation of the second curve in which each of its points describes a curve congruent with the first curve. For this reason the surface is called a *surface of translation*. From this method of generation, as also from equations (32), it follows that the tangents to the curves of one family at their points of intersection with a curve of the second family are parallel to one another. Hence we have the theorem of Lie *:

The developable enveloping a surface of translation along a generating curve is a cylinder.

* *Math. Annalen*, Vol. XIV (1879), pp. 332–387.

Lie has observed that the surface defined by (32) is the locus of the mid-points of the joins of points on the curves

$$x_1 = 2U_1, \qquad y_1 = 2U_2, \qquad z_1 = 2U_3,$$
$$x_2 = 2V_1, \qquad y_2 = 2V_2, \qquad z_2 = 2V_3.$$

It may be that these two sets of equations define the same curve in terms of different parameters. In this case the surface is the locus of the mid-points of all chords of the curve. These results are only a particular case of the following theorem, whose proof is immediate:

The locus of the point which divides in constant ratio the joins of points on two curves, or all the chords of one curve, is a surface of translation; in the latter case the curve is an asymptotic line of the surface.

When the equations of a surface of translation are of the form

$$x = U, \qquad y = V, \qquad z = U_1 + V_1,$$

the generators are plane curves whose planes are perpendicular. We leave it to the reader to show that in this case the asymptotic lines can be found by quadratures.

82. Isothermal-conjugate systems. When the asymptotic lines upon a surface are parametric, the second quadratic form may be written $\lambda\, dudv$. When the surface is real, so also is this quadratic form. Therefore, according as the curvature of the surface is positive or negative, the parameters u and v are conjugate-imaginary or real.

We consider the former case and put

$$u = u_1 + iv_1, \qquad v = u_1 - iv_1,$$

when u_1 and v_1 are real. In terms of these parameters the second quadratic form is $\lambda(du_1^2 + dv_1^2)$. Hence the curves $u_1 =$ const., $v_1 =$ const. form a conjugate system, for which

$$D = D'', \qquad D' = 0. \tag{33}$$

Bianchi * has called a system of this sort *isothermal-conjugate*. Evidently such a system bears to the second quadratic form an analytical relation similar to that of an isothermal-orthogonal system

* Vol. I, p. 167.

to the first quadratic form. In the latter case it was only necessary that $EG - F^2$ be positive, and the analogous requirement, namely $DD'' - D'^2 > 0$, is satisfied by surfaces of positive curvature. Hence all the theorems for isothermal-orthogonal systems (§§ 40, 41) are translated into theorems concerning isothermal-conjugate systems by substituting D, D', D'' for E, F, G respectively in the formulas. In particular, we remark that if the curves $u = \text{const.}$, $v = \text{const.}$ on a surface form an isothermal-conjugate system, all other real isothermal-conjugate systems are given by $u_1 = \text{const.}$, $v_1 = \text{const.}$, the quantities u_1 and v_1 being defined by

$$u_1 + iv_1 = \phi(u \pm iv),$$

where ϕ is any analytic function.

When the curvature of the surface is negative and we put

$$u = u_1 + v_1, \qquad v = u_1 - v_1,$$

in the second quadratic form $\lambda\, du dv$, it becomes $\lambda(du_1^2 - dv_1^2)$. In this case

$$D = -D'', \qquad D' = 0. \tag{34}$$

Hence the curves $u_1 = \text{const.}$ and $v_1 = \text{const.}$ form a conjugate system which may be called *isothermal-conjugate*. With each change of the parameters u and v of the asymptotic lines there is obtained a new isothermal-conjugate system. Hence if u and v are parameters of an isothermal-conjugate system upon a surface of negative curvature, the parameters of all such systems are given by

$$u_1 = \phi(u \pm v) + \psi(u \mp v),$$
$$v_1 = \phi(u \pm v) - \psi(u \mp v),$$

where ϕ and ψ denote arbitrary functions.

It is evident that if the parameters for a surface are such that

$$\frac{D}{D''} = \frac{U}{V}, \qquad D' = 0, \tag{35}$$

where U and V are functions of u and v respectively, then by a change of parameters which does not change the parametric curves we can reduce (35) to one of the forms (33) or (34). Hence equations (35) are a necessary and sufficient condition that the parametric curves form an isothermal-conjugate system. Referring to

§ 77, we see that the lines of curvature upon a surface of constant total curvature form an isothermal-conjugate system.

When equation (35) is of the form (33) or (34), we say that the parameters u and v are isothermal-conjugate.

83. Spherical representation of conjugate systems. When the parametric curves are conjugate, equations (IV, 69) reduce to

$$\mathcal{E}=\frac{GD^2}{H^2}, \qquad \mathcal{F}=-\frac{FDD''}{H^2}, \qquad \mathcal{G}=\frac{ED''^2}{H^2}.$$

From these equations and (III, 15) it follows that the angle ω' between the parametric curves on the sphere is given by

$$\cos\omega'=\frac{\mathcal{F}}{\sqrt{\mathcal{E}\mathcal{G}}}=\mp\frac{F}{\sqrt{EG}}=\mp\cos\omega,$$

where the upper sign corresponds to the case of an elliptic point and the lower to a hyperbolic point. Hence we have the theorem:

The angles between two conjugate directions at a point on a surface, and between the corresponding directions on the sphere, are equal or supplementary, according as the point is hyperbolic or elliptic.

When the parametric curves form a conjugate system, the Codazzi equations (V, 27) reduce to

$$\text{(36)} \qquad \left\{\begin{aligned} \frac{\partial D}{\partial v} &= \begin{Bmatrix}12\\1\end{Bmatrix}' D - \begin{Bmatrix}11\\2\end{Bmatrix}' D'', \\ \frac{\partial D''}{\partial u} &= \begin{Bmatrix}12\\2\end{Bmatrix}' D'' - \begin{Bmatrix}22\\1\end{Bmatrix}' D, \end{aligned}\right.$$

and equations (V, 26) become

$$\text{(37)} \qquad \left\{\begin{aligned} \frac{\partial x}{\partial u} &= \frac{D}{\mathcal{H}^2}\left(-\mathcal{G}\frac{\partial X}{\partial u}+\mathcal{F}\frac{\partial X}{\partial v}\right), \\ \frac{\partial x}{\partial v} &= \frac{D''}{\mathcal{H}^2}\left(\mathcal{F}\frac{\partial X}{\partial u}-\mathcal{E}\frac{\partial X}{\partial v}\right). \end{aligned}\right.$$

Hence, when a system of curves upon the sphere is given, the problem of finding the surfaces with this representation of a conjugate system reduces to the solution of equations (36) and quadratures of the form (37), after X, Y, Z have been determined by the solution of a Riccati equation. By the elimination of D or D'' from equations (36) we obtain a partial differential equation of the second order.

From the general equations (V, 28) we derive the following, when the parametric curves form a conjugate system:*

$$(38)\qquad \left\{\begin{aligned} &\begin{Bmatrix}11\\1\end{Bmatrix}=\frac{\partial\log D}{\partial u}-\begin{Bmatrix}11\\1\end{Bmatrix}', &&\begin{Bmatrix}22\\2\end{Bmatrix}=\frac{\partial\log D''}{\partial v}-\begin{Bmatrix}22\\2\end{Bmatrix}',\\ &\begin{Bmatrix}12\\1\end{Bmatrix}=-\frac{D''}{D}\begin{Bmatrix}11\\2\end{Bmatrix}', &&\begin{Bmatrix}12\\2\end{Bmatrix}=-\frac{D}{D''}\begin{Bmatrix}22\\1\end{Bmatrix}',\\ &\begin{Bmatrix}22\\1\end{Bmatrix}=-\frac{D''}{D}\begin{Bmatrix}12\\2\end{Bmatrix}', &&\begin{Bmatrix}11\\2\end{Bmatrix}=-\frac{D}{D''}\begin{Bmatrix}12\\1\end{Bmatrix}'.\end{aligned}\right.$$

84. Tangential coördinates. Projective transformations. The problem of finding the surfaces with a given representation of a conjugate system is treated more readily from the point of view of tangential coördinates. For, from (V, 22) and (V, 34) it is seen that X, Y, Z, and W are particular solutions of the equation

$$(39)\qquad \frac{\partial^2\theta}{\partial u\,\partial v}-\begin{Bmatrix}12\\1\end{Bmatrix}'\frac{\partial\theta}{\partial u}-\begin{Bmatrix}12\\2\end{Bmatrix}'\frac{\partial\theta}{\partial v}+\mathfrak{F}\theta=0.$$

Hence every solution of this equation linearly independent of X, Y, Z determines a surface with the given representation of a conjugate system, and the calculation of the coördinates x, y, z does not involve quadratures (§ 67).

Conversely, it is readily seen that if the tangential coördinates satisfy an equation of the form

$$\frac{\partial^2\theta}{\partial u\,\partial v}+a\frac{\partial\theta}{\partial u}+b\frac{\partial\theta}{\partial v}+c\theta=0,$$

the coördinate lines form a conjugate system on the surface.

As an example, we determine the surfaces whose lines of curvature are represented on the sphere by a family of curves of constant geodesic curvature and their orthogonal trajectories. If the former family be the curves $v=$ const., and if the linear element on the sphere be written $d\sigma^2=E\,du^2+G\,dv^2$, we must have (IV, 50)

$$\frac{1}{\sqrt{EG}}\frac{\partial\sqrt{E}}{\partial v}=\phi(v),$$

where $\phi(v)$ is a function of v alone. By a change of the parameter v this may be made equal to unity. In this case equation (39) is reducible to

$$\frac{\partial}{\partial u}\left(\frac{\partial\theta}{\partial v}-\sqrt{G}\theta\right)=\frac{\partial\log\sqrt{G}}{\partial u}\left(\frac{\partial\theta}{\partial v}-\sqrt{G}\theta\right).$$

* Cf. Bianchi, Vol. I, p. 167.

The general integral of this equation is

$$\theta = e^{\int_{v_0}^{v}\sqrt{G}\,dv}\left[U + \int_{v_0}^{v} V\sqrt{G}e^{-\int_{v_0}^{v}\sqrt{G}\,dv}dv\right],$$

where v_0 denotes a constant value of v, and U and V are arbitrary functions of u and v respectively. Hence:

The determination of all the surfaces whose lines of curvature are represented on the sphere by a family of curves of constant geodesic curvature and their orthogonal trajectories, requires two quadratures.

In order that among all the surfaces with the same representation of a conjugate system there may be a surface for which the system is isothermal-conjugate, and the parameters be isothermal-conjugate, it is necessary that equations (36) be satisfied by $D'' = \pm D$, according as the total curvature is positive or negative. In this case equations (36) are

$$\frac{\partial \log D}{\partial v} = \begin{Bmatrix} 12 \\ 1 \end{Bmatrix}' \mp \begin{Bmatrix} 11 \\ 2 \end{Bmatrix}', \qquad \frac{\partial \log D}{\partial u} = \begin{Bmatrix} 12 \\ 2 \end{Bmatrix}' \mp \begin{Bmatrix} 22 \\ 1 \end{Bmatrix}'.$$

The condition of integrability is

$$\frac{\partial}{\partial u}\left[\begin{Bmatrix} 12 \\ 1 \end{Bmatrix}' \mp \begin{Bmatrix} 11 \\ 2 \end{Bmatrix}'\right] = \frac{\partial}{\partial v}\left[\begin{Bmatrix} 12 \\ 2 \end{Bmatrix}' \mp \begin{Bmatrix} 22 \\ 1 \end{Bmatrix}'\right]. \tag{40}$$

When this is satisfied D may be found by quadratures, and then the coördinates, by (37). Hence we have the theorem:

A necessary and sufficient condition that a family of curves upon the sphere represent an isothermal-conjugate system on a surface, and that u and v be isothermal-conjugate parameters, is that $\mathcal{E}$, $\mathcal{F}$, $\mathcal{G}$ satisfy (40)*; then the surface is unique to within its homothetics, and its coördinates are given by quadratures.*

The following theorem concerning the invariance of conjugate directions and asymptotic lines is due to Darboux:

When a surface is subjected to a projective transformation or a transformation by reciprocal polars, conjugate directions and asymptotic lines are preserved.

We prove this theorem geometrically. Consider a curve C on a surface S and the developable D circumscribing the surface along C. When a projective transformation is effected upon S we obtain a surface S_1, corresponding point with point to S, and C goes into a curve C_1 upon S_1, and D into a developable D_1 circumscribing S_1 along

C_1; moreover, the tangents to C and C_1 correspond, as do the generators of D and D_1. Since the generators are in each case tangent to the curves conjugate to C and C_1 respectively, the theorem is proved.

In the case of a polar reciprocal transformation a plane corresponds to a point and vice versa, in such a way that a plane and a point of it go into a point and a plane through it. Hence S goes into S_1, C into D_1, D into C_1, and the tangents to C and generators of D into the generators of D_1 and tangents to C_1. Hence the theorem is proved.

EXAMPLES

1. Show that the parametric curves on the surface

$$x = \frac{U_1 + V_1}{U + V}, \qquad y = \frac{U_2 + V_2}{U + V}, \qquad z = \frac{U_3 + V_3}{U + V},$$

where the U's are functions of u alone and the V's of v alone, form a conjugate system.

2. On the surface $x = U_1 V_1$, $y = U_2 V_1$, $z = V_2$, where U_1, U_2 are functions of u alone and V_1, V_2 of v alone, the parametric curves form a conjugate system and the asymptotic lines can be found by quadratures.

3. The generators of a surface of translation form an equidistantial system (cf. Ex. 10, p. 187).

4. Show that a paraboloid is a surface of translation in more than one way.

5. The locus of the mid-points of the chords of a circular helix is a right helicoid.

6. Discuss the surface of translation which is the locus of points dividing in constant ratio the chords of a twisted cubic.

7. From (28) it follows that

$$dx_1^2 + dy_1^2 + dz_1^2 = \frac{c^4 (dx^2 + dy^2 + dz^2)}{(x^2 + y^2 + z^2)^2};$$

consequently the transformation by reciprocal radii is conformal.

8. Determine the condition to be satisfied by the function ω so that a surface with the linear element
$$ds^2 = a^2 (\cos^2\omega \, du^2 + \sin^2\omega \, dv^2)$$

shall have the total curvature $-1/a^2$. Show that if the parametric curves are the lines of curvature, they form an isothermal-conjugate system.

9. A necessary and sufficient condition that the linear element of a surface referred to a conjugate system can be written

$$ds^2 = \rho^2 (\mathcal{E} du^2 - 2 \mathcal{F} dudv + \mathcal{G} dv^2),$$

is that the parametric curves be the characteristic lines. Find the condition imposed upon the curves on the unit sphere in order that they may represent these lines.

10. Conjugate systems and asymptotic lines are transformed into curves of the same sort when a surface is transformed by the general projective transformation

$$x = \frac{A}{D}, \qquad y = \frac{B}{D}, \qquad z = \frac{C}{D},$$

where A, B, C, D are linear functions of the new coördinates x_1, y_1, z_1.

85. Equations of geodesic lines. We have defined a geodesic to be a curve whose geodesic curvature is zero at every point; consequently its osculating plane at any point is perpendicular to the tangent plane to the surface.

From (IV, 49) it follows that every geodesic upon a surface is an integral curve of the differential equation

$$\begin{aligned}(41)\quad &\left(E\frac{du}{ds}+F\frac{dv}{ds}\right)\left(F\frac{d^2u}{ds^2}+G\frac{d^2v}{ds^2}\right)-\left(F\frac{du}{ds}+G\frac{dv}{ds}\right)\left(E\frac{d^2u}{ds^2}+F\frac{d^2v}{ds^2}\right)\\ &+\left(E\frac{du}{ds}+F\frac{dv}{ds}\right)\left[\left(\frac{\partial F}{\partial u}-\frac{1}{2}\frac{\partial E}{\partial v}\right)\left(\frac{du}{ds}\right)^2+\frac{\partial G}{\partial u}\frac{du}{ds}\frac{dv}{ds}+\frac{1}{2}\frac{\partial G}{\partial v}\left(\frac{dv}{ds}\right)^2\right]\\ &-\left(F\frac{du}{ds}+G\frac{dv}{ds}\right)\left[\frac{1}{2}\frac{\partial E}{\partial u}\left(\frac{du}{ds}\right)^2+\frac{\partial E}{\partial v}\frac{du}{ds}\frac{dv}{ds}+\left(\frac{\partial F}{\partial v}-\frac{1}{2}\frac{\partial G}{\partial u}\right)\left(\frac{dv}{ds}\right)^2\right]=0.\end{aligned}$$

If the fundamental identity

$$E\left(\frac{du}{ds}\right)^2+2F\frac{du}{ds}\frac{dv}{ds}+G\left(\frac{dv}{ds}\right)^2=1,$$

which gives the relation between u, v, s along the curve, be differentiated with respect to s, we have

$$\begin{aligned}&2\frac{du}{ds}\left(E\frac{d^2u}{ds^2}+F\frac{d^2v}{ds^2}\right)+2\frac{dv}{ds}\left(F\frac{d^2u}{ds^2}+G\frac{d^2v}{ds^2}\right)+\frac{\partial E}{\partial u}\left(\frac{du}{ds}\right)^3\\ &+\left(\frac{\partial E}{\partial v}+2\frac{\partial F}{\partial u}\right)\left(\frac{du}{ds}\right)^2\frac{dv}{ds}+\left(2\frac{\partial F}{\partial v}+\frac{\partial G}{\partial u}\right)\frac{du}{ds}\left(\frac{dv}{ds}\right)^2+\frac{\partial G}{\partial v}\left(\frac{dv}{ds}\right)^3=0.\end{aligned}$$

If this equation and (41) be solved with respect to $\left(E\frac{d^2u}{ds^2}+F\frac{d^2v}{ds^2}\right)$ and $\left(F\frac{d^2u}{ds^2}+G\frac{d^2v}{ds^2}\right)$, we obtain

$$\begin{aligned}&E\frac{d^2u}{ds^2}+F\frac{d^2v}{ds^2}+\frac{1}{2}\frac{\partial E}{\partial u}\left(\frac{du}{ds}\right)^2+\frac{\partial E}{\partial v}\frac{du}{ds}\frac{dv}{ds}+\left(\frac{\partial F}{\partial v}-\frac{1}{2}\frac{\partial G}{\partial u}\right)\left(\frac{dv}{ds}\right)^2=0,\\ &F\frac{d^2u}{ds^2}+G\frac{d^2v}{ds^2}+\left(\frac{\partial F}{\partial u}-\frac{1}{2}\frac{\partial E}{\partial v}\right)\left(\frac{du}{ds}\right)^2+\frac{\partial G}{\partial u}\frac{du}{ds}\frac{dv}{ds}+\frac{1}{2}\frac{\partial G}{\partial v}\left(\frac{dv}{ds}\right)^2=0.\end{aligned}$$

If these equations be solved with respect to $\frac{d^2u}{ds^2}$ and $\frac{d^2v}{ds^2}$, we have, in consequence of (V, 2),

$$(42)\quad \left\{\begin{aligned}&\frac{d^2u}{ds^2}+\begin{Bmatrix}11\\1\end{Bmatrix}\left(\frac{du}{ds}\right)^2+2\begin{Bmatrix}12\\1\end{Bmatrix}\frac{du}{ds}\frac{dv}{ds}+\begin{Bmatrix}22\\1\end{Bmatrix}\left(\frac{dv}{ds}\right)^2=0,\\ &\frac{d^2v}{ds^2}+\begin{Bmatrix}11\\2\end{Bmatrix}\left(\frac{du}{ds}\right)^2+2\begin{Bmatrix}12\\2\end{Bmatrix}\frac{du}{ds}\frac{dv}{ds}+\begin{Bmatrix}22\\2\end{Bmatrix}\left(\frac{dv}{ds}\right)^2=0.\end{aligned}\right.$$

Every pair of solutions of these equations of the form $u = f_1(s)$, $v = f_2(s)$, determines a geodesic on the surface, and s is its arc.

But a geodesic may be defined in terms of u and v alone, without the introduction of the parameter s. If $v = \phi(u)$ defines such a curve, then

$$\frac{dv}{ds} = \phi' \frac{du}{ds}, \qquad \frac{d^2v}{ds^2} = \phi''\left(\frac{du}{ds}\right)^2 + \phi' \frac{d^2u}{ds^2}.$$

Substituting these expressions in (42) and eliminating $\frac{d^2u}{ds^2}$, we have, to within the factor $(du/ds)^2$,

$$\text{(43)} \qquad \phi'' - \begin{Bmatrix} 22 \\ 1 \end{Bmatrix} \phi'^3 + \left(\begin{Bmatrix} 22 \\ 2 \end{Bmatrix} - 2\begin{Bmatrix} 12 \\ 1 \end{Bmatrix}\right)\phi'^2 + \left(2\begin{Bmatrix} 12 \\ 2 \end{Bmatrix} - \begin{Bmatrix} 11 \\ 1 \end{Bmatrix}\right)\phi' + \begin{Bmatrix} 11 \\ 2 \end{Bmatrix} = 0.$$

From (42) it follows that when du/ds is zero,

$$\text{(44)} \qquad \begin{Bmatrix} 22 \\ 1 \end{Bmatrix} = \frac{1}{2H^2}\left(2G\frac{\partial F}{\partial v} - F\frac{\partial G}{\partial v} - G\frac{\partial G}{\partial u}\right) = 0.$$

Hence, when this condition is not satisfied, equation (43) defines the geodesics on a surface; and when it is satisfied, equations (43) and $u = \text{const.}$ define them.

From the theory of differential equations it follows that there exists a unique integral of (43) which takes a given value for $u = u_0$, and whose first derivative takes a given value for $u = u_0$. Hence we have the fundamental theorem:

Through every point on a surface there passes a unique geodesic with a given direction.

As an example, we consider the geodesics on a surface of revolution. We have found (§ 46) that the linear element of such a surface referred to its meridians and parallels is of the form

$$\text{(45)} \qquad ds^2 = (1 + \phi'^2)\,du^2 + u^2dv^2,$$

where $z = \phi(u)$ is the equation of the meridian curve. If we put

$$\text{(46)} \qquad u_1 = \int \sqrt{1 + \phi'^2}\,du,$$

and indicate the inverse of this equation by $u = \psi(u_1)$, we have

$$\text{(47)} \qquad ds^2 = du_1^2 + \psi^2 dv^2,$$

and the meridians and parallels are still the parametric curves. For this case equations (42) are

$$\text{(48)} \qquad \frac{d^2u_1}{ds^2} - \psi\psi'\left(\frac{dv}{ds}\right)^2 = 0. \qquad \frac{d^2v}{ds^2} + \frac{2\psi'}{\psi}\frac{du_1}{ds}\frac{dv}{ds} = 0.$$

The first integral of the second is

$$\psi^2 \frac{dv}{ds} = c,$$

where c is a constant. Eliminating ds from this equation and (47), and integrating, we have

$$c\int \frac{du_1}{\psi\sqrt{\psi^2 - c^2}} = \pm v + c_1, \tag{49}$$

where c_1 is a constant. The meridians $v = \text{const.}$ correspond to the case $c = 0$. Hence we have the theorem:

The geodesics upon a surface of revolution referred to its meridians and parallels can be found by quadratures.

It should be remarked that equation (49) defines the geodesics upon any surface applicable to a surface of revolution.

86. Geodesic parallels. Geodesic parameters. From (43) it follows that a necessary and sufficient condition that the curves $v = \text{const.}$ on a surface be geodesics is that

$$\begin{Bmatrix} 11 \\ 2 \end{Bmatrix} = \frac{1}{2H^2}\left(2E\frac{\partial F}{\partial u} - F\frac{\partial E}{\partial u} - E\frac{\partial E}{\partial v}\right) = 0. \tag{50}$$

If the parametric system be orthogonal, this condition makes it necessary that E be a function of u alone, say $E = U^2$. By replacing $\int U\,du$ by u we do not change the parametric lines, and E becomes equal to unity. And the linear element has the form

$$ds^2 = du^2 + G\,dv^2, \tag{51}$$

where in general G is a function of both u and v. From this it follows that the length of the segment of a curve $v = \text{const.}$ between the curves $u = u_0$ and $u = u_1$ is given by

$$\int_{u_0}^{u_1} ds_u = \int_{u_0}^{u_1} du = u_1 - u_0.$$

Since this length is independent of v, it follows that the segments of all the geodesics $v = \text{const.}$ included between any two orthogonal trajectories are of equal length. In consequence of the fundamental theorem, we have that there is a unique family of geodesics which are the orthogonal trajectories of a given curve C. The above results enable us to state the following theorem of Gauss *:

If geodesics be drawn orthogonal to a curve C, and equal lengths be measured upon them from C, the locus of their ends is an orthogonal trajectory of the geodesics.

* *L.c.*, p. 25.

This gives us a means of finding all the orthogonal trajectories of a family of geodesics, when one of them is known. And it suggests the name *geodesic parallels* for these trajectories. Referring to § 37, we see that these are the curves there called parallels, and so the theorem of § 37 may be stated thus:

A necessary and sufficient condition that the curves $\phi = const.$ be geodesic parallels is that

$$\Delta_1\phi = f(\phi), \tag{52}$$

where the differential parameter is formed with respect to the linear element of the surface, and f denotes any function. In order that ϕ be the length of the geodesic curves measured from the curve $\phi = 0$, it is necessary and sufficient that

$$\Delta_1\phi = 1. \tag{53}$$

Moreover, we have seen that when a function ϕ satisfies (52), a new function satisfying (53) can be found by quadrature. When this function is taken as u, the linear element has the form (51). In this case we shall call u and v *geodesic parameters.*

87. Geodesic polar coördinates. The following theorem, due to Gauss,* suggests an important system of geodesic parameters:

If equal lengths be laid off from a point P on the geodesics through P, the locus of the end points is an orthogonal trajectory of the geodesics.

In proving the theorem we take the geodesics for the curves $v = \text{const.}$, and let u denote distances measured along these geodesics from P. The points of a curve $u = \text{const.}$ are consequently at the same *geodesic distance* from P, and so we call them *geodesic circles.* It is our problem to show that this parametric system is orthogonal.

From the choice of u we know that $E = 1$, and hence from (50) it follows that F is independent of u. At P, that is for $u = 0$, the derivatives $\frac{\partial x}{\partial v}, \frac{\partial y}{\partial v}, \frac{\partial z}{\partial v}$ are zero. Consequently F and G are zero for $u = 0$, and the former, being independent of u, is always zero. Hence the theorem is proved.

We consider such a system and two points $M_0(u, 0)$, $M_1(u, v_1)$ on the geodesic circle of radius u. The length of the arc M_0M_1

* *L.c.*, p. 24.

is given by $\int_0^{v_1}\sqrt{G}\,dv$. As u approaches zero the ratio M_1M_0/u approaches the angle between the tangents at P to the geodesics $v=0$ and $v=v_1$. If θ denotes this angle, we have

$$\theta = \lim_{u=0}\frac{\int_0^{v_1}\sqrt{G}\,dv}{u} = \int_0^{v_1}\left[\frac{\partial\sqrt{G}}{\partial u}\right]_{u=0} dv.$$

In order that v be θ, it is necessary and sufficient that $\left[\frac{\partial\sqrt{G}}{\partial u}\right]_{u=0}=1$. These particular geodesic coördinates are similar to polar coördinates in the plane, and for this reason are called *geodesic polar coördinates*. The above results may now be stated thus:

The necessary and sufficient conditions that a system of geodesic coördinates be polar are

$$\left[\sqrt{G}\right]_{u=0}=0, \qquad \left[\frac{\partial\sqrt{G}}{\partial u}\right]_{u=0}=1. \tag{54}$$

It should be noticed, however, that it may be necessary to limit the part of the surface under consideration in order that there be a one-to-one correspondence between a point and a pair of coördinates. For, it may happen that two geodesics starting from P meet again, in which case the second point of meeting would be defined by two sets of coördinates.* For example, the helices are geodesics on a cylinder (§ 12), and it is evident that any number of them can be made to pass through two points at a finite distance from one another by varying the angle under which they cut the elements of the cylinder. Hence, in using a system of geodesic polar coördinates with pole at P, we consider the portion of the surface inclosed by a geodesic circle of radius r, where r is such that no two geodesics through P meet within the circle.†

When the linear element is in the form (51), the equation of Gauss (V, 12) reduces to

$$\frac{1}{\sqrt{G}}\frac{\partial^2\sqrt{G}}{\partial u^2}=-K. \tag{55}$$

If K_0 denotes the total curvature of the surface at the pole P, which by hypothesis is not a singular point, from (54) and (55) it follows that

$$\left[\frac{\partial^2\sqrt{G}}{\partial u^2}\right]_{u=0}=0, \qquad \left[\frac{\partial^3\sqrt{G}}{\partial u^3}\right]_{u=0}=-K_0.$$

* Notice that the pole is a singular point for such a system, because $H^2=0$ for $u=0$.

† Darboux (Vol. II, p. 408) shows that such a function r exists; this is suggested also by § 94.

Therefore, for sufficiently small values of u, we have

$$\sqrt{G} = u - \frac{K_0 u^3}{6} + \cdots.$$

Hence the circumference and area of a geodesic circle of radius u have the values *

$$C = \int_0^{2\pi} \sqrt{G}\, dv = 2\pi u - \frac{\pi K_0 u^3}{3} + \epsilon_1,$$

$$A = \int_0^u \int_0^{2\pi} \sqrt{G}\, du dv = \pi u^2 - \frac{\pi K_0 u^4}{12} + \epsilon_2,$$

where ϵ_1 and ϵ_2 denote terms of orders higher than the third and fourth respectively.

EXAMPLES

1. Find the geodesics of an ellipsoid of revolution.

2. The equations $x = u$, $y = v$ define a representation of a surface with the linear element $ds^2 = v(du^2 + dv^2)$ upon the xy-plane in such a way that geodesics on the former are represented by parabolas on the latter.

3. Find the total curvature of a surface with the linear element

$$ds^2 = R^2 \frac{(a^2 - v^2)\,du^2 + 2\,uv\,dudv + (a^2 - u^2)\,dv^2}{(a^2 - u^2 - v^2)^2},$$

where R and a are constants and integrate the equation of geodesics for the surface.

4. A twisted curve is a geodesic on its rectifying developable.

5. The evolutes of a twisted curve are geodesics on its polar developable.

6. Along a geodesic on a surface of revolution the product of the radius of the parallel through a point and the sine of the angle of inclination of the geodesic with the meridian is constant.

7. Upon a surface of revolution a curve cannot be a geodesic and loxodromic at the same time unless the surface be cylindrical.

8. Upon a helicoid the orthogonal trajectories of the helices are geodesics and the other geodesics can be found by quadratures.

9. If a family of geodesics and their orthogonal trajectories on a surface form an isothermal system, the surface is applicable to a surface of revolution.

10. The radius of curvature of a geodesic on a cone of revolution at a point P varies as the cube of the distance of P from the vertex.

88. Area of a geodesic triangle. With the aid of geodesic polar coördinates Gauss proved the following important theorem †:

The excess over 180° of the sum of the angles of a triangle formed by geodesics on a surface of positive curvature, or the deficit from 180°

* Bertrand, *Journal de Mathématiques*, Ser. 1, Vol. XIII (1848), pp. 80–86. † *L.c.*, p. 30.

of the sum of the angles of such a triangle on a surface of negative curvature, is measured by the area of the part of the sphere which represents that triangle.

In the proof of this theorem Gauss made use of the equation of geodesic lines in the form

$$(56)\quad d\theta+\frac{1}{H}\left(\frac{\partial F}{\partial u}-\frac{F}{2E}\frac{\partial E}{\partial u}-\frac{1}{2}\frac{\partial E}{\partial v}\right)du+\frac{1}{2H}\left(\frac{\partial G}{\partial u}-\frac{F}{E}\frac{\partial E}{\partial v}\right)dv=0,$$

where θ denotes the angle which the tangent to a geodesic at a point makes with the curve $v=$ const. through the point. This equation is an immediate consequence of formula (V, 81). When the parametric system is polar geodesic, this becomes

$$(57)\qquad d\theta=-\frac{\partial\sqrt{G}}{\partial u}dv.$$

Let ABC be a triangle whose sides are geodesics, and let α, β, γ denote the included angles. From (IV, 73) it follows that the inclosed area on the sphere is given by

$$(58)\qquad \mathcal{A}=\iint \mathcal{H}\,dudv=\epsilon\iint KH\,dudv,$$

where ϵ is ± 1 according as the curvature is positive or negative, and the double integrals are taken over the respective areas.

Let A be the pole of a polar geodesic system and AB the curve $v=0$. From (55) and (58) we have

$$\mathcal{A}=-\epsilon\int_0^{\alpha}\int_0^{u}\frac{\partial^2\sqrt{G}}{\partial u^2}dvdu.$$

In consequence of (54) we have, upon integration with respect to u,

$$\mathcal{A}=\epsilon\int_0^{\alpha}\left(1-\frac{\partial\sqrt{G}}{\partial u}\right)dv,$$

which, by (57), is equivalent to

$$\mathcal{A}=\epsilon\int_0^{\alpha}dv+\epsilon\int_{\pi-\beta}^{\gamma}d\theta.$$

For, at B the geodesic BC makes the angle $\pi-\beta$ with the curve $v=0$, and at C it makes the angle γ with the curve $v=\alpha$. Hence we have

$$\mathcal{A}=\epsilon(\alpha+\beta+\gamma-\pi),$$

which proves the theorem.

Because of the form of the second part of (58) $\mathcal{A}$ may be said to measure the total curvature of the geodesic triangle, so that the above theorem may also be stated thus:

The total curvature of a geodesic triangle is equal to the excess over 180°, or deficit from 180°, of the sum of the angles of the triangle, according as the curvature is positive or negative.

The extension of these theorems to the case of geodesic polygons is straightforward.

In the preceding discussion it has been tacitly assumed that all the points of the triangle ABC can be uniquely defined by polar coördinates with pole at A. We shall show that this theorem is true, even if this assumption is not made.

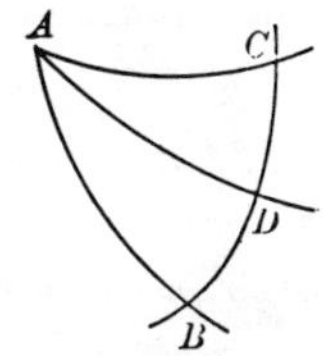

FIG. 18

If the theorem is not true for ABC, it cannot be true for both of the triangles ABD and ACD obtained by joining A and the middle point of BC with a geodesic AD (fig. 18). For, by adding the results for the two triangles, we should have the theorem holding for ABC. Suppose that it is not true for ABD. Divide the latter into two triangles and apply the same reasoning. By continuing this process we should obtain a triangle as small as we please, inside of which a polar geodesic system would not uniquely determine each point. But a domain can be chosen about a point so that a unique geodesic passes through the given point and any other point of the domain.* Consequently the above theorem is perfectly general.

By means of the above result we prove the theorem:

Two geodesics on a surface of negative curvature cannot meet in two points and inclose a simply connected area.

Suppose that two geodesics through a point A pass through a second point B, the two geodesics inclosing a simply connected portion of the surface (fig. 19). Take any geodesic cutting these two segments AB in points C and D. Since the four angles ACD, ADC, BCD, BDC are together equal to four right angles, the sum of the angles of the two triangles ADC and BDC exceed four right angles by the sum of the angles at A and B. Therefore, in consequence of the above theorem of Gauss, the total curvature of the surface cannot be negative at all points of the area $ADBC$.

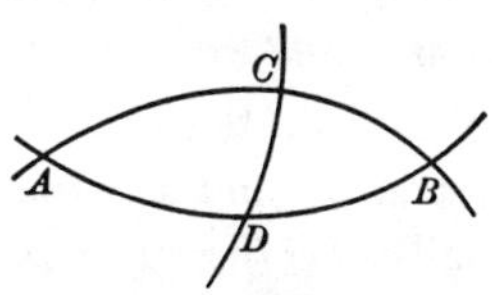

FIG. 19

On the contrary, it can be shown that for a surface of positive curvature geodesics through a point meet again in general. In

* Darboux, Vol. II, p. 408; cf. § 94.

fact, the exceptional points, if there are any, lie in a finite portion of the surface, which may consist of one or more simply connected parts.* For example, the geodesics on a sphere are great circles, and all of these through a point pass through the diametrically opposite point. Again, the helices are geodesics on a cylinder (§ 12), and it is evident that any number of them can be made to pass through two points at a finite distance from one another by varying the angle under which they cut the elements of the cylinder. Hence the domain of a system of polar geodesic coördinates is restricted on a surface of positive or zero curvature.

89. Lines of shortest length. Geodesic curvature. We are now in a position to prove the theorem:

If two points on a surface are such that only one geodesic passes through them, the segment of the geodesic measures the shortest distance on the surface between the two points.

Take one of the points for the pole of a polar geodesic system and the geodesic for the curve $v=0$. The coördinates of the second point are $(u_1, 0)$. The parametric equation of any other curve through the two points is of the form $v=\phi(u)$, and the length of its arc is

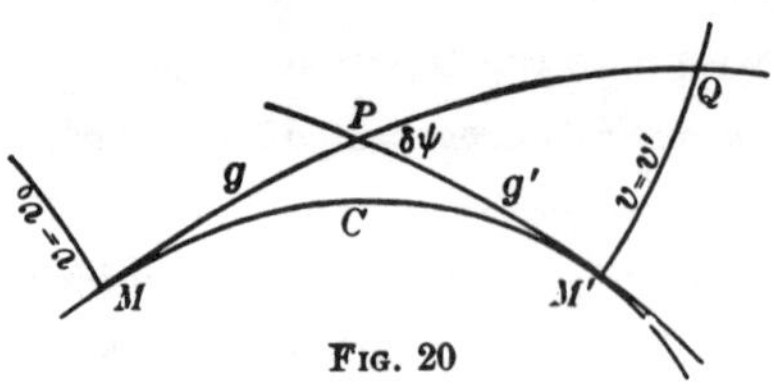

FIG. 20

$$\int_0^{u_1} \sqrt{1+G\phi'^2}\,du.$$

Since $G>0$, the value of this integral is necessarily greater than u_1, and the theorem is proved.

By means of equation (57) we derive another definition of geodesic curvature. Consider two points M and M' upon a curve C, and the unique geodesics g, g' tangent to C at these points (fig. 20). Let P denote the point of intersection of g and g', and $\delta\psi$ the angle under which they cut. Liouville † has called $\delta\psi$ *the angle of geodesic contingence*, because of its analogy to the ordinary angle of contingence. Now we shall prove the theorem:

The limit of the ratio $\delta\psi/\delta s$, as M' approaches M, is the geodesic curvature of C at M.

* For a proof of this the reader is referred to a memoir by H. v. Mangoldt, in *Crelle*, Vol. XCI (1881), pp. 23–53.

† *Journal de Mathématiques*, Vol. XVI (1851), p. 132.

In the proof of this theorem we take for parametric curves the given curve C, its geodesic parallels and their geodesic orthogonals, the parameter u being the distance measured along the latter from C. Since the geodesic g meets the curve $v = v_0$ orthogonally, the angle under which it meets $v = v'$ may be denoted by $\pi/2 + \delta\theta$. As M' approaches M, $\delta\theta$ approaches $d\theta$ given by (57), and the sum of the angles of the triangle $M'PQ$ approaches 180°. Hence $\delta\psi$ approaches $-d\theta$, so that we have

$$\lim \frac{\delta\psi}{\delta s} = -\frac{d\theta}{ds} = \frac{1}{\sqrt{G}}\frac{\partial\sqrt{G}}{\partial u},$$

which is the expression for the geodesic curvature of the curve C.

90. Geodesic ellipses and hyperbolas. An important system of parametric lines for a surface is formed by two families of geodesic parallels. Such a system may be obtained by constructing the geodesic parallels of two curves C_1 and C_2, which are not themselves geodesic parallels of one another, or by taking the two families of geodesic circles with centers at any two points F_1 and F_2. Let u and v measure the geodesic distances from C_1 and C_2, or from F_1 and F_2. They must be solutions of (53). Consequently, in terms of them, we must have

$$\frac{E}{EG - F^2} = \frac{G}{EG - F^2} = 1.$$

If, as usual, ω denotes the angle between these parametric lines, we have, from (III, 15, 16),

$$E = G = \frac{1}{\sin^2\omega}, \qquad F = \frac{\cos\omega}{\sin^2\omega},$$

so that the linear element has the following form, due to Weingarten:

$$ds^2 = \frac{du^2 + 2\cos\omega\, du dv + dv^2}{\sin^2\omega}. \tag{59}$$

Conversely, when the linear element is reducible to this form, u and v are solutions of (53), and consequently the parametric curves are geodesic parallels.

In terms of the parameters u_1 and v_1, defined by $u = u_1 + v_1$ and $v = u_1 - v_1$, the linear element (59) has the form

$$ds^2 = \frac{du_1^2}{\sin^2\frac{\omega}{2}} + \frac{dv_1^2}{\cos^2\frac{\omega}{2}}. \tag{60}$$

The geometrical significance of the curves of parameter u_1 and v_1 is seen when the above equations are written

$$(61)\qquad u_1=\tfrac{1}{2}(u+v),\qquad v_1=\tfrac{1}{2}(u-v).$$

The curves $u_1=$ const. and $v_1=$ const. are respectively the loci of points the sum and difference of whose geodesic distances from C_1 and C_2, or from F_1 and F_2, are constant. In the latter case these curves are analogous to ellipses and hyperbolas in the plane, the points F_1 and F_2 corresponding to the foci. For this reason they are called *geodesic ellipses and hyperbolas*, which names are given likewise to the curves $u_1=$ const., $v_1=$ const., when the distances are measured from two curves, C_1 and C_2. From (60) follows at once the theorem of Weingarten:*

A system of geodesic ellipses and hyperbolas is orthogonal.

By means of (61) equation (60) can be transformed into (59), thus proving that when the linear element of a surface is in the form (60), the parametric curves are geodesic ellipses and hyperbolas.

If θ denotes the angle which the tangent to the curve $v_1=$ const. through a point makes with the curve $v=$ const., it follows from (III, 23) that

$$\cos\theta=\cos\frac{\omega}{2},\qquad \sin\theta=\sin\frac{\omega}{2}.$$

Hence we have the theorem:

Given any two systems of geodesic parallels upon a surface; the corresponding geodesic ellipses and hyperbolas bisect the angles included by the former.

91. Surfaces of Liouville. Dini † inquired whether there were any surfaces with an isothermal system of geodesic ellipses and hyperbolas. A necessary and sufficient condition that such a surface exist is that the coefficients of (60) satisfy a condition of the form (§ 41)

$$V_1\sin^2\frac{\omega}{2}=U_1\cos^2\frac{\omega}{2},$$

where U_1 and V_1 denote functions of u_1 and v_1 respectively. In this case the linear element may be written

$$(62)\qquad ds^2=(U_1+V_1)\left(\frac{du_1^2}{U_1}+\frac{dv_1^2}{V_1}\right).$$

* Ueber die Oberflächen für welche einer der beiden Hauptkrümmungshalbmesser eine Function des anderen ist, *Crelle*, Vol. LXII (1863), pp. 160–173.

† *Annali*, Ser. 2, Vol. III (1869), pp. 269–293.

By the change of parameters defined by

$$u_2 = \int \frac{du_1}{\sqrt{U_1}}, \qquad v_2 = \int \frac{dv_1}{\sqrt{V_1}},$$

this linear element is transformed into

$$ds^2 = (U_2 + V_2)(du_2^2 + dv_2^2), \tag{63}$$

where U_2 and V_2 are functions of u_2 and v_2 respectively, such that

$$U_1(u_1) = U_2(u_2), \qquad V_1(v_1) = V_2(v_2).$$

Conversely, if the linear element is in the form (63), it may be changed into (62) by the transformation of coördinates

$$u_1 = \int \sqrt{U_2}\, du_2, \qquad v_1 = \int \sqrt{V_2}\, dv_2.$$

Surfaces whose linear element is reducible to the form (63) were first studied by Liouville, and on that account are called ***surfaces of Liouville.**** To this class belong the surfaces of revolution and the quadrics (§§ 96, 97). We may state the above results in the form:

When the linear element of a surface is in the Liouville form, the parametric curves are geodesic ellipses and hyperbolas; these systems are the only isothermal orthogonal families of geodesic conics.†

92. Integration of the equation of geodesic lines. Having thus discussed the various properties of geodesic lines, and having seen the advantage of knowing their equations in finite form, we return to the consideration of their differential equation and derive certain theorems concerning its integration.

Suppose, in the first place, that we know a particular first integral of the general equation, that is, a family of geodesics defined by an equation of the form

$$M\,du + N\,dv = 0. \tag{64}$$

From (IV, 58) it follows that M and N must satisfy the equation

$$\frac{\partial}{\partial u}\left(\frac{FN - GM}{\sqrt{EN^2 - 2FMN + GM^2}}\right) + \frac{\partial}{\partial v}\left(\frac{FM - EN}{\sqrt{EN^2 - 2FMN + GM^2}}\right) = 0.$$

* *Journal de Mathématiques*, Vol. XI (1846), p. 345.

† The reader is referred to Darboux, Vol. II, p. 208, for a discussion of the conditions under which a surface is of the Liouville type.

In consequence of this equation we know that there exists a function ϕ defined by

$$(65)\quad \frac{\partial\phi}{\partial u}=\frac{EN-FM}{\sqrt{EN^2-2FMN+GM^2}},\qquad \frac{\partial\phi}{\partial v}=\frac{FN-GM}{\sqrt{EN^2-2FMN+GM^2}}.$$

Moreover, we find that

$$(66)\qquad \Delta_1\phi=1.$$

From (III, 31) and (65) it follows that the curves $\phi=\text{const.}$ are the orthogonal trajectories of the given geodesics, and from (66) it is seen that ϕ measures distance along the geodesics from the curve $\phi=0$. Hence we have the theorem of Darboux *:

When a one-parameter family of geodesics is defined by a differential equation of the first order, the finite equation of their orthogonal trajectories can be obtained by a quadrature, which gives the geodesic parameter at the same time.

Therefore, when the general first integral of the equation of geodesics is known, all the geodesic parallels can be found by quadratures.

We consider now the converse problem of finding the geodesics when the geodesic parallels are known. Suppose that we have a solution of equation (66) involving an arbitrary constant a, which is not additive. If this equation be differentiated with respect to a, we get

$$(67)\qquad \Delta_1\left(\phi,\frac{\partial\phi}{\partial a}\right)=0,$$

where the differential parameter is formed with respect to the linear element. But this is a necessary and sufficient condition (§ 37) that the curves $\phi=\text{const.}$ and the curves

$$(68)\qquad \frac{\partial\phi}{\partial a}=\text{const.}=a'$$

form an orthogonal system. Hence the curves defined by (68) are geodesics. In general, this equation involves two arbitrary constants, a and a', which, as will now be shown, enter in such a way that this equation gives the general integral of the differential equation of geodesic lines.

* *Leçons*, Vol. II, p. 430; cf. also Bianchi, Vol. I, p. 202.

Suppose that a appears in equation (68), and write the latter thus:

$$\psi(u, v, a) = a', \tag{69}$$

in which case equation (67) becomes

$$\Delta_1(\phi, \psi) = 0. \tag{70}$$

The direction of each of the curves (69) is given by $\frac{\partial \psi}{\partial u} \Big/ \frac{\partial \psi}{\partial v}$. If this ratio be independent of a, so also by (70) is the ratio $\frac{\partial \phi}{\partial u} \Big/ \frac{\partial \phi}{\partial v}$. Write the latter in the form

$$\frac{\partial \phi}{\partial u} = f(u, v) \frac{\partial \phi}{\partial v}.$$

If this equation and (66) be solved for $\frac{\partial \phi}{\partial u}$ and $\frac{\partial \phi}{\partial v}$, we obtain values independent of a, so that a would have been additive. Hence f involves a, and so also does $\frac{\partial \psi}{\partial u} \Big/ \frac{\partial \psi}{\partial v}$, and therefore a direction at a point (u_0, v_0) determines the value of a; call it a_0. If then a'_0 be such that

$$\psi(u_0, v_0, a_0) = a'_0,$$

the geodesic $\psi(u, v, a_0) = a'_0$ passes through the point (u_0, v_0) and has the given direction at the point. Hence all the geodesics are defined by equation (68), and we have the theorem:

Given a solution of the equation $\Delta_1 \phi = 1$, *involving an arbitrary constant* a, *in such a way that* $\frac{\partial \phi}{\partial a}$ *involves* a; *the equation*

$$\frac{\partial \phi}{\partial a} = a',$$

for all values of a' *is the finite equation of the geodesics, and the arc of the geodesics is measured by* ϕ.*

By means of this result we establish the following theorem due to Jacobi:

If a first integral of the differential equation of geodesic lines be known, the finite equation can be found by one quadrature.

Such an integral is of the form

$$\frac{dv}{du} = \psi(u, v, a),$$

* Cf. Darboux, Vol. II, p. 429.

where a is an arbitrary constant. As this equation is of the form (64), the function ϕ, defined by

$$\phi = \int \frac{(E + F\psi)\,du + (F + G\psi)\,dv}{\sqrt{E + 2F\psi + G\psi^2}},$$

is a solution of equation (66). As ϕ involves a in the manner specified in the preceding theorem, the finite equation of the geodesics is $\dfrac{\partial\phi}{\partial a} = a'$.

93. Geodesics on surfaces of Liouville. The surfaces of Liouville (§ 91) afford an excellent application of the theorem of Jacobi. We take the linear element in the form *

$$ds^2 = (U - V)(U_1^2 du^2 + V_1^2 dv^2), \tag{71}$$

which evidently is no more general than (63). In this case equation (66) becomes

$$\Delta_1\phi = \frac{1}{U - V}\left[\frac{1}{U_1^2}\left(\frac{\partial\phi}{\partial u}\right)^2 + \frac{1}{V_1^2}\left(\frac{\partial\phi}{\partial v}\right)^2\right] = 1.$$

When this equation is written in the form

$$U - \frac{1}{U_1^2}\left(\frac{\partial\phi}{\partial u}\right)^2 = V + \frac{1}{V_1^2}\left(\frac{\partial\phi}{\partial v}\right)^2,$$

one sees that it belongs to the class of partial differential equations admitting an integral which is the sum of functions of u and v alone.† In order to obtain this integral, we put each side equal to a constant a and integrate. This gives

$$\phi = \int U_1\sqrt{U - a}\,du \pm \int V_1\sqrt{a - V}\,dv. \tag{72}$$

Hence the equation of geodesics is

$$\frac{\partial\phi}{\partial a} = -\frac{1}{2}\int\frac{U_1}{\sqrt{U - a}}\,du \pm \frac{1}{2}\int\frac{V_1}{\sqrt{a - V}}\,dv = a'. \tag{73}$$

If θ denotes the angle which a geodesic through a point makes with the line $v = \text{const.}$ through the point, it follows from (III, 24) and (71) that

$$\tan\theta = \frac{V_1}{U_1}\frac{dv}{du}.$$

* Cf. Darboux, Vol. III, p. 9.

† Forsyth, *Differential Equations* (1888), p. 310.

If the value of dv/du from equation (73) be substituted in this equation, we obtain the following first integral of the Gauss equation (56):

$$U \sin^2\theta + V \cos^2\theta = a. \tag{74}$$

This equation is due to Liouville.*

EXAMPLES

1. On a surface of constant curvature the area of a geodesic triangle is proportional to the difference between the sum of the angles of the triangle and two right angles.

2. Show that for a developable surface the first integral of equation (56) can be found by quadratures.

3. Given any curve C upon a surface and the developable surface which is the envelope of the tangent planes to the surface along C; show that the geodesic curvature of C is equal to the curvature of the plane curve into which C is transformed when the developable is developed upon a plane.

4. When the plane is referred to a system of confocal ellipses and hyperbolas whose foci are at the distance $2c$ apart, the linear element can be written

$$ds^2 = (u^2 - v^2)\left(\frac{du^2}{u^2 - c^2} + \frac{dv^2}{c^2 - v^2}\right).$$

5. A necessary and sufficient condition that ϕ be a solution of $\Delta_1\phi = 1$ is that $ds^2 - d\phi^2$ be a perfect square.

6. If $\phi = \theta_1 a + \theta_2$, where θ_1 and θ_2 are functions of u and v, is a solution of $\Delta_1\phi = 1$, the curves $\theta_1 =$ const. are lines of length zero, and the curves $\theta_1 a + \theta_2 =$ const. are their orthogonal trajectories.

7. When the linear element of a spiral surface is in the form $ds^2 = e^{2v}(du^2 + U^2dv^2)$, the equation $\Delta_1\phi = 1$ admits the solution e^vU_1, where U_1 is a function of u, which satisfies an equation of the first order whose integration gives thus all the geodesics on the surface.

8. For a surface with the linear element

$$ds^2 = V[du^2 + (u + V_1)^2dv^2],$$

where V and V_1 are functions of v alone, the equation $\Delta_1\phi = 1$ admits the solution $\phi = u\psi_1(v) + \psi_2(v)$, the determination of the functions ψ_1 and ψ_2 requiring the solution of a differential equation of the first order and quadratures.

9. If ϕ denotes a solution of $\Delta_1\phi = 1$ involving a nonadditive constant a, the linear element of the surface can be written

$$ds^2 = d\phi^2 + \frac{\Theta\left(\phi, \dfrac{\partial\phi}{\partial a}\right)}{\Theta\left(\dfrac{\partial\phi}{\partial a}, \dfrac{\partial^2\phi}{\partial a^2}\right)}\left(d\frac{\partial\phi}{\partial a}\right)^2,$$

where $\Theta(\phi, \psi)$ indicates the mixed differential parameter (III, 48).

* *L.c.*, p. 348.

94. Lines of shortest length. Envelope of geodesics. We can go a step farther than the first theorem of § 89 and show that whether one or more geodesics pass through two points M_1 and M_2 on a surface, the shortest distance on the surface between these points, if it exists, is measured along one of these geodesics.

Thus, let $v=f(u)$ and $v=f_1(u)$ define two curves C and C_1 passing through the points M_1, M_2, the parametric values of u at the points being u_1 and u_2. The arc of C between these points has the length

$$s=\int_{u_1}^{u_2}\sqrt{E+2Fv'+Gv'^2}\,du, \tag{75}$$

where v' denotes the derivative of v with respect to u. For convenience we write the above thus:

$$s=\int_{u_1}^{u_2}\phi(u,\,v,\,v')\,du. \tag{76}$$

Furthermore, we put

$$f_1(u)=f(u)+\epsilon\,\omega(u),$$

where $\omega(u)$ is a function of u vanishing when u is equal to u_1 and u_2, and ϵ is a constant whose absolute value may be taken so small that the curve C_1 will lie in any prescribed neighborhood of C. Hence the length of the arc M_1M_2 of C_1 is

$$s_1=\int_{u_1}^{u_2}\phi(u,\,v+\epsilon\omega,\,v'+\epsilon\omega')\,du.$$

Thus s_1 is a function of ϵ, reducing for $\epsilon=0$ to s. Hence, in order that the curve C be the shortest of all the near-by curves which pass through M_1 and M_2, it is necessary that the derivative of s_1 with respect to ϵ be zero for $\epsilon=0$. This gives

$$\int_{u_1}^{u_2}\left(\frac{\partial\phi}{\partial v}\omega+\frac{\partial\phi}{\partial v'}\omega'\right)du=0.$$

On the assumption that ω admits a continuous first derivative in the interval (u_1, u_2), and ϕ continuous first and second derivatives, the left-hand member of this equation may be integrated by parts with the result

$$\int_{u_1}^{u_2}\omega\left(\frac{\partial\phi}{\partial v}-\frac{d}{du}\frac{\partial\phi}{\partial v'}\right)du=0;$$

for ω vanishes when u equals u_1 and u_2. As the function ω is arbitrary except for the above conditions upon it, this equation is equivalent to the following equation of Euler *:

$$\frac{\partial\phi}{\partial v}-\frac{d}{du}\frac{\partial\phi}{\partial v'}=0. \tag{77}$$

When this result is applied to the particular form of ϕ in equation (75), we have

$$\frac{d}{du}\left(\frac{F+Gv'}{\sqrt{E+2Fv'+Gv'^2}}\right)-\frac{\dfrac{\partial E}{\partial v}+2\dfrac{\partial F}{\partial v}v'+\dfrac{\partial G}{\partial v}v'^2}{2\sqrt{E+2Fv'+Gv'^2}}=0,$$

which is readily reducible to equation (43).

Hence the shortest distance between two points, if existent, is measured along a geodesic through the points. This geodesic is unique if the surface has negative total curvature at all points. For other surfaces more than one geodesic may pass through the points if the latter are sufficiently far apart. We shall now investigate the nature of this problem.

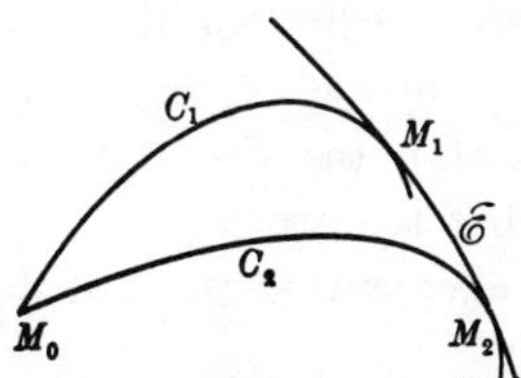

FIG. 21

Let $v=f(u, \alpha)$ define the family of geodesics through a point $M_0(u_0, v_0)$, and let $v=g(u)$ be the equation of their envelope $\mathscr{E}$. We consider two of the geodesics C_1 and C_2 (fig. 21), and let $M_1(u_1, v_1)$ and $M_2(u_2, v_2)$ denote their points of contact with the envelope. Suppose that the arc M_0M_2 is greater than M_0M_1. The distance from M_0 to M_2, measured along C_1 and $\mathscr{E}$ is equal to

$$D=\int_{u_0}^{u_1}\phi(u, f, f')\,du+\int_{u_1}^{u_2}\phi(u, g, g')\,du.$$

If M_2 is considered fixed and M_1 variable, the position of the latter is determined by α. The variation of D with M_1 is given by

$$\frac{\partial D}{\partial\alpha}=\int_{u_0}^{u_1}\left(\frac{\partial\phi}{\partial f}\frac{\partial f}{\partial\alpha}+\frac{\partial\phi}{\partial f'}\frac{\partial f'}{\partial\alpha}\right)du+\left[\phi(u,f,f')\frac{du_1}{d\alpha}-\phi(u,g,g')\frac{du_1}{d\alpha}\right]_{u=u_1}.$$

* *Methodus inveniendi lineas curvas maximi minimive proprietate gaudentes*, chap. ii § 21 (Lausanne, 1744); cf. Bolza, *Lectures on the Calculus of Variations*, p. 22 (Chicago, 1904).

But for $u = u_1$, $f = g$ and $f' = g'$; consequently the last term is zero. Integrating the first member by parts, and noting that $\frac{\partial f}{\partial \alpha}$ is zero for $u = u_0$ and $u = u_1$ (§ 26), we have

$$\frac{\partial D}{\partial \alpha} = \int_{u_0}^{u_1} \frac{\partial f}{\partial \alpha}\left(\frac{\partial \phi}{\partial f} - \frac{d}{du}\frac{\partial \phi}{\partial f'}\right) du.$$

Since C_1 is a geodesic, the expression in parenthesis is zero, and hence D does not vary with M_1. This shows that the envelope of the geodesics through a point bears to them the relation which the evolute of a curve does to a family of normals to the curve. Moreover, the curve $\mathcal{E}$ is not a geodesic, for at each point of it there is tangent a geodesic. Hence there is an arc connecting M_1 and M_2 which is shorter than the arc of $\mathcal{E}$. In this way, by taking different points M_1 on $\mathcal{E}$, we obtain any number of arcs connecting M_0 and M_2 which are shorter than the arc of C_2, each consisting of an arc of a geodesic such as C_1 and the geodesic distance M_1M_2. It is then necessarily true that the shortest distance from M_0 to a point M of C_2 beyond M_2 is not measured along C_2. However, when M lies within the arc M_0M_2, a domain can be chosen about C_2 so small that the arc M_0M of C_2 is shorter than the arc M_0M of any other curve within the domain and passing through these points.*

Another historical problem associated with this problem is the following: †

Given an arc C_0 joining two points A, B on a surface; to find the curve of shortest length joining A and B, and inclosing with C_0 a given area.

The area is given by $\iint H\,dudv$. It is evident that two functions M and N can be found in an infinity of ways such that

$$H = \frac{\partial N}{\partial u} - \frac{\partial M}{\partial v}.$$

By the application of Green's theorem we have

$$\iint H\,dudv = \iint\left(\frac{\partial N}{\partial u} - \frac{\partial M}{\partial v}\right) dudv = \int M du + N dv,$$

where the last integral is curvilinear and is taken around the contour of the area. Since C_0 is fixed, our problem reduces to the determination of a curve C along which the integral $\int_A^B M du + N dv$ is constant, and whose arc AB, that is, the

* For a more complete discussion of this problem the reader is referred to Darboux, Vol. III, pp. 86–112; Bolza, chap. v.

† In fact, it was in the solution of this problem that Minding (*Crelle*, Vol. V (1830), p. 297) discovered the function to which Bonnet (*Journal de l'École Polytechnique*, Vol. XIX (1848), p. 44) gave the name *geodesic curvature*.

integral $\int_A^B \sqrt{E + 2\,Fv' + Gv'^2}\,du$, is a minimum. From the calculus of variations we know that, so far as the differential equations of the solution is concerned, this is the same problem as finding the curve C along which the integral

$$\int_A^B \sqrt{E + 2\,Fv' + Gv'^2}\,du + c\,(M + Nv')\,du$$

is a minimum, c being a constant. Euler's equation for this integral is

$$\frac{d}{du}\left(\frac{F + Gv'}{\sqrt{E + 2\,Fv' + Gv'^2}}\right) - \frac{\dfrac{\partial E}{\partial v} + 2\,v'\dfrac{\partial F}{\partial v} + v'^2\dfrac{\partial G}{\partial v}}{\sqrt{E + 2\,Fv' + Gv'^2}} = cH.$$

Comparing this result with the formula of Bonnet (IV, 56), we see that C has constant geodesic curvature $1/c$, and c evidently depends upon the magnitude of the area between the curves. Hence we have the theorem of Minding:*

In order that a curve C joining two points shall be the shortest which, together with a given curve through these points, incloses a portion of the surface with a given area, it is necessary that the geodesic curvature of C be constant.

GENERAL EXAMPLES

1. When the parametric curves on the unit sphere satisfy the condition

$$\frac{\partial}{\partial u}\left\{\begin{matrix}12\\1\end{matrix}\right\}' = \frac{\partial}{\partial v}\left\{\begin{matrix}12\\2\end{matrix}\right\}' = 2\left\{\begin{matrix}12\\1\end{matrix}\right\}'\left\{\begin{matrix}12\\2\end{matrix}\right\}',$$

they represent the asymptotic lines on a surface whose total curvature is

$$K = -\frac{1}{[\phi(u) + \psi(v)]^2}.$$

2. When the equations of the sphere have the form (III, 35), the parametric curves are asymptotic and the equation (22) is $(1 + uv)^2\dfrac{\partial^2\theta}{\partial u\,\partial v} = -\,2\,\theta$, of which the general integral is

$$\theta = 2\,\frac{v\phi(u) + u\psi(v)}{1 + uv} - \phi'(u) - \psi'(v),$$

where $\phi(u)$ and $\psi(v)$ denote arbitrary functions.

3. The sections of a surface by all the planes through a fixed line L in space, and the curves of contact of the tangent cones to the surface whose vertices are on L, form a conjugate system.

4. Given a surface of translation $x = u,\ y = v,\ z = f(u) + \phi(v)$. Determine the functions f and ϕ so that $(\rho_1 + \rho_2)\,Z = \text{const.}$, where Z denotes the cosine of the angle which the normal makes with the z-axis, and determine the lines of curvature on the surface.

5. Determine the relations between the exponents m_i and n_i in the equations

$$x = U^{m_1}V^{n_1}, \qquad y = U^{m_2}V^{n_2}, \qquad z = U^{m_3}V^{n_3},$$

so that on the surface so defined the parametric curves shall form a conjugate system, and show that the asymptotic lines can be found by quadratures.

* L.c., p. 297.

6. The envelope of the family of planes

$$(U_1 + V_1)x + (U_2 + V_2)y + (U_3 + V_3)z + (U_4 + V_4) = 0,$$

where the U's are functions of u alone and the V's of v, is a surface upon which the parametric curves are plane, and form a conjugate system.

7. The condition that the parametric curves form a conjugate system on the envelope of the plane

$$x \cos u + y \sin u + z \cot v = f(u, v),$$

is that f be the sum of a function of u alone and of v alone; in this case these curves are plane lines of curvature.

8. Find the geodesics on the surface of Ex. 7, p. 219, and determine the expressions for the radii of curvature and torsion of a geodesic.

9. A representation of two surfaces upon one another is said to be *conformal-conjugate* when it is at the same time conformal, and every conjugate system on one surface corresponds to a conjugate system on the other. Show that the lines of curvature correspond and that the characteristic lines also correspond.

10. Given a surface of revolution $x = u \cos v$, $y = u \sin v$, $z = f(u)$, and the function ϕ defined by

$$\text{(i)} \qquad \frac{1}{\phi'} + \sqrt{\frac{1}{\phi'^2} + 1} = A\left\{\frac{1}{f'} + \sqrt{\frac{1}{f'^2} + 1}\right\}^{\frac{1}{c}},$$

where A and c are constants; a conformal-conjugate representation of the surface upon a second surface $x_1 = u_1 \cos v_1$, $y_1 = u_1 \sin v_1$, $z_1 = \phi(u_1)$ is defined by

$$v = cv_1, \qquad c \log u_1 = \int \frac{\sqrt{1 + f'^2}}{\sqrt{1 + F'^2}} \frac{du}{u},$$

where F' denotes the function of u found by solving (i) for ϕ'.

11. If two families of geodesics cut under constant angle, the surface is developable.

12. If a surface with the linear element

$$ds^2 = (au^2 - bv^2 - c)(du^2 + dv^2),$$

where a, b, c are constants, is represented on the xy-plane by $u = x$, $v = y$, the geodesics correspond to the Lissajous figures defined by

$$\sqrt{b} \sin^{-1}\frac{x}{A} - i\sqrt{a} \sin^{-1}\frac{y}{B} = C,$$

where A, B, C are constants.

13. When there is upon a surface more than one family of geodesics which, together with their orthogonal trajectories, form an isothermal system, the curvature of the surface is constant.

14. If the principal normals of a curve meet a fixed straight line, the curve is a geodesic on a surface of revolution whose axis is this line. Examine the case where the principal normals meet the line under constant angle.

15. A representation of two surfaces upon one another is said to be a *geodesic representation* when to a geodesic on one surface there corresponds a geodesic on the other. Show that the representation is geodesic when points with the same parametric values correspond on surfaces with the linear elements

$$ds^2 = (U - V)(U_1^2\,du^2 + V_1^2\,dv^2), \quad ds_1^2 = \left(\frac{1}{V+h} - \frac{1}{U+h}\right)\left(\frac{U_1^2\,du^2}{U+h} + \frac{V_1^2\,dv^2}{V+h}\right),$$

where the U's are functions of u alone, the V's of v alone, and h is a constant.

16. A surface with the linear element

$$ds^2 = (u^4 - v^4)\left[\phi\left(\frac{1}{u}\right)du^2 + \phi(v)\,dv^2\right],$$

where ϕ is any function whatever, admits of a geodesic representation upon itself.

17. A necessary and sufficient condition that an orthogonal system upon a surface may be regarded as geodesic ellipses and hyperbolas in two ways, is that when the curves are parametric the linear element be of the Liouville form; in this case these curves may be so regarded in an infinity of ways.

18. Of all the curves of equal length joining two points, the one which, together with a fixed curve through the points, incloses the area of greatest extent, has constant geodesic curvature.

19. Let Γ be any curve upon a surface, and at two near-by points P, P' draw the geodesics g, g' perpendicular to Γ; let C be the curve through P conjugate to g, P'' the point where it meets g', and Q the intersection of the tangents to g and g' at P and P''; the limiting position of Q, as P' approaches P, is the center of geodesic curvature of Γ at P.

20. Show that if a surface S admits of geodesic representation upon a plane in such a way that four families of geodesics are represented by four families of parallel lines, each geodesic on the surface is represented by a straight line (cf. Ex. 3, p. 209).

CHAPTER VII

QUADRICS. RULED SURFACES. MINIMAL SURFACES

95. Confocal quadrics. Elliptic coördinates. Two quadrics are *confocal* when the foci, real or imaginary, of their principal sections coincide. Hence a family of confocal quadrics is defined by the equation

$$\text{(1)} \qquad \frac{x^2}{a^2-u}+\frac{y^2}{b^2-u}+\frac{z^2}{c^2-u}=1,$$

where u is the parameter of the family and a, b, c are constants, such that

$$\text{(2)} \qquad a^2>b^2>c^2.$$

For each value of u, positive or negative, less than a^2, equation (1) defines a quadric which is

$$\text{(3)} \qquad \begin{cases} \text{an ellipsoid when } c^2>u>-\infty, \\ \text{an hyperboloid of one sheet when } b^2>u>c^2, \\ \text{an hyperboloid of two sheets when } a^2>u>b^2. \end{cases}$$

As u approaches c^2 the smallest axis of the ellipsoid approaches zero. Hence the surface $u=c^2$ is the portion of the xy-plane, counted twice, bounded by the ellipse

$$\text{(4)} \qquad \frac{x^2}{a^2-c^2}+\frac{y^2}{b^2-c^2}=1, \qquad z=0.$$

Again, the surface $u=b^2$ is the portion of the xz-plane, counted twice, bounded by the hyperbola

$$\text{(5)} \qquad \frac{x^2}{a^2-b^2}-\frac{z^2}{b^2-c^2}=1, \qquad y=0,$$

which contains the center of the curve. Equations (4) and (5) define the *focal ellipse* and *focal hyperbola* of the system.

Through each point (x, y, z) in space there pass three quadrics of the family; they are determined by the values of u, which are roots of the equation

$$(6)\qquad \phi(u) = (a^2-u)(b^2-u)(c^2-u) - x^2(b^2-u)(c^2-u) - y^2(a^2-u)(c^2-u) - z^2(a^2-u)(b^2-u) = 0.$$

Since $\quad \phi(a^2) < 0, \quad \phi(b^2) > 0, \quad \phi(c^2) < 0, \quad \phi(-\infty) > 0,$

the roots of equation (6), denoted by u_1, u_2, u_3, are contained in the following intervals:

$$(7)\qquad a^2 > u_1 > b^2, \quad b^2 > u_2 > c^2, \quad c^2 > u_3 > -\infty.$$

From (3) it is seen that the surfaces corresponding to u_1, u_2, u_3 are respectively hyperboloids of two and one sheets and an ellipsoid.

Fig. 22 represents three confocal quadrics; the curves on the ellipsoid are lines of curvature, and on the hyperboloid of one sheet they are asymptotic lines.

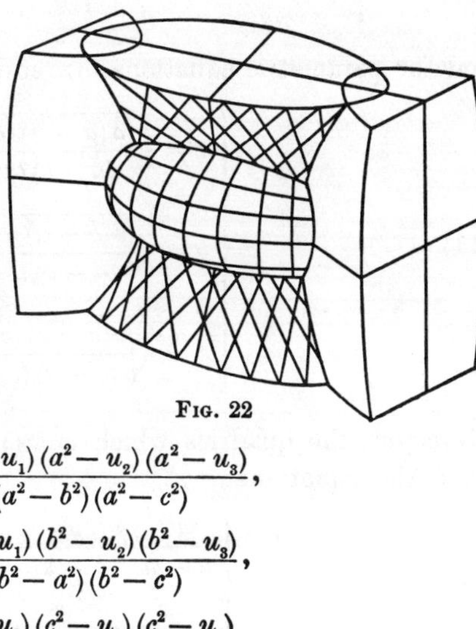

FIG. 22

From the definition of u_1, u_2, u_3 it follows that $\phi(u)$ is equal to $(u_1-u)(u_2-u)(u_3-u)$. When ϕ in (6) is replaced by this expression and u is given successively the values a^2, b^2, c^2, we obtain *

$$(8)\qquad \begin{cases} x^2 = \dfrac{(a^2-u_1)(a^2-u_2)(a^2-u_3)}{(a^2-b^2)(a^2-c^2)}, \\[2ex] y^2 = \dfrac{(b^2-u_1)(b^2-u_2)(b^2-u_3)}{(b^2-a^2)(b^2-c^2)}, \\[2ex] z^2 = \dfrac{(c^2-u_1)(c^2-u_2)(c^2-u_3)}{(c^2-a^2)(c^2-b^2)}. \end{cases}$$

These formulas express the Cartesian coördinates of a point in space in terms of the parameters of the three quadrics which pass through the point. These parameters are called the *elliptic coördinates* of the point. It is evident that to each set of these

* Kirchhoff, *Mechanik*, p. 203. Leipsic, 1877.

coördinates there correspond eight points in space, one in each of the eight compartments bounded by the coördinate planes.

If one of the parameters u_i in (8) be made constant, and the others u_j, u_k, where $i \neq j \neq k$, be allowed to vary, these equations define in parametric form the surface, also defined by equation (1), in which u has this constant value u_i. The parametric curves $u_j = \text{const.}$, $u_k = \text{const.}$ are the curves of intersection of the given quadric and the double system of quadrics corresponding to the parameters u_j and u_k.

If we put

$$(9) \quad a^2 - u_i = a, \quad b^2 - u_i = b, \quad c^2 - u_i = c, \quad u_j - u_i = u, \quad u_k - u_i = v,$$

the equation of the surface becomes

$$(10) \qquad \frac{x^2}{a} + \frac{y^2}{b} + \frac{z^2}{c} = 1,$$

and the parametric equations (8) reduce to

$$(11) \qquad \begin{cases} x = \sqrt{\dfrac{a(a-u)(a-v)}{(a-b)(a-c)}}, \\[2ex] y = \sqrt{\dfrac{b(b-u)(b-v)}{(b-a)(b-c)}}, \\[2ex] z = \sqrt{\dfrac{c(c-u)(c-v)}{(c-a)(c-b)}}. \end{cases}$$

Moreover, the quadrics which cut (10) in the parametric curves have the equations:

$$(12) \qquad \begin{cases} \dfrac{x^2}{a-u} + \dfrac{y^2}{b-u} + \dfrac{z^2}{c-u} = 1, \\[2ex] \dfrac{x^2}{a-v} + \dfrac{y^2}{b-v} + \dfrac{z^2}{c-v} = 1. \end{cases}$$

In consequence of (3) and (9) we have that equations (10) or (11) define

$$(13) \qquad \begin{cases} \text{an ellipsoid when } a > u > b > v > c > 0, \\ \text{an hyperboloid of one sheet when } a > u > b > 0 > c > v, \\ \text{an hyperboloid of two sheets when } a > 0 > b > u > c > v. \end{cases}$$

96. Fundamental quantities for central quadrics. By direct calculation we find from (11)

$$E=\frac{u(u-v)}{f(u)},\qquad F=0,\qquad G=\frac{v(v-u)}{f(v)}, \tag{14}$$

where for the sake of brevity we have put

$$f(\theta)=4(a-\theta)(b-\theta)(c-\theta). \tag{15}$$

We derive also the following:

$$\begin{cases} X=\sqrt{\dfrac{bc(a-u)(a-v)}{uv(a-b)(a-c)}}, \\ Y=\sqrt{\dfrac{ca(b-u)(b-v)}{uv(b-a)(b-c)}}, \\ Z=\sqrt{\dfrac{ab(c-u)(c-v)}{uv(c-a)(c-b)}}, \end{cases} \tag{16}$$

and

$$D=-\sqrt{\frac{abc}{uv}}\frac{u-v}{f(u)},\qquad D'=0,\qquad D''=\sqrt{\frac{abc}{uv}}\frac{u-v}{f(v)}. \tag{17}$$

Since F and D' are zero, the parametric curves are lines of curvature. And since the change of parameters (9) did not change the parametric curves, we have the theorem:

The quadrics of a confocal system cut one another along lines of curvature, and the three surfaces through a point cut one another orthogonally at the point.

This result is illustrated by fig. 22.

From (14) and (17) we have

$$\frac{1}{\rho_1}=-\sqrt{\frac{abc}{u^3v}},\qquad \frac{1}{\rho_2}=-\sqrt{\frac{abc}{uv^3}},\qquad \frac{1}{\rho_1\rho_2}=\frac{abc}{u^2v^2}. \tag{18}$$

Hence the ellipsoid and hyperboloid of two sheets have positive curvature at all points, whereas the curvature is negative at all points of the hyperboloid of one sheet.

If formulas (16) be written

$$X=\sqrt{\frac{abc}{uv}}\frac{x}{a},\qquad Y=\sqrt{\frac{abc}{uv}}\frac{y}{b},\qquad Z=\sqrt{\frac{abc}{uv}}\frac{z}{c},$$

the distance W from the center to the tangent plane is

$$W=\sum Xx=\sqrt{\frac{abc}{uv}}. \tag{19}$$

Hence:

The tangent planes to a central quadric along a curve, at points of which the total curvature of the surface is the same, are equally distant from the center.

From (18) we see that the umbilical points correspond to the values of the parameters such that $u = v$. The conditions (13) show that this common value of u and v for an ellipsoid is b, and c for an hyperboloid of two sheets, whereas there are no real umbilical points for the hyperboloid of one sheet. When these values are substituted in (11), we have as the coördinates of these points on the ellipsoid

$$(20) \qquad x = \pm\sqrt{\frac{a(a-b)}{(a-c)}}, \qquad y = 0, \qquad z = \pm\sqrt{\frac{c(b-c)}{(a-c)}},$$

and on the hyperboloid of two sheets

$$(21) \qquad x = \pm\sqrt{\frac{a(a-c)}{(a-b)}}, \qquad y = \pm\sqrt{\frac{b(b-c)}{(b-a)}}, \qquad z = 0.$$

It should be noticed that these points lie on the focal hyperbola and focal ellipse respectively.

97. Fundamental quantities for the paraboloids. The equation of a paraboloid

$$(22) \qquad 2z = ax^2 + by^2$$

may be replaced by

$$(23) \qquad x = \sqrt{u_1}, \qquad y = \sqrt{v_1}, \qquad z = \frac{1}{2}(au_1 + bv_1).$$

Hence the paraboloids are surfaces of translation (§ 81) whose generating curves are parabolas which lie in perpendicular planes. By direct calculation we find

$$E = \frac{1}{4}\left(\frac{1}{u_1} + a^2\right), \qquad F = \frac{1}{4}ab, \qquad G = \frac{1}{4}\left(\frac{1}{v_1} + b^2\right);$$

$$D = \frac{1}{4}\frac{a}{u_1\sqrt{a^2u_1 + b^2v_1 + 1}}, \qquad D' = 0, \qquad D'' = \frac{1}{4}\frac{b}{v_1\sqrt{a^2u_1 + b^2v_1 + 1}},$$

so that the equation of the lines of curvature is

$$v_1 = u_1\frac{dv_1}{du_1} + \left[\frac{b-a}{a^2b}\frac{dv_1}{du_1} - \frac{b}{a}v_1\frac{dv_1}{du_1} + \frac{b}{a}u_1\left(\frac{dv_1}{du_1}\right)^2\right].$$

The general integral of this equation is

$$(24)\qquad v_1 = c\left[u_1 + \frac{b-a}{a^2b} - \frac{b}{a}v_1\right] + \frac{b}{a}u_1c^2,$$

where c is an arbitrary constant.

When u_1 and v_1 in (24) are given particular values, equation (24) determines two values of c, c_1 and c_2, in general distinct. If these latter values be substituted in (24) successively, we obtain in finite form the equations of the two lines of curvature through the point (u_1, v_1). If c_1 and c_2 be replaced by $-\left(\frac{1+au}{bu}\right)$ and $-\left(\frac{1+av}{bv}\right)$ respectively, we have, in consequence of (23), the two equations

$$(25)\qquad \begin{cases} buy^2 + (1+au)x^2 = u(1+au)\dfrac{b-a}{ab}, \\ bvy^2 + (1+av)x^2 = v(1+av)\dfrac{b-a}{ab}. \end{cases}$$

When these equations are solved for x^2 and y^2, we find that equation (22) can be replaced by

$$(26)\qquad \begin{cases} x^2 = \dfrac{a-b}{b}uv, \\ y^2 = \dfrac{b-a}{ab^2}(1+au)(1+av), \\ z = \dfrac{1}{2}\dfrac{b-a}{ab}(1+au+av), \end{cases}$$

and the parametric curves are the lines of curvature.

Now we have

$$(27)\qquad \begin{cases} E = \dfrac{a-b}{4b^2}(u-v)\dfrac{a(a-b)u-b}{u(1+au)}, \qquad F = 0, \\ G = \dfrac{b-a}{4b^2}(u-v)\dfrac{a(a-b)v-b}{v(1+av)}, \end{cases}$$

$$(28)\quad X, Y, Z = \frac{a^{\frac{3}{2}}\sqrt{a-b}\sqrt{uv},\ \sqrt{b}\sqrt{b-a}\sqrt{(1+au)(1+av)},\ -\sqrt{ab}}{\sqrt{[a(a-b)u-b][a(a-b)v-b]}},$$

and

$$(29)\qquad \begin{cases} D = \dfrac{1}{4}\sqrt{\dfrac{a^3}{b}}\dfrac{(a-b)(u-v)}{\sqrt{[a(a-b)u-b][a(a-b)v-b]}}\cdot\dfrac{1}{u(1+au)}, \\ D' = 0, \\ D'' = -\dfrac{1}{4}\sqrt{\dfrac{a^3}{b}}\dfrac{(a-b)(u-v)}{\sqrt{[a(a-b)u-b][a(a-b)v-b]}}\cdot\dfrac{1}{v(1+av)}. \end{cases}$$

From (27), (28), and (29) we obtain

$$(30) \qquad W=\sum Xx=\frac{\sqrt{ab}\,z}{[a(a-b)u-b]^{\frac{1}{2}}[a(a-b)v-b]^{\frac{1}{2}}},$$

and

$$(31) \qquad \begin{cases} \dfrac{1}{\rho_1}=\sqrt{a^3b^3}\,[a(a-b)u-b]^{-\frac{3}{2}}[a(a-b)v-b]^{-\frac{1}{2}}, \\ \dfrac{1}{\rho_2}=\sqrt{a^3b^3}\,[a(a-b)u-b]^{-\frac{1}{2}}[a(a-b)v-b]^{-\frac{3}{2}}. \end{cases}$$

From these results we find that the ratio W/z is constant along the curves for which the total curvature is constant.

We suppose that b is positive and greater than a. From the first of (26) it follows that u and v at a real point differ in sign, or one is equal to zero. We consider the points at which both u and v are equal to zero. There are two such points, and their coördinates are

$$(32) \qquad x=0, \qquad y=\pm\frac{1}{b}\sqrt{\frac{b-a}{a}}, \qquad z=\frac{1}{2}\frac{b-a}{ab}.$$

Evidently these points are real only on the elliptic paraboloid. From (31) it follows that ρ_1 and ρ_2 are then equal, and consequently these are the umbilical points. Since at points other than these u and v must differ in sign, we may assume that u is always positive and v negative. Moreover, from (26) it is seen that u and v are unrestricted except in the case of the elliptic paraboloid, when v must be greater than $-1/a$.

98. Lines of curvature and asymptotic lines on quadrics. From (14), (27), and § 91 we have the theorem:

The lines of curvature of a quadric surface form an isothermal system of the Liouville type.

Bonnet* has shown that this property is characteristic of the quadrics. There are, however, many surfaces whose lines of curvature form an isothermal system. They are called *isothermic surfaces.* The complete determination of all such surfaces has never been accomplished (cf. Ex. 3, § 65).

* Mémoire sur la théorie des surfaces applicables sur une surface donnée, *Journal de l'École Polytechnique,* Vol. XXV (1867), pp. 121–132.

From (17), (29), and § 82 follows the theorem:

The lines of curvature of a quadric surface form an isothermal-conjugate system, and consequently the asymptotic lines can be found by quadratures.

We shall find the expressions for the coördinates in terms of the latter in another way.

Equation (10) is equivalent to the pair of equations

$$(33) \qquad \left(\frac{x}{\sqrt{a}}+i\frac{z}{\sqrt{c}}\right)=u\left(1-\frac{y}{\sqrt{b}}\right), \quad \left(\frac{x}{\sqrt{a}}-i\frac{z}{\sqrt{c}}\right)=\frac{1}{u}\left(1+\frac{y}{\sqrt{b}}\right),$$

or the pair

$$(34) \qquad \left(\frac{x}{\sqrt{a}}+i\frac{z}{\sqrt{c}}\right)=\frac{1}{v}\left(1+\frac{y}{\sqrt{b}}\right), \quad \left(\frac{x}{\sqrt{a}}-i\frac{z}{\sqrt{c}}\right)=v\left(1-\frac{y}{\sqrt{b}}\right),$$

where u and v are undetermined. For each value of u equations (33) define a line all of whose points lie on the surface. And to each point on the surface there corresponds a value of u determining a line through the point. Hence the surface is ruled, and it is nondevelopable, as seen from (18). Again, for each value of v equations (34) define a line whose points lie on the surface (10), and these lines are different from those of the other system. Hence the central quadrics are doubly ruled. These lines are necessarily the asymptotic lines. Consequently, if equations (33), (34) be solved for x, y, z, thus:

$$(35) \qquad \frac{x}{\sqrt{a}}=\frac{u+v}{uv+1}, \qquad \frac{y}{\sqrt{b}}=\frac{uv-1}{uv+1}, \qquad \frac{z}{\sqrt{c}}=i\frac{v-u}{uv+1},$$

we have the surface defined in terms of parameters referring to the asymptotic lines.

In like manner equation (22) may be replaced by

$$\sqrt{a}x+i\sqrt{b}y=2uz, \qquad \sqrt{a}x-i\sqrt{b}y=\frac{1}{u},$$

or

$$\sqrt{a}x+i\sqrt{b}y=\frac{1}{v}, \qquad \sqrt{a}x-i\sqrt{b}y=2vz.$$

Solving these, we have

$$(36) \qquad \sqrt{a}x=\frac{u+v}{2uv}, \quad \sqrt{b}y=\frac{i(v-u)}{2uv}, \quad z=\frac{1}{2uv}.$$

As in the preceding case, we see that the surface is doubly ruled,* and the parameters in (36) refer to the asymptotic system of straight lines. Hence:

The asymptotic lines on any quadric are straight lines.

EXAMPLES

1. The focal conics of a family of confocal quadrics meet the latter in the umbilical points.

2. Find the characteristic lines on the quadrics of positive curvature.

3. The normal section of an ellipsoid at a point in the direction of the curve along which the total curvature is constant is an ellipse with one of its vertices at the point.

4. Find the equation of the form $\frac{\partial^2\theta}{\partial u\,\partial v} = M\theta$ (cf. § 79) when the corresponding surface is a hyperboloid of one sheet; when a hyperbolic paraboloid.

5. Find the evolute of the hyperboloid of one sheet and derive the following properties:

(*a*) the surface is algebraic of the twelfth order;

(*b*) the section by a principal plane of the hyperboloid consists of a conic and the evolute of a conic;

(*c*) these sections are edges on the surface;

(*d*) the curve of intersection of the two sheets of the surface is cut by each of the principal planes in four ordinary points, four double points, and four cusps, and consequently is of the twenty-fourth order.

6. Determine for the evolute of a hyperbolic paraboloid the properties analogous to those for the surface of Ex. 5.

7. Deduce the equations of the surfaces parallel to a central quadric; determine their order and the character of the sections of the surface by the principal planes of the quadric; find the normal curvature of the curves corresponding to the asymptotic lines on the quadric.

99. Geodesics on quadrics. Since the quadrics are isothermic surfaces of the Liouville type, the finite equation of the geodesics can be found by quadratures (§ 93). From (VI, 74), (14) and (27),

* Moreover, *the quadrics are the only doubly ruled surfaces.* For consider such a surface, and denote by a, b, c three of the generators in one system. A plane α through a meets b and c in unique points B and C, and the line BC meets a in a point A. The line ABC is a generator of the second system, and the only one of this system in the plane α. The other lines of this system meet α in the line a. On this account the plane α cuts the surface in two lines, a and ABC, that is, in a degenerate conic. Hence the surface is of the second degree.

it follows that the first integral of the differential equation of geodesics on any one of the quadrics is

$$u \sin^2\theta + v \cos^2\theta = \alpha, \tag{37}$$

where α is a constant of integration and θ measures the angle which a geodesic, determined by a value of α, makes with the lines of curvature $v = \text{const.}$ We recall that in equations (11) and (26) the parameter u is greater than v, except at the umbilical points, where they are equal. We shall discuss the general case first.

Consider a particular point $M'(u', v')$. According as α is given the value u' or v', equation (37) defines the geodesic tangent at M' to the line of curvature $u = u'$ or $v = v'$ respectively. It is readily seen that the other values of α, determining other geodesics through M', lie in the interval between u' and v'. Moreover, to each value of α in this domain there correspond two geodesics through M' whose tangents are symmetrically placed with respect to the directions of the lines of curvature. From this result it follows also that the whole system of geodesics is defined by (37), when α is given the limiting values of u and v and all the intermediate values.

We write equation (37) in the form

$$(u - \alpha)\sin^2\theta + (v - \alpha)\cos^2\theta = 0, \tag{38}$$

and consider the geodesics on a central quadric defined by this equation when α has a particular value α'. Suppose, first, that α' is in the domain of the values of u. Then at each point of these geodesics $v < \alpha'$ and consequently from (38) $u > \alpha'$. We have seen that these geodesics are tangent to the line of curvature $u = \alpha'$. From (11) it follows that they lie within the zone of the surface bounded by the two branches of the curve $u = \alpha'$. When, now, α' is in the domain of the values of v, $u - \alpha'$ is positive, and consequently from (38) $v < \alpha'$. Hence the geodesics tangent to the curve $v = \alpha'$ lie outside the zone bounded by the two branches of the line of curvature $v = \alpha'$. Similar results are true for the paraboloids, with the difference, as seen from (26), that the geodesics tangent to $u = \alpha'$ lie outside the region bounded by this curve, whereas the curves tangent to $v = \alpha'$ lie inside the region bounded by $v = \alpha'$.

100. Geodesics through the umbilical points. There remains for consideration the case where α takes the unique value which u and v have at the umbilical points. Let it be denoted by α_0, so that the curves defined by

$$(u-\alpha_0)\sin^2\theta+(v-\alpha_0)\cos^2\theta=0 \tag{39}$$

are the *umbilical geodesics*. We have, at once, the theorem:

Through each point on a quadric with real umbilical points there pass two umbilical geodesics which are equally inclined to the lines of curvature through the point.

Hence two diametrically opposite umbilical points of an ellipsoid are joined by an infinity of geodesics, and no two geodesics through the same umbilical point meet again except at the diametrically opposite point. These properties are possessed also by a family of great circles on a sphere through two opposite points. On the elliptic paraboloid and on each sheet of the hyperboloid of two sheets there are two families of umbilical geodesics, but no two of the same family meet except at the umbilical point common to all curves of the family.

For the ellipsoid (11) $\alpha_0=b$ and equations (VI, 72, 73) become

$$\phi=\frac{1}{2}\int\sqrt{\frac{u}{(a-u)(u-c)}}\,du\pm\frac{1}{2}\int\sqrt{\frac{v}{(a-v)(v-c)}}\,dv,$$

$$\frac{\partial\phi}{\partial b}=-\frac{1}{4}\int\sqrt{\frac{u}{(a-u)(u-c)}}\frac{du}{u-b}\mp\frac{1}{4}\int\sqrt{\frac{v}{(a-v)(v-c)}}\frac{dv}{v-b}.$$

Similar results hold for the hyperboloid of two sheets and the elliptic paraboloid. Hence the distances of a point P from two umbilical points (not diametrically opposite) are of the form

$$\phi_1=f_1(u)+f_2(v),\qquad \phi_2=f_1(u)-f_2(v).$$

Hence we have:

The lines of curvature on the quadrics with real umbilical points are geodesic ellipses and hyperbolas with the umbilical points for foci.

101. Ellipsoid referred to a polar geodesic system. A family of umbilical geodesics and their orthogonal trajectories constitute an excellent system for polar geodesic coördinates, because the domain is unrestricted (§ 87) except in the case of the ellipsoid, and then only the diametrically opposite point must be excluded. We consider such a system on the ellipsoid, and let O denote the pole of the system and O', O'', O''' the other umbilical points (fig. 23).

If we put

$$(40)\quad \begin{cases} \phi = \dfrac{1}{2}\displaystyle\int\sqrt{\frac{u}{(a-u)(u-c)}}\,du - \frac{1}{2}\int\sqrt{\frac{v}{(a-v)(v-c)}}\,dv, \\ \psi = \dfrac{1}{2}\displaystyle\int\sqrt{\frac{u}{(a-u)(u-c)}}\,\frac{du}{u-b} - \frac{1}{2}\int\sqrt{\frac{v}{(a-v)(v-c)}}\,\frac{dv}{v-b}, \end{cases}$$

it is readily found that

$$\Delta_1\phi = 1, \qquad \Delta_1(\phi, \psi) = 0, \qquad \Delta_1\psi = -\frac{1}{(b-u)(b-v)}.$$

By means of (11) we may reduce the linear element to the form

$$(41)\qquad ds^2 = d\phi^2 + \frac{(a-b)(b-c)}{b}\,y^2 d\psi^2.$$

In order that the coördinates be polar geodesic, ψ must be replaced by another parameter measuring the angles between the geodesics. For the ellipsoid equation (39) is

$$(42)\quad (u-b)\sin^2\theta + (v-b)\cos^2\theta = 0.$$

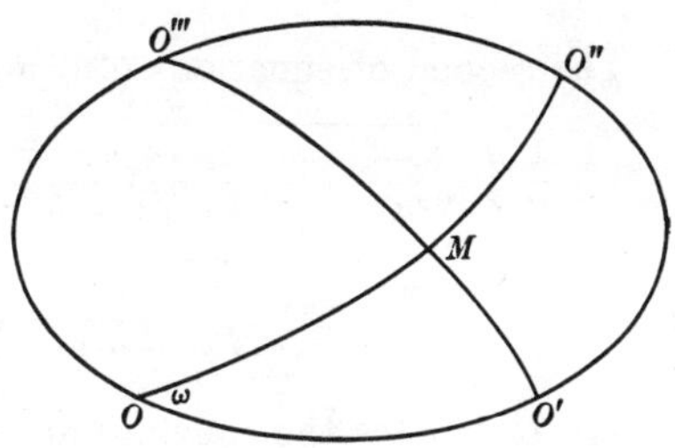

FIG. 23

As previously seen, θ is half of one of the angles between the two geodesics through a point M. As M approaches O along the geodesic joining these two points, the geodesic $O'MO'''$ approaches the section $y=0$. Consequently the angle $2\,\theta$ approaches the angle MOO', denoted by ω, or its supplementary angle. Hence we have from (42)

$$(43)\qquad \lim_{u=b,\, v=b}\left(\frac{b-v}{u-b}\right) = \tan^2\frac{\omega}{2}.$$

We take ω in place of ψ and indicate the relation between them by $\psi = f(\omega)$. From (41) we have

$$\sqrt{G} = \sqrt{(u-b)(b-v)}\,f'(\omega).$$

This expression satisfies the first of conditions (VI, 54). The second is

$$(44)\qquad \lim_{\substack{u=b\\ v=b}} \frac{1}{2}\frac{f'(\omega)}{\sqrt{(u-b)(b-v)}}\left[(b-v)\frac{\partial u}{\partial \phi}-(u-b)\frac{\partial v}{\partial \phi}\right]=1.$$

If we make use of the formulas (III, 11) and (40), we find

$$\frac{\partial u}{\partial \phi}=2\sqrt{\frac{(a-u)(u-c)}{u}}\frac{u-b}{u-v},\quad \frac{\partial v}{\partial \phi}=-2\sqrt{\frac{(a-v)(v-c)}{v}}\frac{b-v}{u-v},$$

so that equation (44) reduces to

$$\lim_{u=b,\, v=b} f'(\omega)\frac{\sqrt{(u-b)(b-v)}}{u-v}\left[\sqrt{\frac{(a-u)(u-c)}{u}}+\sqrt{\frac{(a-v)(v-c)}{v}}\right]=1.$$

By means of (43) we pass from this to

$$(45)\qquad f'(\omega)=\sqrt{\frac{b}{(a-b)(b-c)}}\frac{1}{\sin\omega}.$$

Hence the linear element has the following form due to Roberts *:

$$(46)\qquad ds^2=d\phi^2+\frac{y^2}{\sin^2\omega}d\omega^2.$$

The second of equations (40) may now be put in the form

$$\frac{1}{2}\int\sqrt{\frac{u}{(a-u)(u-c)}}\frac{du}{u-b}-\frac{1}{2}\int\sqrt{\frac{v}{(a-v)(v-c)}}\frac{dv}{v-b}$$
$$=\sqrt{\frac{b}{(a-b)(b-c)}}\log\tan\frac{\omega}{2}+C,$$

where C denotes the constant of integration. In order to evaluate this constant, we consider the geodesic through the point $(0, \sqrt{b}, 0)$. At this point the parameters have the values $u=a$, $v=c$, and the angle ω has a definite value $\overline{\omega}$. Hence the above equation may be replaced by

$$\frac{1}{2}\int_a^u\sqrt{\frac{u}{(a-u)(u-c)}}\frac{du}{u-b}-\frac{1}{2}\int_c^v\sqrt{\frac{v}{(a-v)(v-c)}}\frac{dv}{v-b}$$
$$=\sqrt{\frac{b}{(a-b)(b-c)}}\log\left(\frac{\tan\frac{\omega}{2}}{\tan\frac{\overline{\omega}}{2}}\right).$$

* *Journal de Mathématiques*, Vol. XIII (1848), pp. 1–11.

In like manner, for the umbilical geodesics through one of the other points (not diametrically opposite) we have

$$\frac{1}{2}\int_a^u \sqrt{\frac{u}{(a-u)(u-c)}}\frac{du}{u-b} + \frac{1}{2}\int_c^v \sqrt{\frac{v}{(a-v)(v-c)}}\frac{dv}{v-b}$$
$$= \sqrt{\frac{b}{(a-b)(b-c)}}\log\left(\frac{\tan\frac{\omega'}{2}}{\tan\frac{\overline{\omega}}{2}}\right).$$

It follows at once from these formulas that if M is any point on a line of curvature $u = \text{const.}$ or $v = \text{const.}$, we have respectively

$$\tan\frac{MOO'}{2}\cdot\tan\frac{MO'O}{2} = \text{const.},\qquad \tan\frac{MOO'}{2}\cdot\cot\frac{MO'O}{2} = \text{const.}$$

102. Properties of quadrics. From (18) it follows that for the central quadrics Euler's equation (IV, 34) takes the form

$$\frac{1}{R} = -(v\cos^2\theta + u\sin^2\theta)\sqrt{\frac{abc}{u^3v^3}}.$$

By means of (19) and (37) this reduces to

$$\frac{1}{R} = -\alpha\frac{W^3}{abc}. \tag{47}$$

In like manner, we have for the paraboloids

$$\frac{1}{R} = -\frac{W^3}{z^3}[b + a\alpha(b-a)]. \tag{48}$$

Hence we have:

Along a geodesic or line of curvature on a central quadric the product RW^3 is constant, and on a paraboloid the ratio RW^3/z^3.

Consider any point P on a central quadric and a direction through P. Let α, β, γ be the direction-cosines of the latter. The semi-diameter of the ellipsoid (10) parallel to this direction is given by

$$\frac{1}{\rho^2} = \frac{\alpha^2}{a} + \frac{\beta^2}{b} + \frac{\gamma^2}{c}. \tag{49}$$

By definition $$\alpha = \frac{\cos\theta}{\sqrt{E}}\frac{\partial x}{\partial u} + \frac{\sin\theta}{\sqrt{G}}\frac{\partial x}{\partial v},$$

and similarly for β and γ. When the values of x, y, z, E, G from (11) and (14) are substituted, equation (49) reduces to

$$\frac{1}{\rho^2} = \frac{\cos^2\theta}{u} + \frac{\sin^2\theta}{v}.$$

By means of (19) and (37) this may be reduced to

$$a\rho^2 W^2 = abc. \tag{50}$$

From this follows the theorem of Joachimsthal:

Along a geodesic or a line of curvature on a central quadric the product of the semi-diameter of the quadric parallel to the tangent to the curve at a point P and the distance from the center to the tangent plane at P is constant.

From (47) and (50) we obtain the equation

$$\rho^2 = -RW,$$

for all points on the quadric. Since W is the same for all directions at a point, the maximum and minimum values of ρ and R correspond. Hence we have the theorem:

*In the central section of a quadric parallel to the tangent plane at a point P the principal axes are parallel to the directions of the lines of curvature at P.**

EXAMPLES

1. On a hyperbolic paraboloid, of which the principal parabolas are equal, the locus of a point, the sum or difference of whose distances from the generators through the vertex of the paraboloid is constant, is a line of curvature.

2. Find the radii of curvature and torsion, at the extremity of the mean diameter of an ellipsoid, of an umbilical geodesic through the point.

3. Find the surfaces normal to the tangents to a family of umbilical geodesics on an ellipsoid, and determine the complementary surface (cf. § 76).

4. The geodesic distance of two diametrically opposite umbilical points on an ellipsoid is equal to one half the length of the principal section through the umbilical points.

5. Find the form of the linear element of the hyperboloid of two sheets or the elliptic paraboloid, when the parametric system is polar geodesic with an umbilical point for pole.

6. If M_1 and M_2 are two points of intersection of a geodesic through the umbilical point O with a line of curvature $v = \text{const.}$, then

$$\tan\frac{M_1O'O}{2}\cot\frac{M_2O'O}{2} = \text{const.}$$

* For a more complete discussion of the geodesics on quadrics, the reader is referred to a memoir by v. Braunmühl, in *Math. Annalen*, Vol. XX (1882), pp. 556–586.

7. Given a line of curvature on an ellipsoid and the geodesics tangent to it; the points of intersection of pairs of these geodesics, meeting orthogonally, lie on a sphere.

8. Given the geodesics tangent to two lines of curvature; the points of intersection of pairs of these geodesics, meeting orthogonally, lie on a sphere.

103. Equations of a ruled surface. A surface which can be generated by the motion of a straight line is called a *ruled surface.* Developables are ruled surfaces for which the lines, called the *generators*, are tangent to a curve. As a general thing, ruled surfaces do not possess this property, and in this case they are called *skew surfaces.* Now we make a direct study of ruled surfaces, particularly those of the skew type, limiting our discussion to the case where the generators are real.*

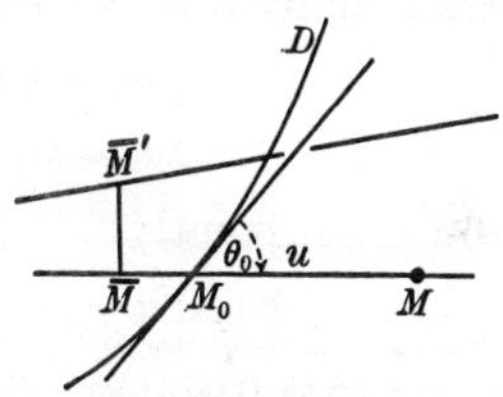

FIG. 24

A ruled surface is completely determined by a curve upon it and the direction of the generators at their points of meeting with the curve. We call the latter the *directrix* D, and the cone formed by drawing through a point lines parallel to the generators the *director-cone.* If the coördinates of a point M_0 of D are x_0, y_0, z_0, expressed in terms of the arc v measured from a point of it, and l, m, n are the direction-cosines of the generator through M_0, the equations of the surface are

$$x = x_0 + lu, \qquad y = y_0 + mu, \qquad z = z_0 + nu, \tag{51}$$

where u is the distance from M_0 to a point M on the generator through M_0. If θ_0 denotes the angle which the generator through M_0 makes with the tangent at M_0 to D, then

$$\cos\theta_0 = x_0'l + y_0'm + z_0'n, \tag{52}$$

where the accent indicates differentiation with respect to v (fig. 24). From (51) we find for the linear element the expression

$$ds^2 = du^2 + 2\cos\theta_0\, du dv + (a^2u^2 + 2bu + 1)dv^2, \tag{53}$$

where we have put for the sake of brevity

$$\begin{cases} a^2 = l'^2 + m'^2 + n'^2, \\ b = l'x_0' + m'y_0' + n'z_0'. \end{cases} \tag{54}$$

* We shall use the term *ruled* to specify the surfaces of the skew type, and *developable* for the others.

Since the generators are geodesics, their orthogonal trajectories can be found by quadratures (§ 92). We arrive at this result directly by remarking that the equation of these trajectories is (III, 26)

$$du + \cos\theta_0 dv = 0,$$

and that θ_0 is a function of v alone.

104. Line of striction. Developable surfaces. We shall now consider the quantities which determine the relative positions of the generators of a ruled surface.

Let g and g' be two generators determined by parametric values v and $v+\delta v$, and let λ, μ, ν denote the direction-cosines of their common perpendicular. If the direction-cosines of g and g' be denoted by l, m, n; $l+\delta l$, $m+\delta m$, $n+\delta n$ respectively, we have

$$(55)\qquad \begin{cases} l\lambda + m\mu + n\nu = 0, \\ (l+\delta l)\lambda + (m+\delta m)\mu + (n+\delta n)\nu = 0, \end{cases}$$

and consequently

$$(56)\qquad \lambda : \mu : \nu = (m\delta n - n\delta m) : (n\delta l - l\delta n) : (l\delta m - m\delta l).$$

From (54) it follows that

$$(mn' - nm')^2 + (nl' - ln')^2 + (lm' - ml')^2 = a^2,$$

and by Taylor's theorem,

$$(57)\qquad l + \delta l = l + l'\delta v + \frac{l''}{\underline{|2}}\delta v^2 + \cdots.$$

Hence equations (56) may be replaced by

$$(58)\qquad \begin{cases} \lambda = \dfrac{1}{a}(mn' - m'n) + \epsilon_1, \\ \mu = \dfrac{1}{a}(nl' - n'l) + \epsilon_2, \\ \nu = \dfrac{1}{a}(lm' - l'm) + \epsilon_3, \end{cases}$$

where $\epsilon_1, \epsilon_2, \epsilon_3$ denote expressions of the first and higher orders in δv.

If $\overline{M}(\overline{x}, \overline{y}, \overline{z})$ and $\overline{M}'(\overline{x}+\delta\overline{x}, \overline{y}+\delta\overline{y}, \overline{z}+\delta\overline{z})$ are the points of meeting of this common perpendicular with g and g' respectively (fig. 24), the length $\overline{M}\overline{M}'$, denoted by Δ, is given by

$$(59)\qquad \Delta = \frac{\delta\overline{x}}{\lambda} = \frac{\delta\overline{y}}{\mu} = \frac{\delta\overline{z}}{\nu},$$

or

$$(60)\qquad \Delta = \lambda\delta\overline{x} + \mu\delta\overline{y} + \nu\delta\overline{z}.$$

From (51), after the manner of (57), we obtain

$$\delta\bar{x} = (x_0' + \bar{u}l')\,\delta v + l\delta\bar{u} + \sigma,$$

where σ involves the second and higher powers of δv. When this and similar values for $\delta\bar{y}$ and $\delta\bar{z}$ are substituted in (60), we have

$$\frac{\Delta}{\delta v} = p + \epsilon, \tag{61}$$

where

$$p = \frac{1}{a}\begin{vmatrix} x_0' & y_0' & z_0' \\ l & m & n \\ l' & m' & n' \end{vmatrix}, \tag{62}$$

and ϵ involves first and higher powers of δv. In consequence of (52) and (54) we have

$$p^2 = \frac{a^2 \sin^2\theta_0 - b^2}{a^2}. \tag{63}$$

As δv approaches zero, the point $\bar{M}$ approaches a limiting position C, which is called the *central point* of the generator. Let α denote the value of u for this point. In order to find its value we remark that it follows from the equations (55) and (59) that

$$\frac{\delta\bar{x}}{\delta v}\frac{\delta l}{\delta v} + \frac{\delta\bar{y}}{\delta v}\frac{\delta m}{\delta v} + \frac{\delta\bar{z}}{\delta v}\frac{\delta n}{\delta v} = 0.$$

If the above expressions for these quantities be substituted in this equation, we have in the limit, as δv approaches zero,

$$a^2\bar{u} + b = 0. \tag{64}$$

Consequently

$$\alpha = -\frac{b}{a^2}. \tag{65}$$

The locus of the central points is called the *line of striction.* Its parametric equation is (64). Evidently $b = 0$ is a necessary and sufficient condition that the line of striction be the directrix.

From (61) and (63) it is seen that the distance between near-by generators is of the second order when

$$a^2 \sin^2\theta_0 - b^2 = 0. \tag{66}$$

Without loss of generality we may take the line of striction for directrix, in which case we may have $\sin\theta_0 = 0$, that is, the generators are tangent to the directrix. Another possibility is afforded by $a = 0$. From (54) it is seen that the only real surfaces satisfying this condition are cylinders. Hence (cf. § 4):

A necessary and sufficient condition that a ruled surface, other than a cylinder, be developable is that the distance between near-by generators be of the second or higher orders; in this case the edge of regression is the line of striction.

105. Central plane. Parameter of distribution. The tangent plane to a ruled surface at a point M necessarily contains the generator through M. It has been found (§ 25) that for a developable surface this plane is tangent at all points of the generator.

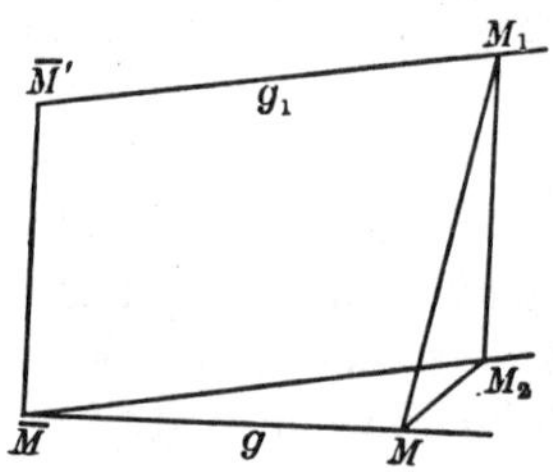

FIG. 25

We shall see that in the case of skew surfaces the tangent plane varies as M moves along the generator. We determine the character of this variation by finding the angle which the tangent plane at M makes with the tangent plane at the central point C of the generator through M. The tangent plane at C is called the *central plane.*

Let g and g_1 be two generators, and $\overline{M}\overline{M}'$ their common perpendicular (fig. 25). Through the point M of g draw the plane normal to g; it meets g_1 in M_1, and the line through $\overline{M}$ parallel to g_1 in M_2. The limiting positions of the planes $M_1M\overline{M}$ and $\overline{M}'\overline{M}M$, as g_1 approaches g, are the tangent planes at M and at C, the limiting position of $\overline{M}$. The angle between these planes, denoted by ϕ', is equal to MM_1M_2, and the angle between g and g_1, denoted by σ, is equal to $M\overline{M}M_2$. By construction MM_2M_1 and $\overline{M}MM_2$ are right angles. Hence

$$\tan\phi' = \frac{MM_2}{M_1M_2} = \frac{M\overline{M}\tan\sigma}{\overline{M}\overline{M}'}.$$

In the limit $\overline{M}$ is the central point C, and so we have

$$\tan\phi = \lim\tan\phi' = \frac{(u-\alpha)\,d\sigma}{p\,dv} = \pm\frac{(u-\alpha)\,a}{p};$$

for we have $\quad d\sigma^2 = \lim(\delta l^2 + \delta m^2 + \delta n^2) = a^2dv^2.$

It is customary to write the above equation in the form

$$(67)\qquad \tan\phi = \frac{u-\alpha}{\beta}.$$

The function β thus defined is called the *parameter of distribution*. It is the limit of the ratio of the shortest distance between two generators and their included angle. As it is independent of the parameter u, we have the theorem:

The tangent of the angle between the tangent plane to a ruled surface at a point M and the central plane is proportional to the distance of M from the central point.

From this it follows that as M moves along a generator from $-\infty$ to $+\infty$, ϕ varies from $-\pi/2$ to $\pi/2$. Hence the tangent planes at the infinitely distant points are perpendicular to the central plane. Since $\beta = 0$ is the condition that a surface be developable, the tangent plane is the same at all points of the generator.

We shall now derive equation (67) analytically. From (51) we find that the direction-cosines of the normal to the surface are of the form

$$(68)\qquad X = \frac{(mz_0' - ny_0') + (mn' - m'n)u}{(a^2u^2 + 2\,bu + \sin^2\theta_0)^{\frac{1}{2}}};$$

the expressions for Y and Z are similar to the above. The direction-cosines X_0, Y_0, Z_0 of the normal at the central point are obtained from these by replacing u by α. From this we have

$$(69)\qquad \cos\phi = \Sigma X X_0$$

$$= \frac{\Sigma(mz_0' - ny_0')^2 + \Sigma(mz_0' - ny_0')(mn' - m'n)(u+\alpha) + a^2u\alpha}{(a^2u^2 + 2\,bu + \sin^2\theta_0)^{\frac{1}{2}}(a^2\alpha^2 + 2\,b\alpha + \sin^2\theta_0)^{\frac{1}{2}}},$$

which leads to

$$\tan^2\phi = \frac{a^4(u-\alpha)^2}{a^2\sin^2\theta_0 - b^2}.$$

From this equation and (67) we have

$$(70)\qquad \beta = \pm\frac{\sqrt{a^2\sin^2\theta_0 - b^2}}{a^2} = \pm\frac{1}{a^2}\begin{vmatrix} x_0' & y_0' & z_0' \\ l & m & n \\ l' & m' & n' \end{vmatrix}.$$

When the surface is defined by its linear element, β is thus determined only to within an algebraic sign. We shall find, however, that this is not the case when the surface is defined by equations of the form (51).

To this end we take a particular generator g for the z-axis. Then for g we have

$$x_0 = y_0 = 0, \qquad l = m = 0, \qquad n = 1, \qquad n' = 0.$$

Let also the central plane be taken for the xz-plane and the central point for the origin. From (68) it follows that $y_0' = 0$. Since the origin is the central point, $b = 0$ and consequently $l' = 0$. Hence the equation of the tangent plane at a point of g has the simple form

$$m'u\xi - x_0'\eta = 0, \tag{71}$$

ξ and η being current coördinates. If the coördinate axes have the usual orientation, and the angle ϕ is measured positively in the direction from the positive x-axis to the positive y-axis, from equation (71) we have

$$\tan\phi = \frac{m'u}{x_0'}. \tag{72}$$

Comparing this with equation (67), we find for β the value x_0'/m'. In order to obtain the same value from (70) for these particular values, we must take the negative sign. Hence we have, in general,

$$\beta = -\frac{1}{a^2}\begin{vmatrix} x_0' & y_0' & z_0' \\ l & m & n \\ l' & m' & n' \end{vmatrix}. \tag{73}$$

It is seen from (72) that, as a point moves along a generator in the direction of u increasing, the motion of the tangent plane is that of a right-handed or left-handed screw, according as β is negative or positive.

EXAMPLES

1. Show that for the ruled surface defined by

$$x = \frac{1}{2}\int (1 - u^2)\phi\, du + \frac{1}{2}(1 - \psi^2)v,$$

$$y = \frac{i}{2}\int (1 + u^2)\phi\, du + \frac{i}{2}(1 + \psi^2)v,$$

$$z = \int u\phi\, du + \psi v,$$

where ϕ and ψ are any functions of u, the directrix and the generators are minimal. Determine under what condition the curvature of the surface is constant.

2. Determine the condition that the directrix of a ruled surface be a geodesic.

3. Prove, by means of (62), that the lines of curvature of a surface $F(x, y, z) = 0$ are defined by

$$\begin{vmatrix} dx, & dy, & dz \\ \frac{\partial F}{\partial x}, & \frac{\partial F}{\partial y}, & \frac{\partial F}{\partial z} \\ d\frac{\partial F}{\partial x}, & d\frac{\partial F}{\partial y}, & d\frac{\partial F}{\partial z} \end{vmatrix} = 0.$$

4. The right helicoid is the only ruled surface whose generators are the principal normals of their orthogonal trajectories. Find the parameter of distribution.

5. Prove for the hyperboloid of revolution of one sheet that:

(*a*) the minimum circle is the line of striction and a geodesic;

(*b*) the parameter of distribution is constant.

6. With every point P on a ruled surface there is associated another point P' on the same generator, such that the tangent planes at these points are perpendicular. Prove that the product $OP \cdot OP'$, where O denotes the central point, has the same value for all points P on the same generator.

7. The normals to a ruled surface along a generator form a hyperbolic paraboloid.

8. The cross-ratio of four tangent planes to a ruled surface at points of a generator is equal to the cross-ratio of the points.

9. If two ruled surfaces are symmetric with respect to a plane, the values of the parameter of distribution for homologous generators differ only in sign.

106. Particular form of the linear element. A number of properties of ruled surfaces are readily obtained when the linear element is given a particular form, which we will now deduce.

Let an orthogonal trajectory of the generators be taken for the directrix. In this case

$$(74) \qquad \theta_0 = \frac{\pi}{2}, \qquad \beta^2 = \frac{a^2 - b^2}{a^4}, \qquad \alpha = -\frac{b}{a^2}.$$

If we make the change of parameters,

$$(75) \qquad u = u, \qquad v_1 = \int_0^v a\, dv,$$

the linear element (53) is reducible to

$$(76) \qquad ds^2 = du^2 + [(u - \alpha)^2 + \beta^2]\, dv_1^2.$$

The angle θ which a curve $v_1 = f(u)$ makes with the generators is given by

$$(77) \qquad \tan\theta = \sqrt{(u - \alpha)^2 + \beta^2}\, f'.$$

Also the expression for the total curvature is

$$(78) \qquad K = -\frac{\beta^2}{[(u - \alpha)^2 + \beta^2]^2}.$$

Hence a real ruled surface has no elliptic points. All the points are hyperbolic except along the generators for which $\beta = 0$, and at the infinitely distant points on each generator. Consequently the linear element of a developable surface may be put in the form

$$ds^2 = du^2 + (u - \alpha)^2 dv^2. \tag{79}$$

Also, in the region of the infinitely distant points of a ruled surface the latter has the character of a developable surface. As another consequence of (78) we have that, for the points of a generator the curvature is greatest in absolute value at the central point, and that at points equally distant from the latter it has the same value.

When the linear element is in the form (76), the Gauss equation of geodesics (VI, 56) has the form

$$\sqrt{(u-\alpha)^2 + \beta^2}\, d\theta + (u - \alpha)\, dv_1 = 0.$$

An immediate consequence is the theorem of Bonnet:

If a curve upon a ruled surface has two of the following properties, it has the third also, namely that it cut the generators under constant angle, that it be a geodesic and that it be the line of striction.

A surface of this kind is formed by the family of straight lines which cut a twisted curve under constant angle and are perpendicular to its principal normals. A particular case is the surface formed of the binormals of a curve. It is readily shown from (73) that the parameter of distribution of this surface is equal to the radius of torsion of the curve.

107. Asymptotic lines. Orthogonal parametric systems. The generators are necessarily asymptotic lines on a ruled surface. We consider now the other family of these lines. From (51) and (68) we find

$$D = 0, \quad D' = \frac{1}{H}\begin{vmatrix} l' & m' & n' \\ l & m & n \\ x_0' & y_0' & z_0' \end{vmatrix}, \quad D'' = \frac{1}{H}\begin{vmatrix} x_0'' + l''u & l & x_0' + l'u \\ y_0'' + m''u & m & y_0' + m'u \\ z_0'' + n''u & n & z_0' + n'u \end{vmatrix}. \tag{80}$$

Hence the differential equation of the other family of asymptotic lines is of the form

$$\frac{du}{dv} + Lu^2 + Mu + N = 0,$$

where L, M, N are functions of v. As this is an equation of the Riccati type, we have, from § 14, the theorem of Serret:

The four points in which each generator of a ruled surface is cut by four curved asymptotic lines are in constant cross-ratio.

From § 14 it follows also that when one of these asymptotic lines is known the others can be found by quadratures.

When the surface is referred to an orthogonal system and the linear element is in the form (76), written

$$ds^2 = du^2 + a^2[(u-\alpha)^2 + \beta^2]\,dv^2, \tag{81}$$

the expressions (80) can be given a simpler form.

From (73) and (81) we have

$$H = a[(u-\alpha)^2 + \beta^2]^{\frac{1}{2}}, \qquad D' = \frac{a^2\beta}{H}.$$

From the equations $\quad \Sigma x_0' l = 0, \quad \Sigma x_0'^2 = 1, \qquad \Sigma l^2 = 1,$

and (54) we obtain, by differentiation,

$$\Sigma x_0' x_0'' = 0, \qquad \Sigma l l' = 0, \qquad \Sigma x_0'' l = -b,$$
$$\Sigma l' l'' = a a', \quad \Sigma l l'' = -a^2, \quad \Sigma l' x_0'' = b' - t,$$

where t is defined by $\qquad \Sigma l'' x_0' = t.$

If the expression for D'' in (80) be multiplied by the determinant of the right-hand member of (73), and the result be divided by its equal, $-a^2\beta$, we have, in consequence of the above identities,

$$D'' = -\frac{1}{Ha^2\beta}[u^2(ta^2 - aa'b) + u(2tb - aa' - bb') + t - b'].$$

If equations (74) be solved for a and b as functions of α and β, and the resulting expressions be substituted in this equation, we have

$$D'' = -\frac{a^3}{H}\{r[(u-\alpha)^2 + \beta^2] + \beta'(u-\alpha) + \beta\alpha'\},$$

where the primes indicate differentiation with respect to v_1, given by (75), and r is defined by

$$a^3\beta r = t - \frac{a'}{a}b.$$

From the above equations it follows that the mean curvature (cf. § 52) is expressible in the form

$$\frac{1}{\rho_1} + \frac{1}{\rho_2} = -\frac{r[(u-\alpha)^2 + \beta^2] + \beta'(u-\alpha) + \beta\alpha'}{[(u-\alpha)^2 + \beta^2]^{\frac{3}{2}}}. \tag{82}$$

EXAMPLES

1. When the linear element of a ruled surface is in the form (76), the direction-cosines of the limiting position of the common perpendicular to two generators are given by

$$\frac{x_0' + \alpha l'}{a\beta}, \qquad \frac{y_0' + \alpha m'}{a\beta}, \qquad \frac{z_0' + \alpha n'}{a\beta}.$$

2. Prove that the developable surfaces are the only ruled surfaces with real generators whose total curvature is constant.

3. Show that the perpendicular upon the z-axis from any point of the cubic $x = u$, $y = u^2$, $z = u^3$ lies in the osculating plane at the point, and find the asymptotic lines on the ruled surface generated by this perpendicular.

4. Determine the function ϕ in the equations

$$x = u, \qquad y = u^n, \qquad z = \phi(u),$$

so that the osculating plane at any point M of this curve shall pass through the projection P of M on the y-axis. Find the asymptotic lines on the surface generated by the line MP.

5. Show that the equations

$$x = u \sin\theta \cos\psi, \qquad y = u \sin\theta \sin\psi, \qquad z = v + u\cos\theta,$$

where θ and ψ are functions of v, define the most general ruled surface with a rectilinear directrix, and prove that the equation of asymptotic lines can be integrated by two quadratures. Discuss the case where θ is constant.

6. Concerning the curved asymptotic lines on a ruled surface the following are to be proved:

(a) if one of them is an orthogonal trajectory of the generators, the determination of the rest reduces to quadratures;

(b) if two of them are orthogonal trajectories, they are curves of Bertrand;

(c) if all of them are orthogonal trajectories, the surface is a right helicoid.

7. Determine the condition that the line of striction be an asymptotic line, and show that in this case the other curved asymptotic lines can be found by quadratures.

8. Find a ruled surface of the fourth degree which is generated by a line passing through the two lines $x = 0$, $y = 0$; $z = 0$, $x + y + z = 1$. Show that these lines and the line $x = 0$, $x + y + z = 1$ are double lines. Find the line of striction.

9. The right helicoid is the only ruled surface each of whose lines of curvature cuts the generators under constant angle; however, on any other ruled surface there are in general four lines of curvature which have this property.

108. Minimal surfaces. In 1760 Lagrange extended to double integrals the Euler theorems about simple integrals in the calculus of variations, and as an example he proposed the following problem*:

Given a closed curve C and a connected surface S bounded by the curve; to determine S so that the inclosed area shall be a minimum.

If the surface be defined by the equation

$$z = f(x, y),$$

the problem requires the determination of $f(x, y)$ so that the integral (cf. Ex. 1, p. 77)

$$\iint \sqrt{1 + p^2 + q^2}\, dx dy$$

* *Œuvres de Lagrange*, Vol. I, pp. 354–357. Paris, 1867.

extended over the portion of the surface bounded by C shall be a minimum. As shown by Lagrange, the condition for this is

$$\frac{\partial}{\partial x}\left(\frac{p}{\sqrt{1+p^2+q^2}}\right)+\frac{\partial}{\partial y}\left(\frac{q}{\sqrt{1+p^2+q^2}}\right)=0, \tag{83}$$

or, in other form,

$$(1+q^2)r-2pqs+(1+p^2)t=0. \tag{84}$$

Lagrange left the solution of the problem in this form, and Meusnier,* sixteen years later, proved that this equation is equivalent to the vanishing of the mean curvature (§ 52), thus showing that the surfaces furnishing the solution of Lagrange's problem are characterized by the geometrical property which now is usually taken as the definition of *minimal surfaces;* however, the name indicates the connection with the definition of Lagrange.†

In what follows we purpose giving a discussion of minimal surfaces from the standpoint of their definition as the surfaces whose mean curvature is zero at all the points. At each point of such a surface the principal radii differ only in sign, and so every point is a hyperbolic point and its Dupin indicatrix is an equilateral hyperbola. Consequently minimal surfaces are characterized by the property that their asymptotic lines form an orthogonal system. Moreover, the tangents to the two asymptotic lines at a point bisect the angles between the lines of curvature at the point, and vice versa.

We recall the formulas giving the relations between the fundamental quantities of a surface and its spherical representation (IV, 70):

$$\mathcal{E}=K_mD-KE,\quad \mathcal{F}=K_mD'-KF,\quad \mathcal{G}=K_mD''-KG. \tag{85}$$

From these we have at once the theorem:

The necessary and sufficient condition that the spherical representation of a surface be conformal is that it be minimal or a sphere.

* Mémoire sur la courbure des surfaces, *Mémoires des Savants étrangers,* Vol. X (1785), p. 477.

† For a historical sketch of the development of the theory of minimal surfaces and a complete discussion of them the reader is referred to the *Leçons* of Darboux (Vol. I, pp. 267 *et seq.*). The questions in the calculus of variations involved in the study of minimal surfaces are treated by Riemann, *Gesammelte Werke*, p. 287 (Leipzig, 1876); and by Schwarz, *Gesammelte Abhandlungen*, Vol. I, pp. 223, 270 (Berlin, 1890).

Hence isothermal orthogonal systems on the surface are represented by similar systems on the sphere, and conversely. All the isothermal orthogonal systems on the sphere are known (§§ 35, 40). Suppose that one of these systems is parametric and that the linear element is

$$d\sigma^2 = \lambda\,(du^2 + dv^2).$$

From the general condition for minimal surfaces (IV, 77), namely

$$\mathcal{E}D'' + \mathcal{G}D - 2\mathcal{F}D' = 0, \tag{86}$$

it follows that in this case

$$D + D'' = 0.$$

In consequence of this the Codazzi equations (V, 27) are reducible to

$$\frac{\partial D}{\partial v} - \frac{\partial D'}{\partial u} = 0, \qquad \frac{\partial D}{\partial u} + \frac{\partial D'}{\partial v} = 0. \tag{87}$$

By eliminating D or D' we find that both D and D' are integrals of the equation

$$\frac{\partial^2 \theta}{\partial u^2} + \frac{\partial^2 \theta}{\partial v^2} = 0.$$

Hence the most general form of D' is

$$D' = \phi\,(u + iv) + \psi\,(u - iv), \tag{88}$$

where ϕ and ψ are arbitrary functions. Then from (87) we have

$$D = -D'' = -i\,(\phi - \psi) + c, \tag{89}$$

where c is the constant of integration. To each pair of functions ϕ, ψ there corresponds a minimal surface whose Cartesian coördinates are given by the quadratures (V, 26), namely

$$\frac{\partial x}{\partial u} = -\frac{1}{\lambda}\left(D\frac{\partial X}{\partial u} + D'\frac{\partial X}{\partial v}\right), \qquad \frac{\partial x}{\partial v} = -\frac{1}{\lambda}\left(D'\frac{\partial X}{\partial u} + D''\frac{\partial X}{\partial v}\right), \tag{90}$$

and similar expressions in y and z. Evidently the surface is real only when ϕ and ψ are conjugate functions.

In obtaining the preceding results we have tacitly assumed that neither D nor D' is zero. We notice that either may be zero and then the other is a constant, which is zero only for the plane. These results may be stated thus:

Every isothermal system on the sphere is the representation of the lines of curvature of a unique minimal surface and of the asymptotic lines of another minimal surface.

The converse also is true, namely:

The spherical representations of the lines of curvature and of the asymptotic lines of a minimal surface are isothermal systems.

For, if the lines of curvature are parametric, equation (86) may be replaced by

$$D = \rho\mathcal{E}, \qquad D'' = -\rho\mathcal{G},$$

where ρ is equal to either principal radius to within its algebraic sign. When these values and $D' = \mathcal{F} = 0$ are substituted in the Codazzi equations (V, 27), we obtain

$$\frac{\partial}{\partial v}(\rho\mathcal{E}) = 0, \qquad \frac{\partial}{\partial u}(\rho\mathcal{G}) = 0,$$

so that $\mathcal{E}/\mathcal{G} = U/V$, which proves the first part of the theorem (§ 41).

When the asymptotic lines are parametric, we have $D = D'' = \mathcal{F} = 0$, and equations (V, 27) reduce to

$$\frac{\partial}{\partial u}\left(\sqrt{\frac{\mathcal{G}}{\mathcal{E}}}\, D'\right) = 0, \qquad \frac{\partial}{\partial v}\left(\sqrt{\frac{\mathcal{E}}{\mathcal{G}}}\, D'\right) = 0,$$

from which it follows that $\mathcal{E}/\mathcal{G} = U/V$.

109. Lines of curvature and asymptotic lines. Adjoint minimal surfaces. We return to the consideration of equations (87) and investigate first the minimal surface with its lines of curvature represented by an isothermal system. Without loss of generality,* we may take

$$D = -D'' = 1, \qquad D' = 0. \tag{91}$$

From (IV, 77) it follows that

$$\frac{1}{\rho_1\rho_2} = -\frac{1}{\rho_1^2} = -\lambda^2, \qquad E = G = \rho,$$

where

$$\rho = |\rho_1| = |\rho_2|.$$

Hence we have the theorem:

The parameters of the lines of curvature of a minimal surface may be so chosen that the linear elements of the surface and of its spherical representation have the respective forms

$$ds^2 = \rho\,(du^2 + dv^2), \qquad d\sigma^2 = \frac{1}{\rho}(du^2 + dv^2),$$

where ρ is the absolute value of each principal radius.

* Any other value of the constant leads to homothetic surfaces.

In like manner we may take, for the solution of equations (87),

$$D = D'' = 0, \qquad D' = 1. \tag{92}$$

Again we find

$$\frac{1}{\rho_1 \rho_2} = -\frac{1}{\rho_1^2} = -\lambda^2, \qquad E = G = \rho,$$

so that we have a result similar to the above:

The parameters of the asymptotic lines of a minimal surface may be so chosen that the linear elements of the surface and of its spherical representation have the respective forms

$$ds^2 = \rho\,(du^2 + dv^2), \qquad d\sigma^2 = \frac{1}{\rho}\,(du^2 + dv^2),$$

where ρ is the absolute value of each principal radius.

From the symmetric form of equations (87) it follows that if (88) and (89) represent one set of solutions, another set is given by

$$D_1 = -D_1'' = \phi + \psi, \qquad D_1' = i\,(\phi - \psi) - c.$$

These values are such that

$$DD_1'' + D''D_1 - 2\,D'D_1' = 0,$$

which is the condition that asymptotic lines on either surface correspond to a conjugate system on the other (§ 56). When this condition is satisfied by two minimal surfaces, and the tangent planes at corresponding points are parallel, the two surfaces are said to be the *adjoints* of one another. Hence a pair of functions ϕ, ψ determines a pair of *adjoint minimal surfaces.* When, in particular, the asymptotic lines on one surface are parametric, the functions have the values (92), and on the other the values (91). It follows, then, from (90), that between the Cartesian coördinates of a minimal surface and its adjoint the following relations hold:

$$\frac{\partial x_1}{\partial u} = \frac{\partial x}{\partial v}, \qquad \frac{\partial x_1}{\partial v} = -\frac{\partial x}{\partial u}; \tag{93}$$

and similar expressions in the y's and z's, when the parametric curves are asymptotic on the locus of (x, y, z).

110. Minimal curves on a minimal surface. The lines of length zero upon a minimal surface are of fundamental importance. When they are taken for parametric curves, the equations of the surface take a simple form, which we shall now obtain.

Since the lines of length zero, or minimal lines, are parametric, we have

$$E = G = 0. \tag{94}$$

From (85) it follows that the parametric lines on the sphere also are minimal lines, that is, the imaginary rectilinear generators. And from (86) we find that D' is zero. Conversely, when the latter is zero, and the parametric lines are minimal curves, it follows from (IV, 33) that K_m is equal to zero. Hence:

*A necessary and sufficient condition that a surface be minimal is that the lines of length zero form a conjugate system.**

In consequence of (94) and (VI, 26) the point equation of a minimal surface, referred to its minimal lines, is

$$\frac{\partial^2 \theta}{\partial u \partial v} = 0.$$

Hence the finite equations of the surface are of the form

$$x = U_1 + V_1, \qquad y = U_2 + V_2, \qquad z = U_3 + V_3, \tag{95}$$

where U_1, U_2, U_3 are functions of u alone, and V_1, V_2, V_3 are functions of v alone, satisfying the conditions

$$U_1'^2 + U_2'^2 + U_3'^2 = 0, \qquad V_1'^2 + V_2'^2 + V_3'^2 = 0. \tag{96}$$

From (95) it is seen that minimal surfaces are surfaces of translation (§ 81), and from (96) that the generators are minimal curves (§ 22). In consequence of the second theorem of § 81 we may state this result thus:

A minimal surface is the locus of the mid-points of the joins of points on two minimal curves.

In § 22 we found that the Cartesian coördinates of any minimal curve are expressible in the form

$$\int (1 - u^2) F(u)\, du, \qquad i \int (1 + u^2) F(u)\, du, \qquad 2 \int u F(u)\, du. \tag{97}$$

* This follows also from the fact that an equilateral hyperbola is the only conic for which the directions with angular coefficients $\pm i$ are conjugate.

Hence by the above theorem the following equations, due to Enneper *, define a minimal surface referred to its minimal lines:

$$(98)\qquad \begin{cases} x = \dfrac{1}{2}\displaystyle\int (1-u^2)\,F(u)\,du + \dfrac{1}{2}\displaystyle\int (1-v^2)\,\Phi(v)\,dv, \\ y = \dfrac{i}{2}\displaystyle\int (1+u^2)\,F(u)\,du - \dfrac{i}{2}\displaystyle\int (1+v^2)\,\Phi(v)\,dv, \\ z = \displaystyle\int u\,F(u)\,du + \displaystyle\int v\,\Phi(v)\,dv, \end{cases}$$

where F and Φ are any analytic functions whatever. Moreover, any minimal surface can be defined by equations of this form. For, the only apparent lack of generality is due to the fact that the algebraic signs of the expressions (98) are not determined by equations (96), and consequently the signs preceding the terms in the right-hand members of equations (98) could be positive or negative. But it can be shown that by a suitable change of the parameters and of the functions F and Φ all of these cases reduce to (98). Thus, for example, we consider the surface defined by the equations which result when the second terms of the right-hand members of (98) are replaced by

$$\frac{1}{2}\int (1-v_1^2)\,\Phi_1(v_1)\,dv_1, \qquad \frac{i}{2}\int (1+v_1^2)\,\Phi_1(v_1)\,dv_1, \qquad \int v_1\Phi_1(v_1)\,dv_1.$$

In order that the surface thus defined can be brought into coincidence, by a translation, with the surface (98), we must have

$$(1-v_1^2)\Phi_1 dv_1 = (1-v^2)\,\Phi\,dv, \qquad (1+v_1^2)\,\Phi_1 dv_1 = -(1+v^2)\,\Phi\,dv,$$
$$v_1\Phi_1 dv_1 = v\Phi\,dv.$$

Dividing these equations, member by member, we have

$$\frac{1-v_1^2}{1-v^2} = -\frac{1+v_1^2}{1+v^2} = \frac{v_1}{v},$$

from which it follows that

$$v_1 = -\frac{1}{v}.$$

Substituting this value in the last of the above equations, we find

$$\Phi_1(v_1) = -v^4\,\Phi(v),$$

* *Zeitschrift für Mathematik und Physik*, Vol. IX (1864), p. 107.

and this value satisfies the other equations. Similar results follow when another choice of signs is made. The reason for the particular choice made in (98) will be seen when we discuss the reality of the surfaces.

Incidentally we have proved the theorem:

When a minimal surface is defined by equations (98), *the necessary and sufficient condition that the two generating curves be congruent is that*

$$F(u) = -\frac{1}{u^4}\Phi\left(-\frac{1}{u}\right). \tag{99}$$

From (98) we obtain

$$E = 0, \qquad F = \frac{1}{2}(1+uv)^2 F(u)\Phi(v), \qquad G = 0,$$

so that the linear element is

$$ds^2 = (1+uv)^2 F(u)\Phi(v)\,dudv. \tag{100}$$

We find for the expressions of the direction-cosines of the normal

$$X = \frac{u+v}{1+uv}, \qquad Y = i\frac{v-u}{1+uv}, \qquad Z = \frac{uv-1}{1+uv}, \tag{101}$$

and the linear element of the sphere is

$$d\sigma^2 = \frac{4\,dudv}{(1+uv)^2}.$$

Also we have

$$D = -F(u), \qquad D' = 0, \qquad D'' = -\Phi(v), \tag{102}$$

so that the equations of the lines of curvature and of the asymptotic lines are respectively

$$F(u)\,du^2 - \Phi(v)\,dv^2 = 0, \tag{103}$$

$$F(u)\,du^2 + \Phi(v)\,dv^2 = 0. \tag{104}$$

These equations are of such a form that we have the theorem:

When a minimal surface is referred to its minimal lines, the finite equations of the lines of curvature and asymptotic lines are given by quadratures, which are the same in both cases.

In order that a surface be real its spherical representation must be real. Consequently u and v must be conjugate imaginaries, as is seen from (101) and § 13, and the functions F and Φ must be conjugate imaginary. Hence if $R\theta$ denotes the real part of a function θ, all real minimal surfaces are defined by

$$x = R\int(1-u^2)F(u)\,du, \qquad y = R\int i(1+u^2)F(u)\,du,$$
$$z = R\int 2\,uF(u)\,du,$$

where $F(u)$ is any function whatever of a complex variable u. In like manner the equations of the lines of curvature may be written in the form

$$(105) \qquad R\int\sqrt{F(u)}\,du = \text{const.}, \qquad R\int i\sqrt{F(u)}\,du = \text{const.}$$

111. Double minimal surfaces. It is natural to inquire whether the same minimal surface can be defined in more than one way by equations of the form (98). We assume that this is possible, and indicate by u_1, v_1 and $F_1(u_1)$, $\Phi_1(v_1)$ the corresponding parameters and functions. As the parameters u_1, v_1 refer to the lines of length zero on the surface, each is a function of either u or v. In order to determine the forms of the latter we make use of the fact that the positive directions of the normal to the surface in the two forms of parametric representation may have the same or opposite senses. When they have the same sense, the expressions (101) and similar ones in u_1 and v_1 must be equal respectively. In this case

$$(106) \qquad u_1 = u, \qquad v_1 = v.$$

If the senses are opposite, the respective expressions are equal to within algebraic signs. From the resulting equations we find

$$(107) \qquad u_1 = -\frac{1}{v}, \qquad v_1 = -\frac{1}{u}.$$

When we compare equations (98) with analogous equations in u_1 and v_1, we find that for the case (106) we must have

$$F_1(u_1) = F(u), \qquad \Phi_1(v_1) = \Phi(v),$$

and for the case (107)

$$F_1(u_1) = -v^4\Phi(v), \qquad \Phi_1(v_1) = -u^4F(u).$$

Hence we have the theorem:

A necessary and sufficient condition that two minimal surfaces, determined by the pairs of functions F, Φ and F_1, Φ_1, be congruent is that

$$(108) \qquad F_1(u) = -\frac{1}{u^4}\Phi\left(-\frac{1}{u}\right), \qquad \Phi_1(v) = -\frac{1}{v^4}F\left(-\frac{1}{v}\right);$$

to the point (u, v) on one surface corresponds the point $\left(-\frac{1}{v}, -\frac{1}{u}\right)$ on the other, and the normals at these points are parallel but of different sense.

In general, the functions F and F_1 as given by (108) are not the same. If they are, so also are Φ and Φ_1. In this case the right-hand members of equations (98) are unaltered when u and v are replaced by $-1/v$ and $-1/u$ respectively. Hence the Cartesian coördinates of the points (u, v) and $\left(-\frac{1}{v}, -\frac{1}{u}\right)$ differ at most by constants. And so the regions of the surface about these points either coincide or can be brought into coincidence by a translation. In the latter case the surface is periodic and consequently transcendental.

Suppose that it is not periodic, and consider a point $P_0(u_0, v_0)$. As u varies continuously from u_0 to $-1/v_0$, v varies from v_0 to $-1/u_0$, and the point describes a closed curve on the surface by returning to P_0. But now the positive normal is on the other side of the surface. Hence these surfaces have the property that a point can pass continuously from one side to the other without going through the surface. On this account they were called *double minimal surfaces* by Lie,* who was the first to study them.

From the third theorem of § 110 it follows that double minimal surfaces are characterized by the property that the minimal curves in both systems are congruent. The equations of such a surface may be written

$$x = \tfrac{1}{2}[f_1(u) + f_1(v)], \quad y = \tfrac{1}{2}[f_2(u) + f_2(v)], \quad z = \tfrac{1}{2}[f_3(u) + f_3(v)].$$

The surface is consequently the locus of the mid-points of the chords of the curve

$$\xi = f_1(u), \qquad \eta = f_2(u), \qquad \zeta = f_3(u),$$

which lies upon the surface and is the envelope of the parametric curves.

* *Math. Annalen*, Vol. XIV (1878), pp. 345–350.

EXAMPLES

1. The focal sheets of a minimal surface are applicable to one another and to the surface of revolution of the evolute of the catenary about the axis of the latter.

2. Show that there are no minimal surfaces with the minimal lines in one family straight.

3. If two minimal surfaces correspond with parallelism of tangent planes, the minimal curves on the two surfaces correspond.

4. If two minimal surfaces correspond with parallelism of tangent planes, and the joins of corresponding points be divided in the same ratio, the locus of the points of division is a minimal surface.

5. Show that the right helicoid is defined by $F(u) = im/2u^2$, where m is a real constant.

6. The surface for which $F(u) = \dfrac{2}{1-u^4}$ is called the *surface of Scherk*. Find its equation in the Monge form $z = f(x, y)$. Show that it is doubly periodic and that it is a surface of translation with real generators which are in perpendicular planes.

7. By definition a *meridian curve* on a surface is one whose spherical representation is a great circle on the unit sphere. Show that the surface of Scherk possesses two families of plane meridian curves.

112. Algebraic minimal surfaces. Weierstrass* remarked that formulas (98) can be put in a form free of all quadratures. This is done by replacing $F(u)$ and $\Phi(v)$ by $f'''(u)$ and $\phi'''(v)$, where the accents indicate differentiation, and then integrating by parts. This gives

$$(109)\quad \begin{cases} x = \dfrac{1-u^2}{2}f''(u) + uf'(u) - f(u) + \dfrac{1-v^2}{2}\phi''(v) + v\phi'(v) - \phi(v), \\ y = i\dfrac{1+u^2}{2}f''(u) - iuf'(u) + if(u) - i\dfrac{1+v^2}{2}\phi''(v) + iv\phi'(v) \\ \qquad - i\phi(v), \\ z = uf''(u) - f'(u) + v\phi''(v) - \phi'(v). \end{cases}$$

It is clear that the surface so defined is real when f and ϕ are conjugate imaginary functions. In this case the above formulas may be written:

$$(110)\quad \begin{cases} x = R[(1-u^2)f''(u) + 2uf'(u) - 2f(u)], \\ y = Ri[(1+u^2)f''(u) - 2uf'(u) + 2f(u)], \\ z = R[2uf''(u) - 2f'(u)]. \end{cases}$$

* *Monatsberichte der Berliner Akademie* (1866), p. 619.

However, it is not necessary, as Darboux * has pointed out, that f and ϕ be conjugate imaginaries in order that the surface be real. For, equations (109) are unaltered if f and ϕ be replaced by

$$f_1(u) = f(u) + A(1-u^2) + Bi(1+u^2) + 2\,Cu,$$
$$\phi_1(v) = \phi(v) - A(1-v^2) + Bi(1+v^2) - 2\,Cv,$$

where A, B, C are any constants whatever. Evidently, if f and ϕ are conjugate imaginaries, the same is not true in general of f_1 and ϕ_1; but the surface was real for the former and consequently is real for the latter also. It is readily found that f_1 and ϕ_1 are conjugate imaginary functions only in case A, B, C are pure imaginaries.

Formulas (109) are of particular value in the study of algebraic surfaces. Thus, it is evident that the surface is algebraic when f and ϕ are algebraic. Conversely, every algebraic minimal surface is determined by algebraic functions f and ϕ. In proving this we follow the method suggested by Weierstrass.†

We establish first the following lemma:

Given a function $\Phi(\xi + i\eta)$ and let $\Psi(\xi, \eta)$ denote the real part of Φ; if in a certain domain an algebraic relation exists between Ψ, ξ, and η, Φ is an algebraic function of $\xi + i\eta$.

If the point $\xi = 0$, $\eta = 0$ does not lie within the domain under consideration, this can be effected by a change of variables without vitiating the argument. Assuming that this has been done, we develop the function Φ in a power series, thus:

$$\Phi = a_0 + ib_0 + (a_1 + ib_1)(\xi + i\eta) + (a_2 + ib_2)(\xi + i\eta)^2 + \cdots,$$

where the a's and b's are real constants. Evidently Ψ is given by

$$\Psi = a_0 + \tfrac{1}{2}(a_1 + ib_1)(\xi + i\eta) + \tfrac{1}{2}(a_2 + ib_2)(\xi + i\eta)^2 + \cdots$$
$$\tfrac{1}{2}(a_1 - ib_1)(\xi - i\eta) + \tfrac{1}{2}(a_2 - ib_2)(\xi - i\eta)^2 + \cdots.$$

Let $F(\Psi, \xi, \eta) = 0$ denote a rational integral relation between Ψ, ξ, and η. When Ψ has been replaced by the above value, and the resulting expression is arranged in powers of ξ and η, the coefficient of every term is identically zero. They will continue to be zero when ξ and η have been replaced by two complex quantities

* Vol. I, p. 293. † *Monatsberichte der Berliner Akademie* (1867), pp. 511–518.

α and β, provided that the development remains convergent. The condition for the latter is that the moduli of α and β be each one half the modulus of $\xi + i\eta$. This condition is satisfied if we take

$$\alpha = \frac{\xi + i\eta}{2}, \qquad \beta = \frac{\xi + i\eta}{2i}.$$

Now we have

$$\Psi(\alpha, \beta) = \frac{1}{2}(a_0 - ib_0) + \frac{1}{2}\Phi(\xi + i\eta),$$

so that

$$F[\Psi(\alpha, \beta), \alpha, \beta] = F_1[\Phi(\xi + i\eta), \xi + i\eta] = 0,$$

which proves the lemma.

In applying this lemma to real minimal surfaces we note from (101) that

$$\frac{X}{1-Z} = \frac{u+v}{2}, \qquad \frac{Y}{1-Z} = \frac{u-v}{2i};$$

consequently the left-hand members of these equations are equal to u_1 and v_1 respectively, where $u = u_1 + iv_1$. When the surface is algebraic there exists an algebraic relation between the functions $\frac{X}{1-Z}, \frac{Y}{1-Z}$ and each of the Cartesian coördinates.* Since, then, there is an algebraic relation between u_1, v_1, and each of the coördinates given by (110), it follows from the lemma that each of the three expressions

$$\phi_1(u) = (1 - u^2)f''(u) + 2uf'(u) - 2f(u),$$

$$\phi_2(u) = i(1 + u^2)f''(u) - 2iuf'(u) + 2if(u),$$

$$\phi_3(u) = 2uf''(u) - 2f'(u)$$

are algebraic functions of u, and so also is $f(u)$; for,

$$f(u) = \frac{1}{4}(u^2 - 1)\phi_1(u) - \frac{i}{4}(u^2 + 1)\phi_2(u) - \frac{1}{2}u\phi_3(u).$$

Hence we have demonstrated the theorem of Weierstrass:

The necessary and sufficient condition that equation (110) *define an algebraic surface is that* $f(u)$ *be algebraic.*

* For, if the surface is defined by $F(x, y, z) = 0$, the direction-cosines of the normal are functions of x, y, z. Eliminating two of the latter between $\frac{X}{1-Z}, \frac{Y}{1-Z}$, and $F(x, y, z) = 0$, we have a relation of the kind described.

113. Associate surfaces. When the equations of a minimal surface S are written in the abbreviated form (95), the linear element is

$$ds^2 = 2(dU_1 dV_1 + dU_2 dV_2 + dU_3 dV_3).$$

This is the linear element also of a surface defined by

$$x_\alpha = e^{i\alpha} U_1 + e^{-i\alpha} V_1, \quad y_\alpha = e^{i\alpha} U_2 + e^{-i\alpha} V_2, \quad z_\alpha = e^{i\alpha} U_3 + e^{-i\alpha} V_3, \tag{111}$$

where α is any constant. There are an infinity of such surfaces, called *associate minimal surfaces*. It is readily found that the direction-cosines of the normal to any one have the values (101). Hence any two associate minimal surfaces defined by (111) have their tangent planes at corresponding points parallel, and are applicable.

Of particular interest is the surface S_1 for which $\alpha = \pi/2$. Its equations are

$$\left\{\begin{aligned} x_1 &= \frac{i}{2}\int (1-u^2)F(u)\,du - \frac{i}{2}\int (1-v^2)\Phi(v)\,dv, \\ y_1 &= -\frac{1}{2}\int (1+u^2)F(u)\,du - \frac{1}{2}\int (1+v^2)\Phi(v)\,dv, \\ z_1 &= i\int uF(u)\,du - i\int v\Phi(v)\,dv. \end{aligned}\right. \tag{112}$$

In order to show that S_1 is the adjoint (§ 109) of S, we have only to prove that the asymptotic lines on either surface correspond to the lines of curvature on the other. For S_1 the equations of the lines of curvature and asymptotic lines are

$$iF(u)\,du^2 + i\Phi(v)\,dv^2 = 0,$$
$$iF(u)\,du^2 - i\Phi(v)\,dv^2 = 0,$$

respectively. Comparing these with (103) and (104), we see that the desired condition is satisfied.

From (98) and (112) we obtain the identities

$$\left\{\begin{aligned} &dx^2 + dy^2 + dz^2 = dx_1^2 + dy_1^2 + dz_1^2, \\ &dx\,dx_1 + dy\,dy_1 + dz\,dz_1 = 0. \end{aligned}\right. \tag{113}$$

The latter has the following interpretation:

On two adjoint minimal surfaces at points corresponding with parallelism of tangent planes the tangents to corresponding curves are perpendicular.

From (105) it follows that if we put

$$\bar{u} + i\bar{v} = \int \sqrt{F(u)}\, du,$$

the curves $\bar{u} = \text{const.}$ and $\bar{v} = \text{const.}$ on the surface are its lines of curvature. Moreover, for an associate surface the lines of curvature are given by

$$R\left[e^{\frac{i\alpha}{2}}(\bar{u} + i\bar{v})\right] = \text{const.}, \qquad R\left[ie^{\frac{i\alpha}{2}}(\bar{u} + i\bar{v})\right] = \text{const.},$$

or

$$\bar{u} \cos\frac{\alpha}{2} - \bar{v} \sin\frac{\alpha}{2} = \text{const.}, \qquad \bar{u} \sin\frac{\alpha}{2} + \bar{v} \cos\frac{\alpha}{2} = \text{const.}$$

From this result follows the theorem:

The lines of curvature on a minimal surface associate to a surface S correspond to the curves on S which cut its lines of curvature under the constant angle $\alpha/2$.

Since equations (111) may be written

$$\text{(114)} \qquad \begin{cases} x_\alpha = x \cos\alpha + x_1 \sin\alpha, \\ y_\alpha = y \cos\alpha + y_1 \sin\alpha, \\ z_\alpha = z \cos\alpha + z_1 \sin\alpha, \end{cases}$$

the plane determined by the origin of coördinates, a point P on a minimal surface and the corresponding point on its adjoint, contains the point P_α corresponding to P on every associate minimal surface. Moreover, the locus of these points P_α is an ellipse with its center at the origin. Combining this result and the first one of this section, we have

A minimal surface admits of a continuous deformation into a series of minimal surfaces, and each point of the surface describes an ellipse whose plane passes through a fixed point which is the center of the ellipse.

114. Formulas of Schwarz. Since the tangent planes to a minimal surface and its adjoint at corresponding points are parallel, we have

$$X\,dx_1 + Y\,dy_1 + Z\,dz_1 = 0.$$

From this and the second of (113) we obtain the proportion

$$\frac{dx_1}{Z\,dy - Y\,dz} = \frac{dy_1}{X\,dz - Z\,dx} = \frac{dz_1}{Y\,dx - X\,dy}.$$

In consequence of the first of (113) the sums of the squares of the numerators and of the denominators are equal. And so the common ratio is $+1$ or -1. If the expressions for the various quantities be substituted from (98), (101), and (112), it is found that the value is -1. Hence we have

$$(115)\quad dx_1 = Y\,dz - Z\,dy,\quad dy_1 = Z\,dx - X\,dz,\quad dz_1 = X\,dy - Y\,dx.$$

From these equations and the formulas (95), (112) we have

$$(116)\quad \begin{cases} 2U_1 = x - ix_1 = x + i\int Z\,dy - Y\,dz, \\ 2U_2 = y - iy_1 = y + i\int X\,dz - Z\,dx, \\ 2U_3 = z - iz_1 = z + i\int Y\,dx - X\,dy, \end{cases}$$

and

$$(117)\quad \begin{cases} 2V_1 = x + ix_1 = x - i\int Z\,dy - Y\,dz, \\ 2V_2 = y + iy_1 = y - i\int X\,dz - Z\,dx, \\ 2V_3 = z + iz_1 = z - i\int Y\,dx - X\,dy. \end{cases}$$

These equations are known as the *formulas of Schwarz.** Their importance is due to their ready applicability to the solution of the problem:

To determine a minimal surface passing through a given curve and admitting at each point of the curve a given tangent plane.†

In solving this problem we let C be a curve whose coördinates x, y, z are analytic functions of a parameter t, and let X, Y, Z be analytic functions of t satisfying the conditions

$$X^2 + Y^2 + Z^2 = 1,\qquad X\,dx + Y\,dy + Z\,dz = 0.$$

* *Crelle*, Vol. LXXX (1875), p. 291.

† This problem is a special case of the more general one solved by Cauchy: *To determine an integral surface of a differential equation passing through a curve and admitting at each point of the curve a given tangent plane.* For minimal surfaces the equation is (84). Cauchy showed that such a surface exists in general, and that it is unique unless the curve is a characteristic for the equation. His researches are inserted in Vols. XIV, XV of the *Comptes Rendus*. The reader may consult also Kowalewski, Theorie der partiellen Differentialgleichungen, *Crelle*, Vol. LXXX (1875), p. 1; and Goursat, *Cours d'Analyse Mathématique*, Vol. II, pp. 563–567 (Paris, 1905).

If x_u, y_u, z_u denote the values of x, y, z when t is replaced by a complex variable u, and x_v, y_v, z_v the values when t is replaced by v, the equations

$$\text{(118)} \qquad \left\{ \begin{aligned} \bar{x} &= U_1 + V_1 = \frac{x_u + x_v}{2} + \frac{i}{2}\int_v^u (Z\,dy - Y\,dz), \\ \bar{y} &= U_2 + V_2 = \frac{y_u + y_v}{2} + \frac{i}{2}\int_v^u (X\,dz - Z\,dx), \\ \bar{z} &= U_3 + V_3 = \frac{z_u + z_v}{2} + \frac{i}{2}\int_v^u (Y\,dx - X\,dy) \end{aligned} \right.$$

define a minimal surface which passes through C and admits at each point for tangent plane the plane through the point with direction-cosines X, Y, Z. For, when u and v are replaced by t, these equations define C. And the conditions (96) and

$$\sum X\,dU_1 = 0, \qquad \sum X\,dV_1 = 0$$

are satisfied. Furthermore, the surface defined by (118) affords the unique solution, as is seen from (116) and (117).

When, in particular, C and t are real, the equations of the real minimal surface, satisfying the conditions of the problem, may be put in the form

$$\bar{x} = R\left[x + i\int_t^u (Z\,dy - Y\,dz)\right],$$

$$\bar{y} = R\left[y + i\int_t^u (X\,dz - Z\,dx)\right],$$

$$\bar{z} = R\left[z + i\int_t^u (Y\,dx - X\,dy)\right].$$

As an application of these formulas, we consider minimal surfaces containing a straight line. If we take the latter for the z-axis, and let ϕ denote the angle which the normal to the surface at a point of the line makes with the x-axis, we have

$$x = y = 0, \quad z = t, \quad X = \cos\phi, \quad Y = \sin\phi, \quad Z = 0.$$

Hence the equations of the surface are

$$\bar{x} = -Ri\int_t^u \sin\phi\,dt, \qquad \bar{y} = Ri\int_t^u \cos\phi\,dt, \qquad \bar{z} = R(u).$$

Here ϕ is an analytic function of t, whose form determines the character of the surface. For two points corresponding to conjugate values of u, the $\bar{z}$-coördinates are equal, and the $\bar{x}$- and $\bar{y}$-coördinates differ in sign. Hence:

Every straight line upon a minimal surface is an axis of symmetry.

EXAMPLES

1. The tangents to corresponding curves on two associate minimal surfaces make a constant angle.

2. If corresponding directions on two applicable surfaces make a constant angle, the latter are associate minimal surfaces.

3. Show that the catenoid and the right helicoid are adjoint surfaces and determine the function $F(u)$ which defines the former.

4. Let C be a geodesic on a minimal surface S. Show that

(*a*) the equations of the surface may be put in the form

$$x = R\left[\xi + i\int \lambda\, ds\right], \quad y = R\left[\eta + i\int \mu\, ds\right], \quad z = R\left[\zeta + i\int \nu\, ds\right],$$

where ξ, η, ζ are the coördinates of a point on C, and λ, μ, ν the direction-cosines of its binormal;

(*b*) if C' denotes the curve on the adjoint S_1 corresponding to C, the radii of first and second curvature of C' are the radii of second and first curvature of C;

(*c*) if C is a plane curve, the surface is symmetric with respect to its plane.

5. The surface for which $F(u) = 1 - \dfrac{1}{u^4}$ is called the *surface of Henneberg;* it is a double algebraic surface of the fifteenth order and fifth class.

GENERAL EXAMPLES

1. The edge of regression of the developable surface circumscribed to two confocal quadrics has for projections on the three principal planes the evolutes of the focal conics.

2. By definition a *tetrahedral surface* is one whose equations are of the form

$$x = A(u-a)^m(v-a)^n, \quad y = B(u-b)^m(v-b)^n, \quad z = C(u-c)^m(v-c)^n,$$

where A, B, C, m, n are any constants. Show that the parametric curves are conjugate, and that the asymptotic lines can be found by quadratures; also that when $m = n$, the equation of the surface is

$$\left(\frac{x}{A}\right)^{\frac{1}{m}}(b-c) + \left(\frac{y}{B}\right)^{\frac{1}{m}}(c-a) + \left(\frac{z}{C}\right)^{\frac{1}{m}}(a-b) = (a-b)(b-c)(a-c).$$

3. Determine the tetrahedral surfaces, defined as in Ex. 2, upon which the parametric curves are the lines of curvature.

4. Find the surfaces normal to the tangents to a family of umbilical geodesics on an elliptic paraboloid, and find the complementary surface.

5. At every point of a geodesic circle with center at an umbilical point on the ellipsoid (10)

$$abc = \rho^2 W^2 (a + c - r^2),$$

where r is the radius vector of the point (cf. § 102).

6. The tangent plane to the director-cone of a ruled surface along a generator is parallel to the tangent plane to the surface at the infinitely distant point on the corresponding generator.

7. Upon the hyperboloid of one sheet, and likewise upon the hyperbolic paraboloid, find the two lines of striction.

8. The line of striction of a ruled surface is an orthogonal trajectory of the generators only in case the latter are the binormals of a curve or the surface is a right conoid.

9. Determine for a geodesic on a developable surface the relation existing between the curvature, torsion, and angle of inclination of the geodesic with the generators.

10. If h denotes the shortest distance and α the angle between two lines l_1 and l_2, and the latter revolves about the former with a helicoidal motion of parameter a (cf. § 62), the locus of l_2 is a developable surface if $a = h \cot \alpha$. If $a = h \tan \alpha$, the surface is the locus of the binormals of a circular helix.

11. If the lines of curvature in one family upon a ruled surface are such that the segments of the generators between two curves of the family are of the same length, the parameter of distribution is constant and the line of striction is a line of curvature.

12. If two ruled surfaces meet one another in a generator, they are tangent to one another at two points of the generator or at every point; in the latter case the central point for the common generator is the same, and the parameter of distribution has the same value.

13. If tangents be drawn to a ruled surface at points of the line of striction and in directions perpendicular to the generators, these tangents form the *conjugate* ruled surface. It has the same line of striction as the given surface. Moreover, a generator of the given surface, the normal to the surface at the central point C of this generator, and the generator of the conjugate surface through C are parallel to the tangent, principal normal, and binormal of a twisted curve.

14. Let C be a curve on a surface S, and Σ the ruled surface formed by the normals to S along C. Derive the following results:

(*a*) the distance between near-by generators of Σ is of the first order unless C is a line of curvature;

(*b*) if r denotes the distance from the central point of a generator to the point of intersection with S,

$$r\Sigma (dX)^2 = - \Sigma\, dx dX;$$

(*c*) the tangent to C at a point M is conjugate to the tangent to the surface at M parallel to the line of shortest distance;

(*d*) the maximum and minimum values of r are the principal radii of S, ρ_1, and ρ_2, and the above equation may be written $r = \rho_1 \sin^2 \phi + \rho_2 \cos^2 \phi$, where ϕ is the angle which the corresponding line of shortest distance makes with the tangent to the line of curvature corresponding to ρ_2.

15. If C and C' are two orthogonal curves on a surface, then at the point of intersection (cf. Ex. 14)

$$\frac{1}{rR} + \frac{1}{r'R'} = \frac{1}{\rho_1^2} + \frac{1}{\rho_2^2}.$$

16. If C and C' are two conjugate curves on a surface, then at the point of intersection (cf. Ex. 14)

$$\frac{1}{r} + \frac{1}{r'} = \frac{1}{\rho_1} + \frac{1}{\rho_2}, \qquad \frac{r}{r'} = \frac{R}{R'}.$$

17. If two surfaces are applicable, and the radii of first and second curvature of every geodesic on one surface are equal to the radii of second and first curvature of the corresponding geodesic on the other, the surfaces are minimal.

18. The surface for which F in (98) is constant, say 3, is called the *minimal surface of Enneper;* it possesses the following properties:

(*a*) it is an algebraic surface of the ninth degree whose equation is unaltered when x, y, z are replaced by y, x, $-z$ respectively;

(*b*) it meets the plane $z = 0$ in two orthogonal straight lines;

(*c*) if we put $u = \alpha - i\beta$, the equations of the surface are

$$x = 3\alpha + 3\alpha\beta^2 - \alpha^3, \qquad y = 3\beta + 3\alpha^2\beta - \beta^3, \qquad z = 3\alpha^2 - 3\beta^2,$$

and the curves $\alpha =$ const., $\beta =$ const. are the lines of curvature;

(*d*) the lines of curvature are rectifiable unicursal curves of the third order and they are plane curves, the equations of the planes being

$$x + \alpha z - 3\alpha - 2\alpha^3 = 0, \qquad y - \beta z - 3\beta - 2\beta^3 = 0;$$

(*e*) the lines of curvature are represented on the unit sphere by a double family of circles whose planes form two pencils with perpendicular axes which are tangent to the sphere at the same point;

(*f*) the asymptotic lines are twisted cubics;

(*g*) the sections of the surface by the planes $x = 0$ and $y = 0$ are cubics, which are double curves on the surface and the locus of the double points of the lines of curvature;

(*h*) the associate minimal surfaces are positions of the original surface rotated through the angle $-\alpha/2$ about the z-axis, where α has the same meaning as in § 113;

(*i*) the surface is the envelope of the plane normal, at the mid-point, to the join of any two points, one on each of the focal parabolas

$$x = 4\alpha, \quad y = 0, \quad z = 2\alpha^2 - 1; \qquad x = 0, \quad y = 4\beta, \quad z = 1 - 2\beta^2;$$

the planes normal to the two parabolas at the extremities of the join are the planes of the lines of curvature through the point of contact of the first plane.

19. Find the equations of Schwarz of a minimal surface when the given curve is an asymptotic line.

20. Let S and S' be two surfaces, and let the points at which the normals are parallel correspond; for convenience let S and S' be referred to their common conjugate system. Show that if the correspondence is conformal, either S and S' are homothetic; or both are minimal surfaces; or the parametric curves are the lines of curvature on both surfaces, and form an isothermal system.

21. Find the coördinates of the surface which corresponds to the ellipsoid after the manner of Ex. 20. Show that the surface is periodic, and investigate the points corresponding to the umbilical points on the ellipsoid.

22. When the equations of an ellipsoid are in the form (11), the curves $u + v =$ const. lie on spheres whose centers coincide with the origin; and at all points of such a curve the product ρW is constant (§ 102).

CHAPTER VIII

SURFACES OF CONSTANT TOTAL CURVATURE. W-SURFACES. SURFACES WITH PLANE OR SPHERICAL LINES OF CURVATURE

115. Spherical surfaces of revolution. Surfaces whose total curvature K is the same at all points are called *surfaces of constant curvature.* When this constant value is zero, the surface is developable (§ 64). The nondevelopable surfaces of this kind are called *spherical* or *pseudospherical,* according as K is positive or negative. We consider these two kinds and begin our study of them with the determination of surfaces of revolution of constant curvature.

When upon a surface of revolution the curves $v = \text{const.}$ are the meridians and $u = \text{const.}$ the parallels, the linear element is reducible to the form

$$ds^2 = du^2 + G\,dv^2, \tag{1}$$

where G is a function of u alone (§ 46). In this case the expression for the total curvature (V, 12) is

$$K = -\frac{1}{\sqrt{G}}\frac{\partial^2\sqrt{G}}{\partial u^2}. \tag{2}$$

For spherical surfaces we have $K=1/a^2$, where a is a real constant. Substituting this value in equation (2) and integrating, we have

$$\sqrt{G} = c\cos\left(\frac{u}{a}+b\right), \tag{3}$$

where b and c are constants of integration. From (1) it is seen that a change in b means simply a different choice of the parallel $u = 0$. If we take $b = 0$, the linear element is

$$ds^2 = du^2 + c^2\cos^2\frac{u}{a}\,dv^2. \tag{4}$$

From (III, 99, 100) it follows that the equations of the meridian curve are

$$r = c\cos\frac{u}{a}, \qquad z = \int\sqrt{1-\frac{c^2}{a^2}\sin^2\frac{u}{a}}\,du, \tag{5}$$

and that v measures the angle between the meridian planes. There are three cases to be considered, according as c is equal to, greater than, or less than, a.

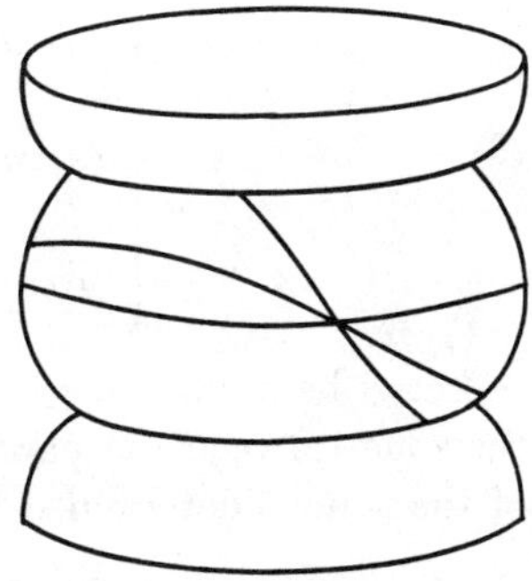

FIG. 26

CASE I. $c = a$. Now

$$r = a\cos\frac{u}{a}, \qquad z = a\sin\frac{u}{a},$$

and consequently the surface is a sphere.

CASE II. $c > a$. From the expression for z it follows that $\sin^2\frac{u}{a} < 1$ and consequently $r > 0$. Hence the surface is made up of zones bounded by minimum parallels whose radii are equal to the minimum value of $\cos\frac{u}{a}$, and the greatest parallel of each zone is of radius c; as in fig. 26, where the curves represent geodesics.

CASE III. $c < a$. Now r varies from 0 to c, the former corresponding to the value $u = ma\pi/2$, where m is any odd integer. At these points on the axis the meridians meet the latter under the angle $\sin^{-1}\frac{c}{a}$. Hence the surface is made up of a series of spindles (fig. 27). For the cases II and III the expression for z can be integrated in terms of elliptic functions.*

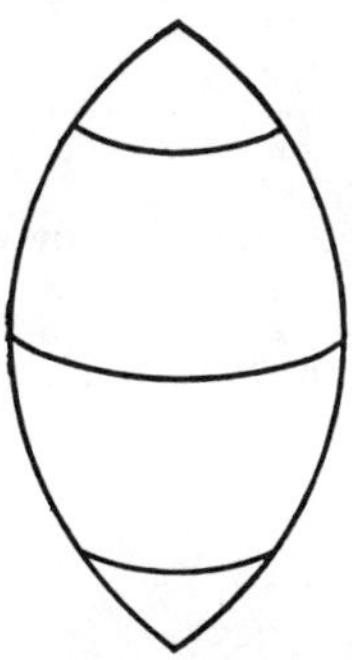

FIG. 27

It is readily found that these two surfaces are applicable to the sphere with the meridians and parallels of each in correspondence. Thus, if we write the linear element of the sphere in the form

$$ds^2 = d\bar{u}^2 + a^2\cos^2\frac{\bar{u}}{a}\,d\bar{v}^2,$$

it follows from (4) that the equations

$$\bar{u} = u, \qquad \bar{v} = \frac{c}{a}v$$

determine the correspondence desired.

It is evident that for values of b other than zero we should be brought to the same results. However, for the sake of future

*Cf. Bianchi, Vol. I, p. 233.

reference we write down the expressions for the linear element when $b = -\pi/2$ and $-\pi/4$ together with (4), thus:

$$
(6) \qquad \begin{cases} \text{(i)} \ ds^2 = du^2 + c^2 \cos^2 \dfrac{u}{a} dv^2, \\[2ex] \text{(ii)} \ ds^2 = du^2 + c^2 \sin^2 \dfrac{u}{a} dv^2, \\[2ex] \text{(iii)} \ ds^2 = du^2 + c^2 \cos^2 \left(\dfrac{u}{a} - \dfrac{\pi}{4}\right) dv^2. \end{cases}
$$

Let S be a surface with the linear element (6, i), and consider the zone between the parallels $u_0 = \text{const.}$ and $u_1 = \text{const.}$ A point of the zone is determined by values of u and v such that

$$u_1 \geq u \geq u_0, \qquad 2\pi \geq v \geq 0.$$

The parametric values of the corresponding point on the sphere are such that

$$u_1 \geq \bar{u} \geq u_0, \qquad \frac{2\pi c}{a} \geq \bar{v} \geq 0.$$

Hence when $c < a$, the given zone on S does not cover the zone on the sphere between the parallels $u_0 = \text{const.}$ and $u_1 = \text{const.}$; but when $c > a$ it not only covers it, but there is an overlapping.

116. Pseudospherical surfaces of revolution. In order to find the pseudospherical surfaces of revolution we replace K in (2) by $-1/a^2$ and integrate. This gives

$$\sqrt{G} = c_1 \cosh \frac{u}{a} + c_2 \sinh \frac{u}{a},$$

where c_1 and c_2 are constants of integration. We consider first the particular forms of the linear element arising when either of these constants is zero or both are equal. They may be written

$$
(7) \qquad \begin{cases} \text{(i)} \ ds^2 = du^2 + c^2 \cosh^2 \dfrac{u}{a} dv^2, \\[2ex] \text{(ii)} \ ds^2 = du^2 + c^2 \sinh^2 \dfrac{u}{a} dv^2, \\[2ex] \text{(iii)} \ ds^2 = du^2 + c^2 e^{\frac{2u}{a}} dv^2. \end{cases}
$$

Any case other than these may be obtained by taking for $\sqrt{G}$ either of the values $\cosh\left(\dfrac{u}{a} + b\right)$ or $\sinh\left(\dfrac{u}{a} + b\right)$, where b is a constant.

By a change of the parameter u the corresponding linear elements are reducible to (i) or (ii). Hence the forms (7) are the most general. The corresponding meridian curves are defined by

$$(8) \qquad \begin{cases} \text{(i)} \; r = c\cosh\dfrac{u}{a}, & z = \displaystyle\int\sqrt{1-\frac{c^2}{a^2}\sinh^2\frac{u}{a}}\,du; \\[2ex] \text{(ii)} \; r = c\sinh\dfrac{u}{a}, & z = \displaystyle\int\sqrt{1-\frac{c^2}{a^2}\cosh^2\frac{u}{a}}\,du; \\[2ex] \text{(iii)} \; r = ce^{\frac{u}{a}}, & z = \displaystyle\int\sqrt{1-\frac{c^2}{a^2}e^{\frac{2u}{a}}}\,du. \end{cases}$$

We consider these three cases in detail.

CASE I. The maximum and minimum values of $\sinh^2\dfrac{u}{a}$ are a^2/c^2 and 0. Hence the maximum and minimum values of r are $\sqrt{a^2+c^2}$ and c. At points of a maximum parallel the tangents to the meridians are perpendicular to the axis, and at points of a minimum parallel they are parallel to the axis. Hence the former is a cuspidal edge, and the latter a circle of gorge, so that the surface is made up of spool-like sections. It is represented by fig. 28, upon which the closed curves are geodesic circles and the other curves are geodesics. These pseudospherical surfaces are said to be of the *hyperbolic type.* *

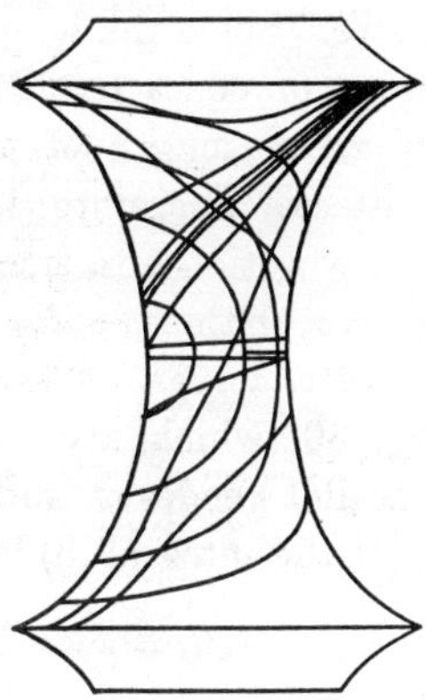

FIG. 28

CASE II. In order that the surface be real c^2 cannot be greater than a^2, a restriction not necessary in either of the other cases. If we put $c = a\sin\alpha$,† the maximum and minimum values of $\cosh^2\dfrac{u}{a}$ are $\operatorname{cosec}^2\alpha$ and 1, and the corresponding values of r are $a\cos\alpha$ and 0. The tangents to the meridians at points of the former circle are perpendicular to the axis, and at the points for which r is zero they meet the axis under the angle α. Hence the surface is made up of a series of parts similar in shape

* Cf. Bianchi, Vol. I, p. 223.

† Cf. Bianchi, Vol. I, p. 226.

to hour-glasses. Fig. 29 represents one half of such a part; one of the curves is an asymptotic line and the others are parallel geodesics. The surface is called a *pseudospherical surface of the elliptic type.*

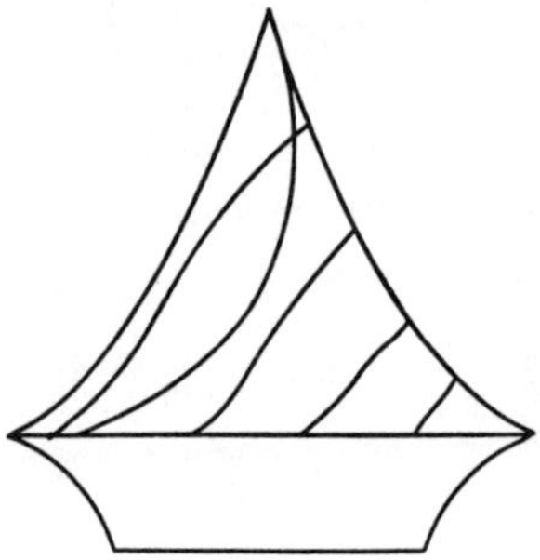

FIG. 29

CASE III. In the preceding cases the equations of the meridian curve can be expressed without the quadrature sign by means of elliptic functions.* In this case the same can be done by means of trigonometric functions. For, if we put

$$\sin\phi = \frac{c}{a} e^{\frac{u}{a}},$$

equations (iii) of (8) become

$$(9) \quad r = a\sin\phi, \quad z = a\left(\log\tan\frac{\phi}{2} + \cos\phi\right).$$

We find that ϕ is the angle which the tangent to a meridian at a point makes with the axis. Hence the axis is an asymptote to the curve. Since the length of the segment of a tangent between the point of contact and the intersection with the axis is $r \operatorname{cosec} \phi$ or a, the length of the segment is independent of the point of contact. Therefore the meridian curve is a *tractrix*. The surface of revolution of a tractrix about its asymptote is called the *pseudosphere*, or the *pseudospherical surface of the parabolic type*. The surface is shown in fig. 30, which also pictures a family of parallel geodesics and an asymptotic line.

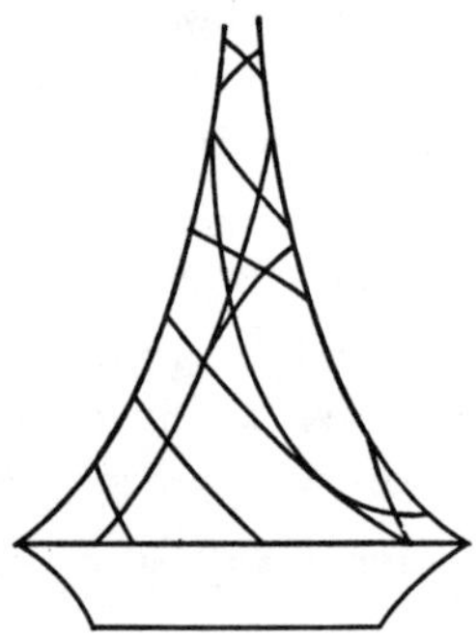

FIG. 30

If the integral (3) be written in the form

$$\sqrt{G} = c_1 \cos\frac{u}{a} + c_2 \sin\frac{u}{a},$$

the cases (i), (ii), (iii) of (6) are seen to correspond to the similar cases of (7). We shall find other marks of similarity between these cases, but now we desire to call attention to differences.

Each of the three forms (7) determines a particular kind of pseudospherical surface of revolution, and c is restricted in value

* Cf. Bianchi, Vol. I, pp. 226–228.

only for the second case. On the contrary each of the three forms (6) serves to define any of the three types of spherical surfaces of revolution according to the magnitude of c.

From (IV, 51) we find that the geodesic curvature of the parallels on the surfaces with the linear elements (7) is measured by the expressions

$$\frac{1}{a}\tanh\frac{u}{a}, \qquad \frac{1}{a}\coth\frac{u}{a}, \qquad \frac{1}{a}.$$

Since no two of these expressions can be transformed into the other if u be replaced by u plus a real constant, it follows that two pseudospherical surfaces of revolution of different types are not applicable to one another with meridians in correspondence.

117. Geodesic parametric systems. Applicability. Now we shall show that in corresponding cases of (6) and (7) the parametric geodesic systems are of the same kind, and then we shall prove that when such a geodesic system is chosen for any surface of constant curvature, not necessarily one of revolution, the linear element can be brought to the corresponding form of (6) or (7).

In the first place we recall that when on any surface the curves $v = \text{const.}$ are geodesics, and $u = \text{const.}$ their orthogonal trajectories, the linear element is reducible to the form (1), where G is, in general, a function of both u and v; and the geodesic curvature of the curves $u = \text{const.}$ is given by (IV, 51), namely

$$\frac{1}{\rho_g} = \frac{1}{\sqrt{EG}}\frac{\partial\sqrt{G}}{\partial u}. \tag{10}$$

When, in particular, the curvature of the surface is constant, $\sqrt{G}$ is given by equation (2) in which K may by replaced by $\pm 1/a^2$. Hence, for spherical surfaces, the general form of $\sqrt{G}$ is

$$\sqrt{G} = \phi(v)\cos\frac{u}{a} + \psi(v)\sin\frac{u}{a}, \tag{11}$$

and for pseudospherical surfaces

$$\sqrt{G} = \phi(v)\cosh\frac{u}{a} + \psi(v)\sinh\frac{u}{a}, \tag{12}$$

where ϕ and ψ are, at most, functions of v. We consider now the three cases of (6) and (7).

CASE I. From the forms (i) of (6) and (7), and from (10), it follows that the curve $u=0$ is a geodesic and that its arc is measured by cv. Moreover, a necessary and sufficient condition that the curve $u=0$ on any surface with the linear element (1) satisfy these conditions is

$$\left(\sqrt{G}\right)_{u=0} = c, \qquad \left(\frac{1}{\sqrt{G}}\frac{\partial\sqrt{G}}{\partial u}\right)_{u=0} = 0.$$

Applying these conditions to (11) and (12), we are brought to the forms (i) of (6) and (7) respectively.

CASE II. The forms (ii) of (6) and (7) satisfy the conditions

$$\left(\sqrt{G}\right)_{u=0} = 0, \qquad \left(\frac{\partial}{\partial u}\sqrt{G}\right)_{u=0} = \frac{c}{a},$$

which are necessary and sufficient that the parametric system be geodesic polar, in which cv measures angles (cf. VI, 54). When these conditions are applied to (11) and (12), we obtain (ii) of (6) and of (7) respectively.

CASE III. For (iii) of (6) the curve $u=0$ has constant geodesic curvature $1/a$, and for (iii) of (7) all of the curves $u=\text{const.}$ have the same geodesic curvature $1/a$. Conversely, we find from (11) and (12) that when this condition is satisfied on any surface of constant curvature the linear element is reducible to one of the forms (iii). We gather these results together into the theorem:

The linear element of any surface of constant curvature is reducible to the forms (i), (ii), (iii) *of* (6) *or* (7) *according as the parametric geodesics are orthogonal to a geodesic, pass through a point, or are orthogonal to a curve of constant geodesic curvature.*

When the linear element of a surface of constant curvature is in one of the forms (i), (ii), (iii) of (6) and (7), it is said to be of the *hyperbolic, elliptic, or parabolic* type accordingly.

The above theorem may be stated as follows:

Any spherical surface of curvature $1/a^2$ is applicable to a sphere of radius a in such a way that to a family of great circles with the same diameter there correspond the geodesics orthogonal to a *given geodesic on the surface, or all the geodesics through any point of it, or those which are orthogonal to a curve of geodesic curvature $1/a$.*

Any pseudospherical surface of curvature $-1/a^2$ is applicable to a pseudospherical surface of revolution of any of the three types; according as the latter surface is of the hyperbolic, elliptic, or parabolic type, to its meridians correspond on the given surface geodesics which are orthogonal to a geodesic, or pass through a point, or are orthogonal to a curve of geodesic curvature $1/a$.

In the case of spherical surfaces one system of geodesics can satisfy all three conditions; for in the case of the sphere the great circles with the same diameter are orthogonal to the equator, pass through both poles, and are orthogonal to two small circles of radius $a/\sqrt{2}$, whose geodesic curvature is $1/a$. But on a pseudospherical surface a geodesic system can satisfy only one of these conditions. Otherwise it would be possible to apply two surfaces of revolution of different types in such a way that meridians and parallels correspond.

From the foregoing theorems it follows that, in order to carry out the applicability of a surface of constant curvature upon any one of the surfaces of revolution, it is only necessary to find the geodesics on the given surface. The nature of this problem is set forth in the theorem:

The determination of the geodesic lines on a surface of constant curvature requires the solution of a Riccati equation.

In proving this theorem we consider first a spherical surface defined in terms of any parametric system. It is applicable to a sphere of the same curvature with center at the origin. The coördinates of this sphere, expressed as functions of the parameters u, v, can be found by the solution of a Riccati equation (§ 65). To great circles on the sphere correspond geodesic lines on the spherical surface; hence the finite equation of the geodesics is $a\bar{x} + b\bar{y} + c\bar{z} = 0$, where a, b, c are arbitrary constants.

When the surface is pseudospherical we use an imaginary sphere of the same curvature, and the analysis is similar.

118. Transformation of Hazzidakis. Let a spherical surface of curvature $1/a^2$ be defined in terms of isothermal-conjugate parameters. Then *

$$(13) \qquad \frac{D}{H} = \frac{D''}{H} = \frac{1}{a},$$

and the Codazzi equations (V, 13′) reduce to

$$(14) \qquad \left\{ \begin{aligned} &\frac{\partial E}{\partial u} - \frac{\partial G}{\partial u} + 2\frac{\partial F}{\partial v} = 0, \\ &\frac{\partial E}{\partial v} - \frac{\partial G}{\partial v} - 2\frac{\partial F}{\partial u} = 0. \end{aligned} \right.$$

From these equations and §§ 77, 82 follows the theorem:

The lines of curvature of a spherical surface form an isothermal-conjugate system.

For, a solution of these equations is

$$E - G = \text{const.}, \qquad F = 0.$$

When this constant is zero the surface is a sphere because of (13). Excluding this case, we replace the above by

$$(15) \qquad E = a^2\cosh^2\omega, \qquad F = 0, \qquad G = a^2\sinh^2\omega.$$

Now

$$(16) \qquad D = D'' = a\sinh\omega\cosh\omega.$$

When these values are substituted in the Gauss equation (V, 12), namely

$$(17) \quad \frac{1}{2H}\left\{\frac{\partial}{\partial v}\left[\frac{2}{H}\frac{\partial F}{\partial u} - \frac{1}{H}\frac{\partial E}{\partial v} - \frac{F}{HE}\frac{\partial E}{\partial u}\right] + \frac{\partial}{\partial u}\left[\frac{F}{HE}\frac{\partial E}{\partial v} - \frac{1}{H}\frac{\partial G}{\partial u}\right]\right\} = \frac{1}{a^2},$$

it is found that ω must satisfy the equation

$$(18) \qquad \frac{\partial^2\omega}{\partial u^2} + \frac{\partial^2\omega}{\partial v^2} + \sinh\omega\cosh\omega = 0.$$

Conversely, for each solution of this equation the quantities (15) and (16) determine a spherical surface.

If equations (14) be differentiated with respect to u and v respectively, and the resulting equations be added, we have

$$(19) \qquad \frac{\partial^2 E}{\partial u^2} + \frac{\partial^2 E}{\partial v^2} = \frac{\partial^2 G}{\partial u^2} + \frac{\partial^2 G}{\partial v^2}.$$

* The ambiguity of sign may be neglected, as a change of sign gives a surface symmetrical with respect to the origin.

In consequence of (14) equation (17) is reducible to

$$(20)\quad \frac{1}{4H^4}\left\{E\left[\left(\frac{\partial G}{\partial u}\right)^2+\left(\frac{\partial G}{\partial v}\right)^2\right]+2F\left[\frac{\partial E}{\partial u}\frac{\partial G}{\partial v}-\frac{\partial E}{\partial v}\frac{\partial G}{\partial u}\right]\right.$$
$$\left.+G\left[\left(\frac{\partial E}{\partial u}\right)^2+\left(\frac{\partial E}{\partial v}\right)^2\right]\right\}-\frac{1}{2H^2}\left(\frac{\partial^2 G}{\partial u^2}+\frac{\partial^2 G}{\partial v^2}\right)=\frac{1}{a^2}.$$

Equations (14) are unaltered if E and G be interchanged and the sign of F be changed. The same is true of (17) because of (19) and (20). Hence we have:

If the linear element of a spherical surface referred to an isothermal-conjugate system of parameters be

$$ds^2 = E\,du^2 + 2F\,dudv + G\,dv^2,$$

there exists a second spherical surface of the same curvature referred to a similar parametric system with the linear element

$$ds_1^2 = G\,du^2 - 2F\,dudv + E\,dv^2,$$

and with the same second quadratic form as the given surface; moreover, the lines of curvature correspond on the two surfaces.

The latter fact is evident from the equation of the lines of curvature (IV, 26), which reduces to $F\,du^2+(G-E)\,dudv - F\,dv^2 = 0$. From (IV, 69) it is seen that the linear elements of the spherical representation of the respective surfaces are

$$d\sigma^2 = \frac{1}{a^2}(G\,du^2 - 2F\,dudv + E\,dv^2),$$
$$d\sigma_1^2 = \frac{1}{a^2}(E\,du^2 + 2F\,dudv + G\,dv^2).$$

In particular we have the theorem:

Each solution ω of equation (18) determines two spherical surfaces of curvature $1/a^2$; the linear elements of the surfaces are

$$ds^2 = a^2(\cosh^2\omega\,du^2 + \sinh^2\omega\,dv^2),$$
$$ds_1^2 = a^2(\sinh^2\omega\,du^2 + \cosh^2\omega\,dv^2),$$

and of their spherical representations

$$(21)\quad \begin{cases} d\sigma^2 = \sinh^2\omega\,du^2 + \cosh^2\omega\,dv^2, \\ d\sigma_1^2 = \cosh^2\omega\,du^2 + \sinh^2\omega\,dv^2; \end{cases}$$

moreover, their principal radii are respectively

$$\rho_1 = a\coth\omega, \qquad \rho_2 = a\tanh\omega,$$
$$\rho_1' = a\tanh\omega, \qquad \rho_2' = a\coth\omega.$$

Bianchi * has given the name *Hazzidakis transformation* to the relation between these two surfaces. It is evident that the former theorem defines this transformation in a more general way.

119. Transformation of Bianchi. We consider now a pseudospherical surface of curvature $-1/a^2$, defined in terms of isothermal-conjugate parameters. We have

$$\frac{D}{H} = -\frac{D''}{H} = -\frac{1}{a},\dagger$$

and the Codazzi equations reduce to

$$\frac{\partial E}{\partial u} + \frac{\partial G}{\partial u} - 2\frac{\partial F}{\partial v} = 0, \qquad \frac{\partial E}{\partial v} + \frac{\partial G}{\partial v} - 2\frac{\partial F}{\partial u} = 0.$$

These equations are satisfied by the values (cf. §§ 77, 82)

$$E = a^2 \cos^2\omega, \qquad F = 0, \qquad G = a^2 \sin^2\omega, \tag{22}$$

where ω is a function which, because of the Gauss equation (V, 12), must satisfy the equation

$$\frac{\partial^2\omega}{\partial u^2} - \frac{\partial^2\omega}{\partial v^2} = \sin\omega\cos\omega. \tag{23}$$

Conversely, every solution of this equation determines a pseudospherical surface whose fundamental quantities are given by (22) and by

$$D = -D'' = -a\sin\omega\cos\omega. \tag{24}$$

Moreover, the linear element of the spherical representation is

$$d\sigma^2 = \sin^2\omega\, du^2 + \cos^2\omega\, dv^2. \tag{25}$$

There is not a transformation for pseudospherical surfaces similar to the Hazzidakis transformation of spherical surfaces, but there are transformations of other kinds which are of great importance. One of these is involved in the following theorem of Ribaucour:

If in the tangent planes to a pseudospherical surface of curvature $-1/a^2$ circles of radius a be described with centers at the points of contact, these circles are the orthogonal trajectories of an infinity of surfaces of curvature $-1/a^2$.

* Vol. II, p. 437.

† This choice of sign is made so that the following formulas may have the customary form.

In proving this theorem we imagine the given surface S referred to its lines of curvature, and we associate with it the moving trihedral whose axes are tangent to the parametric lines. From (22) and (V, 75, 76) it follows that

$$p = 0, \quad p_1 = \cos\omega, \qquad q = \sin\omega, \quad q_1 = 0, \qquad r = \frac{\partial\omega}{\partial v}, \quad r_1 = \frac{\partial\omega}{\partial u},$$

$$\xi = a\cos\omega, \qquad \eta_1 = a\sin\omega, \qquad \xi_1 = \eta = 0.$$

In the tangent xy-plane we draw from the origin M a segment of length a, and let θ denote its angle of inclination with the x-axis. The coördinates of the other extremity M_1 with respect to these axes are $a\cos\theta$, $a\sin\theta$, 0, and the projections upon these axes of a displacement of M_1 as M moves over S are, by (V, 51),

$$a\left[-\sin\theta\,d\theta + \cos\omega\,du - \left(\frac{\partial\omega}{\partial v}du + \frac{\partial\omega}{\partial u}dv\right)\sin\theta\right],$$

$$a\left[\cos\theta\,d\theta + \sin\omega\,dv + \left(\frac{\partial\omega}{\partial v}du + \frac{\partial\omega}{\partial u}dv\right)\cos\theta\right],$$

$$a\left[\cos\omega\sin\theta\,dv - \sin\omega\cos\theta\,du\right].$$

We seek now the conditions which θ must satisfy in order that the line MM_1 be tangent to the locus of M_1 denoted by S_1, and that the tangent plane to S_1 at M_1 be perpendicular to the tangent plane to S at M. Under these conditions the direction-cosines of the tangent plane to S_1 with reference to the moving trihedral are

$$\sin\theta, \qquad -\cos\theta, \qquad 0, \tag{26}$$

and since the tangent to the above displacement must be in this plane, we have

$$d\theta + \left(\frac{\partial\omega}{\partial v} - \sin\theta\cos\omega\right)du + \left(\frac{\partial\omega}{\partial u} + \cos\theta\sin\omega\right)dv = 0. \tag{27}$$

As this equation must hold for all displacements of M, it is equivalent to

$$\left\{\begin{aligned} &\frac{\partial\theta}{\partial u} + \frac{\partial\omega}{\partial v} = \cos\omega\sin\theta, \\ &\frac{\partial\theta}{\partial v} + \frac{\partial\omega}{\partial u} = -\sin\omega\cos\theta. \end{aligned}\right. \tag{28}$$

These equations satisfy the condition of integrability in consequence of (23). Moreover, θ is a solution of equation (23), as is seen by differentiating equations (28) with respect to u and v respectively and subtracting.

By means of (28) the above expressions for the projections of a displacement of M_1 can be put in the form

$$a\cos\theta\,(\cos\omega\cos\theta\,du + \sin\omega\sin\theta\,dv),$$
$$a\sin\theta\,(\cos\omega\cos\theta\,du + \sin\omega\sin\theta\,dv),$$
$$a\,(\cos\omega\sin\theta\,dv - \sin\omega\cos\theta\,du).$$

From these it follows that the linear element of S_1 is

$$ds_1^2 = a^2(\cos^2\theta\,du^2 + \sin^2\theta\,dv^2).$$

In order to prove that S_1 is a pseudospherical surface referred to its lines of curvature, it remains for us to show that the spherical representation of these curves forms an orthogonal system. We obtain this representation with the aid of a trihedral whose vertex is fixed, and which rotates so that its axes are always parallel to the corresponding axes of the trihedral for S. The point whose coördinates with reference to the new trihedral are given by (26) serves for the spherical representation of S_1. The projections upon these axes of a displacement of this point are reducible, by means of (28), to

$$\cos\theta\,(\cos\omega\sin\theta\,du - \sin\omega\cos\theta\,dv),$$
$$\sin\theta\,(\cos\omega\sin\theta\,du - \sin\omega\cos\theta\,dv),$$
$$-\sin\omega\sin\theta\,du - \cos\omega\cos\theta\,dv,$$

from which it follows that the linear element is

$$d\sigma_1^2 = \sin^2\theta\,du^2 + \cos^2\theta\,dv^2.$$

Since θ is a solution of (23), the surface S_1 is pseudospherical, of curvature $-1/a^2$, and the lines of curvature are parametric. To each solution θ of equations (28) there corresponds a surface S_1. Darboux * has called this process of finding S_1 the *transformation of Bianchi.* As the complete integral of equations (28) involves an arbitrary constant, there are an infinity of surfaces S_1, as remarked by Ribaucour. Moreover, if we put

$$\phi = \tan\frac{\theta}{2}, \tag{29}$$

these equations are of the Riccati type in ϕ. Hence, by § 14,

When one transform of Bianchi of a pseudospherical surface is known, the determination of the others requires only quadratures.

* Vol. III, p. 422.

From (III, 24) it follows that the differential equation of the curves to which the lines joining corresponding points on S and S_1 are tangent is

(30) $$\cos\omega \sin\theta\, du - \sin\omega \cos\theta\, dv = 0.$$

Hence, along such a curve, equation (27) reduces to

$$d\theta + \frac{\partial \omega}{\partial v} du + \frac{\partial \omega}{\partial u} dv = 0.$$

But from (VI, 56) it is seen that this is the Gauss equation of geodesics upon a surface whose first fundamental coefficients have the values (22). Hence:

The curves on S to which the lines joining corresponding points on S and S_1 are tangent are geodesics.

The orthogonal trajectories of the curves (30) are defined by

(31) $$\cos\omega \cos\theta\, du + \sin\omega \sin\theta\, dv = 0.$$

In consequence of (28) the left-hand member of this equation is an exact differential. If we put

$$d\xi = -a(\cos\omega \cos\theta\, du + \sin\omega \sin\theta\, dv),$$

the quantity $e^{-\xi/a}$ is an integrating factor of the left-hand member of (30). Consequently we may define a function η thus:

$$d\eta = ae^{-\xi/a}(\cos\omega \sin\theta\, du - \sin\omega \cos\theta\, dv).$$

In terms of ξ and η the linear element of S is expressible in the parabolic form (7),

(32) $$ds^2 = d\xi^2 + e^{2\xi/a} d\eta^2.$$

Equation (31) defines also the orthogonal trajectories of the curves on S_1 to which the lines MM_1 are tangent, and the equation of the latter curves is

$$\sin\omega \cos\theta\, du - \cos\omega \sin\theta\, dv = 0.$$

The quantity $e^{\xi/a}$ is an integrating factor of this equation, and if we put accordingly

$$d\zeta = ae^{\xi/a}(\sin\omega \cos\theta\, du - \cos\omega \sin\theta\, dv),$$

the linear element of S_1 may be expressed in the parabolic form

(33) $$ds_1^2 = d\xi^2 + e^{-2\xi/a} d\zeta^2.$$

As the expressions (32) and (33) are of the form of the linear element of a surface of revolution, the finite equations of the geodesics can be found by quadratures. Hence:

When a Bianchi transformation is known for a surface, the finite equation of its geodesics can be found by quadratures.

This follows also from the preceding theorem and the last one of § 117.

120. Transformation of Bäcklund. The transformation of Bianchi is only a particular case of a transformation discovered by Bäcklund,* by means of which from one pseudospherical surface S another S_1, of the same curvature, can be found. Moreover, on these two surfaces the lines of curvature correspond, the join of corresponding points is tangent at these points to the surfaces and is of constant length, and the tangent planes at corresponding points meet under constant angle.

We refer S to the same moving trihedral as in the preceding case, and let λ and θ denote the length of MM_1 and the angle which the latter makes with the x-axis. The coördinates of M_1 are $\lambda \cos\theta$, $\lambda \sin\theta$, 0, and the projections of a displacement of M_1 are

$$(34)\qquad \begin{cases} -\lambda\sin\theta\, d\theta + a\cos\omega\, du - \lambda\sin\theta\left(\dfrac{\partial\omega}{\partial v}du + \dfrac{\partial\omega}{\partial u}dv\right), \\ \lambda\cos\theta\, d\theta + a\sin\omega\, dv + \lambda\cos\theta\left(\dfrac{\partial\omega}{\partial v}du + \dfrac{\partial\omega}{\partial u}dv\right), \\ \lambda(\cos\omega\sin\theta\, dv - \sin\omega\cos\theta\, du). \end{cases}$$

If σ denotes the constant angle between the tangent planes to S and S_1 at M and M_1 respectively, since these planes are to intersect in MM_1, the direction-cosines of the normal to S_1 are

$$\sin\sigma\sin\theta, \qquad -\sin\sigma\cos\theta, \qquad \cos\sigma.$$

Hence θ must satisfy the condition

$$\begin{aligned} \lambda\sin\sigma\, d\theta - a\sin\sigma(\cos\omega\sin\theta\, du - \sin\omega\cos\theta\, dv) & \\ + \lambda\sin\sigma\left(\frac{\partial\omega}{\partial v}du + \frac{\partial\omega}{\partial u}dv\right) & \\ + \lambda\cos\sigma(\sin\omega\cos\theta\, du - \cos\omega\sin\theta\, dv) &= 0. \end{aligned}$$

Since this condition must be satisfied for every displacement, it is equivalent to

$$\lambda\sin\sigma\left(\frac{\partial\theta}{\partial u} + \frac{\partial\omega}{\partial v}\right) = a\sin\sigma\cos\omega\sin\theta - \lambda\cos\sigma\sin\omega\cos\theta,$$

$$\lambda\sin\sigma\left(\frac{\partial\theta}{\partial v} + \frac{\partial\omega}{\partial u}\right) = -a\sin\sigma\sin\omega\cos\theta + \lambda\cos\sigma\cos\omega\sin\theta.$$

*Om ytor med konstant negativ krökning, *Lunds Universitets Arsskrift*, Vol. XIX (1883). An English translation of this memoir has been made by Miss Emily Coddington of New York, and privately printed.

If these equations be differentiated with respect to v and u respectively, and the resulting equations be subtracted, we have

$$a^2 \sin^2 \sigma - \lambda^2 = 0,$$

on the assumption that λ is a constant. Without loss of generality we take $\lambda = a \sin \sigma$. If this value be substituted in the above equations, we have

$$(35) \qquad \begin{cases} \sin \sigma \left(\dfrac{\partial \theta}{\partial u} + \dfrac{\partial \omega}{\partial v} \right) = \sin \theta \cos \omega - \cos \sigma \cos \theta \sin \omega, \\ \sin \sigma \left(\dfrac{\partial \theta}{\partial v} + \dfrac{\partial \omega}{\partial u} \right) = - \cos \theta \sin \omega + \cos \sigma \sin \theta \cos \omega, \end{cases}$$

and these equations satisfy the condition of integrability. If they be differentiated with respect to u and v respectively, and the resulting equations be subtracted, it is found that θ is a solution of (23).

In consequence of (35) the expressions (34) reduce to

$$\begin{aligned} &a \cos \theta (\cos \omega \cos \theta + \cos \sigma \sin \omega \sin \theta)\, du \\ &\qquad\qquad + a \sin \theta (\sin \omega \cos \theta - \cos \sigma \cos \omega \sin \theta)\, dv, \\ &a \cos \theta (\cos \omega \sin \theta - \cos \sigma \sin \omega \cos \theta)\, du \\ &\qquad\qquad + a \sin \theta (\sin \omega \sin \theta + \cos \sigma \cos \omega \cos \theta)\, dv, \\ &a \sin \sigma (\cos \omega \sin \theta\, dv - \sin \omega \cos \theta\, du), \end{aligned}$$

and the linear element of S_1 is

$$ds_1^2 = a^2 (\cos^2 \theta\, du^2 + \sin^2 \theta\, dv^2).$$

In a manner similar to that of § 119 it can be shown that the spherical representation of the parametric curves is orthogonal, and consequently these curves are the lines of curvature on S_1.

Equations (35) are reducible to the Riccati form by the change of variable (29). Moreover, the general solution of these equations involves two constants, namely σ and the constant of integration. Hence we have the theorem:

By the integration of a Riccati equation a double infinity of pseudospherical surfaces can be obtained from a given surface of this kind.

We refer to this as the *transformation of Bäcklund*, and indicate it by B_σ, thus putting in evidence the constant σ.

121. Theorem of permutability. Let S_1 be a transform of S by means of the functions (θ_1, σ_1). Since conversely S is a transform of S_1, and the equations for the latter similar to (35) are reducible to the Riccati type, all the transforms of S_1 can be found by quadratures. But even these quadratures can be dispensed with because of the following *theorem of permutability* due to Bianchi*:

If S_1 and S_2 are transforms of S by means of the respective pairs of functions (θ_1, σ_1) and (θ_2, σ_2), a function ϕ can be found without quadratures which is such that by means of the pairs (ϕ, σ_2) and (ϕ, σ_1) the surfaces S_1 and S_2 respectively are transformable into a pseudospherical surface S'.

By hypothesis ϕ is a solution of the equations

$$(36)\qquad \begin{cases} \sin\sigma_2\left(\dfrac{\partial\phi}{\partial u}+\dfrac{\partial\theta_1}{\partial v}\right)=\sin\phi\cos\theta_1-\cos\sigma_2\cos\phi\sin\theta_1, \\[2ex] \sin\sigma_2\left(\dfrac{\partial\phi}{\partial v}+\dfrac{\partial\theta_1}{\partial u}\right)=-\cos\phi\sin\theta_1+\cos\sigma_2\sin\phi\cos\theta_1, \end{cases}$$

and also of the equations

$$(37)\qquad \begin{cases} \sin\sigma_1\left(\dfrac{\partial\phi}{\partial u}+\dfrac{\partial\theta_2}{\partial v}\right)=\sin\phi\cos\theta_2-\cos\sigma_1\cos\phi\sin\theta_2, \\[2ex] \sin\sigma_1\left(\dfrac{\partial\phi}{\partial v}+\dfrac{\partial\theta_2}{\partial u}\right)=-\cos\phi\sin\theta_2+\cos\sigma_1\sin\phi\cos\theta_2. \end{cases}$$

The projections of the line M_1M' on the tangents to the lines of curvature of S_1 and on its normal, where M_1 and M' are corresponding points on S_1 and S', are

$$(38)\qquad a\sin\sigma_2\cos\phi, \qquad a\sin\sigma_2\sin\phi, \qquad 0.$$

The direction-cosines of the tangents to the lines of curvature of S_1 with respect to the line MM_1, the line MQ_1 perpendicular to the latter and in the tangent plane at M, and the normal to S are

$$\begin{array}{lll} \cos\omega, & -\cos\sigma_1\sin\omega, & -\sin\sigma_1\sin\omega, \\ \sin\omega, & \cos\sigma_1\cos\omega, & \sin\sigma_1\cos\omega. \end{array}$$

From these and (38) it follows that the coördinates of M' with respect to MM_1, MQ_1, and the normal to S are

$$a[\sin\sigma_1+\sin\sigma_2\cos(\phi-\omega)], \qquad a[\sin\sigma_2\cos\sigma_1\sin(\phi-\omega)],$$
$$a[\sin\sigma_1\sin\sigma_2\sin(\phi-\omega)].$$

* Vol. II, p. 418.

Hence the coördinates of M' with respect to the axes of the moving trihedral for S are

$$(39)\quad \begin{cases} \dfrac{x'}{a} = \cos\theta_1 \sin\sigma_1 + \cos\theta_1 \sin\sigma_2 \cos(\phi-\omega) \\ \qquad\quad - \sin\theta_1 \sin\sigma_2 \cos\sigma_1 \sin(\phi-\omega), \\ \dfrac{y'}{a} = \sin\theta_1 \sin\sigma_1 + \sin\theta_1 \sin\sigma_2 \cos(\phi-\omega) \\ \qquad\quad + \cos\theta_1 \sin\sigma_2 \cos\sigma_1 \sin(\phi-\omega), \\ \dfrac{z'}{a} = \sin\sigma_1 \sin\sigma_2 \sin(\phi-\omega). \end{cases}$$

If S_2 be transformed by means of σ_1 and the same function ϕ, the coördinates x'', y'', z'' of the resulting surface can be obtained from (39) by interchanging the subscripts 1 and 2. Evidently z' and z'' are equal. A necessary and sufficient condition that x', y' be equal to x'', y'' respectively is

$$\cos\theta_1(x'-x'') + \sin\theta_1(y'-y'') = 0,$$
$$\cos\theta_2(x'-x'') + \sin\theta_2(y'-y'') = 0.$$

If the above values be substituted in these equations, we obtain

$$[\sin\sigma_1 \cos(\theta_2-\theta_1) - \sin\sigma_2]\cos(\phi-\omega)$$
$$\quad - \sin\sigma_1 \cos\sigma_2 \sin(\theta_2-\theta_1)\sin(\phi-\omega) = \sin\sigma_1 - \sin\sigma_2\cos(\theta_2-\theta_1),$$
$$[\sin\sigma_2 \cos(\theta_2-\theta_1) - \sin\sigma_1]\cos(\phi-\omega)$$
$$\quad + \sin\sigma_2 \cos\sigma_1 \sin(\theta_2-\theta_1)\sin(\phi-\omega) = \sin\sigma_2 - \sin\sigma_1\cos(\theta_2-\theta_1).$$

Solving these equations with respect to $\sin(\phi-\omega)$ and $\cos(\phi-\omega)$, we get

$$\sin(\phi-\omega) = \frac{(\cos\sigma_2 - \cos\sigma_1)\sin(\theta_2-\theta_1)}{\sin\sigma_1\sin\sigma_2\cos(\theta_2-\theta_1) + \cos\sigma_1\cos\sigma_2 - 1},$$
$$\cos(\phi-\omega) = \frac{\sin\sigma_1\sin\sigma_2 + (\cos\sigma_1\cos\sigma_2 - 1)\cos(\theta_2-\theta_1)}{\sin\sigma_1\sin\sigma_2\cos(\theta_2-\theta_1) + \cos\sigma_1\cos\sigma_2 - 1}.$$

These two expressions satisfy the condition that the sum of their squares be unity, and the function ϕ satisfies equations (36) and (37). Hence our hypotheses are consistent and the theorem of permutability is demonstrated.

We may replace the above equations by

$$(40)\quad \tan\left(\frac{\phi-\omega}{2}\right) = \frac{\sin\left(\dfrac{\sigma_1+\sigma_2}{2}\right)}{\sin\left(\dfrac{\sigma_1-\sigma_2}{2}\right)}\tan\left(\frac{\theta_1-\theta_2}{2}\right).$$

The preceding result may be expressed in the following form:

When the transforms of a given pseudospherical surface are known, all the transformations of the former can be effected by algebraic processes and differentiation.

Thus, suppose that the complete integral of equations (35) is

$$(41) \qquad \theta = f(u, v, \sigma, c),$$

and that a particular integral is

$$\theta_1 = f(u, v, \sigma_1, c_1),$$

corresponding to particular values of the constants, and let S_1 denote the transform of S by means of θ_1 and σ_1. All the transformations of S_1 are determined by the functions ϕ and σ, where

$$(42) \qquad \tan\left(\frac{\phi - \omega}{2}\right) = \frac{\sin\left(\frac{\sigma + \sigma_1}{2}\right)}{\sin\left(\frac{\sigma - \sigma_1}{2}\right)} \tan\left(\frac{f - \theta_1}{2}\right).$$

Exceptional cases arise when σ has the value σ_1. For all values of c other than c_1 formula (42) gives $\phi = \omega + m\pi$, where m is an odd integer. When this is substituted in equations (36) they reduce to (35). In this case S' coincides with S.

We consider now the remaining case where c has the value c_1, whereupon the right-hand member of (42) is indeterminate. In order to handle this case we consider c in (41) to be a function of σ, reducing to c_1 for $\sigma = \sigma_1$. If we apply the ordinary methods to the function $\tan\left(\frac{f - \theta_1}{2}\right) \Big/ \sin\left(\frac{\sigma - \sigma_1}{2}\right)$, which becomes indeterminate for $\sigma = \sigma_1$, differentiating numerator and denominator with respect to σ, we have

$$\tan\left(\frac{\phi - \omega}{2}\right) = \sin \sigma_1 \left(\frac{\partial f}{\partial \sigma} + \frac{\partial f}{\partial c}\frac{\partial c}{\partial \sigma}\right)_{\sigma = \sigma_1},$$

or

$$\tan\left(\frac{\phi - \omega}{2}\right) = \sin \sigma_1 \left(\frac{\partial f}{\partial \sigma} + c' \frac{\partial f}{\partial c}\right)_{\sigma = \sigma_1}$$

where c' is an arbitrary constant. It is necessary to verify that this value of ϕ satisfies the equations (36), which is easily done.*

* Cf. Bianchi, Vol. II, p. 418.

122. Transformation of Lie. Another transformation of pseudospherical surfaces which, however, is analytical in character was discovered by Lie.* It is immediate when the surface is referred to its asymptotic lines, or to any isothermal-conjugate system of lines.

Since the parameters in terms of which the surface is defined in § 119 are isothermal-conjugate, the parameters of the asymptotic lines may be given by

$$u + v = 2\,\alpha, \qquad u - v = 2\,\beta. \tag{43}$$

In terms of these curvilinear coördinates the linear elements of the surface and its spherical representation have the forms

$$ds^2 = a^2(d\alpha^2 + 2\cos 2\,\omega\, d\alpha d\beta + d\beta^2),$$
$$d\sigma^2 = d\alpha^2 - 2\cos 2\,\omega\, d\alpha d\beta + d\beta^2,$$

and equation (23) takes the form

$$\frac{\partial^2 \omega}{\partial\alpha\,\partial\beta} = \sin\omega\cos\omega.$$

From the form of this equation it is evident that if $\omega = \phi(\alpha, \beta)$ be a solution, so also is $\omega_1 = \phi(\alpha m, \beta/m)$, where m is any constant. Hence from one pseudospherical surface we can obtain an infinity of others by the *transformation of Lie.* It should be remarked, however, that only the fundamental quantities of the new surfaces are thus given, and that the determination of the coördinates requires the solution of a Riccati equation which may be different from that for the given surface.

Lie has called attention to the fact that every Bäcklund transformation is a combination of transformations of Lie and Bianchi.† In order to prove this we effect the change of parameters (43) upon equations (35) and obtain

$$\left\{\begin{aligned} &\frac{\partial}{\partial\alpha}(\theta + \omega) = \frac{1 + \cos\sigma}{\sin\sigma}\sin(\theta - \omega), \\ &\frac{\partial}{\partial\beta}(\theta - \omega) = \frac{1 - \cos\sigma}{\sin\sigma}\sin(\theta + \omega). \end{aligned}\right. \tag{44}$$

* *Archiv for Mathematik og Naturvidenskab*, Vol. IV (1879), p. 150.

† Cf. Bianchi, Vol. II, p. 434; Darboux, Vol. III, p. 432.

In particular, for a transformation of Bianchi we have

$$\frac{\partial}{\partial\alpha}(\theta+\omega)=\sin(\theta-\omega), \qquad \frac{\partial}{\partial\beta}(\theta-\omega)=\sin(\theta+\omega).$$

Suppose that we have a pair of functions θ and ω satisfying these equations, and that we effect upon them the Lie transformation for which m has the value $(1+\cos\sigma)/\sin\sigma$. This gives

$$\omega_1=\omega\left(\frac{1+\cos\sigma}{\sin\sigma}\alpha,\ \frac{1-\cos\sigma}{\sin\sigma}\beta\right),$$
$$\theta_1=\theta\left(\frac{1+\cos\sigma}{\sin\sigma}\alpha,\ \frac{1-\cos\sigma}{\sin\sigma}\beta\right).$$

As these functions satisfy (44), they determine a transformation B_σ. But θ_1 may be obtained from ω_1 by effecting upon the latter an inverse Lie transformation, denoted by L_σ^{-1}, upon this result a Bianchi transformation, $B_{\pi/2}$, and then a direct Lie transformation, L_σ. Hence we may write symbolically

$$B_\sigma=L_\sigma^{-1}B_{\pi/2}L_\sigma,$$

which may be expressed thus:

*A Bäcklund transformation B_σ is the transform of a Bianchi transformation by means of a Lie transformation L_σ.**

EXAMPLES

1. The asymptotic lines on a pseudospherical surface are curves of constant torsion.

2. Every surface whose asymptotic lines are of the same length as their spherical images is a pseudospherical surface of curvature -1.

3. Show that on the pseudosphere, defined by (9), the curves

$$b\cos\phi\,d\phi+\sin^2\phi\sqrt{a^2\sin^2\phi-b^2}\,dv=0,$$

where b is a constant, are geodesics, and find the radius of curvature of these curves.

4. When the linear element of a pseudospherical surface is in the parabolic form (iii) of (7), the surface defined by

$$x'=x-a\frac{\partial x}{\partial u}, \qquad y'=y-a\frac{\partial y}{\partial u}, \qquad z'=z-a\frac{\partial z}{\partial u}$$

is pseudospherical (cf. § 76); it is a Bianchi transform of the given surface.

* Spherical surfaces admit of transformations similar to those of Lie and Bäcklund. The latter are imaginary, but such combinations of them can be made that the resulting surface is real. For a complete discussion of these the reader is referred to chap. v. of the *Lezioni* of Bianchi.

5. The helicoids

$$x = u\cos v,\qquad y = u\sin v,\qquad z = \int\sqrt{\frac{1}{a - k^2u^2} - \frac{h^2}{u^2} - 1}\,du + hv,$$

where a, h, k are constants, are spherical surfaces.

6. The helicoid whose meridian curve is the tractrix is called the *surface of Dini*. Find its equations when $\sin\sigma$ denotes the helicoidal parameter and $\cos\sigma$ the constant length of the segment of the tangent between the curve and its axis. Show that the surface is pseudospherical.

7. The curves tangent to the joins of corresponding points on a pseudospherical surface and on a Bäcklund transform are geodesics only when $\sigma = \pi/2$.

8. Let S be a pseudospherical surface and S_1 a Bianchi transform by means of a function θ (§ 119). Show that

$$X_1' = \cos\omega(\cos\theta X_1 + \sin\theta X_2) - \sin\omega X,$$
$$X_2' = \sin\omega(\cos\theta X_1 + \sin\theta X_2) + \cos\omega X,$$
$$X' = \sin\theta X_1 - \cos\theta X_2,$$

where X_1, X_2, X are direction-cosines, with respect to the x-axis, of the tangents to the lines of curvature on S and of the normal to S, and X_1', X_2', X' are the similar functions for S_1.

123. W-surfaces. Fundamental quantities. Minimal surfaces and surfaces of constant curvature possess, in common with a great many other surfaces, the property that each of the principal radii is a function of the other. Surfaces of this kind were first studied in detail by Weingarten,* and, in consequence, are called *Weingarten surfaces*, or simply *W-surfaces*. Since the principal radii of surfaces of revolution and of the general helicoids are functions of a single parameter (§§ 46, 62), these are W-surfaces. We shall find other surfaces of this kind, but now we consider the properties which are common to W-surfaces.

When a surface S is referred to its lines of curvature, the Codazzi equations may be given the form

$$\text{(45)}\qquad \frac{\partial\log\sqrt{\mathcal{E}}}{\partial v} = \frac{1}{\rho_2 - \rho_1}\frac{\partial\rho_1}{\partial v},\qquad \frac{\partial\log\sqrt{\mathcal{G}}}{\partial u} = \frac{1}{\rho_1 - \rho_2}\frac{\partial\rho_2}{\partial u}.$$

If a relation exists between ρ_1 and ρ_2, as

$$\text{(46)}\qquad f(\rho_1, \rho_2) = 0,$$

the integration of equations (45) is reducible to quadratures, thus:

$$\sqrt{\mathcal{E}} = Ue^{\int\frac{d\rho_1}{\rho_2 - \rho_1}},\qquad \sqrt{\mathcal{G}} = Ve^{\int\frac{d\rho_2}{\rho_1 - \rho_2}},$$

* Crelle, Vol. LXII (1863), pp. 160–173.

where U and V are functions of u and v respectively. Without changing the parametric lines the parameters can be so chosen that the above expressions reduce to

$$(47)\qquad \sqrt{\mathcal{E}} = e^{\int \frac{d\rho_1}{\rho_2 - \rho_1}}, \qquad \sqrt{\mathcal{G}} = e^{\int \frac{d\rho_2}{\rho_1 - \rho_2}}.$$

Thus $\mathcal{E}$ and $\mathcal{G}$ are expressible as functions of ρ_1 or ρ_2, and consequently they are functions of one another. This relation becomes more clear when we introduce an additional parameter κ defined by

$$(48)\qquad \kappa = e^{\int \frac{d\rho_1}{\rho_1 - \rho_2}}.$$

By the elimination of ρ_2 from this equation and (46) we have a relation of the form

$$\rho_1 = \phi(\kappa).$$

When this value is substituted in (48) we obtain

$$\rho_2 = \phi(\kappa) - \kappa\phi'(\kappa),$$

where the accent indicates differentiation with respect to κ. From (47) it follows that

$$\sqrt{\mathcal{E}} = \frac{1}{\kappa}, \qquad \sqrt{\mathcal{G}} = \frac{1}{\phi'}.$$

When these values are substituted in the Gauss equation for the sphere (V, 24), the latter becomes

$$\frac{\partial}{\partial u}\left(\frac{\kappa\phi''}{\phi'^2}\frac{\partial\kappa}{\partial u}\right) + \frac{\partial}{\partial v}\left(\frac{\phi'}{\kappa^2}\frac{\partial\kappa}{\partial v}\right) - \frac{1}{\kappa\phi'} = 0.$$

This equation places a restriction upon the forms of κ and $\phi(\kappa)$, but it is the only restriction, for the Codazzi equations (45) are satisfied. Hence we have the theorem of Weingarten:*

When one has an orthogonal system on the unit sphere for which the linear element is reducible to the form

$$(49)\qquad d\sigma^2 = \frac{du^2}{\kappa^2} + \frac{dv^2}{\phi'^2(\kappa)},$$

there exists a W-surface whose lines of curvature are represented by this system and whose principal radii are expressed by

$$(50)\qquad \rho_1 = \phi(\kappa), \qquad \rho_2 = \phi(\kappa) - \kappa\phi'(\kappa).$$

* *L.c.*, p. 163.

If the coördinates of the sphere, namely X, Y, Z, are known functions of u and v, the determination of the W-surface with this representation reduces to quadratures. For, from the formulas of Rodrigues (IV, 32) we have

$$x = -\int \rho_1 \frac{\partial X}{\partial u} du + \rho_2 \frac{\partial X}{\partial v} dv,$$

$$y = -\int \rho_1 \frac{\partial Y}{\partial u} du + \rho_2 \frac{\partial Y}{\partial v} dv,$$

$$z = -\int \rho_1 \frac{\partial Z}{\partial u} du + \rho_2 \frac{\partial Z}{\partial v} dv.$$

The right-hand members of these equations are exact differentials, since the Codazzi equations (45) have been satisfied. If X, Y, Z are not known, their determination requires the solution of a Riccati equation. The relation between the radii of the form (46) is obtained by eliminating κ from equations (50).

We find readily that the fundamental quantities for the surface have the values

$$(51) \qquad \begin{cases} E = \dfrac{\phi^2}{\kappa^2}, & F = 0, \qquad G = \left(\dfrac{\phi - \kappa\phi'}{\phi'}\right)^2, \\ D = \dfrac{\phi}{\kappa^2}, & D' = 0, \qquad D'' = \dfrac{\phi - \kappa\phi'}{\phi'^2}. \end{cases}$$

And from (48), (50), and (51) we obtain

$$(52) \qquad \sqrt{E} = \rho_1 e^{\int \frac{d\rho_1}{\rho_2 - \rho_1}}, \qquad \sqrt{G} = \rho_2 e^{\int \frac{d\rho_2}{\rho_1 - \rho_2}}.$$

Consider the quadratic form

$$(53) \quad \frac{1}{H}\left[(ED' - FD)\,du^2 + (ED'' - GD)\,du dv + (FD'' - GD')\,dv^2\right],$$

which when equated to zero defines the lines of curvature. When these lines are parametric, this quadratic form is reducible by means of (IV, 74) to

$$\sqrt{\mathcal{E}\mathcal{G}}(\rho_1 - \rho_2)\,du dv.$$

But in consequence of (47) this is further reducible for W-surfaces to $du dv$. Since the curvature of this latter form is zero, the curvature of (53) also is zero, and consequently (§ 135) the form (53) is reducible by quadratures to $du dv$. Hence we have the theorem of Lie:

The lines of curvature of a W-surface can be found by quadratures.

124. Evolute of a *W*-surface. The evolute of a *W*-surface possesses several properties which are characteristic. Referring to the results of § 75, we see that by means of (52) the linear elements of the sheets of the evolute of a *W*-surface are reducible to the form

$$(54)\qquad \begin{cases} ds_1^2 = d\rho_1^2 + e^{2\int \frac{d\rho_1}{\rho_1 - \rho_2}} dv^2, \\ ds_2^2 = d\rho_2^2 + e^{2\int \frac{d\rho_2}{\rho_2 - \rho_1}} du^2, \end{cases}$$

or, in terms of κ,

$$(55)\qquad \begin{cases} ds_1^2 = \phi'^2 d\kappa^2 + \kappa^2 dv^2, \\ ds_2^2 = \kappa^2 \phi''^2 d\kappa^2 + \phi'^2 du^2. \end{cases}$$

From these results and the remarks of § 46 we obtain at once the following theorem of Weingarten:

Each surface of center of a W-surface is applicable to a surface of revolution whose meridian curve is determined by the relation between the radii of the given surface.

We have also the converse theorem, likewise due to Weingarten:

If a surface S_1 be applicable to a surface of revolution, the tangents to the geodesics on S_1 corresponding to the meridians of the surface of revolution are normal to a family of parallel W-surfaces; if S_1 be deformed in any manner whatever, the relation between the radii of these W-surfaces is unaltered.

In proving this theorem we apply the results of § 76. If the linear element of S_1 be

$$ds_1^2 = du^2 + U^2 dv^2,$$

the principal radii of S are given by

$$(56)\qquad \rho_1 = u, \qquad \rho_2 = u - \frac{U}{U'}.$$

Since both are functions of a single parameter, a relation exists between them which depends upon U alone, and consequently is unaltered in the deformation of S_1.

From (V, 99) the projections upon the moving trihedral for S_1 of a displacement of a point on the complementary surface S_2 are

$$d\left(u - \frac{U}{U'}\right), \qquad 0, \qquad (q\,du + q_1 dv)\frac{U}{U'}.$$

In consequence of formulas (V, 48, 75) the expression $U(q\,du + q_1 dv)$ is an exact differential, which will be denoted by dw. Hence the linear element of S_2 is

$$ds_2^2 = U^2\left(\frac{1}{U'}\right)'^2 du^2 + \frac{1}{U'^2}dw^2, \tag{57}$$

from which it follows that S_2 also is applicable to a surface of revolution.*

The last theorem of § 75 may be stated thus:

A necessary and sufficient condition that the asymptotic lines on the surfaces of center S_1, S_2 of a surface S correspond is that S be a W-surface; in this case to every conjugate system on S_1 or S_2 there corresponds a conjugate system on the other.

From (V, 98, 98′) it follows that when S is a W-surface, and only in this case, we have

$$K_1 K_2 = \frac{1}{(\rho_1 - \rho_2)^4}. \tag{58}$$

Hence at corresponding points the curvature is of the same kind.

An exceptional form of equation (46) is afforded by the case where one or both of the principal radii is constant. For the plane both radii are infinite; for a circular cylinder one is infinite and the other has a finite constant value. The sphere is the only surface with both radii finite and constant. For, if ρ_1 and ρ_2 are different constants, from (45) it follows that $\mathcal{E}$ and $\mathcal{G}$ are functions of u and v respectively, which is impossible for the sphere. When one of the radii is infinite, the surface is developable. There remains the case where one has a finite constant value; then S is a canal surface (§ 29).

In considering the last case we take

$$\rho_2 = a;$$

then, from (48), we have

$$\rho_1 = \kappa + a,$$

and the linear element of the sphere is

$$d\sigma^2 = \frac{du^2}{\kappa^2} + dv^2.$$

Conversely, when the linear element of the sphere is reducible to this form, the curves on the sphere represent the lines of curvature on an infinity of parallel canal surfaces.

* Cf. Darboux, Vol. III, p. 329.

125. Surfaces of constant mean curvature. For surfaces of contant total curvature the relation (46) may be written

$$\rho_2 = \frac{c}{\rho_1},$$

where c denotes a constant. When this value is substituted in (48) we have, by integration,

$$\rho_1 = \sqrt{\kappa^2 + c}, \qquad \rho_2 = \frac{c}{\sqrt{\kappa^2 + c}}, \tag{59}$$

so that the linear element of the sphere is

$$d\sigma^2 = \frac{du^2 + (\kappa^2 + c)\,dv^2}{\kappa^2}. \tag{60}$$

Conversely, when we have an orthogonal system on the sphere for which the linear element is reducible to the form (60), it serves for the representation of the lines of curvature of a surface of constant curvature, and of an infinity of parallel surfaces.

When c is positive, two of these parallel surfaces have constant mean curvature, as follows from the theorem of Bonnet (§ 73). In fact, the radii of these surfaces are

$$\rho_1 = \sqrt{\kappa^2 + c} \pm \sqrt{c}, \qquad \rho_2 = \frac{c}{\sqrt{\kappa^2 + c}} \pm \sqrt{c}. \tag{61}$$

If we put

$$c = a^2, \qquad \kappa = a \operatorname{csch} \omega, \tag{62}$$

and replace u by au, the linear element (60) becomes

$$d\sigma^2 = \sinh^2 \omega \, du^2 + \cosh^2 \omega \, dv^2.$$

In like manner, if we replace u by iau, v by iv, and take

$$c = a^2, \qquad \kappa = ai \operatorname{sech} \omega, \tag{63}$$

the linear element of the sphere is

$$d\sigma_1^2 = \cosh^2 \omega \, du^2 + \sinh^2 \omega \, dv^2.$$

For the values (62) we have, from (61),

$$\rho_1 = \frac{ae^{\pm\omega}}{\sinh \omega}, \quad \rho_2 = \pm \frac{ae^{\pm\omega}}{\cosh \omega}, \tag{64}$$

and the linear elements of the corresponding surfaces are

$$ds^2 = a^2 e^{\pm 2\omega} (du^2 + dv^2). \tag{65}$$

Moreover, for the values (63) the radii have the values

$$\rho_1 = \pm\frac{ae^{\pm\omega}}{\cosh\omega}, \qquad \rho_2 = \frac{ae^{\pm\omega}}{\sinh\omega}, \tag{66}$$

but the linear elements are the same (65). In each case the mean curvature is $\pm 1/a$. We state these results in the following form:

The lines of curvature upon a surface of constant mean curvature form an isothermic system, the parameters of which can be chosen so that the linear element has one of the forms (65), *where* ω *is a solution of the equation*

$$\frac{\partial^2\omega}{\partial u^2} + \frac{\partial^2\omega}{\partial v^2} + \sinh\omega\cosh\omega = 0. \tag{67}$$

Conversely, each solution of this equation determines two pairs of applicable surfaces of constant mean curvature $\pm 1/a$, *whose lines of curvature correspond, and for which the radii* ρ_1, ρ_2 *of one surface are equal to the radii of* ρ_2, ρ_1 *of the applicable surface.*

It can be shown that if $\omega = \phi(u, v)$ is a solution of equation (67), so also is

$$\omega_1 = \phi(u\cos\sigma - v\sin\sigma,\ u\sin\sigma + v\cos\sigma), \tag{68}$$

where σ is any constant whatever. Hence there exists for spherical surfaces a transformation analogous to the Lie transformation of pseudospherical surfaces. This transformation can be given a geometrical interpretation if it is considered in connection with the surfaces of constant mean curvature parallel to the spherical surfaces.

Let S_1 denote the surface with the linear element

$$ds_1^2 = a^2 e^{2\omega_1}(du^2 + dv^2). \tag{69}$$

If we put

$$u_1 = u\cos\sigma - v\sin\sigma, \qquad v_1 = u\sin\sigma + v\cos\sigma, \tag{70}$$

the solution (68) becomes $\omega_1 = \phi(u_1, v_1)$, and (69) reduces to

$$ds_1^2 = a^2 e^{2\omega_1}(du_1^2 + dv_1^2).$$

Hence if we make a point (u, v) on S with the linear element (65), in which the positive sign is taken, correspond to the point (u_1, v_1) on S_1, the surfaces are applicable, and to the lines of curvature $u = \text{const.}$, $v = \text{const.}$ on S correspond on S_1 the curves

$$u\cos\sigma - v\sin\sigma = \text{const.}, \qquad u\sin\sigma + v\cos\sigma = \text{const.}$$

But the latter cut the lines of curvature $u = \text{const.}$, $v = \text{const.}$ on S_1 under the angle σ. Moreover, the corresponding principal radii of S and S_1 are equal at corresponding points. Hence we have the following theorem of Bonnet:*

A surface of constant mean curvature admits an infinity of applicable surfaces of the same kind with preservation of the principal radii at corresponding points, and the lines of curvature on one surface correspond to lines on the other which cut the lines of curvature under constant angle.

Weingarten has considered the W-surfaces whose lines of curvature are represented on the sphere by geodesic ellipses and hyperbolas. In this case the linear element of the sphere is reducible to the form (§ 90)

$$d\sigma^2 = \frac{du^2}{\sin^2\frac{\omega}{2}} + \frac{dv^2}{\cos^2\frac{\omega}{2}}.$$

Comparing this with (49), we have

$$\kappa = \sin\frac{\omega}{2}, \qquad \phi' = \cos\frac{\omega}{2},$$

from which it follows that

$$\phi = \frac{\omega + \sin\omega}{4}.$$

Hence

$$\rho_1 = \frac{\omega + \sin\omega}{4}, \qquad \rho_2 = \frac{\omega - \sin\omega}{4}, \tag{71}$$

and the relation between the radii is found, by the elimination of ω, to be

$$2(\rho_1 - \rho_2) = \sin 2(\rho_1 + \rho_2).\dagger \tag{72}$$

* Mémoire sur la théorie des surfaces applicables sur une surface donnée, *Journal de l'École Polytechnique*, Cahier 42 (1867), pp. 72 *et seq.* In this memoir Bonnet solves completely the problem of finding applicable surfaces with corresponding principal radii equal. When a surface possesses an infinity of applicable surfaces of this kind, its lines of curvature form an isothermal system.

† Darboux (Vol. III, p. 373) proves that these surfaces may be generated as follows: Let C and C_1 be two curves of constant torsion, differing only in sign. The locus of the mid-points M of the join of any points P and P_1 of these curves is a surface of translation. If a line be drawn through M parallel to the intersection of the osculating planes of C and C_1 at P and P_1, this line is normal to a W-surface of the above type for all positions of M.

126. Ruled *W*-surfaces. We conclude the present study of *W*-surfaces with the solution of the problem:

To determine the W-surfaces which are ruled.

This problem was proposed and solved simultaneously by Beltrami * and Dini.† We follow the method of the latter.

In §§ 106, 107 we found that when the linear element of a ruled surface is in the form

$$ds^2 = du^2 + [(u-\alpha)^2 + \beta^2]\,dv^2,$$

the expressions for the total and mean curvatures are

$$K = -\frac{\beta^2}{g^4}, \qquad K_m = -\frac{rg^2 + \beta'(u-\alpha) + \beta\alpha'}{g^3},$$

where r is a function of v at most, and

$$g^2 = (u-\alpha)^2 + \beta^2.$$

In order that a relation exist between the principal radii it is necessary and sufficient that the equation

$$\frac{\partial}{\partial u}K \cdot \frac{\partial}{\partial v}K_m - \frac{\partial}{\partial v}K \cdot \frac{\partial}{\partial u}K_m = 0$$

be satisfied identically. If the above values be substituted, the resulting equation reduces to

$$\frac{2(u-\alpha)}{\beta}\frac{\partial}{\partial v}\left[\frac{rg^2 + \beta'(u-\alpha) + \beta\alpha'}{g^3}\right] - \left\{\frac{\beta^2\beta' - (u-\alpha)[rg^2 + 2\beta'(u-\alpha) + 3\beta\alpha']}{g^5}\right\}\frac{\partial}{\partial v}\frac{g^2}{\beta} = 0.$$

As this is an identical equation, it is true when $u = \alpha$, in which case it reduces to $\beta' = 0$. Hence β is a constant and the above equation becomes

$$r'(u-\alpha)^2 + r'\beta^2 + \beta\alpha'' = 0.$$

Since this equation must be true independently of the value of u, both r' and α'' are zero. Therefore we have

$$(73) \qquad \alpha = cv + d, \qquad \beta = e, \qquad r = k,$$

where c, d, e, k are constants.

The linear element is

$$ds^2 = du^2 + [(u - cv - d)^2 + e^2]\,dv^2.$$

* *Annali*, Vol. VII (1865), pp. 139–150. † *Annali*, Vol. VII (1865), pp. 205–210.

In order to interpret this result we calculate the expression for the tangent of the angle which the generators $v = \text{const.}$ make with the line of striction

$$u - cv - d = 0.$$

From (III, 24) we have

$$\tan\theta = \frac{e}{c};$$

consequently the angle is constant. Conversely, if θ and the parameter of distribution β be constant, α has the form (73). Hence we have the theorem:

A necessary and sufficient condition that a ruled surface be a W-surface is that the parameter of distribution be constant and that the generators be inclined at a constant angle to the line of striction, which consequently is a geodesic.

EXAMPLES

1. Show that the helicoids are W-surfaces.

2. Find the form of equation (49), when the surface is minimal, and show that each conformal representation of the sphere upon the plane determines a minimal surface.

3. Show that the tangents to the curves $v = \text{const.}$ on a spherical surface with the linear element (i) of (6) are normal to a W-surface for which

$$\rho_2 - \rho_1 = a \cot\frac{\rho_1}{a}.$$

4. The helicoids are the only W-surfaces which are such that the curves $\rho_1 = \text{const.}$ meet the lines of curvature under constant angle (cf. Ex. 23, p. 188).

5. The asymptotic lines on the surfaces of center of a surface for which $\rho_1 + \rho_2 = \text{const.}$ correspond to the minimal lines on the spherical representation of the surface; and, when $\rho_1 - \rho_2 = \text{const.}$, to a rectangular system on the sphere.

127. Spherical representation of surfaces with plane lines of curvature in both systems. Surfaces whose lines of curvature in one or both systems are plane curves have been an object of study by many geometers. Since the tangents to a line of curvature and to its spherical representation at corresponding points are parallel, a plane line of curvature is represented on the sphere by a plane curve, that is, a circle; and conversely, a line of curvature is plane when its spherical representation is a circle.

We consider first the determination of surfaces with plane lines of curvature in both systems from the point of view of their spherical representation.* To this end we must find orthogonal systems of circles on the sphere. If two circles cut one another orthogonally, the plane of each must pass through the pole of the plane of the other. Hence the planes of the circles of one system pass through a point in the plane of each circle of the second system, and consequently the planes of each family form a pencil, the two axes being polar reciprocal with respect to the sphere.†

We consider separately the two cases: I, when one axis is tangent to the sphere, and therefore the other is tangent at the same point and perpendicular to it; II, when neither is tangent.

CASE I. We take the center of the unit sphere for origin O, the x- and y-axes parallel to the axes of the pencils, and let the coördinates of the point of contact be (0, 0, 1). The equations of the pencils of planes may be put in the form

$$x + u(z-1) = 0, \qquad y + v(z-1) = 0, \tag{74}$$

where u and v are the parameters of the respective families. If these equations be solved simultaneously with the equation of the sphere, and, as usual, X, Y, Z denote coördinates of the latter, we have

$$X = \frac{2u}{u^2+v^2+1}, \qquad Y = \frac{2v}{u^2+v^2+1}, \qquad Z = \frac{u^2+v^2-1}{u^2+v^2+1}. \tag{75}$$

Now the linear element of the sphere is

$$d\sigma^2 = \frac{4(du^2+dv^2)}{(u^2+v^2+1)^2}. \tag{76}$$

CASE II. As in the preceding case, we take for the z-axis the common perpendicular to the axes of the pencils, and for the x- and y-axes we take lines through O parallel to the axes of the pencils. The coördinates of the points of meeting of the latter with the z-axis are of the form $(0, 0, a)$, $(0, 0, 1/a)$. The equations of the two pencils of planes could be written in forms

* Bianchi, Vol. II, p. 256; Darboux, Vol. I, p. 128, and Vol. IV, p. 180.

† Bonnet, *Journal de l'École Polytechnique*, Vol. XX (1853), pp. 136, 137.

similar to (74), but the expressions for X, Y, Z will be found to be of a more suitable form if the equations of the families of planes be written

$$x - \frac{\tan u}{\sqrt{1-a^2}}(z-a) = 0, \qquad y - \frac{a \tanh v}{\sqrt{1-a^2}}\left(z - \frac{1}{a}\right) = 0.$$

Proceeding as in Case I, we find

$$(77) \qquad \begin{cases} X = \dfrac{\sqrt{1-a^2}\sin u}{\cosh v + a\cos u}, \\ Y = -\dfrac{\sqrt{1-a^2}\sinh v}{\cosh v + a\cos u}, \\ Z = \dfrac{\cos u + a\cosh v}{\cosh v + a\cos u}, \end{cases}$$

and the linear element is

$$(78) \qquad d\sigma^2 = \frac{(1-a^2)(du^2+dv^2)}{(\cosh v + a\cos u)^2}.$$

From the preceding discussion we have tacitly excluded the system of meridians and parallels. As before, the planes of the two families of circles form pencils, but now the axis of one pencil passes through the center of the sphere and the other is at infinity. Hence this case corresponds to the value zero for a in Case II. In fact, if we put $a = 0$ in (77), the resulting equations define a sphere referred to a system of meridians and parallels, namely

$$(79) \qquad X = \frac{\sin u}{\cosh v}, \qquad Y = -\frac{\sinh v}{\cosh v}, \qquad Z = \frac{\cos u}{\cosh v}.$$

Since the planes of the lines of curvature on a surface are parallel to the planes of their spherical images, the curves $v = \text{const.}$ on a surface with the representation (79) lie in parallel planes, and the planes of the curves $u = \text{const.}$ envelop a cylinder. These surfaces are called the *molding surfaces.** We shall consider them later.

128. Surfaces with plane lines of curvature in both systems. By a suitable choice of coördinate axes and parameters the expressions for the direction-cosines of the normal to a surface with plane lines of curvature in both systems can be given one

* These surfaces were first studied by Monge, *Application de L'Analyse à la Géométrie*, § 17. Paris, 1849.

of the forms (75) or (77). For the complete determination of all surfaces of this kind it remains then for us to find the expression for the other tangential coördinate W, that is, the distance from the origin to the tangent plane. The linear element of the sphere in both cases is of the form

$$d\sigma^2 = \frac{du^2 + dv^2}{\lambda^2},$$

where λ is such that

$$\frac{\partial^2 \lambda}{\partial u \, \partial v} = 0. \tag{80}$$

From (VI, 39) we see that the equation satisfied by W is

$$\frac{\partial^2 \theta}{\partial u \, \partial v} + \frac{\partial \log \lambda}{\partial v} \frac{\partial \theta}{\partial u} + \frac{\partial \log \lambda}{\partial u} \frac{\partial \theta}{\partial v} = 0.$$

In consequence of (80), if we change the unknown function in accordance with $\theta_1 = \lambda \theta$, the equation in θ_1 is of the form (80). Hence the most general value for W is

$$W = \frac{U + V}{\lambda},$$

where U and V are arbitrary functions of u and v respectively.

Hence any surface with plane lines of curvature in both systems is the envelope of a family of planes whose equation is of the form

$$2\,ux + 2\,vy + (u^2 + v^2 - 1)\,z = 2\,(U + V), \tag{81}$$

or

$$\begin{aligned} &\sqrt{1 - a^2} \sin u x - \sqrt{1 - a^2} \sinh v y + (\cos u + a \cosh v)\,z \\ &\qquad = (U + V)\sqrt{1 - a^2}. \end{aligned} \tag{82}$$

The expressions for the Cartesian coördinates of these surfaces can be found without quadrature by the methods of § 67. Thus, for the surface envelope of (81) we have to solve for x, y, z equation (81) and its derivatives with respect to u and v. The latter are

$$x + uz = U', \qquad y + vz = V', \tag{83}$$

where the accents indicate differentiation. We shall not carry out this solution, but remark that as each of these equations contains a single parameter they define the planes of the lines of curvature.

From the form of (83) it is seen that these planes in each system envelop a cylinder, and that the axes of these two cylinders are perpendicular. This fact was remarked by Darboux, who also observed that equation (81) defines the radical plane of the two spheres

$$\left\{\begin{aligned} &x^2+y^2+z^2+2\,ux+(u^2-\tfrac{1}{2})z=2\,U, \\ &x^2+y^2+z^2-2\,vy-(v^2-\tfrac{1}{2})z=-2\,V. \end{aligned}\right. \tag{84}$$

These are the equations of two one-parameter families of spheres, whose centers lie on the focal parabolas

$$\left\{\begin{aligned} &x_1=-u, \quad y_1=0, \quad z_1=\tfrac{1}{2}(\tfrac{1}{2}-u^2), \\ &x_2=0, \quad y_2=v, \quad z_2=\tfrac{1}{2}(v^2-\tfrac{1}{2}), \end{aligned}\right. \tag{85}$$

and whose radii are determined by the arbitrary functions U and V. The characteristics of each family are defined by its equation and the corresponding equation of the pair (83). Consequently the original surface is the locus of the point of intersection of the planes of these characteristics and the radical planes of the spheres.

Similar results follow for the equation (82), which defines the radical planes of two families of spheres whose centers are on the focal ellipse and hyperbola

$$\left\{\begin{aligned} &x_1=\tfrac{1}{2}\sqrt{1-a^2}\sin u, \quad y_1=0, \quad z_1=\tfrac{1}{2}\cos u, \\ &x_2=0, \quad y_2=\tfrac{1}{2}\sqrt{1-a^2}\sinh v, \quad z_2=-\tfrac{1}{2}\,a\cosh v. \end{aligned}\right. \tag{86}$$

When in particular $a=0$, these curves of center are a circle and its axis.

From the foregoing results it follows that these surfaces may be generated by the following geometrical method due to Darboux:*

Every surface with plane lines of curvature in two systems can be obtained from two singly infinite families of spheres whose centers lie on focal conics and whose radii vary according to an arbitrary law. The surface is the envelope of the radical plane of two spheres S and Σ, belonging to two different families. If one associate with S and Σ two infinitely near spheres S' and Σ', the radical center of these four spheres describes the surface; and the radical planes of S and S' and of Σ and Σ' are the planes of the lines of curvature.

* Vol. I, p. 132.

129. Surfaces with plane lines of curvature in one system. Surfaces of Monge. When the lines of curvature in one system are plane, the curves on the sphere are a family of circles and their orthogonal trajectories; and conversely. Every system of this kind may be obtained from a system of circles and their orthogonal trajectories in a plane by a stereographic projection. The determination of such a system in the plane reduces to the integration of a Riccati equation (Ex. 11, p. 50). Since the circles are curves of constant geodesic curvature we have, in consequence of the first theorem of § 84, the theorem:

The determination of all the surfaces with plane lines of curvature in one system requires the solution of a Riccati equation and quadratures.

We shall discuss at length several kinds of surfaces with plane lines of curvature in one system, and begin with the case where these curves are geodesics. They are consequently normal sections of the surface. Their planes envelop a developable surface, called the *director-developable*, and the lines of curvature in the other system are the orthogonal trajectories of these planes. Conversely, the locus of any simple infinity of the orthogonal trajectories of a one-parameter system of planes is a surface of the kind sought. For, the planes cut the surface orthogonally, and consequently they are lines of curvature and geodesics (§ 59). Since these planes are the osculating planes of the edge of regression of the developable, the orthogonal trajectories can be found by quadratures (§ 17).

Suppose that we have such a surface, and that C denotes one of the orthogonal trajectories of the family of plane lines of curvature. Let the coördinates of C be expressed in terms of the arc of the curve from a point of it, which will be denoted by v. As the plane of each plane line of curvature Γ is normal to C at its point of meeting with the latter, the coördinates of a point P of Γ with reference to the moving trihedral of C are $0, \eta, \zeta$. Since P describes an orthogonal trajectory of the planes, we must have (I, 82)

$$\frac{d\eta}{dv} + \frac{\zeta}{\tau} = 0, \qquad \frac{d\zeta}{dv} - \frac{\eta}{\tau} = 0,$$

where τ denotes the radius of torsion of C. If we change the parameter of C in accordance with the equation

$$(87) \qquad v_1 = \int_0^v \frac{dv}{\tau},$$

the above equations become

$$\frac{d\eta}{dv_1} + \zeta = 0, \qquad \frac{d\zeta}{dv_1} - \eta = 0.$$

The general integral of these equations is

$$(88) \qquad \eta = U_1 \cos v_1 - U_2 \sin v_1, \qquad \zeta = U_1 \sin v_1 + U_2 \cos v_1,$$

where U_1 and U_2 are functions of the parameter u of points of Γ. When $v = 0$ we have $v_1 = 0$, and so the curve Γ in the plane through the point $v = 0$ of C has the equations $\eta = U_1$, $\zeta = U_2$. Hence the character of the functions U_1 and U_2 is determined by the form of the curve; and conversely, the functions U_1 and U_2 determine the character of the curve.

By definition (87) the function v_1 measures the angle swept out in the plane normal to C by the binormal of the latter, as this plane moves from $v = 0$ to any other point. Hence equations (88) define the same curve, in this moving plane, for each value of v_1, but it is defined with respect to axes which have rotated through the angle v_1. Hence we have the theorem:

Any surface whose lines of curvature in one system are geodesics can be generated by a plane curve whose plane rolls, without slipping, over a developable surface.

These surfaces are called the *surfaces of Monge*, by whom they were first studied. He proposed the problem of finding a surface with one sheet of the evolute a developable. It is evident that the above surfaces satisfy this condition. Moreover, they furnish the only solution. For, the tangents to a developable along an element lie in the plane tangent along this element, and if these tangents are normals to a surface, the latter is cut normally by this plane, and consequently the curve of intersection is a line of curvature. In particular, a molding surface (§ 127) is a surface of Monge with a cylindrical director-developable.

Since every curve in the moving plane of the lines of curvature generates a surface of Monge, a straight line in this plane generates a developable surface of Monge. For, all the normals to the surface along a generator lie in a plane (§ 25). Hence:

A necessary and sufficient condition that a curve Γ in a plane normal to a curve C at a point Q generate a surface of Monge as the plane moves, remaining normal to the curve, is that the line joining a point of Γ to Q generate a developable.

130. Molding surfaces. When the orthogonal trajectory C is a plane curve, the planes of the curves Γ are perpendicular to the plane of C, and consequently the director-developable is a cylinder whose right section is the plane evolute of C. The surface is a *molding surface* (§ 127), and all the lines of curvature of the second system are plane curves, — involutes of the right section of the cylinder. Hence a molding surface may be generated by a plane curve whose plane rolls without slipping over a cylinder. We shall apply the preceding formulas to this particular case.

Since $1/\tau$ is equal to zero, it follows from (88) that η and ζ are functions of u alone. If u be taken as a measure of the arc of the curve Γ, we have, in all generality,

$$\eta = U, \qquad \zeta = \int \sqrt{1 - U'^2}\, du,$$

where the function U determines the form of Γ. If we take the plane of the curve C for $z = 0$, and x_0, y_0 denote the coördinates of a point of C, the equations of the surface may be written

$$x = x_0 + U\cos v, \qquad y = y_0 + U\sin v, \qquad z = \int \sqrt{1 - U'^2}\, du,$$

where v denotes the angle which the principal normal to C makes with the x-axis. Since
$$\frac{dx_0}{ds_0} = \sin v \qquad \frac{dy_0}{ds_0} = -\cos v,$$

if V denote the radius of curvature of C, then $ds_0 = V\,dv$, and the equations of the surface can be put in the following form, given by Darboux *:

$$(89) \qquad \begin{cases} x = U\cos v + \displaystyle\int V \sin v\, dv, \\ y = U\sin v - \displaystyle\int V\cos v\, dv, \\ z = \displaystyle\int \sqrt{1 - U'^2}\, du. \end{cases}$$

* Vol. I, p. 105.

The equations of the right section of the cylinder are

$$\bar{x} = x_0 + V \cos v = \int V' \cos v \, dv,$$

$$\bar{y} = y_0 + V \sin v = \int V' \sin v \, dv.$$

In passing, we remark that surfaces of revolution are molding surfaces, whose director-cylinder is a line; this corresponds to the case $V = 0$.

EXAMPLES

1. When the spherical representation of the lines of curvature of a surface is isothermal and the curves in one family on the sphere are circles, the curves in the other family also are circles.

2. If the lines of curvature in one system on a minimal surface are plane, those in the other system also are plane.

3. Show that the surface

$$x = au + \sin u \cosh v, \quad y = v + a \cos u \sinh v, \quad z = \sqrt{1 - a^2} \cos u \cosh v,$$

is minimal and that its lines of curvature are plane. Find the spherical representation of these curves and determine the form of the curves.

4. Show that the surface of Ex. 3 and the Enneper surface (Ex. 18, p. 209) are the only minimal surfaces with plane lines of curvature.

5. When the lines of curvature in one system lie in parallel planes, the surface is of the molding type.

6. A necessary and sufficient condition that the lines of curvature in one system on a surface be represented on the unit sphere by great circles is that it be a surface of Monge.

7. Derive the expressions for the point coördinates of a molding surface by the method of § 67.

131. Surfaces of Joachimsthal. Another interesting class of surfaces with plane lines of curvature in one system are those for which all the planes pass through a straight line. Let one of these lines of curvature be denoted by Γ, and one of the other system by C. The developable enveloping the surface along the latter has for its elements the tangents to the curves Γ at their points of intersection with C. Since these elements lie in the planes of the curves Γ, the developable is a cone with its vertex on the line D, through which all these planes pass. This cone is tangent to the surface along C, and its elements are orthogonal to the latter. Consequently C is the intersection of the surface and a sphere with center at the vertex of the cone which cuts the surface orthogonally. Hence we have the following result, due to Joachimsthal*:

When the lines of curvature in one system lie in planes passing through a line D, the lines of curvature in the second system lie on spheres whose centers are on D and which cut the surface orthogonally.

Such surfaces are called *surfaces of Joachimsthal.* Each of the curves of the first system is an orthogonal trajectory of the circles in which the spheres are cut by its plane. Therefore, in order to derive the equations of such a surface, we consider first the orthogonal trajectories of a family of circles whose centers are on a line. If the latter be taken for the η-axis, the circles are defined by

$$\xi = r \sin\theta, \qquad \eta = r\cos\theta + u,$$

where r denotes the radius, θ the angle which the latter makes with the η-axis, and u the distance of the center from the origin. Now r is a function of u, and θ is independent of u. In order that these same equations may define an orthogonal trajectory of the circles, θ must be such a function of u that

$$\cos\theta \frac{\partial \xi}{\partial u} - \sin\theta \frac{\partial \eta}{\partial u} = 0,$$

or

$$r\frac{d\theta}{du} - \sin\theta = 0.$$

By integration we have

$$\tan\frac{\theta}{2} = Ve^{\int \frac{du}{r}}, \tag{90}$$

where V denotes the constant of integration.

Since each section of a surface of Joachimsthal by a plane through its axis is an orthogonal trajectory of a family of circles whose centers are on this axis, the equations of the most general surface of this kind are of the form

$$x = r\sin\theta\cos v, \qquad y = r\sin\theta\sin v, \qquad z = u + r\cos\theta,$$

where v denotes the angle which the plane through a point and the axis makes with the plane $y = 0$, and θ is given by (90), in which now V is a function of v.

* *Crelle*, Vol. LIV (1857), pp. 181–192.

When V is constant θ is a function of u alone, and the surface is one of revolution. For other forms of V the geometrical generation of the surfaces is given by the theorem:

Given the orthogonal trajectories of a family of circles whose centers lie on a right line D; if they be rotated about D through different angles, according to a given law, the locus of the curves is a surface of Joachimsthal.

132. Surfaces with circular lines of curvature. We consider next surfaces whose lines of curvature in one system are circles. Let σ denote the constant angle between the plane of the circle C and the tangent planes to the surface along C (cf. § 59), ρ the radius of normal curvature in the direction of C, and r the radius of the latter. Now equation (IV, 17) may be written

$$r = \rho \sin \sigma. \tag{91}$$

As an immediate consequence we have the theorem:

A necessary and sufficient condition that a plane line of curvature be a circle is that the normal curvature of the surface in its direction be the same at all of its points.

Since the normals to the surface along C are inclined to its plane under constant angle, they form a right circular cone whose vertex is on the axis of C. Moreover, the cone cuts the surface at right angles, and consequently the sphere of radius ρ and center at the vertex of the cone is tangent to the surface along C. Hence the surface is the envelope of a family of spheres of variable or constant radius, whose centers lie on a curve.

Conversely, we have seen in § 29 that the characteristics of the family of spheres

$$(X-x)^2+(Y-y)^2+(Z-z)^2=R^2,$$

where x, y, z are the coördinates of a curve expressed in terms of its arc, and R is a function of the same parameter, are circles of radius

$$r = R(1-R'^2)^{\frac{1}{2}}, \tag{92}$$

whose axes are tangent to the curve of centers and whose centers have the coördinates

$$x_1 = x - \alpha RR', \qquad y_1 = y - \beta RR', \qquad z_1 = z - \gamma RR', \tag{93}$$

where α, β, γ are the direction-cosines of the axis, and the accent indicates differentiation. The normals to the envelope along a characteristic form a cone, and consequently these circles are lines of curvature upon it. Hence:

A necessary and sufficient condition that the lines of curvature in one family be circles is that the surface be the envelope of a single infinity of spheres, the locus of whose centers is a curve, the radii being determined by an arbitrary law.

From equations (91), (92) it follows that $R' = \cos\sigma$. Hence the circles are geodesics only when R is constant, that is, for canal surfaces (§ 29). In this case, as is seen from (92), all the circles are equal.

The circles are likewise of equal radius a when

$$R^2 = (s+c)^2 + a^2,$$

where s is the arc of the curve of centers and c is a constant of integration. Now equations (93) become

$$x_1 = x - (s+c)\alpha. \qquad y_1 = y - (s+c)\beta, \qquad z_1 = z - (s+c)\gamma,$$

which are the equations also of an involute of the curve of centers (§ 21). This result may be stated thus*:

If a string be unwound from a curve in such a way that its moving extremity M generates an involute of the curve, and if at M a circle be constructed whose center is M and whose plane is normal to the string, then as the string is unwound this circle generates a surface with a family of equal circles for lines of curvature.

The locus of the centers of the spheres enveloped by a surface is evidently one sheet of the evolute of the surface, and the radius of the sphere is the radius of normal curvature in the direction of the circle. Consequently this radius is a function of the parameter of the spheres. Conversely, from § 75, we have that when S_2 is a curve $H_2 = 0$, and consequently

$$(\rho_1 - \rho_2)\frac{\partial \rho_2}{\partial v} = 0.$$

* Cf. Bianchi, Vol. II, p. 272.

Excluding the case of the sphere, we have that ρ_2 is a function of u alone. From the formulas of Rodrigues (IV, 32),

$$\frac{\partial x}{\partial v} = -\rho_2 \frac{\partial X}{\partial v}, \qquad \frac{\partial y}{\partial v} = -\rho_2 \frac{\partial Y}{\partial v}, \qquad \frac{\partial z}{\partial v} = -\rho_2 \frac{\partial Z}{\partial v},$$

we have, by integration,

$$x = -\rho_2 X + U_1, \qquad y = -\rho_2 Y + U_2, \qquad z = -\rho_2 Z + U_3.$$

Hence the points of the surface lie on the spheres

$$(x - U_1)^2 + (y - U_2)^2 + (z - U_3)^2 = \rho_2^2,$$

and the spheres are tangent to the surface.

Since the normals to a surface along a circular line of curvature form a cone of revolution, the second sheet of the evolute is the envelope of a family of such cones. The characteristics of such a family are conics. Hence we have the theorem:

A necessary and sufficient condition that one sheet of the evolute of a surface be a curve is that the surface be the envelope of a single infinity of spheres; the second focal sheet is the locus of a family of conics.

133. Cyclides of Dupin. From the preceding theorem it results that if also the second sheet of the evolute of a surface be a curve, it is a conic, and then the first sheet also is a conic. Moreover, these conics are so placed that the cone formed by joining any point on one conic to all the points of the other is a cone of revolution. A pair of focal conics is characterized by this property. And so we have the theorem:

*A necessary and sufficient condition that the lines of curvature in both families be circles is that the sheets of the evolute be a pair of focal conics.**

These surfaces are called the *cyclides of Dupin.* They are the envelopes of two one-parameter families of spheres, and all such envelopes are cyclides of Dupin. A sphere of one family touches each sphere of the other family. Consequently the spheres of which the cyclide is the envelope are tangent to three spheres.

We shall prove the converse theorem of Dupin †:

The envelope of a family of spheres tangent to three fixed spheres is a cyclide.

* Cf. Ex. 19, p. 188.

† *Applications de géométrie et de méchanique*, pp. 200–210. Paris, 1822.

The plane determined by the centers of the three spheres cuts the latter in three circles. If any point on the circumference C, orthogonal to these circles, be taken for the pole of a transformation by reciprocal radii (cf. § 80), C is transformed into a straight line L. Since angles are preserved in this transformation, the three fixed spheres are changed into three spheres whose centers are on L. Evidently the envelope of a family of spheres tangent to these three spheres is a tore with L as axis. Hence the given envelope is transformed into a tore. However, the latter surface is the envelope of a second family of spheres whose centers lie on L. Therefore, if the above transformation be reversed, we have a second family of spheres tangent to the envelope, and so the latter is a cyclide of Dupin. We shall now find the equations of these surfaces.

Let (x_1, y_1, z_1) and (x_2, y_2, z_2) denote the coördinates of the points on the focal conics which are the curves of centers of the spheres, and R_1, R_2 the radii of the spheres. The condition of tangency is

$$(x_1 - x_2)^2 + (y_1 - y_2)^2 + (z_1 - z_2)^2 = (R_1 + R_2)^2. \tag{94}$$

We consider first the case where the evolute curves are the focal parabolas defined by (85). Now equation (94) reduces to

$$\frac{1}{4}(u^2 + v^2 + 1)^2 = (R_1 + R_2)^2.$$

Since R_1 and R_2 are functions of u and v respectively, this equation is equivalent to

$$R_1 = \frac{1}{2}\left(u^2 + \frac{1}{2} + a\right), \quad R_2 = \frac{1}{2}\left(v^2 + \frac{1}{2} - a\right), \tag{95}$$

where a is an arbitrary constant whose variation gives parallel surfaces.

By the method of § 132 we find that the coördinates (ξ, η, ζ) of the centers of the circular lines of curvature $u = \text{const.}$ and the radius ρ are

$$\xi = \frac{u}{2(1+u^2)}\left[u^2 + \frac{1}{2} + a - 2(1+u^2)\right],$$

$$\eta = 0,$$

$$\zeta = \frac{1}{2(1+u^2)}\left[u^2\left(u^2 + \frac{1}{2} + a\right) - \left(u^2 - \frac{1}{2}\right)(1+u^2)\right],$$

$$\rho = \frac{1}{2}\left(u^2 + \frac{1}{2} + a\right)(1+u^2)^{-\frac{1}{2}}.$$

Hence if P be a point on the circle and θ denote the angle which the radius to P makes with the positive direction of the normal to the parabola (85), the coördinates of P are

$$\bar{x} = \xi + \frac{\rho u}{\sqrt{1+u^2}} \cos\theta, \qquad \bar{y} = \rho \sin\theta, \qquad \bar{z} = \zeta - \frac{\rho}{\sqrt{1+u^2}} \cos\theta.$$

This surface is algebraic and of the third order.

If the evolute curves are the focal ellipse and hyperbola (86), we have

$$(96) \qquad R_1 = \frac{1}{2}(a \cos u + \kappa), \qquad R_2 = \frac{1}{2}(\cosh v - \kappa),$$

where κ is an arbitrary constant whose variation gives parallel surfaces. This cyclide of Dupin is of the fourth degree. When in particular the constant a is zero, the surface is the ordinary tore, or anchor ring.*

134. Surfaces with spherical lines of curvature in one system. Surfaces with circular lines of curvature in one system belong evidently to the general class of surfaces with spherical lines of curvature in one system. We consider now surfaces of the latter kind.

Let S be such a surface referred to its lines of curvature, and in particular let the lines $v =$ const. be spherical. The coördinates of the centers of the spheres as well as their radii are functions of v alone. They will be denoted by (V_1, V_2, V_3) and R. By Joachimsthal's theorem (§ 59) each sphere cuts the surface under the same angle at all its points. Hence for the family of spheres the expression for the angle is a function of v alone; we call it V.

Since the direction-cosines of the tangent to a curve $u =$ const. are

$$\frac{1}{\sqrt{\mathcal{G}}}\frac{\partial X}{\partial v}, \qquad \frac{1}{\sqrt{\mathcal{G}}}\frac{\partial Y}{\partial v}, \qquad \frac{1}{\sqrt{\mathcal{G}}}\frac{\partial Z}{\partial v},$$

when the linear element of the spherical representation is written $d\sigma^2 = \mathcal{E}\,du^2 + \mathcal{G}\,dv^2$, the coördinates of S are of the form

$$(97) \qquad \begin{cases} x = V_1 + \dfrac{R \sin V}{\sqrt{\mathcal{G}}}\dfrac{\partial X}{\partial v} + XR\cos V, \\ y = V_2 + \dfrac{R \sin V}{\sqrt{\mathcal{G}}}\dfrac{\partial Y}{\partial v} + YR\cos V, \\ z = V_3 + \dfrac{R \sin V}{\sqrt{\mathcal{G}}}\dfrac{\partial Z}{\partial v} + ZR\cos V. \end{cases}$$

* For other geometrical constructions of the cyclides of Dupin the reader is referred to the article in the *Encyklopädie der Math. Wissenschaften*, Vol. III, 3, p. 290.

By hypothesis X, Y, Z are the direction-cosines of the normal to S; consequently we must have

$$\sum X\frac{\partial x}{\partial u}=0, \qquad \sum X\frac{\partial x}{\partial v}=0.$$

If the values of the derivatives obtained from (97) be reduced by means of (V, 22), and the results substituted in the above equations, the first vanishes identically and the second reduces to

$$(98) \qquad XV_1'+YV_2'+ZV_3'+(R\cos V)'-R\sin V\sqrt{\mathcal{G}}=0,$$

where the primes indicate differentiation with respect to v. Conversely, when this condition is satisfied, equations (97) define a surface on which the curves $v=$ const. are spherical. Hence:

A necessary and sufficient condition that the curves $v=$ const. of an orthogonal system on the unit sphere represent spherical lines of curvature upon a surface is that five functions of v, namely V_1, V_2, V_3, R, V, can be found which satisfy the corresponding equation (98).

We note that V_1, V_2, V_3, and $R\cos V$ are determined by (98) only to within additive constants. A change of these constants for the first three gives a translation of the surface. If $R\cos V$ be increased by a constant, we have a new surface parallel to the other one. Hence *:

If the lines of curvature in one system upon a surface be spherical, the same is true of the corresponding system on each parallel surface.

Since equation (98) is homogeneous in the quantities V_1', V_2', V_3', $(R\cos V)'$, $R\sin V$, the latter are determined only to within a factor which may be a function of v. This function may be chosen so that all the spheres pass through a point. From these results we have the theorem of Dobriner †:

With each surface with spherical lines of curvature in one system there is associated an infinity of nonparallel surfaces of the same kind with the same spherical representation of these lines of curvature. Among these surfaces there is at least one for which all the spheres pass through a point. At corresponding points of the loci of the centers of spheres of two surfaces of the family the tangents are parallel.

* Cf. Bianchi, Vol. II, p. 303. † *Crelle*, Vol. XCIV (1883), pp. 118, 125.

If the values of x, y, z from (97) be substituted in the formulas of Rodrigues (IV, 32),

$$\frac{\partial x}{\partial u} = -\rho_1 \frac{\partial X}{\partial u}, \qquad \frac{\partial x}{\partial v} = -\rho_2 \frac{\partial X}{\partial v}, \tag{99}$$

and similarly for y and z, we obtain by means of (V, 22),

$$-\rho_1 = R\cos V + \frac{R\sin V}{\sqrt{\mathcal{E}\mathcal{G}}}\frac{\partial\sqrt{\mathcal{E}}}{\partial v},$$

$$-\rho_2 = R\cos V + \frac{(R\sin V)'}{\sqrt{\mathcal{G}}} + \frac{1}{\mathcal{G}}\sum V_1'\frac{\partial X}{\partial v}.$$

Conversely, when for a surface referred to its lines of curvature the principal radius ρ_1 is of the form

$$-\rho_1 = \phi_1(v) + \frac{\phi_2(v)}{\sqrt{\mathcal{E}\mathcal{G}}}\frac{\partial\sqrt{\mathcal{E}}}{\partial v}, \tag{100}$$

where ϕ_1 and ϕ_2 are any functions whatever of v, the curves $v = \text{const.}$ are spherical. For, by (V, 22),

$$\frac{\partial}{\partial u}\left(\frac{1}{\sqrt{\mathcal{G}}}\frac{\partial X}{\partial v}\right) = \frac{1}{\sqrt{\mathcal{E}\mathcal{G}}}\frac{\partial\sqrt{\mathcal{E}}}{\partial v}\frac{\partial X}{\partial u}.$$

Consequently, from the first of (99), in which ρ_1 is given the above value, we obtain by integration

$$x = \phi_1 X + \phi_2 \frac{1}{\sqrt{\mathcal{G}}}\frac{\partial X}{\partial v} + V_1,$$

where V_1 is a function of v alone. Similar results follow for y and z. As these expressions are of the form (97), we have the theorem:

A necessary and sufficient condition that the lines of curvature $v = \text{const.}$ be spherical is that ρ_1 be of the form (100).

EXAMPLES

1. If the lines of curvature in one system are plane and one is a circle, all are circles.

2. When the lines of curvature in one family on a surface are circles, their spherical images are circles whose spherical centers constitute the spherical indicatrix of the tangents to the curve of centers of the spheres which are enveloped by the given surface. Show also that each one-parameter system of circles on the unit sphere represents the circular lines of curvature on an infinity of surfaces, for one of which the circles are equal.

3. If the lines of curvature of a surface are parametric, and the curves $v = \text{const.}$ are spherical, we have

$$\frac{1}{\rho_{gu}} = \frac{1}{R \sin V} + \frac{\cot V}{\rho_1},$$

where ρ_{gu}, ρ_1, R denote the radii of geodesic curvature and normal curvature in the direction $v = \text{const.}$ and of the sphere respectively, and V denotes the angle under which the sphere cuts the surface.

4. When a line of curvature is spherical, the developable circumscribing the surface along this line of curvature also circumscribes a sphere; and conversely, if such a developable circumscribes a sphere, the line of curvature lies on a sphere concentric with the latter (cf. Ex. 7, p. 149).

5. Let S be a pseudospherical surface with the spherical representation (25) of its lines of curvature. Show that a necessary and sufficient condition that the curves $v = \text{const.}$ be plane is

$$\frac{\partial}{\partial u}\left(\frac{1}{\sin \omega}\frac{\partial \omega}{\partial v}\right) = 0;$$

show also that in this case ω is given by

$$\cos \omega = \frac{V' - U'}{U^2 - V^2 - 1},$$

where U and V are functions of u and v respectively, which satisfy the conditions

$$U'^2 = U^4 + (a-2)U^2 + b, \qquad V'^2 = V^4 + aV^2 + (a + b - 1),$$

a and b being constants, and the accent indicating differentiation, unless U' or V' is zero.

6. When the lines of curvature $v = \text{const.}$ upon a pseudospherical surface are plane, the linear element is reducible to the form

$$ds^2 = \frac{a^2 \tanh^2 (u + v)\, du^2}{C - A \cosh 2u + B \sinh 2u} + \frac{a^2 \operatorname{sech}^2 (u + v)\, dv^2}{C + A \cosh 2v + B \sinh 2v - 1},$$

where A, B, C are constants. Find the expressions for the principal radii.

7. When the lines of curvature $v = \text{const.}$ on a spherical surface are plane, the linear element is reducible to

$$ds^2 = \frac{a^2 \cot^2 (u + v)\, du^2}{A \sin 2u + B - 1} + \frac{a^2 \csc^2 (u + v)\, dv^2}{A \sin 2v - B},$$

where A and B are constants. The surfaces of Exs. 5 and 6 are called *the surfaces of Enneper of constant curvature.*

GENERAL EXAMPLES

1. The lines of curvature and the asymptotic lines on a surface of constant curvature can be found by quadratures.

2. When the linear element of a pseudospherical surface is in the form (iii) of (7), the equations $x = cv$, $y = ae^{-\frac{u}{a}}$ determine a conformal representation of the surface upon the plane, which is such that any geodesic on the surface is represented on the plane by a circle with its center on the x-axis, or by a line perpendicular to this axis.

3. When the linear elements of a developable surface, a spherical surface, and a pseudospherical surface are in the respective forms

$$ds^2 = du^2 + u^2 dv^2, \quad ds^2 = a^2(du^2 + \sin^2 u\, dv^2), \quad ds^2 = a^2(du^2 + \sinh^2 u\, dv^2),$$

the finite equations of the geodesics are respectively

$$Au \cos v + Bu \sin v + C = 0, \quad A \tan u \cos v + B \tan u \sin v + C = 0,$$
$$A \tanh u \cos v + B \tanh u \sin v + C = 0,$$

where A, B, C are constants; if the coefficients of A and B are in any case equated to x and y, the resulting equations define a correspondence between the surface and the plane such that geodesics on the former correspond to straight lines on the latter. Find the expression for each linear element in terms of x and y as parameters.

4. Each surface of center of a pseudospherical surface is applicable to the catenoid.

5. The asymptotic lines on the surfaces of center of a surface of constant mean curvature correspond to the minimal lines on the latter.

6. Surfaces of constant mean curvature are characterized by the property that if $u =$ const., $v =$ const. are the minimal curves, then D is a function of u alone and D'' of v alone.

7. Equation (23) admits the solution $\omega = 0$, in which case the surface degenerates into a curve. Show that the general integral of the corresponding equations (35) is $\tan \theta/2 = Ce^{\frac{u + v \cos \sigma}{\sin \sigma}}$; take for S the line $x = 0$, $y = 0$, $z = au$ and derive the equations of the transforms of S; show that the latter are surfaces of Dini (Ex. 6, § 122), or a pseudosphere.

8. Show that the Bäcklund transforms of the surfaces of Dini and of the pseudosphere can be found without integration, and that if the pseudosphere be transformed by the transformation of Bianchi, the resulting surface may be defined by

$$x = \frac{2a \cosh u}{\cosh^2 u + v^2}(\sin v - v \cos v), \qquad y = -\frac{2a \cosh u}{\cosh^2 u + v^2}(\cos v + v \sin v),$$
$$z = a\left(u - \frac{2 \sinh u \cosh u}{\cosh^2 u + v^2}\right),$$

Show that the lines of curvature $v =$ const. lie in planes through the z-axis.

9. The tangents to a family of geodesics of the elliptic or hyperbolic type on a pseudospherical surface are normal to a W-surface; the relations between the radii are respectively

$$\rho_1 - \rho_2 = a \tanh \frac{\rho_1 + c}{a}, \qquad \rho_1 - \rho_2 = a \coth \frac{\rho_1 + c}{a},$$

where a and c are constants (cf. § 76).

10. Show that the linear elements of the second surfaces of center of the W-surfaces of Ex. 9 are reducible to the respective forms

$$ds_2^2 = \tanh^4 \frac{u}{a} du^2 + \operatorname{sech}^2 \frac{u}{a} dv^2, \qquad ds_2^2 = \coth^4 \frac{u}{a} du^2 + \operatorname{csch}^2 \frac{u}{a} dv^2,$$

and that consequently these surfaces are applicable to surfaces of revolution whose meridians are defined by

$$r = \frac{a}{\sqrt{a^2\kappa^2 + 1}} \sin \phi, \qquad z = a\left(\log \tan \frac{\phi}{2} + \cos \phi\right),$$
$$r = \frac{a}{\sqrt{1 - a^2\kappa^2}} \sin \phi, \qquad z = a\left(\log \tan \frac{\phi}{2} + \cos \phi\right),$$

where κ denotes a constant.

11. Determine the particular form of the linear element (49), and the nature of the curves upon the surface to which the asymptotic lines on the sheets of the evolute correspond, when

$$(a)\quad \frac{\rho_1}{\rho_2} = \text{const}; \qquad (b)\quad \frac{1}{\rho_1} - \frac{1}{\rho_2} = \text{const}.$$

12. When a W-surface is of the type (72), the surfaces of center are applicable to one another and to an imaginary paraboloid of revolution.

13. When a W-surface is of the type (72) and the linear element of the sphere has the form (VI, 60), the curves $u + v =$ const. and $u - v =$ const. on the spherical representation are geodesic parallels whose orthogonal trajectories correspond to the asymptotic lines on the surfaces of center; hence on each sheet there is a family of geodesics such that the tangents at their points of meeting with an asymptotic line are parallel to a plane, which varies in general with the asymptotic line.

14. Show that the equations

$$x = aU\cos\frac{v}{a} + \int V\sin\frac{v}{a}\,dv, \qquad y = aU\sin\frac{v}{a} - \int V\cos\frac{v}{a}\,dv,$$
$$z = \int\sqrt{1 - a^2U'^2}\,du,$$

where a denotes an arbitrary constant, define a family of applicable molding surfaces.

15. When the lines of curvature in one system on a surface are plane, and the lines of the second system lie on spheres which cut the surface orthogonally, the latter is a surface of Joachimsthal.

16. The spherical lines of curvature on a surface of Joachimsthal have constant geodesic curvature, the radius of geodesic curvature being the radius of the sphere on which a curve lies.

17. When the lines of curvature in one system on a surface lie on concentric spheres, it is a surface of Monge, whose director-developable is a cone with its vertex at the center of the spheres; and conversely.

18. The sheets of the evolute of a surface of Monge are the director-developable and a second surface of Monge, which has the same director-developable and whose generating curve is the evolute of the generating curve of the given surface.

19. If the lines of curvature in one system on a surface are plane, and two in the second system are plane, then all in the latter system are plane.

20. A surface with plane lines of curvature in both systems, in one of which they are circles, is

(*a*) A surface of Joachimsthal.

(*b*) The locus of the orthogonal trajectories of a family of spheres, with centers on a straight line, which pass through a circle on one of the spheres.

(*c*) The envelope of a family of spheres whose centers lie on a plane curve C, and whose radii are proportional to the distances of these centers from a straight line fixed in the plane of C.

21. If an arbitrary curve C be drawn in a plane, and the plane be made to move in such a way that a fixed line of it envelop an arbitrary space curve Γ, and at the same time the plane be always normal to the principal normal to Γ, the curve C describes a surface of Monge.

22. If all the Bianchi transforms of a pseudospherical surface S are surfaces of Enneper (cf. Ex. 5, § 134), S is a surface of revolution.

23. When ω has the value in Ex. 5, § 134, the surfaces with the spherical representation (25), and with the linear element

$$ds^2 = \left(U_1 \cos\omega + \frac{U_1'}{U}\right)^2 du^2 + U_1^2 \sin^2\omega\, dv^2,$$

where U_1 is an arbitrary function of u, are surfaces of Joachimsthal.

24. If the lines of curvature in both systems be plane for a surface S with the same spherical representation of its lines of curvature as for a pseudospherical surface, S is a molding surface.

25. If S is a pseudospherical surface with the spherical representation (25) of its lines of curvature, and the curves $v = \text{const.}$ are plane, the function θ, given by

$$\sin\theta \frac{\partial^2\omega}{\partial v^2} + \cos\theta \frac{\partial\omega}{\partial u}\frac{\partial\omega}{\partial v} + \sin\omega \frac{\partial\omega}{\partial v} = 0,$$

determines a transformation of Bianchi of S into a surface S_1 for which the lines of curvature $v = \text{const.}$ are plane.

26. A necessary and sufficient condition that the lines of curvature $v = \text{const.}$ on a pseudospherical surface with the representation (25) of its lines of curvature be spherical is that

$$\cot\omega = V_1 + \frac{V}{\sin\omega}\frac{\partial\omega}{\partial v},$$

where V and V_1 are functions of v alone. Show that when ω is a solution of (23) and of

$$\begin{vmatrix} \dfrac{1}{\sin^2\omega}\dfrac{\partial\omega}{\partial u} & \dfrac{\partial}{\partial u}\left(\dfrac{1}{\sin\omega}\dfrac{\partial\omega}{\partial v}\right) \\ \dfrac{\partial}{\partial u}\left(\dfrac{1}{\sin^2\omega}\dfrac{\partial\omega}{\partial u}\right) & \dfrac{\partial^2}{\partial u^2}\left(\dfrac{1}{\sin\omega}\dfrac{\partial\omega}{\partial v}\right) \end{vmatrix} = 0,$$

the curves $v = \text{const.}$ are plane or spherical, and that in the latter case V and V_1 can be found directly.

27. Show that when ω is a solution of (23) and of

$$\frac{\partial\omega}{\partial v}\frac{\partial^3\omega}{\partial u\,\partial v^2} - \frac{\partial^2\omega}{\partial v^2}\frac{\partial^2\omega}{\partial u\,\partial v} + \frac{\partial\omega}{\partial u}\left(\frac{\partial\omega}{\partial v}\right)^3 = 0,$$

and $\dfrac{\partial}{\partial u}\left(\dfrac{1}{\cos\omega}\dfrac{\partial\omega}{\partial v}\right) \neq 0$, the lines of curvature $u = \text{const.}$ are spherical on the pseudospherical surface with the spherical representation (25); and that when ω is such a function, upon the surfaces with the linear element

$$ds^2 = V^2\left(\frac{\partial\omega}{\partial v}\right)^2 du^2 + \left(V\frac{\partial^2\omega}{\partial v^2} + V'\frac{\partial\omega}{\partial v}\right)^2 \frac{dv^2}{\left(\dfrac{\partial\omega}{\partial v}\right)^2},$$

or

$$ds^2 = \left(\sin\omega + V\frac{\partial\omega}{\partial v}\right)^2 du^2 + \left(\cos\omega + V' + V\frac{\dfrac{\partial^2\omega}{\partial v^2}}{\dfrac{\partial\omega}{\partial v}}\right)^2 dv^2,$$

where V is a function of v alone, the curves $v = \text{const.}$ are spherical; in the former case the spheres cut the surface orthogonally.

CHAPTER IX

DEFORMATION OF SURFACES

135. Problem of Minding. Surfaces of constant curvature. According to § 43 two surfaces are *applicable* when a one-to-one correspondence can be established between them which is of such a nature that in the neighborhood of corresponding points corresponding figures are congruent or symmetric. It was seen that two surfaces with the same linear element are applicable, the parametric curves on the two surfaces being in correspondence. But the fact that the linear elements of two surfaces are unlike is not a sufficient condition that they are not applicable; in evidence of this we have merely to recall the effect of a change of parameters, to say nothing of a change of parametric lines. Hence we are brought to the following problem, first proposed by Minding:*

To find a necessary and sufficient condition that two surfaces be applicable.

From the second theorem of § 64 it follows that a necessary condition is that the total curvature of the two surfaces at corresponding points be the same. We shall show that this condition is sufficient for surfaces of constant curvature.

In § 64 we found that when K is zero at all points of a surface, the surface is applicable to the plane. If the plane be referred to the system of straight lines parallel to the rectangular axes, its linear element is

$$ds^2 = dx^2 + dy^2.$$

Hence the analytical problem of the application of a developable surface upon the plane reduces to the determination of orthogonal systems of geodesics such that when these curves are parametric the linear element takes the above form.

* *Crelle*, Vol. XIX (1839), pp. 371–387.

Referring to the results of § 39, we see that in this case the factor tt_1 must equal unity. Consequently we must find a function θ such that the left-hand members of the equations

$$e^{i\theta}\left(\sqrt{E}\,du+\frac{F+iH}{\sqrt{E}}\,dv\right)=d\,(x+iy),$$

$$e^{-i\theta}\left(\sqrt{E}\,du+\frac{F-iH}{\sqrt{E}}\,dv\right)=d\,(x-iy)$$

are exact differentials, in which case these equations give x and y by quadratures. Hence we must have

$$\frac{\partial}{\partial v}(e^{i\theta}\sqrt{E})=\frac{\partial}{\partial u}\left(e^{i\theta}\frac{F+iH}{\sqrt{E}}\right),\quad \frac{\partial}{\partial v}(e^{-i\theta}\sqrt{E})=\frac{\partial}{\partial u}\left(e^{-i\theta}\frac{F-iH}{\sqrt{E}}\right),$$

which are equivalent to

$$\frac{\partial\theta}{\partial u}=\frac{1}{H}\frac{\partial F}{\partial u}-\frac{F}{2\,EH}\frac{\partial E}{\partial u}-\frac{1}{2\,H}\frac{\partial E}{\partial v},$$

$$\frac{\partial\theta}{\partial v}=\frac{1}{2\,H}\frac{\partial G}{\partial u}-\frac{F}{2\,EH}\frac{\partial E}{\partial v}.$$

From (V, 12) it is seen that these equations are consistent when $K=0$. In this case θ, and consequently x and y, can be found by quadratures.

The additive constants of integration are of such a character that if x_0, y_0 are a particular set of solutions, the most general are

$$x=x_0\cos\alpha-y_0\sin\alpha+a,\qquad y=x_0\sin\alpha+y_0\cos\alpha+b,$$

where α, a, b are arbitrary constants.

In the above manner we can effect the isometric representation of any developable surface upon the plane, and consequently upon itself or any other developable. These results may be stated thus:

A developable surface is applicable to itself, or to any other developable, in a triple infinity of ways, and the complete determination of the applicability requires quadratures only.

Incidentally we have the two theorems:

The geodesics upon a developable surface can be found by quadratures.

If the total curvature of a quadratic form be zero, the quadratic form is reducible by quadratures to $d\alpha d\beta$.

Suppose now that the total curvature of two surfaces S, S_1 is $1/a^2$, where a is a real constant. Let P and P_1 be points on S and S_1 respectively, C and C_1 geodesics through these respective points, and take P and P_1 for the poles and C and C_1 for the curves $v = 0$ of a polar geodesic system on these surfaces. The linear elements are accordingly (VIII, 6)

$$ds^2 = du^2 + \sin^2\frac{u}{a}\,dv^2, \qquad ds_1^2 = du_1^2 + \sin^2\frac{u_1}{a}\,dv_1^2.$$

Hence the equations

$$u_1 = u, \qquad v_1 = \pm v$$

determine an isometric representation of one surface upon the other, in which P and C correspond to P_1 and C_1 respectively. According as the upper or lower sign in the second equation is used, corresponding figures are equal or symmetric. Similar results obtain for pseudospherical surfaces. Hence we have:

Any two surfaces of constant curvature, different from zero, are in two ways applicable so that a given point and geodesic through it on one surface correspond to a given point and geodesic through it on the other.

In particular, a surface of constant curvature can be applied to itself so that a given point shall go into any other point and a geodesic through the former into one through the latter. Combining these results with the last theorem of § 117, we have:

A nondevelopable surface of constant curvature can be applied to itself, or to any surface of the same curvature, in a triple infinity of ways, and the complete realization of the applicability requires the solution of a Riccati equation.

136. Solution of the problem of Minding. We proceed to the determination of a necessary and sufficient condition that two surfaces S, S' of variable curvature be applicable. Let their linear elements be

$$ds^2 = E\,du^2 + 2F\,du dv + G\,dv^2, \quad ds'^2 = E'\,du'^2 + 2F'\,du'dv' + G'\,dv'^2.$$

By definition S and S' are applicable if there exist two independent equations

$$(1) \qquad \phi(u, v) = \phi'(u', v'), \qquad \psi(u, v) = \psi'(u', v'),$$

establishing a one-to-one correspondence between the surfaces of such a nature that by means of (1) either of the above quadratic forms can be transformed into the other.

It is evident that if the two surfaces are applicable, the differential parameters formed with respect to the two linear elements are equal. Hence a necessary condition is

(2) $$\Delta_1\phi = \Delta_1'\phi', \quad \Delta_1(\phi, \psi) = \Delta_1'(\phi', \psi'), \quad \Delta_1\psi = \Delta_1'\psi',$$

where the primes indicate functions pertaining to S'. These conditions are likewise sufficient that the transformation (1) change either of the above quadratic forms into the other. For, if the curves $\phi = \text{const.}$, $\psi = \text{const.}$; $\phi' = \text{const.}$, $\psi' = \text{const.}$ be taken for the parametric curves on S and S' respectively, the respective linear elements may be written (cf. § 37)

$$ds^2 = \frac{\Delta_1\psi\, d\phi^2 - 2\,\Delta_1(\phi, \psi)\, d\phi d\psi + \Delta_1\phi\, d\psi^2}{\Delta_1\phi \cdot \Delta_1\psi - \Delta_1^2(\phi, \psi)},$$

$$ds'^2 = \frac{\Delta_1'\psi'\, d\phi'^2 - 2\,\Delta_1'(\phi', \psi')\, d\phi' d\psi' + \Delta_1'\phi'\, d\psi'^2}{\Delta_1'\phi' \cdot \Delta_1'\psi' - \Delta_1'^2(\phi', \psi')}.$$

Hence when equations (1) and (2) hold, the surfaces are applicable.

The next step is the determination of equations of the form (1). Since the curvature of two applicable surfaces at corresponding points is the same, one such equation is afforded by the necessary condition

(3) $$K(u, v) = K'(u', v').$$

The first of equations (2) is

(4) $$\Delta_1 K = \Delta_1' K'.$$

Both members of this equation cannot vanish identically. For, in this case the curves $K = \text{const.}$ and $K' = \text{const.}$ would be minimal (§ 37), and consequently imaginary. If these two equations are independent of one another, that is,

$$\Delta_1 K \neq f(K), \qquad \Delta_1' K' \neq f(K'),$$

they establish a correspondence, and the condition that it be isometric is, as seen from (2),

$$\Delta_1(K, \Delta_1 K) = \Delta_1'(K', \Delta_1' K'), \qquad \Delta_1\Delta_1 K = \Delta_1'\Delta_1' K'.$$

If, however,

(5) $$\Delta_1 K = f(K), \qquad \Delta_1' K' = f(K'),$$

we may take for the second of (1)

(6) $$\Delta_2 K = \Delta_2' K',$$

unless

(7) $$\Delta_2 K = \phi(K), \qquad \Delta_2' K' = \phi(K').$$

If this condition be not satisfied, the conditions that (3), (6) define an isometric correspondence are

$$\Delta_1(K, \Delta_2 K) = \Delta_1'(K', \Delta_2' K'), \qquad \Delta_1\Delta_2 K = \Delta_1'\Delta_2' K'.^*$$

Finally, we consider the case where both (5) and (7) hold. Since the ratio of $\Delta_1 K$ and $\Delta_2 K$ is a function of K, the curves $K =$ const. and their orthogonal trajectories $t =$ const. form an isothermal system of lines on S (§ 41). Moreover, the function t can be found by quadratures, and the linear element is reducible to

$$ds^2 = \frac{1}{f(K)}\left(dK^2 + e^{2\int \frac{\Phi(K)}{f(K)} dK} dt^2\right). \tag{8}$$

When in particular $\Delta_2 K = 0$, the linear element is

$$ds^2 = \frac{1}{f(K)}(dK^2 + dt^2).$$

In like manner the linear element of S' is reducible to

$$ds'^2 = \frac{1}{f(K')}\left(dK'^2 + e^{2\int \frac{\Phi(K')}{f(K')} dK'} dt'^2\right),$$

or, in the particular case $\Delta_2' K' = 0$, to

$$ds'^2 = \frac{1}{f(K')}(dK'^2 + dt'^2).$$

In either case the equations

$$K = K', \qquad t = \pm t' + a,$$

where a is an arbitrary constant, define the applicability of the surfaces.

We have thus treated all possible cases and found that it can be determined without quadrature whether two surfaces are applicable. Moreover, in the first two cases the equations defining the correspondence follow directly, but in the last case the determination requires a quadrature. The last case differs also in this respect: the application can be effected in an infinity of ways, whereas in the first two cases it is unique.

* If the surface be referred to the curves $\sigma =$ const. and their orthogonal trajectories, where $\sigma = \int \frac{dK}{\sqrt{f(K)}}$, equation (6) may be replaced by $\Delta_2\sigma = \Delta_2'\sigma'$, and it can be shown that $\Delta_1(\sigma, \Delta_2\sigma) = \Delta_1'(\sigma', \Delta_2'\sigma')$ is a consequence of the other conditions. Cf. Darboux, Vol. III, p. 227.

Furthermore, we notice from (8) that in the third case the surface S is applicable to a surface of revolution, the parallels of the latter corresponding to the curves $K=$ const. of the former. Conversely, the linear element of every surface applicable to a surface of revolution can be put in the form (8). For, a necessary and sufficient condition that a surface be applicable to a surface of revolution is that its linear element be reducible to

$$ds^2 = du^2 + U^2\, dv^2,$$

where U is a function of u alone (§ 46). Now

$$\Delta_1 u = 1, \qquad K = -\frac{U''}{U}, \qquad \Delta_2 \int \frac{du}{U} = 0.$$

From the second it follows that $u = F(K)$, and consequently $\int \frac{du}{U} = F_1(K)$. When these values are substituted in the above equations, we have, in consequence of Ex. 5, p. 91,

$$(9) \qquad \Delta_1 K = f(K), \qquad \Delta_2 K = \phi(K).$$

Hence we have the theorem:

Equations (9) *constitute a necessary and sufficient condition that a surface be applicable to a surface of revolution.*

The equations

$$K = K', \qquad t = \pm t' + a$$

define an isometric representation of a surface with the linear element (8) upon itself. Therefore we have:

Every surface applicable to a surface of revolution admits of a continuous deformation into itself in such a way that each curve $K=$ const. slides over itself.

Conversely, every surface applicable to itself in an infinity of ways is applicable to a surface of revolution. For, if the curvature is constant, the surface is applicable to a surface of revolution (§ 135), and the only case in which two surfaces of variable curvature are applicable in an infinity of ways is that for which conditions (5) and (7) are satisfied.

137. Deformation of minimal surfaces. These results suggest a means of determining the minimal surfaces applicable to a surface of revolution. In the first place we inquire under what conditions two minimal surfaces are applicable. The latter problem reduces to the determination of two pairs of parameters, u, v and u_1, v_1, and two pairs of functions, $F(u)$, $\Phi(v)$* and $F_1(u_1)$, $\Phi_1(v_1)$, which satisfy the condition

$$(10)\qquad (1+uv)^2 F(u)\,\Phi(v)\,du\,dv = (1+u_1v_1)^2 F_1(u_1)\,\Phi_1(v_1)\,du_1dv_1.$$

From the nature of this equation it follows that the equations which serve to establish the correspondence between the two surfaces are either of the form

$$(11)\qquad u_1 = \phi(u), \qquad v_1 = \psi(v),$$

or

$$(12)\qquad u_1 = \phi(v), \qquad v_1 = \psi(u).$$

If either set of values for u_1 and v_1 be substituted in (10), and if after removing the common factor $dudv$ we take the logarithmic derivative with respect to u and v, we obtain

$$\frac{\phi'\psi'}{(1+u_1v_1)^2} = \frac{1}{(1+uv)^2}.$$

As this may be written

$$(13)\qquad \frac{du_1dv_1}{(1+u_1v_1)^2} = \frac{dudv}{(1+uv)^2},$$

the spherical images of corresponding parts on the two surfaces are equal or symmetric according as (11) or (12) obtains (§ 47). The latter case reduces to the former when the sense of the normal to either surface is changed. When this has been done, corresponding spherical images are equal and can be made to coincide by a rotation of the unit sphere about a diameter. Hence one surface can be so displaced in space that corresponding normals become parallel, in which case the two surfaces have the same representation, that is, $u_1 = u$, $v_1 = v$. Now equation (10) is

$$F(u)\,\Phi(v) = F_1(u)\,\Phi_1(v),$$

which is equivalent to

$$F_1(u) = cF(u), \qquad \Phi_1(v) = \frac{1}{c}\,\Phi(v),$$

* § 110.

where c denotes a constant. If the surfaces are real, c must be of the form $e^{i\alpha}$. Hence, in consequence of § 113, we have the theorem:

A minimal surface admits of a continuous deformation into an infinity of minimal surfaces, which are either associate to it or can be made such by a suitable displacement.

We pass to the determination of a minimal surface which admits of a continuous deformation into itself, and consequently is applicable to a surface of revolution. In consequence of the interpretation of equation (13) it follows that if a minimal surface be deformed continuously into itself, a point p on the sphere tends to move in the direction of the small circle through p, whose axis is the momentary axis of rotation, and consequently each of these small circles moves over itself. From § 47 it follows that if the axis of rotation be taken for the z-axis, these small circles are the curves $uv = \text{const.}$ In the deformation each point of the surface moves along the curve $K = \text{const.}$ through it. Hence K is a function of uv. From (VII, 100, 102) we have

$$K = \frac{-4}{(1+uv)^4 F(u)\,\Phi(v)};$$

consequently $F(u)\,\Phi(v)$ must be a function of uv, and hence

$$\frac{uF'(u)}{F(u)} = \frac{v\Phi'(v)}{\Phi(v)}.$$

The common value of these two terms is a constant. If it be denoted by κ, we have

$$F(u) = cu^\kappa, \qquad \Phi(v) = c_1 v^\kappa,$$

where c and c_1 are constants. Hence from (VII, 98) we have:

Any minimal surface applicable to a surface of revolution can be defined by equations of the form

$$\text{(14)} \qquad \begin{cases} x = \dfrac{1}{2}c\displaystyle\int(1-u^2)\,u^\kappa du + \dfrac{1}{2}c_1\displaystyle\int(1-v^2)\,v^\kappa dv, \\[2ex] y = \dfrac{i}{2}c\displaystyle\int(1+u^2)\,u^\kappa du - \dfrac{i}{2}c_1\displaystyle\int(1+v^2)\,v^\kappa dv, \\[2ex] z = c\displaystyle\int u^{\kappa+1}du + c_1\displaystyle\int v^{\kappa+1}dv, \end{cases}$$

where c, c_1, and κ are arbitrary constants.

Since the curves $K=$ const. are represented on the sphere by the small circles whose axis is the z-axis, in each finite deformation of the surface into itself, as well as in a very small one, the unit sphere undergoes a rotation about this axis. In § 47 it was seen that such a rotation is equivalent to replacing u, v by $ue^{i\alpha}$, $ve^{-i\alpha}$, where α denotes the angle of rotation. Hence the continuous deformation of a surface (14) is defined by the equations resulting from the substitution in (14) of $ue^{i\alpha}$, $ve^{-i\alpha}$ for u, v respectively.

An important property of the surfaces (14) is discovered when we submit such a surface to a rotation of angle α about the z-axis. Let $\bar{S}$ denote the surface in its new position, and write its equations in the form

$$\bar{x}=\tfrac{1}{2}\int(1-\bar{u}^2)\,\bar{F}(\bar{u})\,d\bar{u}+\tfrac{1}{2}\int(1-\bar{v}^2)\,\bar{\Phi}(\bar{v})\,d\bar{v},$$

and similarly for $\bar{y}$ and $\bar{z}$. Between the parameters $\bar{u}$, $\bar{v}$ and u, v the following relations hold:

$$\bar{u}=ue^{i\alpha},\qquad \bar{v}=ve^{-i\alpha},$$

and we have also

$$\bar{x}=x\cos\alpha-y\sin\alpha,\qquad \bar{y}=x\sin\alpha+y\cos\alpha,\qquad \bar{z}=z.$$

Combining these equations with (14), we find

$$\bar{F}(u)=cu^{\kappa}e^{-i\alpha(\kappa+2)},\qquad \bar{\Phi}(v)=c_1v^{\kappa}e^{i\alpha(\kappa+2)}.$$

Hence, for the correspondence defined by $\bar{u}=u$, $\bar{v}=v$, the surface $\bar{S}$ is an associate of S, unless $\kappa+2=0$, in which case it is the same surface.

We consider the latter case, and remark that its equations are (cf. § 110)

$$x=R\left[c\left(\frac{1}{u}+u\right)\right],\quad y=R\left[ic\left(\frac{1}{u}-u\right)\right],\quad z=-2R\left[c\log u\right].$$

If u, v be replaced by $ue^{i\alpha}$, $ve^{-i\alpha}$, and the resulting expressions be denoted by x_1, y_1, z_1, we have

$$(15)\quad x_1=x\cos\alpha-y\sin\alpha,\quad y_1=x\sin\alpha+y\cos\alpha,\quad z_1=z+2R(i\alpha c).$$

Hence, in a continuous deformation, the surface slides over itself with a helicoidal motion. Consequently it is a helicoid. Moreover, it is the only minimal helicoid. For, every helicoid is applicable to a surface of revolution, and each minimal surface applicable to a surface of revolution with the z-axis for the axis of revolution of the sphere is defined by (14). But only when $\kappa = -2$ will the substitution of $ue^{i\alpha}$, $ve^{-i\alpha}$ give a set of equations such as (15). Hence we have:

The helicoidal minimal surfaces are defined by the Weierstrass formulas when $F(u) = c/u^2$.

And we may state the other results thus:

If any nonhelicoidal minimal surface, which is applicable to a surface of revolution, be rotated through any angle about the axis of the unit sphere whose small circles represent the curves $K=$ *const. on the surface, and a correspondence with parallelism of tangent planes be established between the surfaces, they are associate; consequently the associates of such a minimal surface are superposable.*

EXAMPLES

1. Find under what conditions the surfaces, whose equations are

$$x = r\cos v, \qquad y = r\sin v, \qquad z = f(r),$$
$$x_1 = r\cos v, \qquad y_1 = r\sin v, \qquad z_1 = F(r) + av,$$

can be brought into a one-to-one correspondence, so that the total curvature at corresponding points is the same. Determine under what condition the surfaces are applicable.

2. If the tangent planes to two applicable surfaces at corresponding points are parallel, the surfaces are associate minimal surfaces.

3. Show that the equations

$$x = e^{\alpha}u, \qquad y = e^{-\alpha}v, \qquad z = ae^{\alpha}u^2 + be^{-\alpha}v^2,$$

where α is a real parameter, and a and b are constants, define a family of paraboloids which have the same total curvature at points with the same curvilinear coördinates. Are these surfaces applicable to one another?

4. Find the geodesics on a surface with the linear element

$$ds^2 = \frac{du^2 - 4\,v\,du\,dv + 4\,u\,dv^2}{4\,(u - v^2)}.$$

Show that the surface is applicable to a surface of revolution, and determine the form of a meridian of the latter.

5. Determine the values of the constants a and b in

$$ds^2 = du^2 + [(u + av)^2 + b^2]\,dv^2,$$

so that a surface with this linear element shall be applicable to

(*a*) the right helicoid.
(*b*) the ellipsoid of revolution.

6. A necessary and sufficient condition that a surface be applicable to a surface of revolution is that each curve of a family of geodesic parallels have constant geodesic curvature.

7. Show that the helicoidal minimal surfaces are applicable to the catenoid and to the right helicoid.

138. Second general problem of deformation. We have seen that it can always be determined whether or not two given surfaces are applicable to one another. The solution of this problem was an important contribution to the theory of deformation. An equally important problem, but a more difficult one, is the following:

To determine all the surfaces applicable to a given one.

This problem was proposed by the French Academy in 1859, and has been studied by the most distinguished geometers ever since. Although it has not been solved in the general case, its profound study has led to many interesting results, some of which we shall derive.

If the linear element of the given surface be

$$ds^2 = E\,du^2 + 2\,F\,du\,dv + G\,dv^2,$$

every surface applicable to it is determined by this form and by a second, namely $D du^2 + 2\,D' du dv + D'' dv^2$, whose coefficients satisfy the Gauss and Codazzi equations (§ 64). Conversely, every set of solutions D, D', D'' of these equations defines a surface applicable to the given one, and the determination of the Cartesian coördinates of the corresponding surface requires the solution of a Riccati equation. But neither the Codazzi equations, nor a Riccati equation, can be integrated in the general case with our present knowledge of differential equations. Later we shall make use of this method in the study of particular cases, but for the present we proceed to the exposition of another means of attacking the general problem.

When the values of D, D', D'' obtained from the Gauss equations (V, 7) are substituted in the equation $H^2K = DD'' - D'^2$, the resulting equation is reducible, in consequence of the identity $\Delta_1 x = 1 - X^2$ (cf. Ex. 6, p. 120), to

$$(16)\quad KH^2(1-\Delta_1 x)$$
$$= \left(\frac{\partial^2 x}{\partial u^2} - \begin{Bmatrix}11\\1\end{Bmatrix}\frac{\partial x}{\partial u} - \begin{Bmatrix}11\\2\end{Bmatrix}\frac{\partial x}{\partial v}\right)\left(\frac{\partial^2 x}{\partial v^2} - \begin{Bmatrix}22\\1\end{Bmatrix}\frac{\partial x}{\partial u} - \begin{Bmatrix}22\\2\end{Bmatrix}\frac{\partial x}{\partial v}\right)$$
$$- \left(\frac{\partial^2 x}{\partial u\,\partial v} - \begin{Bmatrix}12\\1\end{Bmatrix}\frac{\partial x}{\partial u} - \begin{Bmatrix}12\\2\end{Bmatrix}\frac{\partial x}{\partial v}\right)^2.$$

This equation, which is satisfied also by y and z, involves only E, F, G and their derivatives, and consequently its integration will give the complete solution of the problem. It is linear in $\left[\frac{\partial^2 x}{\partial u^2}\frac{\partial^2 x}{\partial v^2}-\left(\frac{\partial^2 x}{\partial u\,\partial v}\right)^2\right]$, $\frac{\partial^2 x}{\partial u^2}$, $\frac{\partial^2 x}{\partial u\,\partial v}$, $\frac{\partial^2 x}{\partial v^2}$, and therefore is of the form studied by Ampère. Hence we have the theorem:

The determination of all surfaces applicable to a given one requires the integration of a partial differential equation of the second order of the Ampère type.

In consequence of (16) and (V, 36) we have that the coördinates of a surface with the linear element

$$ds^2 = E\,du^2 + 2F\,du\,dv + G\,dv^2 \tag{17}$$

are integrals of

$$\Delta_{22}\theta = (1-\Delta_1\theta)\,K, \tag{18}$$

the differential parameters being formed with respect to (17). We shall find that when one of these coördinates is known the other two can be found by quadratures.

Our general problem may be stated thus:

Given three functions E, F, G of u and v; to find all functions x, y, z of u and v which satisfy the equation

$$dx^2 + dy^2 + dz^2 = E\,du^2 + 2F\,du\,dv + G\,dv^2,$$

where du and dv may be chosen arbitrarily.

Darboux * observed that as the equation may be written

$$dx^2 + dy^2 = E\,du^2 + 2F\,du\,dv + G\,dv^2 - dz^2, \tag{19}$$

whose left-hand member is the linear element of the plane, or of a developable surface, the total curvature of the quadratic form

$$\left[E-\left(\frac{\partial z}{\partial u}\right)^2\right]du^2 + 2\left[F-\frac{\partial z}{\partial u}\frac{\partial z}{\partial v}\right]du\,dv + \left[G-\left(\frac{\partial z}{\partial v}\right)^2\right]dv^2 \tag{20}$$

must be zero (§ 64).

In order to find the condition for this, we assume that z is known, and take for parametric lines the curves $z = \text{const.}$ and their

* *L.c.*, Vol. III, p. 253.

orthogonal trajectories for $v=$const. With this choice of parameters the right-hand member of (19) reduces to $(E-1)dz^2+G\,dv^2$. The condition that the curvature of this form be zero is

$$(E-1)\left[\frac{\partial E}{\partial v}\frac{\partial G}{\partial v}+\left(\frac{\partial G}{\partial z}\right)^2\right]+G\left[\frac{\partial E}{\partial z}\frac{\partial G}{\partial z}+\left(\frac{\partial E}{\partial v}\right)^2\right]$$
$$-2(E-1)G\left[\frac{\partial^2 E}{\partial v^2}+\frac{\partial^2 G}{\partial z^2}\right]=0,$$

or

$$\frac{\partial E}{\partial z}\frac{\partial G}{\partial z}+\left(\frac{\partial E}{\partial v}\right)^2+4\,E^2G(E-1)K=0,$$

where K denotes the curvature of the surface. But this is the condition also that z be a solution of (18) when the differential parameters are formed with respect to $E\,dz^2+G\,dv^2$. However, the members of equation (18) are differential parameters; consequently z is a solution of this equation whatever be the parametric curves.

By reversing the above steps we prove the theorem:

When z is any integral of the equation (18), *the quadratic form* (20) *has zero curvature.*

When such a solution is known we can find by quadratures (cf. § 135) two functions x, y such that the quadratic form (20) is equal to dx^2+dy^2, provided that

$$\left[E-\left(\frac{\partial z}{\partial u}\right)^2\right]\left[G-\left(\frac{\partial z}{\partial v}\right)^2\right]-\left[F-\frac{\partial z}{\partial u}\frac{\partial z}{\partial v}\right]^2>0,$$

that is, $\Delta_1 z<1$. Hence we have the theorem:

If z be a solution of $\Delta_{22}\theta=(1-\Delta_1\theta)K$ such that $\Delta_1 z<1$, it is one of the rectangular coördinates of a surface with the given linear element, and the other two coördinates can be obtained by quadratures.

139. Deformations which change a curve on the surface into a given curve in space. We consider the problem:

Can a surface be deformed in such a manner that a given curve C upon it comes into coincidence with a given curve Γ in space?

Let the surface be referred to a family of curves orthogonal to C and to their orthogonal trajectories, C being the curve $v=0$, and its arc being the parameter u, so that $E=1$ for $v=0$. The same conditions hold for Γ on the deform.

Since the geodesic curvature of C is unaltered in the deformation (§ 58), it follows from the equation (IV, 47) for the new surface, namely

$$\rho = \rho_g \sin \varpi, \tag{21}$$

that the deformation is impossible, if the curvature of Γ at any point is less than the geodesic curvature of C at the corresponding point. Since both ρ and ρ_g are known, equation (21) determines ϖ, and consequently the direction of the normal to the new surface along Γ is fixed. This being the case, the direction of the tangents to the curves $u=$ const. on the new surface at points of Γ can be found, and so we have the values of $\frac{1}{\sqrt{G}}\frac{\partial x}{\partial v}, \frac{1}{\sqrt{G}}\frac{\partial y}{\partial v}, \frac{1}{\sqrt{G}}\frac{\partial z}{\partial v}$ for $v=0$, as well as $\frac{\partial x}{\partial u}, \frac{\partial y}{\partial u}, \frac{\partial z}{\partial u}$ for $v=0$, the latter being the direction-cosines of the tangent to Γ. If these expressions be differentiated with respect to u, we obtain the values of $\frac{\partial^2 x}{\partial u^2}, \frac{\partial^2 y}{\partial u^2}, \frac{\partial^2 z}{\partial u^2}; \frac{\partial^2 x}{\partial u \partial v}, \frac{\partial^2 y}{\partial u \partial v}, \frac{\partial^2 z}{\partial u \partial v}$ for $v=0$. Since $F=0$ and $E=1$ for $v=0$, the Gauss equations (V, 7) for $v=0$ are

$$\left\{\begin{aligned} \frac{\partial^2 x}{\partial u^2} &= -\frac{1}{2G}\frac{\partial E}{\partial v}\frac{\partial x}{\partial v} + DX, \\ \frac{\partial^2 x}{\partial u \partial v} &= \frac{1}{2}\frac{\partial E}{\partial v}\frac{\partial x}{\partial u} + \frac{1}{2G}\frac{\partial G}{\partial u}\frac{\partial x}{\partial v} + D'X, \\ \frac{\partial^2 x}{\partial v^2} &= -\frac{1}{2}\frac{\partial G}{\partial u}\frac{\partial x}{\partial u} + \frac{1}{2G}\frac{\partial G}{\partial v}\frac{\partial x}{\partial v} + D''X. \end{aligned}\right. \tag{22}$$

All the terms of the first two equations have been determined except D and D'; hence the latter are given by these equations. Since the total curvature K is unaltered by the deformation, it is known at all points of Γ; consequently D'' is given by $H^2K = DD'' - D'^2$, unless D is zero, in which case Γ is an asymptotic line and $\rho = \rho_g$. When D'' is found we can obtain the value of $\frac{\partial^2 x}{\partial v^2}$ from the last of equations (22). From the method of derivation of equation (16) it follows that the above process is equivalent to finding the value of $\frac{\partial^2 x}{\partial v^2}$ from equation (16), which is possible unless $D=0$. Excluding this exceptional case, we remark that if equations (22) be differentiated with respect to u, we obtain the values of all the derivatives of x of the third order for $v=0$ except $\frac{\partial^3 x}{\partial v^3}\cdot$ The latter may be obtained from the equation which results from the differentiation of equation (16) with respect to v. By continuing this process we obtain the values for $v=0$ of the derivatives of x of all orders, and likewise of y and z. If we indicate by subscript null the values of functions, when $u=u_0$, $v=0$, the expansions

$$(23)\quad x = x_0 + \left(\frac{\partial x}{\partial u}\right)_0 u + \left(\frac{\partial x}{\partial v}\right)_0 v + \frac{1}{2}\left(\frac{\partial^2 x}{\partial u^2}\right)_0 u^2 + \left(\frac{\partial^2 x}{\partial u\, \partial v}\right)_0 uv + \frac{1}{2}\left(\frac{\partial^2 x}{\partial v^2}\right)_0 v^2 + \cdots,$$

and similar expansions for y and z, are convergent in general, as Cauchy has shown,* and x, y, z thus defined are the solutions of equation (16) which for $v=0$ satisfy the given conditions. Hence:

A surface S can be deformed in such a manner that a curve C upon it comes into coincidence with a given curve Γ, provided that the curvature of Γ at each point is greater than the geodesic curvature of C at the corresponding point.

There remains the exceptional case $\rho = \rho_g$. If the desired deformation is possible, Γ is an asymptotic line on the deform, and consequently, by Enneper's theorem (§ 59), its radius of torsion must satisfy the condition $\tau^2 = -1/K$. Hence when C is given, Γ is determined, if it is to be an asymptotic line.

If Γ satisfies these conditions, the value of D'' for $v=0$ is arbitrary, as we have seen. But when it has been chosen, the further determination of the values of the derivatives of x, y, z of higher order for $v=0$ is unique, it being the same as that pursued in the general case. Hence equation (16) admits as solution a family of these surfaces, depending upon an arbitrary function. For all of these surfaces the directions of the tangent planes at each point of Γ are the same. Hence we have the theorem:

Given a curve C upon a surface S; there exists in space a unique curve Γ with which C can be brought into coincidence by a deformation of S in an infinity of ways; moreover, all the new surfaces are tangent to one another along Γ.

* Cf. Goursat, *Leçons sur l'integration des équations aux dérivées partielles du second ordre*, chap. ii. Paris, 1896.

If C is an asymptotic line on S, it may be taken for Γ; hence:

A surface may be subjected to a continuous deformation during which a given asymptotic line is unaltered in form and continues to be an asymptotic line on each deform.

This result suggests the problem:

Can a surface be subjected to a continuous deformation in which a curve other than an asymptotic line is unaltered?

By hypothesis the curvature is not changed and the geodesic curvature is necessarily invariant; hence from (21) we have that $\sin \varpi$ must have the same value for all the surfaces. If ϖ is the same for all surfaces, the tangent plane is the same, and consequently the expansions (23) are the same. Hence all the surfaces coincide in this case. However, there are always two values of ϖ for which $\sin \varpi$ has the same value, unless ϖ is a right angle. Hence it is possible to have two applicable surfaces passing through a curve whose points are self-correspondent, but not an infinity of such surfaces. Therefore:

An asymptotic line is the only curve on a surface which can remain unaltered in a continuous deformation.

140. Lines of curvature in correspondence. We inquire whether a surface S can be deformed in such a manner that a given curve C upon it may become a line of curvature on the new surface. Suppose it is possible, and let Γ denote this line of curvature. The radii of curvature and torsion of Γ must satisfy (21) and $1/\tau - d\varpi/ds = 0$ (cf. § 59), where ρ_g is the same for Γ as for C. If we choose for ϖ any function whatever, the functions ρ and τ are thus determined, and Γ is unique. Since ϖ fixes the direction of the tangent plane to the new surface along Γ, there is only one deform of S of the kind desired for each choice of ϖ (cf. § 139). Hence:

A surface can be deformed in an infinity of ways so that a given curve upon it becomes a line of curvature on the deform.

This result suggests the following problem of Bonnet*:

To determine the surfaces which can be deformed with preservation of their lines of curvature.

* Mémoire sur la théorie des surfaces applicables sur une surface donnée, *Journal de l'École Polytechnique*, Cahier 42 (1867), p. 58.

We follow the method of Bonnet in making use of the fundamental equations in the form (V, 48, 55). We assume that the lines of curvature are parametric. In this case these equations reduce to

$$(24)\qquad \begin{cases} \dfrac{\partial p_1}{\partial u} = -qr_1, \quad \dfrac{\partial q}{\partial v} = rp_1, \quad \dfrac{\partial r}{\partial v} - \dfrac{\partial r_1}{\partial u} = -qp_1, \\ r = -\dfrac{1}{\sqrt{G}}\dfrac{\partial \sqrt{E}}{\partial v}, \quad r_1 = \dfrac{1}{\sqrt{E}}\dfrac{\partial \sqrt{G}}{\partial u}. \end{cases}$$

From these equations it follows that if S and S' are two applicable surfaces referred to corresponding lines of curvature, the functions r and r_1 have the same value for both surfaces, and consequently the same is true of the product qp_1. Hence our problem reduces to the determination of two sets of functions p_1, q; p_1', q', satisfying the above equations. In consequence of the identity

$$(25)\qquad p_1'q' = p_1q,$$

we have from the first two of (24)

$$(26)\qquad p_1'\frac{\partial p_1'}{\partial u} = p_1\frac{\partial p_1}{\partial u}, \quad q'\frac{\partial q'}{\partial v} = q\frac{\partial q}{\partial v},$$

of which the integrals are $p_1'^2 = p_1^2 + f(v)$, $q'^2 = q^2 + \phi(u)$, where $f(v)$ and $\phi(u)$ are functions of v and u respectively. The parameters u, v may be chosen so that these functions become constants α, β, and consequently

$$(27)\qquad p_1'^2 = p_1^2 + \alpha, \quad q'^2 = q^2 + \beta.$$

If these equations be multiplied together, the resulting equation is reducible by means of (25) to either of the forms

$$(28)\qquad p_1^2\beta + q^2\alpha + \alpha\beta = 0, \quad p_1'^2\beta + q'^2\alpha - \alpha\beta = 0.$$

From the first we see that α and β cannot both be positive if S is real, and from the second that they cannot both be negative. We assume that α is negative and β positive, and without loss of generality write

$$(29)\qquad p_1'^2 = p_1^2 - 1, \quad q'^2 = q^2 + 1.$$

The first of (28) reduces to $p_1^2 - q^2 = 1$. In conformity with this we introduce a function ω, thus

$$p_1 = \cosh\omega, \quad q = \sinh\omega.$$

Then equations (29) may be replaced by

$$p_1' = \sinh\omega, \quad q' = \cosh\omega.$$

Moreover, the fundamental equations (24) reduce to

$$\frac{\partial \omega}{\partial v}=r=-\frac{1}{\sqrt{G}}\frac{\partial \sqrt{E}}{\partial v}, \qquad \frac{\partial \omega}{\partial u}=-r_1=-\frac{1}{\sqrt{E}}\frac{\partial \sqrt{G}}{\partial u},$$

$$\frac{\partial^2 \omega}{\partial u^2}+\frac{\partial^2 \omega}{\partial v^2}=-\sinh \omega \cosh \omega.$$

Comparing these results with § 118, we see that the spherical representation of lines of curvature of the surfaces S and S' respectively is the same as of the lines of curvature of a spherical surface and of its Hazzidakis transform. Conversely, we have that every surface of this kind admits of an applicable surface with lines of curvature in correspondence.

The preceding investigation rested on the hypothesis that neither the first nor second of equations (24) vanishes identically. Suppose that the second vanishes; then q is a function of u alone, say $\phi(u)$. Since the product $p_1 q$ differs from the total curvature only by a factor (cf. § 70), p_1 cannot be zero; therefore $r=0$ and $q'=\phi_1(u)$. Equation (25) is now of the form $p_1\phi(u)=p_1'\phi_1(u)$. If p_1' be eliminated from this equation and the first of (27), it is found that p_1 also is a function of u alone. Hence the curves $v=$ const. on the sphere are great circles with a common diameter, and therefore S is a *molding surface* (§ 130). The parameter u may be chosen so that we may take $q=1$ and $p_1=U$; then from (27) and (25) we find $p_1'=\sqrt{U^2+a}$, $q'=U/\sqrt{U^2+a}$, where a is an arbitrary constant. Hence we have the theorem:

*A necessary and sufficient condition that a surface admit of an applicable surface with lines of curvature in correspondence is that the surface have the same spherical representation of its lines of curvature as a spherical surface Σ, or be a molding surface; in the first case there is one applicable surface, and the spherical representation of its lines of curvature is the same as of the Hazzidakis transform of Σ; in the second case there is an infinity of applicable surfaces.**

141. Conjugate systems in correspondence. When two surfaces are applicable to one another, there is a system of corresponding lines which is conjugate for both surfaces (cf. § 56). The results of § 140 show that for a given conjugate system on a surface S

* Cf. Ex. 14, p. 319.

there is not in general a surface S_1 applicable to S with the corresponding system conjugate. We inquire under what conditions a given conjugate system of S possesses this property.

Let S be referred to the given conjugate system. If the corresponding system on an applicable surface S_1 is conjugate, we have

$$D' = D_1' = 0, \qquad D_1 D_1'' = DD'';$$

for the total curvature of the two surfaces is the same. We replace this equation by the two

$$D_1 = \tanh\theta \cdot D, \qquad D_1'' = \coth\theta \cdot D'',$$

thus defining a function θ. The Codazzi equations for S are

$$\frac{\partial D}{\partial v} - \begin{Bmatrix} 12 \\ 1 \end{Bmatrix} D + \begin{Bmatrix} 11 \\ 2 \end{Bmatrix} D'' = 0, \qquad \frac{\partial D''}{\partial u} + \begin{Bmatrix} 22 \\ 1 \end{Bmatrix} D - \begin{Bmatrix} 12 \\ 2 \end{Bmatrix} D'' = 0.$$

Since these equations must be satisfied by D_1 and D_1'', we have

$$(30) \qquad \frac{\partial \theta}{\partial u} = - \begin{Bmatrix} 22 \\ 1 \end{Bmatrix} \frac{D}{D''} \tanh\theta, \qquad \frac{\partial \theta}{\partial v} = - \begin{Bmatrix} 11 \\ 2 \end{Bmatrix} \frac{D''}{D} \coth\theta.$$

The condition of integrability of (30) is reducible to

$$(31) \qquad \cosh^2\theta \left[\frac{\partial}{\partial u}\left(\begin{Bmatrix} 11 \\ 2 \end{Bmatrix} \frac{D''}{D} \right) - \frac{\partial}{\partial v}\left(\begin{Bmatrix} 22 \\ 1 \end{Bmatrix} \frac{D}{D''} \right) \right] + \left[\frac{\partial}{\partial v}\left(\begin{Bmatrix} 22 \\ 1 \end{Bmatrix} \frac{D}{D''} \right) + 2 \begin{Bmatrix} 22 \\ 1 \end{Bmatrix} \begin{Bmatrix} 11 \\ 2 \end{Bmatrix} \right] = 0.$$

As the two roots of this equation differ only in sign, and thus lead to symmetric surfaces, we need consider only one. If it be substituted in (30), we obtain two conditions upon E, F, G; D, D'', which are necessary in order that S admit of an applicable surface of the kind sought. Hence in general there is no solution of the problem. However, if the two expressions in the brackets of (31) vanish identically, the conditions of integrability of equations (30) are completely satisfied, and S admits of an infinity of applicable surfaces upon which the coördinate curves form a conjugate system. Consequently we have the theorem:

If a conjugate system on a surface S corresponds to a conjugate system on more than one surface applicable to S, it corresponds to a conjugate system on an infinity of surfaces applicable to S.

We shall give this result another interpretation by considering the spherical representation of S. From (VI, 38) we have

$$\left\{ {22 \atop 1} \right\} \frac{D}{D''} = -\left\{ {12 \atop 2} \right\}', \qquad \left\{ {11 \atop 2} \right\} \frac{D''}{D} = -\left\{ {12 \atop 1} \right\}', \tag{32}$$

where the symbols $\left\{ {rs \atop t} \right\}'$ are formed with respect to the linear element of the spherical representation of S. If we substitute these values in (30), we get

$$\frac{\partial \theta}{\partial u} = \left\{ {12 \atop 2} \right\}' \tanh \theta, \qquad \frac{\partial \theta}{\partial v} = \left\{ {12 \atop 1} \right\}' \coth \theta,$$

and the condition that these equations have an integral involving a parameter becomes

$$\frac{\partial}{\partial u} \left\{ {12 \atop 1} \right\}' = \frac{\partial}{\partial v} \left\{ {12 \atop 2} \right\}' = 2 \left\{ {12 \atop 1} \right\}' \left\{ {12 \atop 2} \right\}'. \tag{33}$$

The first of these equations is the condition that the curves upon the sphere represent the asymptotic lines upon a certain surface Σ (cf. § 78). Moreover, if K denotes the total curvature of Σ, and we put $K = -1/\rho^2$, we have

$$\frac{\partial \log \rho}{\partial u} = -2 \left\{ {12 \atop 2} \right\}', \qquad \frac{\partial \log \rho}{\partial v} = -2 \left\{ {12 \atop 1} \right\}'. \tag{34}$$

Now equations (33) are equivalent to (34), and

$$\frac{\partial^2 \log \rho}{\partial u \, \partial v} + \frac{\partial \log \rho}{\partial u} \cdot \frac{\partial \log \rho}{\partial v} = 0,$$

which reduces to $\frac{\partial^2 \rho}{\partial u \, \partial v} = 0$. As the general integral of this equation is $\rho = \phi(u) + \psi(v)$, where ϕ and ψ are arbitrary functions of u and v respectively, we have the following theorem due to Bianchi*:

A necessary and sufficient condition that a surface S admit a continuous deformation in which a conjugate system remains conjugate is that the spherical representation of this system be that of the asymptotic lines of a surface whose total curvature, expressed in terms of parameters referring to these lines, is of the form

$$K = \frac{-1}{[\phi(u) + \psi(v)]^2}. \tag{35}$$

* *Annali*, Ser. 2, Vol. XVIII (1890), p. 320; also *Lezioni*, Vol. II, p. 83.

The pseudospherical surfaces afford an example of surfaces with K of this form. In this case ϕ and ψ are constants, so that equations (34) reduce to $\begin{Bmatrix} 12 \\ 1 \end{Bmatrix}' = \begin{Bmatrix} 12 \\ 2 \end{Bmatrix}' = 0$, which, in consequence of (32), are equivalent to $\begin{Bmatrix} 11 \\ 2 \end{Bmatrix} = \begin{Bmatrix} 22 \\ 1 \end{Bmatrix} = 0$. But these are the conditions that the parametric curves on S be geodesics. A surface with a conjugate system of geodesics is called a *surface of Voss.* We state these results thus:

A surface of Voss admits of a continuous deformation in which the geodesic conjugate system is preserved; consequently all the new surfaces are of the same kind.

EXAMPLES

1. Show that every integral of the equation $\Delta_1 \theta = 1$ is an integral of the fundamental equation (18).

2. On a right helicoid the helices are asymptotic lines. Find the surfaces applicable to the helicoid in such a way that one of the helices is unaltered in form and continues to be an asymptotic line.

3. A surface applicable to a surface of revolution with the lines of curvature on the two surfaces in correspondence is a surface of revolution.

4. Show that the equations

$$x = \kappa r \cos \frac{v}{\kappa}, \qquad y = \kappa r \sin \frac{v}{\kappa}, \qquad z = \int \sqrt{1 - \kappa^2 r'^2}\, du,$$

define a family of applicable surfaces of revolution with lines of curvature in correspondence. Discuss the effect of a variation of the parameter κ.

5. Let S denote a surface parallel to a spherical surface Σ. Find the surface applicable to S with preservation of the lines of curvature.

6. If S_1 and S_2 be applicable surfaces referred to the common conjugate system, their coördinates x_1, y_1, z_1; x_2, y_2, z_2 are solutions of the same point equation (cf. VI, 26), and the function $x_1^2 + y_1^2 + z_1^2 - (x_2^2 + y_2^2 + z_2^2)$ also is a solution.

7. Show that the locus of a point which divides in constant ratio the join of corresponding points on the surfaces S_1 and S_2 of Ex. 6 is a surface upon which the parametric lines form a conjugate system. Under what condition is this surface applicable to S_1 and S_2?

8. The tetrahedral surface

$$x = A\,(a + u)^{\frac{3}{2}}\,(a + v)^{\frac{3}{2}}, \quad y = B\,(b + u)^{\frac{3}{2}}\,(b + v)^{\frac{3}{2}}, \quad z = C\,(c + u)^{\frac{3}{2}}\,(c + v)^{\frac{3}{2}},$$

admits of an infinity of deforms

$$x_1 = A_1(a_1 + u)^{\frac{3}{2}}\,(a_1 + v)^{\frac{3}{2}}, \quad y_1 = B_1(b_1 + u)^{\frac{3}{2}}\,(b_1 + v)^{\frac{3}{2}}, \quad z_1 = C_1(c_1 + u)^{\frac{3}{2}}\,(c_1 + v)^{\frac{3}{2}}.$$

The curves $u = v$ upon these surfaces are congruent, and consequently each is an asymptotic line on the surface through it.

9. If the equations of a surface are of the form

$$x = U_1V_1, \qquad y = U_2V_1, \qquad z = V_2,$$

the equations

$$x_1 = \sqrt{U_1^2 + U_2^2 + h - 1}\,V_1 \cos\theta, \quad y_1 = \sqrt{U_1^2 + U_2^2 + h - 1}\,V_1 \sin\theta,$$

$$z_1 = \int \sqrt{V_2'^2 - (h-1)V_1'^2}\,dv, \qquad \theta = \int \frac{\sqrt{(U_1U_2' - U_2U_1')^2 + (h-1)(U_1'^2 + U_2'^2)}}{U_1^2 + U_2^2 + h - 1}\,du,$$

where h denotes a constant, define a family of applicable surfaces upon which the parametric lines form a conjugate system.

10. Show that the equations of the quadrics can be put in the form of Ex. 9, and apply the results to this case.

142. Asymptotic lines in correspondence. Deformation of a ruled surface. We have seen (§ 139) that a surface can be subjected to a continuous deformation in which an asymptotic line remains asymptotic. We ask whether two surfaces are applicable with the asymptotic lines in one system corresponding to asymptotic lines of the other. We assume that there are two such surfaces, S, S_1, and we take the corresponding asymptotic lines for the curves $v =$ const. and their orthogonal trajectories for $u =$ const. In consequence of this choice and the fact that the total curvature of the two surfaces is the same, we have

$$D = D_1 = 0, \qquad F = 0, \qquad D' = D_1'. \tag{36}$$

The Codazzi equations (V, 13′) for S reduce to

$$\left\{\begin{aligned} &\frac{\partial}{\partial u}\left(\frac{D'}{\sqrt{EG}}\right) + \frac{\partial \log G}{\partial u}\frac{D'}{\sqrt{EG}} + \frac{1}{2G}\frac{\partial E}{\partial v}\frac{D''}{\sqrt{EG}} = 0,\\ &\frac{\partial}{\partial u}\left(\frac{D''}{\sqrt{EG}}\right) - \frac{\partial}{\partial v}\left(\frac{D'}{\sqrt{EG}}\right) - \frac{\partial \log E}{\partial v}\frac{D'}{\sqrt{EG}} + \frac{\partial \log\sqrt{E}}{\partial u}\frac{D''}{\sqrt{EG}} = 0. \end{aligned}\right. \tag{37}$$

Because of (36) the Codazzi equation for S_1 analogous to the first of (37) will differ from the latter only in the last term. Hence we must have either $D_1'' = D''$, or $E = f(u)$. In the former case the surfaces S and S_1 are congruent. Hence we are brought to the second, which is the condition that the curves $v =$ const. be geodesics. As the latter are asymptotic lines also, they are straight, and consequently S must be a ruled surface. By changing the parameter u, we have $E = 1$, and equations (37) reduce to

$$\frac{\partial}{\partial u}(\sqrt{G}D') = 0, \qquad \frac{\partial}{\partial u}\left(\frac{D''}{\sqrt{G}}\right) = \frac{\partial}{\partial v}\left(\frac{D'}{\sqrt{G}}\right).$$

By a suitable choice of the parameter v the first of these equations may be replaced by $D'=1/\sqrt{G}$, and the second becomes

$$\frac{D''}{\sqrt{G}}=\int\frac{\partial}{\partial v}\left(\frac{1}{G}\right)du+\phi(v),$$

where ϕ is an arbitrary function. These results establish the following theorem of Bonnet:

A necessary and sufficient condition that a surface admit an applicable surface with the asymptotic lines in one system on each surface corresponding is that the surface be ruled; moreover, a ruled surface admits of a continuous deformation in which the generators remain straight.

To this may be added the theorem:

If two surfaces are applicable and the asymptotic lines in both systems on each surface are in correspondence, the surfaces are congruent, or symmetric.

This is readily proved when the asymptotic lines are taken as parametric.

We shall establish the second part of the above theorem in another manner. For this purpose we take the equations of the ruled surface in the form (§ 103)

$$x=x_0+lu,\qquad y=y_0+mu,\qquad z=z_0+nu, \tag{38}$$

where x_0, y_0, z_0 are the coördinates of the directrix C expressed as functions of its arc v, and l, m, n are the direction-cosines of the generators, also functions of v. They satisfy the conditions

$$x_0'^2+y_0'^2+z_0'^2=1,\qquad l^2+m^2+n^2=1, \tag{39}$$

where the accents indicate differentiation with respect to v. Furthermore, the linear element is

$$ds^2=du^2+2\cos\theta_0\,du\,dv+(a^2u^2+2\,bu+1)\,dv^2, \tag{40}$$

where

$$\begin{cases} a^2=l'^2+m'^2+n'^2, & b=l'x_0'+m'y_0'+n'z_0',\\ \cos\theta_0=x_0'l+y_0'm+z_0'n. \end{cases} \tag{41}$$

Hence if we have a ruled surface with the linear element (40), the problem of finding a ruled surface applicable to it, with the generators of the two surfaces corresponding, reduces to the determination of six functions of v, namely x_0, y_0, z_0; l, m, n, satisfying the five conditions (39), (41). From this it follows that there is an arbitrary function of v involved in the problem, and consequently there is an infinity of ruled surfaces with the linear element (40).

There are two general ways in which the choice of this arbitrary function may be made, — either as determining the form of the director-cone of the required surface, or by a property of the directrix. We consider these two cases.

143. Method of Minding. The first case was studied by Minding.* He took l, m, n in the form

$$(42)\qquad l = \cos\phi\cos\psi, \qquad m = \cos\phi\sin\psi, \qquad n = \sin\phi,$$

which evidently satisfy the second of (39). The first of (41) reduces to

$$(43)\qquad \phi'^2 + \psi'^2\cos^2\phi = a^2.$$

If we solve equations (39) and (41) for x_0', y_0', z_0', the resulting expressions are reducible by means of (VII, 63) to

$$(44)\qquad x_0' = l\cos\theta_0 + \frac{1}{a^2}\left[l'b \pm (mn' - m'n)\sqrt{a^2\sin^2\theta_0 - b^2}\right],$$

and analogous expressions for y_0' and z_0'. Hence, if ϕ be an arbitrary function of v, and ψ be given by

$$(45)\qquad \psi = \int\frac{\sqrt{a^2 - \phi'^2}}{\cos\phi}\,dv,$$

the functions x_0, y_0, z_0, obtained from (44) by quadratures, together with l, m, n from (42), determine a ruled surface with the linear element (40).

Each choice of ϕ gives a different director-cone, which is determined by the curve in which the cone cuts the unit sphere, whose center is at the vertex of the cone. Such a curve is defined by a relation $f(\phi, \psi) = 0$, so that instead of choosing ϕ arbitrarily we may take f as arbitrary; for, by combining equations (43) and $f(\phi, \psi) = 0$, we obtain the expressions for ϕ and ψ as functions of v. Hence:

A ruled surface may be deformed in such a way that the director-cone takes an arbitrary form.

* *Crelle*, Vol. XVIII (1838), pp. 297–302.

When the given ruled surface is nondevelopable, the radicand in (44) is different from zero, and consequently there are two different sets of functions x_0, y_0, z_0. Hence there are two applicable ruled surfaces with the same director-cone. If the parameters of distribution of these two surfaces be calculated by (VII, 73), they are found to differ only in sign. Hence we have the theorem of Beltrami: *

A ruled surface admits of an applicable ruled surface such that corresponding generators are parallel, and the parameters of distribution differ only in sign.

144. Particular deformations of ruled surfaces. By means of the preceding results we prove the theorem:

A ruled surface may be deformed in an infinity of ways so that a given curve becomes plane.

Let the given curve be taken for the directrix of the original surface. Assuming that a deform of the kind desired exists, we take its plane for the xy-plane. From (44) we have

$$a^2 n \cos\theta_0 + bn' \pm (lm' - l'm)\sqrt{a^2 \sin^2\theta_0 - b^2} = 0,$$

which, in consequence of (42) and (43), reduces to

$$b \cos\phi \cdot \phi' + a^2 \sin\phi \cos\theta_0 \pm \cos\phi \sqrt{a^2 - \phi'^2}\sqrt{a^2 \sin^2\theta_0 - b^2} = 0.$$

The integral of this equation involves an arbitrary constant, and thus the theorem is proved.

The preceding example belongs to the class of problems whose general statement is as follows:

To deform a ruled surface into a ruled surface in such a way that the deform of a given curve C on the original surface shall possess a certain property on the resulting surface.

We consider this general problem. Let the deform of C be the directrix of the required surface, and let $\alpha_0, \beta_0, \gamma_0;\ l_0, m_0, n_0;\ \lambda_0, \mu_0, \nu_0$ denote the direction-cosines of its tangent, principal normal, and binormal. If σ denotes the angle between the osculating plane to the curve and the tangent plane to the surface, we have

(46) $$l = \alpha_0 \cos\theta_0 + \sin\theta_0 (l_0 \cos\sigma + \lambda_0 \sin\sigma),$$

* *Annali*, Vol. VII (1865), p. 115.

and similar expressions for m and n. When these values are substituted in the first two of equations (41), the resulting equations are reducible, by means of the Frenet formulas (I, 50), to

$$(47)\qquad \left\{\begin{aligned} &\frac{\cos\sigma}{\rho} = -\left(\theta_0' + \frac{b}{\sin\theta_0}\right),\\ &\left[\frac{\cos\theta_0}{\rho} + (\cos\sigma\sin\theta_0)' + \frac{\sin\sigma\sin\theta_0}{\tau}\right]^2\\ &\qquad + \left[(\sin\sigma\sin\theta_0)' - \frac{\cos\sigma\sin\theta_0}{\tau}\right]^2 = a^2 - b^2.\end{aligned}\right.$$

These are two equations of condition on σ, ρ, τ, as functions of v. Each set of solutions determines a solution of the problem; for, the directrix is determined by expressions for ρ and τ, and equations (46) give the direction-cosines of the generators.

We leave it to the reader to prove the above theorem by this means, and we proceed to the proof of the theorem:

A ruled surface may be deformed in such a manner that a given curve C becomes an asymptotic line on the new ruled surface.

On the deform we must have $\sigma = 0$ or $\sigma = \pi$, so that from (47)

$$\frac{1}{\rho} = \pm\left(\theta_0' + \frac{b}{\sin\theta_0}\right),$$

the sign being fixed by the fact that ρ is necessarily positive. The second of (47) reduces to

$$\frac{1}{\tau} = \pm\frac{\sqrt{a^2\sin^2\theta_0 - b^2}}{\sin^2\theta_0}.$$

If the curve with these intrinsic equations be constructed, and in the osculating plane at each point the line be drawn which makes the angle θ_0 with the tangent, the locus of these lines is a ruled surface satisfying the given conditions.

When the curve C is an orthogonal trajectory of the generators, the same is true of its deform. Hence:

A ruled surface may be deformed in such a way that all the generators become the principal normals of the deform of any one of their orthogonal trajectories.

Having thus considered the deformation of ruled surfaces in which the generators remain straight, we inquire whether two ruled surfaces are applicable with the generators of each corresponding to curves on the other. Assume that it is possible, and let $v=$ const. be the generators of S and $u=$ const. the curves on S corresponding to the generators of S_1. From (V, 13′) it follows that the conditions for this are respectively

$$\frac{\partial}{\partial u}\log\rho = 2\left\{\begin{matrix}12\\2\end{matrix}\right\}, \qquad \frac{\partial}{\partial v}\log\rho = 2\left\{\begin{matrix}12\\1\end{matrix}\right\}, \tag{48}$$

where $K=-1/\rho^2$. But equations (48) are the necessary and sufficient conditions that there be a surface Σ applicable to S and S_1, upon which the asymptotic lines are parametric (cf. VI, 3). But the curves $v=$ const. and $u=$ const. are geodesics on S and S_1, and consequently on Σ. Therefore Σ is doubly ruled. Hence:

If two ruled surfaces S and S_1 are applicable to one another, the generators correspond unless the surfaces are applicable to a quadric with the generators of S and S_1 corresponding to the two different systems of generators of the quadric.

EXAMPLES

1. A ruled surface can be deformed into another ruled surface in such a way that a geodesic becomes a straight line.

2. A ruled surface formed by the binormals of a curve C can be deformed into a right conoid; the latter is the right helicoid when the torsion of C is constant. Prove the converse also.

3. On the hyperboloid of revolution, defined by

$$\frac{x}{c}=\frac{u}{\Delta}\cos\frac{v}{c}+\sin\frac{v}{c}, \qquad \frac{y}{c}=\frac{u}{\Delta}\sin\frac{v}{c}-\cos\frac{v}{c}, \qquad \frac{z}{d}=\frac{u}{\Delta},$$

where $\Delta^2=c^2+d^2$, the circle of gorge is a geodesic, which is met by the generators under the angle $\cos^{-1}c/\Delta$.

4. Show that the ruled surface which results from the deformation of the hyperboloid of Ex. 3, in which the circle of gorge becomes straight, is given by

$$x=\frac{ud}{\Delta}\cos\frac{v}{d}, \qquad y=\frac{ud}{\Delta}\sin\frac{v}{d}, \qquad z=\frac{uc}{\Delta}+v.$$

5. Show that the ruled surface to which the hyperboloid of Ex. 3 is applicable with parallelism of corresponding generators is the helicoid

$$\frac{x}{c}=\frac{u}{\Delta}\cos\frac{v}{c}+\frac{c^2-d^2}{c^2+d^2}\sin\frac{v}{c}, \qquad \frac{y}{c}=\frac{u}{\Delta}\sin\frac{v}{c}-\frac{c^2-d^2}{c^2+d^2}\cos\frac{v}{c}, \qquad \frac{z}{d}=\frac{u}{\Delta}+\frac{2c}{\Delta^2}v,$$

and that the circle of gorge of the former corresponds to a helix upon the latter.

6. When the directrix is a geodesic, equations (47) reduce to

$$\sin\theta_0\cdot\theta_0'+b=0, \qquad \left(\frac{\cos\theta_0}{\rho}\pm\frac{\sin\theta_0}{\tau}\right)^2=a^2-\theta_0'^2.$$

7. When an hyperboloid of revolution of one sheet is deformed into another ruled surface, the circle of gorge becomes a Bertrand curve and the generators are parallel to the corresponding binormals of the conjugate Bertrand curve.

8. A ruled surface can be deformed in such a way that a given curve is made to lie upon a sphere of arbitrary radius.

9. When a ruled surface admits a continuous deformation into itself the total curvature of the surface is constant along the line of striction, the generators meet the latter under constant angle, and the parameter of distribution is constant (cf. § 126).

10. Two applicable ruled surfaces whose corresponding generators are parallel cannot be obtained from one another by a continuous deformation.

GENERAL EXAMPLES

1. Determine the systems of coördinate lines in the plane such that the linear element of the plane is

$$ds^2 = \frac{du^2 + dv^2}{(U+V)^2},$$

where U and V are functions of u and v respectively.

2. Solve for the sphere the problem similar to Ex. 1.

3. Determine the functions $\phi(u)$ and $\psi(u)$ so that the helicoids, defined by

$$x = a\sqrt{U^2 - b^2}\cos\frac{v-\phi}{a}, \quad y = a\sqrt{U^2 - b^2}\sin\frac{v-\phi}{a}, \quad z = bv + \psi,$$

shall be applicable to the surface whose equations are

$$x = U\cos v, \quad y = U\sin v, \quad z = \int\sqrt{1-U'^2}\,du,$$

where U is any function of u.

4. Apply the method of Ex. 3 to find helicoids applicable to the pseudosphere; to the catenoid.

5. The equations

$$x = a\sqrt{2u-2}\cos\frac{v}{a}, \quad y = a\sqrt{2u-2}\sin\frac{v}{a}, \quad z = \frac{a}{2}(u-1)$$

define a paraboloid of revolution. Show that surfaces applicable to it are defined by

$$x = \frac{ia}{2}\left[f_3\phi_2 - f_2\phi_3 + \int(f_2 df_3 - f_3 df_2) - \int(\phi_2 d\phi_3 - \phi_3 d\phi_2)\right],$$

$$y = \frac{ia}{2}\left[f_1\phi_3 - f_3\phi_1 + \int(f_3 df_1 - f_1 df_3) - \int(\phi_3 d\phi_1 - \phi_1 d\phi_3)\right],$$

$$z = \frac{ia}{2}\left[f_2\phi_1 - f_1\phi_2 + \int(f_1 df_2 - f_2 df_1) - \int(\phi_1 d\phi_2 - \phi_2 d\phi_1)\right],$$

where a is a real constant, and the f's and ϕ's are functions of a parameter α and β respectively such that

$$f_1^2 + f_2^2 + f_3^2 = 1, \quad \phi_1^2 + \phi_2^2 + \phi_3^2 = 1, \quad f_1\phi_1 + f_2\phi_2 + f_3\phi_3 = u.$$

6. Investigate the special case of Ex. 5 for which α and β are conjugate imaginary functions, and

$$f_1 = \frac{2 + a - 2\alpha^2}{2\sqrt{2a}}, \quad f_2 = i\frac{2 - a - 2\alpha^2}{2\sqrt{2a}}, \quad f_3 = \alpha,$$

and the ϕ's are functions conjugate imaginary to the f's.

7. Show that the surface of translation

$$x = a(\cos u + \cos v), \quad y = a(\sin u + \sin v), \quad z = c(u + v)$$

is applicable to a surface of revolution.

8. Show that the minimal surfaces applicable to a spiral surface (Ex. 22, p. 151) are determined by the functions $F(u) = cu^{m+in}$, $\Phi(v) = c_1 v^{m-in}$, and that the associate surfaces are similar to the given one.

9. If the coefficients E, F, G of the linear element of a surface are homogeneous functions of u and v of order -2, the surface is applicable to a surface of revolution.

10. If x, y, z are the coördinates of a surface S referred to a conjugate system, the equations

$$\frac{\partial x'}{\partial u} = P\frac{\partial x}{\partial u}, \quad \frac{\partial y'}{\partial u} = P\frac{\partial y}{\partial u}, \quad \frac{\partial z'}{\partial u} = P\frac{\partial z}{\partial u}; \quad \frac{\partial x'}{\partial v} = Q\frac{\partial x}{\partial v}, \quad \frac{\partial y'}{\partial v} = Q\frac{\partial y}{\partial v}, \quad \frac{\partial z'}{\partial v} = Q\frac{\partial z}{\partial v}$$

are integrable if P and Q satisfy the conditions

$$\frac{\partial P}{\partial v} + (P - Q)\begin{Bmatrix} 12 \\ 1 \end{Bmatrix} = 0, \qquad \frac{\partial Q}{\partial u} + (Q - P)\begin{Bmatrix} 12 \\ 2 \end{Bmatrix} = 0,$$

where the Christoffel symbols are formed with respect to the linear element of S. Show that on the surface S', whose coördinates are x', y', z', the parametric curves form a conjugate system, and that the normals to S and S' at corresponding points are parallel.

11. Show that for the surface

$$x = \int \lambda f_1(u)\,du + \phi_1(v), \quad y = \int \lambda f_2(u)\,du + \phi_2(v), \quad z = \int \lambda f_3(u)\,du + \phi_3(v),$$

where λ is any function of u and v, and f_1, f_2, f_3; ϕ_1, ϕ_2, ϕ_3 are functions of u and v respectively, the parametric curves form a conjugate system. Apply the results of Ex. 10 to this surface, and discuss the case for which λ is independent of v.

12. If S and S_1 are two applicable surfaces, and S_1' denotes the surface corresponding to S_1 in the same manner as S' to S in Ex. 10 and by means of the same functions P and Q, then S' and S_1' are applicable surfaces.

13. If x, y, z and x_1, y_1, z_1 are the coördinates of a pair of applicable surfaces S and S_1, a second pair of applicable surfaces S' and S_1' is defined by

$$\begin{aligned} x' &= x + h(z + z_1) - k(y + y_1), & x_1' &= x_1 - h(z + z_1) + k(y + y_1), \\ y' &= y + k(x + x_1) - g(z + z_1), & y_1' &= y_1 - k(x + x_1) + g(z + z_1), \\ z' &= z + g(y + y_1) - h(x + x_1), & z_1' &= z_1 - g(y + y_1) + h(x + x_1), \end{aligned}$$

where g, h, and k are constants. Show that the line segments joining corresponding points of S and S' are equal and parallel to those for S_1 and S_1'; that the lines joining corresponding points on S and S_1 meet the similar lines for S' and S_1'; and that the common conjugate system on S and S_1 corresponds to the common conjugate system on S' and S_1'.

14. Apply the results of Ex. 13 to the surfaces of translation

$$x = u^2 - v^2 + 2\,av, \quad y = 2\,u^2 + v^2 - 2\,av - 2\int \sqrt{b^2 + 3\,u^2}\,du, \quad z = 2\,bu,$$

$$x_1 = u^2 + 2\,v^2 - 2\,av - 2\int \sqrt{b^2 + 3\,u^2}\,du, \quad y_1 = -\,u^2 + v^2 + 2\int \sqrt{b^2 + 3\,u^2}\,du,$$

$$z_1 = 2\int \sqrt{a^2 - 3\,v^2}\,dv.$$

Show that when $g = h = 0$, $k = -1$, the surface S' is an elliptic paraboloid.

15. Show that the equations

$$x = \int \sqrt{1 - \frac{V'^2}{a^2}}\, dv, \quad y = \int \sqrt{1 - a^2 U'^2}\, du, \quad z = aU + \frac{V}{a},$$

where the accent indicates differentiation with respect to the argument, define a family of applicable surfaces of translation. Apply the results of Ex. 12 to this case.

16. Show that when S and S_1 in Exs. 12 and 13 are surfaces of translation, and their generating curves correspond, the same is true of S' and S_1'.

17. If lines be drawn through points of a Bertrand curve parallel to the binormals of the conjugate curve, their locus is applicable to a surface of revolution.

18. If a real ruled surface is applicable to a surface of revolution, it is applicable to the right helicoid or to a hyperboloid of revolution of one sheet (cf. Ex. 9, § 144).

19. A ruled surface can be deformed in an infinity of ways so that a curve not orthogonal to the generators shall be a line of curvature on the new ruled surface, unless the given curve is a geodesic; in the latter case the deformation is unique and the line of curvature is plane.

20. Let P be any point of a twisted curve C, and M_1, M_2 points on the principal normal to C such that

$$PM_1 = -PM_2 = a \sin\left(\int \frac{ds}{\rho} + b\right),$$

where a, b are constants and ρ is the radius of curvature of C. The loci of the lines through M_1 and M_2 parallel to the tangent to C at P are applicable ruled surfaces.

21. On the surface whose equations are

$$x = u, \quad y = f(u)\phi'(v) + \psi'(v), \quad z = f(u)[\phi(v) - v\phi'(v)] + \psi(v) - v\psi'(v),$$

the parametric curves form a conjugate system, the curves $u = \text{const.}$ lie in planes parallel to the yz-plane, and the curves $v = \text{const.}$ in planes parallel to the x-axis; hence the tangents to the curves $u = \text{const.}$ at their points of intersection with a curve $v = \text{const.}$ are parallel.

22. Investigate the character of the surfaces of Ex. 21 in the following cases: (a), $\phi(v) = \sqrt{v^2+1}$; (b), $\phi(v) = \text{const.}$; (c), $\psi(v) = 0$; (d), $f(u) = au + b$.

23. If the equations of Ex. 21 be written

$$x = u, \quad y = f(u)\phi_1(v) + \psi_1(v), \quad z = f(u)\phi_2(v) + \psi_2(v),$$

the most general applicable surfaces of the same kind with parametric curves corresponding are defined by

$$x_1 = \int \sqrt{1 + \kappa f'^2(u)}\, du, \quad y_1 = f(u)\Phi_1(v) + \Psi_1(v), \quad z_1 = f(u)\Phi_2(v) + \Psi_2(v),$$

where κ is a parameter, and the functions Φ_1, Φ_2, Ψ_1, Ψ_2 satisfy the conditions

$$\Phi_1^2 + \Phi_2^2 = \phi_1^2 + \phi_2^2 - \kappa, \qquad \Phi_1'^2 + \Phi_2'^2 = \phi_1'^2 + \phi_2'^2,$$

$$\Psi_2' = \frac{\Phi_1'(\phi_2\psi_2' + \phi_1\psi_1') - \Phi_1(\phi_2'\psi_2' + \phi_1'\psi_1')}{\Phi_2\Phi_1' - \Phi_2'\Phi_1},$$

$$\Psi_1' = \frac{\Phi_2(\phi_2'\psi_2' + \phi_1'\psi_1') - \Phi_2'(\phi_2\psi_2' + \phi_1\psi_1')}{\Phi_2\Phi_1' - \Phi_2'\Phi_1}.$$

Show also that the determination of Φ_1 and Φ_2 requires only a quadrature.

CHAPTER X

DEFORMATION OF SURFACES. THE METHOD OF WEINGARTEN

145. Reduced form of the linear element. Weingarten has remarked that when we reduce the determination of all surfaces applicable to a given one to the solution of the equation (IX, 18), namely

$$(1) \qquad \Delta_{22}\theta = (1 - \Delta_1\theta)\,K,$$

we make no use of our knowledge of the given surface, and in reality are trying to solve the problem of finding all the surfaces with an assigned linear element. In his celebrated memoir, *Sur la déformation des surfaces*,* which was awarded the grand prize of the French Academy in 1894, Weingarten showed that by taking account of the given surface the above equation can be replaced by another which can be solved in several important cases. This chapter is devoted to the exposition of this method. We begin by determining a particular moving trihedral for the given surface.

It follows from (VII, 64) that the necessary and sufficient condition that the directrix of a ruled surface be the line of striction is

$$(2) \qquad b = x_0'l' + y_0'm' + z_0'n' = 0.$$

The functions l', m', n' are proportional to the direction-cosines of the curve in which the director-cone of the surface meets the unit sphere with center at the vertex of the cone. We call this curve the *spherical indicatrix* of the surface. From (2) and the identity

$$ll' + mm' + nn' = 0$$

it is seen that the tangent to the spherical indicatrix is perpendicular to the tangent plane to the surface at the corresponding point of the line of striction. This fact is going to enable us to determine under what conditions a ruled surface Σ, tangent to a curved surface S along a curve C, admits the latter for its line of striction.

* *Acta Mathematica*, Vol. XX (1896), pp. 159–200.

We suppose that the parameters u, v are any whatever, and that the surface is referred to a moving trihedral. We consider the ruled surface formed by the x-axis of the trihedral as the origin of the latter describes the curve C. The point (1, 0, 0) of a second trihedral parallel to this one, but with origin fixed, describes the spherical indicatrix of Σ. From equations (V, 51) we find that the components of a displacement of this point are

$$0, \qquad r\,du + r_1 dv, \qquad -(q\,du + q_1 dv).$$

In order that the displacement be perpendicular to the tangent plane to Σ at the corresponding point of C, that is, perpendicular to the xy-plane of the moving trihedral, we must have

$$r\,du + r_1 dv = 0. \tag{3}$$

Hence if a trihedral T be associated with a surface S in any manner, as the vertex of T describes an integral curve of equation (3), the x-axis of T generates a ruled surface whose line of striction is this curve.

When the parametric lines on S are given, and also the angle U which the x-axis of T makes with the tangent to the curve $v=$ const., the functions r and r_1 are completely determined, as follows from (V, 52, 55). They are

$$r = \frac{H}{E}\begin{Bmatrix}11\\2\end{Bmatrix} - \frac{\partial U}{\partial u}, \qquad r_1 = \frac{H}{E}\begin{Bmatrix}12\\2\end{Bmatrix} - \frac{\partial U}{\partial v}. \tag{4}$$

Hence if U be given the value

$$U = \int \frac{H}{E}\begin{Bmatrix}12\\2\end{Bmatrix} dv + \phi(u), \tag{5}$$

where $\phi(u)$ denotes an arbitrary function of u, the function r_1 is zero, and as the vertex of the trihedral describes a curve $u=$ const., the x-axis describes a ruled surface whose line of striction is this curve.

Suppose now that the trihedral is such that $r_1=0$. From (V, 48, 64) it follows that

$$\frac{\partial r}{\partial v} = HK; \tag{6}$$

consequently

$$r = \int HK\,dv + \psi(u), \tag{7}$$

where ψ is an arbitrary function of u.

Let the right-hand member of (7) be denoted by $f(u, v)$, and change the parameters of the surface in accordance with the equations

$$u_1 = u, \quad v_1 = f(u, v).$$

From § 32 and equation (7) it follows that

$$H = H_1 \frac{\partial f}{\partial v} = HH_1K.$$

Since K is unaltered by the transformation, in terms of the new coördinates H_1K is equal to unity, and hence from (6) we have $r = v_1$. Therefore the coördinate curves and the moving trihedral of a surface can be chosen in such a way that

$$r_1 = 0, \qquad r = v, \qquad HK = 1. \tag{8}$$

In this case we say that the linear element of the surface is in its *reduced form*. It should be remarked that for surfaces of negative curvature the parameters are imaginary.

146. General formulas. If X_1, Y_1, Z_1; X_2, Y_2, Z_2; X, Y, Z denote the direction-cosines of the axes of the moving trihedral with respect to fixed axes, we have, from (V, 47),

$$\left\{\begin{aligned} &\frac{\partial X_1}{\partial u} = X_2 v - Xq, && \frac{\partial X_2}{\partial u} = Xp - X_1 v, && \frac{\partial X}{\partial u} = X_1 q - X_2 p, \\ &\frac{\partial X_1}{\partial v} = -Xq_1, && \frac{\partial X_2}{\partial v} = Xp_1, && \frac{\partial X}{\partial v} = X_1 q_1 - X_2 p_1. \end{aligned}\right. \tag{9}$$

The rotations p, p_1, q, q_1 satisfy equations (V, 48) in the reduced form

$$\frac{\partial p}{\partial v} - \frac{\partial p_1}{\partial u} = -vq_1, \quad \frac{\partial q}{\partial v} - \frac{\partial q_1}{\partial u} = vp_1, \quad pq_1 - p_1 q = 1. \tag{10}$$

The coördinates x, y, z of S with reference to these fixed axes are given by

$$\left\{\begin{aligned} x &= \int (\xi X_1 + \eta X_2)\, du + (\xi_1 X_1 + \eta_1 X_2)\, dv, \\ y &= \int (\xi Y_1 + \eta Y_2)\, du + (\xi_1 Y_1 + \eta_1 Y_2)\, dv, \\ z &= \int (\xi Z_1 + \eta Z_2)\, du + (\xi_1 Z_1 + \eta_1 Z_2)\, dv, \end{aligned}\right. \tag{11}$$

where

$$\xi^2 + \eta^2 = E, \qquad \xi\xi_1 + \eta\eta_1 = F, \qquad \xi_1^2 + \eta_1^2 = G, \tag{12}$$

and

$$\left\{\begin{aligned} &\frac{\partial \xi}{\partial v} - \frac{\partial \xi_1}{\partial u} = -v\eta_1, \quad \frac{\partial \eta}{\partial v} - \frac{\partial \eta_1}{\partial u} = v\xi_1, \\ &p\eta_1 - \eta p_1 + \xi q_1 - q\xi_1 = 0. \end{aligned}\right. \tag{13}$$

Weingarten's method consists in replacing the coefficients of ξ, η, ξ_1, η_1 in the last of equations (13) by differential parameters of u formed with respect to the linear element of the spherical representation of the x-axis of the moving trihedral.

By means of (9) this linear element is reducible to

$$(14)\qquad d\sigma^2 = dX_1^2 + dY_1^2 + dZ_1^2 = (v^2 + q^2)\,du^2 + 2\,qq_1du\,dv + q_1^2dv^2.$$

The differential parameters of u, formed with respect to this form, have the values *

$$(15)\qquad \begin{cases} \Delta_1 u = \dfrac{1}{v^2}, & \Delta_1\Delta_1 u = \dfrac{4\,(v^2 + q^2)}{v^8 q_1^2}, \\ \Delta_1(u,\, \Delta_1 u) = \dfrac{2\,q}{v^5 q_1}, & \Delta_2 u = \dfrac{q}{v^3 q_1} - \dfrac{p_1}{v q_1}. \end{cases}$$

Because of the identity (V, 38)

$$\Delta_{22}\theta = \frac{2\,\Delta_2\theta\cdot\Delta_1(\theta,\,\Delta_1\theta) - \Delta_1\Delta_1\theta}{4\,\Delta_1\theta},$$

we have also

$$(16)\qquad \Delta_{22} u = -\frac{p}{v^4 q_1}.$$

If the last of equations (13) be divided by q_1, and the values of p/q_1, q/q_1, p_1/q_1 obtained from (15) and (16) be substituted, we have

$$(17)\qquad \eta_1\Delta_{22}u - \frac{\eta}{v^3}\Delta_2 u - \frac{\xi}{v^4} + \frac{\eta + \xi_1 v^2}{2\,v}\Delta_1(u,\,\Delta_1 u) = 0.$$

In consequence of the first of equations (15), written

$$(18)\qquad v = \frac{1}{\sqrt{\Delta_1 u}},$$

the coefficients of ξ, η, ξ_1, η_1 in (17) are expressible in terms of differential parameters of u formed with respect to (14), as was to be proved.

An exceptional case is that in which $q_1 = 0$. Under this condition the spherical representation of the x-axis reduces to a curve, as is seen from (14).

* Previously we have indicated by a prime differential parameters formed with respect to the linear element of the spherical representation. For the sake of simplicity we disregard this practice in this chapter.

By means of (9) we find that

$$\Delta_1(X_1, u) = \frac{X_2}{v}, \quad \Delta_1(Y_1, u) = \frac{Y_2}{v}, \quad \Delta_1(Z_1, u) = \frac{Z_2}{v}, \tag{19}$$

and consequently equations (11) may be written

$$x = \int [\xi X_1 + \eta v \Delta_1(X_1, u)]\,du + [\xi_1 X_1 + \eta_1 v \Delta_1(X_1, u)]\,dv, \tag{20}$$

and similarly for y and z.

147. The theorem of Weingarten. Equation (17) is the equation which Weingarten has suggested as a substitute for equation (1). We notice that ξ, η, ξ_1, η_1 are known functions of u and v when the surface S is given. By means of (18) equation (17) can be given a form which involves only u and differential parameters of u formed with respect to (14). On account of the invariant character of these differential parameters this linear element may be expressed in terms of any parameters, say u' and v'. We shall show that each solution of equation (17) determines a surface applicable to S. We formulate the theorem of Weingarten as follows:

Let S be a surface whose linear element in the reduced form is

$$ds^2 = (\xi^2 + \eta^2)\,du^2 + 2(\xi\xi_1 + \eta\eta_1)\,du\,dv + (\xi_1^2 + \eta_1^2)\,dv^2; \tag{21}$$

then ξ, η, ξ_1, η_1 are functions of u and v such that

$$\frac{\partial \xi}{\partial v} - \frac{\partial \xi_1}{\partial u} = -v\eta_1, \qquad \frac{\partial \eta}{\partial v} - \frac{\partial \eta_1}{\partial u} = v\xi_1. \tag{22}$$

Furthermore, let X_1, Y_1, Z_1 be the coördinates of a point on the unit sphere, expressed in terms of any two parameters u' and v', the linear element of the sphere being

$$d\sigma'^2 = \mathcal{E}'\,du'^2 + 2\mathcal{F}'\,du'\,dv' + \mathcal{G}'\,dv'^2. \tag{23}$$

Any integral u_1 of the equation

$$\begin{aligned}\eta_1\left(u, \frac{1}{\sqrt{\Delta_1 u}}\right)\Delta_{22}u - \eta\left(u, \frac{1}{\sqrt{\Delta_1 u}}\right)(\Delta_1 u)^{\frac{3}{2}}\Delta_2 u - \xi\left(u, \frac{1}{\sqrt{\Delta_1 u}}\right)(\Delta_1 u)^2 \\ + \frac{1}{2}\left[\eta\left(u, \frac{1}{\sqrt{\Delta_1 u}}\right)(\Delta_1 u)^{\frac{1}{2}} + \xi_1\left(u, \frac{1}{\sqrt{\Delta_1 u}}\right)(\Delta_1 u)^{-\frac{1}{2}}\right]\Delta_1(u, \Delta_1 u) = 0,\end{aligned} \tag{24}$$

the differential parameters being formed with respect to (23), *renders the following expression and similar ones in y and z total differentials:*

$$(25)\qquad dx=\left[\xi(u_1, v_1)X_1(u', v')+\eta(u_1, v_1)\frac{\Delta_1(X_1, u_1)}{\sqrt{\Delta_1 u_1}}\right]du_1$$

$$+\left[\xi_1(u_1, v_1)X_1(u', v')+\eta_1(u_1, v_1)\frac{\Delta_1(X_1, u_1)}{\sqrt{\Delta_1 u_1}}\right]dv_1,$$

where

$$v_1=\frac{1}{\sqrt{\Delta_1 u_1}};$$

and the surface whose coördinates are the functions x, y, z thus defined has the linear element (21).

Before proving this theorem we remark that the parameters u' and v' may be chosen either as known functions of u and v, or in such a way that the linear element (14) shall have a particular form. In the former case X_1, Y_1, Z_1 are known as functions of u' and v', and in the second their determination requires the solution of a Riccati equation. However, in what follows we assume that X_1, Y_1, Z_1 are known.

Suppose now that u' and v' are any parameters whatever, and that we have a solution u_1 of equation (24), where the differential parameters are formed with respect to (23). Let v_1 denote the quantity $(\Delta_1 u_1)^{-\frac{1}{2}}$. Both u_1 and v_1 are functions of u' and v', and consequently the latter are expressible as functions of the former. We express X_1, Y_1, Z_1 as functions of u_1 and v_1 and determine the corresponding linear element of the unit sphere, which we write

$$(26)\qquad d\sigma_1^2=\mathcal{E}_1\,du_1^2+2\,\mathcal{F}_1\,du_1dv_1+\mathcal{G}_1\,dv_1^2.$$

In terms of u_1 and v_1 we have

$$\frac{1}{v_1^2}=\Delta_1 u_1=\frac{\mathcal{G}_1}{\mathcal{H}_1^2},\qquad \Delta_1(X_1, u_1)=\frac{\mathcal{G}_1\dfrac{\partial X_1}{\partial u_1}-\mathcal{F}_1\dfrac{\partial X_1}{\partial v_1}}{\mathcal{H}_1^2}.$$

From these expressions it follows that if we put

$$X_2=v_1\Delta_1(X_1, u_1),\qquad Y_2=v_1\Delta_1(Y_1, u_1),\qquad Z_2=v_1\Delta_1(Z_1, u_1),$$

we have

$$(27)\qquad \begin{cases}\sum X_2^2=1, & \sum X_1X_2=0,\\[4pt] \sum X_2\dfrac{\partial X_1}{\partial u_1}=v_1, & \sum X_2\dfrac{\partial X_1}{\partial v_1}=0.\end{cases}$$

Hence if we put

$$X = Y_1 Z_2 - Z_1 Y_2, \qquad Y = Z_1 X_2 - X_1 Z_2, \qquad Z = X_1 Y_2 - Y_1 X_2,$$

the functions $X_1, Y_1, \cdots, Z$ satisfy a set of equations similar to equations (V, 47).

In consequence of (27) the corresponding rotations have the values

$$\begin{aligned} \bar{p} &= -\sum X_2 \frac{\partial X}{\partial u_1}, & \bar{q} &= \sum X_1 \frac{\partial X}{\partial u_1}, \\ \bar{p}_1 &= -\sum X_2 \frac{\partial X}{\partial v_1}, & \bar{q}_1 &= \sum X_1 \frac{\partial X}{\partial v_1}, \\ \bar{r} &= v_1, & \bar{r}_1 &= 0. \end{aligned}$$

It is readily shown that these functions satisfy equations similar to (10).

Since the functions ξ, η, ξ_1, η_1 are of the same form in (25) as in (21), equations similar to the first two of equations (13) are necessarily satisfied. Hence the only other equation to be satisfied, in order that the expressions (25) be exact differentials, is

$$\bar{p}\eta_1(u_1, v_1) - \bar{p}_1 \eta(u_1, v_1) + \bar{q}_1 \xi(u_1, v_1) - \bar{q}\xi_1(u_1, v_1) = 0. \tag{28}$$

But it can be shown that the coefficients of (26) are expressible in the form
$$\mathcal{E}_1 = v_1^2 + \bar{q}^2, \qquad \mathcal{F}_1 = \bar{q}\bar{q}_1, \qquad \mathcal{G}_1 = \bar{q}_1^2,$$
so that by means of differential parameters of u_1 formed with respect to (26) the equation (28) can be given the form (17). Hence all the conditions are satisfied, and the theorem of Weingarten has been established.

148. Other forms of the theorem of Weingarten. It is readily found that equations (22) are satisfied by the expressions

$$\begin{cases} \xi = \dfrac{\partial^2 \phi}{\partial u^2} + v^3 \dfrac{\partial \phi}{\partial v}, & \xi_1 = \dfrac{\partial^2 \phi}{\partial u\, \partial v}, \\[2ex] \eta = -v^2 \dfrac{\partial^2 \phi}{\partial u\, \partial v}, & \eta_1 = -3v \dfrac{\partial \phi}{\partial v} - v^2 \dfrac{\partial^2 \phi}{\partial v^2}, \end{cases} \tag{29}$$

where ϕ is any function of u and v. Since now

$$\eta + v^2 \xi_1 = 0, \tag{30}$$

equation (17) reduces to

$$\left(3v \frac{\partial \phi}{\partial v} + v^2 \frac{\partial^2 \phi}{\partial v^2}\right) \Delta_{22} u - \frac{1}{v} \frac{\partial^2 \phi}{\partial u\, \partial v} \Delta_2 u + \frac{1}{v^4}\left(\frac{\partial^2 \phi}{\partial u^2} + v^3 \frac{\partial \phi}{\partial v}\right) = 0. \tag{31}$$

This equation will be simplified still more by the introduction of two new parameters which are suggested by the following considerations.

As previously defined, the functions X_1, Y_1, Z_1 are the direction-cosines of lines tangent to the given surface S in such a way that the ruled surface formed by these tangents at points of a curve $u = \text{const.}$ has this curve for its line of striction. Moreover, from the theorem of Weingarten it follows that the functions X_1, Y_1, Z_1 have the same significance for the surface applicable to S which corresponds to a particular solution of equation (17).

But X_1, Y_1, Z_1 may be taken also as the direction-cosines of the normals to a large group of surfaces, as shown in § 67. In particular, we consider the surface Σ which is the envelope of the plane

$$X_1 x + Y_1 y + Z_1 z = u.$$

Each solution of equation (17) determines such a surface. If $\bar{x}, \bar{y}, \bar{z}$ denote the coördinates of the point of contact of this plane with Σ, we have from (V, 32)

$$\bar{x} = uX_1 + \Delta_1(u, X_1), \tag{32}$$

which, in consequence of (19), may be written

$$\bar{x} = uX_1 + \frac{1}{v} X_2. \tag{32}$$

Hence the point of contact of Σ lies in the plane through the origin parallel to the tangent plane to S at the corresponding point.

If the square of the distance of the point of contact from the origin be denoted by $2q$, and the distance from the origin to the tangent plane by p,* we have

$$2q = \bar{x}^2 + \bar{y}^2 + \bar{z}^2 = u^2 + \frac{1}{v^2}, \qquad p = u. \tag{33}$$

From (V, 35, 37) it follows that the principal radii of Σ are given by

$$\begin{cases} \rho_1 + \rho_2 = -(\Delta_2 u + 2u), \\ \rho_1 \rho_2 = \Delta_{22} u + u\Delta_2 u + u^2, \end{cases} \tag{34}$$

* The reader will observe that the functions p and q thus defined are different from the rotations designated by the same letters. As this notation is generally employed in the treatment of the theorem of Weingarten, it has seemed best to retain it, even at the risk of a confusion of notation.

where the differential parameters are formed with respect to (14). From these equations we have

$$(35)\qquad \begin{cases} \Delta_2 u = -(\rho_1 + \rho_2) - 2u, \\ \Delta_{22} u = \rho_1\rho_2 + (\rho_1 + \rho_2)u + u^2. \end{cases}$$

We shall now effect a change of parameters, using p and q defined by (33) as the new ones. By direct calculation we obtain

$$(36)\qquad \begin{cases} \dfrac{\partial\phi}{\partial u} = \dfrac{\partial\phi}{\partial p} + \dfrac{\partial\phi}{\partial q}p, \qquad \dfrac{\partial\phi}{\partial v} = -\dfrac{1}{v^3}\dfrac{\partial\phi}{\partial q}, \\ \dfrac{\partial^2\phi}{\partial u^2} = \dfrac{\partial^2\phi}{\partial p^2} + 2p\dfrac{\partial^2\phi}{\partial p\,\partial q} + p^2\dfrac{\partial^2\phi}{\partial q^2} + \dfrac{\partial\phi}{\partial q}, \\ \dfrac{\partial^2\phi}{\partial u\,\partial v} = -\dfrac{1}{v^3}\left(\dfrac{\partial^2\phi}{\partial p\,\partial q} + \dfrac{\partial^2\phi}{\partial q^2}p\right), \\ \dfrac{\partial^2\phi}{\partial v^2} = \dfrac{1}{v^6}\dfrac{\partial^2\phi}{\partial q^2} + \dfrac{3}{v^4}\dfrac{\partial\phi}{\partial q}. \end{cases}$$

By means of the equations (33) and (36) the fundamental equation (31) can be reduced to

$$(37)\qquad \frac{\partial^2\phi}{\partial q^2}\rho_1\rho_2 - \frac{\partial^2\phi}{\partial p\,\partial q}(\rho_1 + \rho_2) + \frac{\partial^2\phi}{\partial p^2} = 0.$$

This is the form in which the fundamental equation was first considered by Weingarten.* The method of §§ 146, 147 was a subsequent development.

In terms of the parameters p and q the formulas (29) become

$$(38)\qquad \begin{cases} \xi = \dfrac{\partial^2\phi}{\partial p^2} + 2p\dfrac{\partial^2\phi}{\partial p\,\partial q} + p^2\dfrac{\partial^2\phi}{\partial q^2}, \qquad \eta = \sqrt{2q - p^2}\left(\dfrac{\partial^2\phi}{\partial p\,\partial q} + p\dfrac{\partial^2\phi}{\partial q^2}\right), \\ \xi_1 = -(2q - p^2)^{\frac{3}{2}}\left(\dfrac{\partial^2\phi}{\partial p\,\partial q} + p\dfrac{\partial^2\phi}{\partial q^2}\right), \; \eta_1 = -(2q - p^2)^2\dfrac{\partial^2\phi}{\partial q^2}. \end{cases}$$

If these values and the expression for $\Delta_1(u, X_1)$ given by (32) be substituted in (20), it is reducible to

$$x = \int\left(X_1\frac{\partial^2\phi}{\partial p^2} + \bar{x}\frac{\partial^2\phi}{\partial p\,\partial q}\right)dp + \left(X_1\frac{\partial^2\phi}{\partial p\,\partial q} + \bar{x}\frac{\partial^2\phi}{\partial q^2}\right)dq.$$

* *Comptes Rendus*, Vol. CXII (1891), p. 607.

Hence the equations for S may be written

$$(39)\qquad \begin{cases} dx = X_1 d\left(\dfrac{\partial\phi}{\partial p}\right) + \bar{x} d\left(\dfrac{\partial\phi}{\partial q}\right), \\ dy = Y_1 d\left(\dfrac{\partial\phi}{\partial p}\right) + \bar{y} d\left(\dfrac{\partial\phi}{\partial q}\right), \\ dz = Z_1 d\left(\dfrac{\partial\phi}{\partial p}\right) + \bar{z} d\left(\dfrac{\partial\phi}{\partial q}\right), \end{cases}$$

and consequently the linear element of S is of the form

$$(40)\qquad ds^2 = \left[d\left(\frac{\partial\phi}{\partial p}\right)\right]^2 + 2\,p\,d\left(\frac{\partial\phi}{\partial p}\right)d\left(\frac{\partial\phi}{\partial q}\right) + 2\,q\left[d\left(\frac{\partial\phi}{\partial q}\right)\right]^2.$$

Since these various expressions and equations differ only in form from those which figure in the theorem of Weingarten, the latter is just as true for these new equations. We remark also that the right-hand member of (40) depends only upon the form of ϕ. Hence we have the theorem of Weingarten in the form:

When ϕ in equation (37) *is a definite function of p and q, this equation defines a large group of surfaces with the same spherical representation, the functions ρ_1 and ρ_2 denoting the principal radii, and p and $2\,q$ the distance from the origin to the tangent plane and the square of the distance to the point of contact. Each surface Σ satisfying this condition gives by quadratures* (39) *a surface with the linear element* (40). *Conversely, each surface with this linear element stands in such relation to some surface satisfying the corresponding equation* (37).

As a corollary to the preceding results, we have the theorem:

The linear element of any surface S is reducible to the form

$$(41)\qquad ds^2 = du^2 + 2\frac{\partial\psi}{\partial u}\,du\,dv + 2\frac{\partial\psi}{\partial v}\,dv^2,$$

where ψ is a function of u and v.

For, we have seen that the linear element of any surface is reducible to the form (40). If, then, we change the parameters by means of the equations

$$(42)\qquad u = \frac{\partial\phi}{\partial p}, \qquad v = \frac{\partial\phi}{\partial q},$$

we have

$$(43)\qquad ds^2 = du^2 + 2\,p\,du\,dv + 2\,q\,dv^2.$$

From (42) it follows that

$$du = \frac{\partial^2\phi}{\partial p^2}dp + \frac{\partial^2\phi}{\partial p\,\partial q}dq, \qquad dv = \frac{\partial^2\phi}{\partial p\,\partial q}dp + \frac{\partial^2\phi}{\partial q^2}dq.$$

and consequently

$$(44) \qquad \begin{cases} \dfrac{\partial p}{\partial u} = \dfrac{\partial^2\phi}{\partial q^2}\Delta, & \dfrac{\partial p}{\partial v} = -\dfrac{\partial^2\phi}{\partial p\,\partial q}\Delta, \\[2ex] \dfrac{\partial q}{\partial u} = -\dfrac{\partial^2\phi}{\partial p\,\partial q}\Delta, & \dfrac{\partial q}{\partial v} = \dfrac{\partial^2\phi}{\partial p^2}\Delta, \end{cases}$$

where

$$\frac{1}{\Delta} = \frac{\partial^2\phi}{\partial p^2}\frac{\partial^2\phi}{\partial q^2} - \left(\frac{\partial^2\phi}{\partial p\,\partial q}\right)^2.$$

From (44) it is seen that $\dfrac{\partial p}{\partial v} = \dfrac{\partial q}{\partial u}$, and consequently the inverse of equations (42) are of the form

$$(45) \qquad p = \frac{\partial\psi}{\partial u}, \qquad q = \frac{\partial\psi}{\partial v}.$$

Hence equation (43) is of the form (41), as was to be proved. Moreover, equations (44) reduce to

$$(46) \qquad \frac{\partial^2\psi}{\partial u^2} = \frac{\partial^2\phi}{\partial q^2}\Delta, \qquad \frac{\partial^2\psi}{\partial u\,\partial v} = -\frac{\partial^2\phi}{\partial p\,\partial q}\Delta, \qquad \frac{\partial^2\psi}{\partial v^2} = \frac{\partial^2\phi}{\partial p^2}\Delta.$$

In terms of these parameters u, v equations (39) reduce to

$$(47) \quad dx = X_1du + \bar{x}dv, \quad dy = Y_1du + \bar{y}dv, \quad dz = Z_1du + \bar{z}dv.$$

Hence the coördinates of Σ are given by

$$(48) \qquad \bar{x} = \frac{\partial x}{\partial v}, \qquad \bar{y} = \frac{\partial y}{\partial v}, \qquad \bar{z} = \frac{\partial z}{\partial v},$$

and the direction-cosines of the normal to Σ are

$$(49) \qquad X_1 = \frac{\partial x}{\partial u}, \qquad Y_1 = \frac{\partial y}{\partial u}, \qquad Z_1 = \frac{\partial z}{\partial u},$$

that is, the normals to Σ are parallel to the corresponding tangents to the curves $v = \text{const.}$ on S. Hence we have the following theorem:

When the linear element of a surface is in the form (41), *the surface Σ whose coördinates are given by* (48) *has the same spherical* *representation of its normals as the tangents to the curves $v = \text{const.}$ on S. If p and $2q$ denote the distance from the origin to the tangent plane to Σ and the square of the distance to the point of contact, they have the values* (45). *Moreover, if the change of parameters defined by these equations be expressed in the inverse form*

$$(50) \qquad u = \frac{\partial \phi}{\partial p}, \qquad v = \frac{\partial \phi}{\partial q},$$

the principal radii of Σ satisfy the condition

$$(51) \qquad \rho_1 \rho_2 \frac{\partial^2 \phi}{\partial q^2} - (\rho_1 + \rho_2) \frac{\partial^2 \phi}{\partial p \, \partial q} + \frac{\partial^2 \phi}{\partial p^2} = 0,$$

and the coördinates of S are given by quadratures of the form

$$(52) \qquad dx = X_1 d\left(\frac{\partial \phi}{\partial p}\right) + \bar{x} d\left(\frac{\partial \phi}{\partial q}\right).$$

Moreover, every surface with the same representation as Σ, and whose functions ρ_1, ρ_2, p, q satisfy (51) *for the same ϕ, determines by equations of the form* (52) *a surface applicable to S.**

149. Surfaces applicable to a surface of revolution. When the linear element of a surface applicable to a surface of revolution is written

$$(53) \qquad ds^2 = du_1^2 + \rho^2(u_1)\, dv_1^2,$$

and the x-axis of the moving trihedral is tangent to the curve $v = \text{const.}$, the function r is equal to zero, as follows from (4).

In order to obtain the conditions (8), we effect the transformation of variables

$$\bar{u} = v_1, \qquad \bar{v} = -u_1,$$

so that the linear element becomes

$$(54) \qquad ds^2 = \rho^2\, d\bar{u}^2 + d\bar{v}^2.$$

Now $r = \rho'$, $r_1 = 0$, and consequently in order to have the linear element in the reduced form we must take

$$(55) \qquad u = \bar{u}, \qquad v = \rho'(-\bar{v}).$$

* For a direct proof of this theorem the reader is referred to a memoir by Goursat, Sur un théorème de M. Weingarten, et sur la théorie des surfaces applicables, *Toulouse Annales*, Vol. V (1891); also Darboux, Vol. IV, p. 316, and Bianchi, Vol. II, p. 198.

From these results and (32′) we find that the coördinates of the surface Σ are given by

$$\bar{x} = -\frac{1}{\rho'}\frac{\partial x}{\partial u_1} + \frac{v_1}{\rho}\frac{\partial x}{\partial v_1}, \qquad \bar{y} = -\frac{1}{\rho'}\frac{\partial y}{\partial u_1} + \frac{v_1}{\rho}\frac{\partial y}{\partial v_1},$$
$$\bar{z} = -\frac{1}{\rho'}\frac{\partial z}{\partial u_1} + \frac{v_1}{\rho}\frac{\partial z}{\partial v_1},$$

and the direction-cosines of the normals to Σ are

$$X_1 = \frac{1}{\rho}\frac{\partial x}{\partial v_1}, \qquad Y_1 = \frac{1}{\rho}\frac{\partial y}{\partial v_1}, \qquad Z_1 = \frac{1}{\rho}\frac{\partial z}{\partial v_1}.$$

Also, we have

(56) $$p = \sum \bar{x} X_1 = v_1, \qquad 2q = \sum \bar{x}^2 = v_1^2 + \frac{1}{\rho'^2}.$$

Hence we have the theorem:

To a curve which is the deform of a meridian of a surface of revolution there corresponds on the surface Σ a curve such that the tangent planes to Σ at points of the curve are at a constant distance from the origin, and to a deform of a parallel there corresponds a curve such that the projection of the radius vector upon the tangent plane at a point is constant.

For the present case $\eta = \xi_1 = 0$; consequently we have, from (38),

$$\frac{\partial^2 \phi}{\partial p\, \partial q} + p\frac{\partial^2 \phi}{\partial q^2} = 0.$$

This equation is satisfied by

(57) $$\phi(p, q) = f(2q - p^2),$$

where f is any function whatever. In terms of this function we have, from (38),

(58) $$\xi = -2f', \qquad \eta_1 = -4(2q - p^2)^2 f'',$$

where the accents indicate differentiation with respect to the argument, $2q - p^2$.

By means of (55) the linear element (54) can be transformed into

$$ds^2 = \omega^2(v)\, du^2 + \frac{\omega^{2\prime}(v)}{v^2}\, dv^2,$$

the function $\omega(v)$ being defined by

$$\omega(v) = \rho(-\bar{v}).$$

Since $\eta = \xi_1 = 0$, we have

$$\xi = \omega(v), \qquad \eta_1 = -\frac{\omega'(v)}{v},$$

and we know that $r = v$. Now equations (58) become

$$\omega(v) = -2f', \qquad \omega'(v) = 4v(2q - p^2)^2 f'',$$

and these are consistent because of the relation $2q - p^2 = 1/v^2$, which results from (56). Hence we have the theorem:

When $\phi(p, q)$ is a function of $2q - p^2$, the corresponding surface S is applicable to a surface of revolution, the tangents to the deforms of the parallels being parallel to the corresponding normals to Σ.

If we give ϕ the form (57) and put $\psi = 2f'$, the linear element of S is

$$ds^2 = (2q - p^2)\,d\psi^2 + \psi^2 dp^2, \tag{59}$$

as follows from (40) or (58).

150. Minimal lines on the sphere parametric. In § 147 we remarked that the parametric curves on the sphere may be any whatever. An interesting case is that in which they are the imaginary generatrices. In § 35 we saw that the parameters of these lines, say α and β, can be so chosen that

$$X_1 = \frac{\alpha + \beta}{1 + \alpha\beta}, \qquad Y_1 = i\frac{(\beta - \alpha)}{1 + \alpha\beta}, \qquad Z_1 = \frac{\alpha\beta - 1}{1 + \alpha\beta}. \tag{60}$$

Consequently

$$d\sigma^2 = dX_1^2 + dY_1^2 + dZ_1^2 = \frac{4\,d\alpha\,d\beta}{(1 + \alpha\beta)^2}. \tag{61}$$

From (32) we find that the coördinates of Σ, the envelope of the plane

$$X_1 x + Y_1 y + Z_1 z - p = 0,$$

are

$$\left\{\begin{aligned} \bar{x} &= pX_1 + \frac{1}{2}\left[(1 - \alpha^2)\frac{\partial p}{\partial \alpha} + (1 - \beta^2)\frac{\partial p}{\partial \beta}\right], \\ \bar{y} &= pY_1 + \frac{i}{2}\left[(1 + \alpha^2)\frac{\partial p}{\partial \alpha} - (1 + \beta^2)\frac{\partial p}{\partial \beta}\right], \\ \bar{z} &= pZ_1 + \alpha\frac{\partial p}{\partial \alpha} + \beta\frac{\partial p}{\partial \beta}. \end{aligned}\right. \tag{62}$$

From these we obtain

$$2q = p^2 + (1 + \alpha\beta)^2 \frac{\partial p}{\partial \alpha}\frac{\partial p}{\partial \beta}. \tag{63}$$

By means of (34) the expressions for $\rho_1+\rho_2$ and $\rho_1\rho_2$ in terms of p and its derivatives with respect to α and β can be readily found, and thus the fundamental equation (37) put in a new form. However, it is not with the general case that we shall now concern ourselves, but with a particular form of the function $\phi(p, q)$.

This function has been considered by Weingarten*; it is

$$\phi(p, q)=pq-\frac{p^3}{3}-\overline{\omega}(p). \tag{64}$$

In this case

$$\frac{\partial^2\phi}{\partial p^2}=-2p-\overline{\omega}''(p), \qquad \frac{\partial^2\phi}{\partial p\,\partial q}=1, \qquad \frac{\partial^2\phi}{\partial q^2}=0,$$

$$\Delta_2 p=(1+\alpha\beta)^2\frac{\partial^2 p}{\partial\alpha\,\partial\beta},$$

so that equation (37) reduces to

$$\rho_1+\rho_2=-(2p+\overline{\omega}''(p)), \tag{65}$$

which, in consequence of (34), may be written

$$\frac{\partial^2 p}{\partial\alpha\,\partial\beta}=\frac{\overline{\omega}''(p)}{(1+\alpha\beta)^2}. \tag{66}$$

When the values from (62) are substituted in (52), we obtain

$$\begin{cases} x=u_1X_1-\int u_1dX_1+\frac{1}{2}\left[(1-\alpha^2)\frac{\partial p}{\partial\alpha}+(1-\beta^2)\frac{\partial p}{\partial\beta}\right]dp, \\ y=u_1Y_1-\int u_1dY_1+\frac{i}{2}\left[(1+\alpha^2)\frac{\partial p}{\partial\alpha}-(1+\beta^2)\frac{\partial p}{\partial\beta}\right]dp, \\ z=u_1Z_1-\int u_1dZ_1+\left(\alpha\frac{\partial p}{\partial\alpha}+\beta\frac{\partial p}{\partial\beta}\right)dp, \end{cases} \tag{67}$$

where

$$u_1=q-\frac{p^2}{2}-\overline{\omega}'(p)=\frac{(1+\alpha\beta)^2}{2}\frac{\partial p}{\partial\alpha}\frac{\partial p}{\partial\beta}-\overline{\omega}'(p). \tag{68}$$

From (42) and (64) we have

$$u=q-p^2-\overline{\omega}'(p), \qquad v=p.$$

Hence the linear element (43) of S is, in this case,

$$ds^2=du^2+2\,v\,dudv+2[u+v^2+\overline{\omega}'(v)]\,dv^2. \tag{69}$$

* *Acta Mathematica*, Vol. XX (1896), p. 195.

However, from (68) it is seen that

$$u_1 = u + \frac{v^2}{2}, \tag{70}$$

so that (69) may be written

$$ds^2 = du_1^2 + 2\,[u_1 + \overline{\omega}'(v)]\,dv^2. \tag{71}$$

Gathering together these results, we have the theorem:

The determination of all the surfaces with the linear element (71) *reduces to the integration of the equation*

$$\frac{\partial^2 p}{\partial\alpha\,\partial\beta} = \frac{\overline{\omega}''(p)}{(1+\alpha\beta)^2}. \tag{72}$$

The integral of this equation for $\omega(p)$ arbitrary is not known. However, the integral is known in certain cases. We consider several of these.

151. Surfaces of Goursat. Surfaces applicable to certain paraboloids. When we take

$$\overline{\omega}'(p) = \tfrac{1}{2}\,m\,(1-m)\,p^2, \tag{73}$$

m being any constant, equation (72) becomes

$$\frac{\partial^2 p}{\partial\alpha\,\partial\beta} = \frac{m\,(1-m)\,p}{(1+\alpha\beta)^2}. \tag{74}$$

The general integral of this equation can be found by the method of Laplace,* in finite form or in terms of definite integrals, according as m is integral or not.

The linear element of the surface S is

$$ds^2 = du_1^2 + [2\,u_1 + m\,(1-m)\,v^2]\,dv^2. \tag{75}$$

And the surfaces Σ are such that

$$\rho_1 + \rho_2 + 2\,p = m\,(m-1)\,p, \tag{76}$$

that is, the sum of the principal radii is proportional to the distance of the tangent plane from a fixed point. These surfaces were first studied by Goursat,† and are called, consequently, *the surfaces of Goursat.*

* Darboux, Vol. II, p. 56.

† *American Journal*, Vol. X (1888), p. 187.

Darboux has remarked * that equation (71) is similar to the linear element of ruled surfaces (VII, 53). In fact, if the equations of a ruled surface are written in the form

$$x = x_0 + lu_1, \qquad y = y_0 + mu_1, \qquad z = z_0 + nu_1, \tag{77}$$

where $x_0, \cdots; l, m, n$ are functions of v alone, which now is not necessarily the arc of the directrix, the linear element of the surface will have the form (71), provided that

$$\Sigma l^2 = 1, \quad \Sigma x_0' l = 0, \quad \Sigma x_0'^2 = 2\,\overline{\omega}'(v), \quad \Sigma x_0' l' = 1, \quad \Sigma l'^2 = 0. \tag{78}$$

In consequence of the equations

$$\Sigma l l' = 0, \qquad \Sigma l'^2 = 0,$$

it follows that a ruled surface of this kind admits an isotropic plane director. If this plane be $x + iy = 0$, that is, if

$$l' : m' : n' = 1 : i : 0,$$

we have

$$l = V, \qquad m = iV, \qquad n = 1,$$

where V is a function of v. By means of these values and equations (78), we can put (77) in the form

$$\left\{\begin{aligned} x + iy &= \int \frac{dv}{V'}, \\ x - iy &= 2\,Vu_1 + 2\int V'\overline{\omega}' dv - \int \frac{V^2}{V'} dv, \\ z &= u_1 - \int \frac{V}{V'} dv. \end{aligned}\right. \tag{79}$$

We shall find that among these surfaces there is an imaginary paraboloid to which are applicable certain surfaces to which Weingarten called attention. To this end we consider the function

$$\overline{\omega}'(p) = -\sqrt{\kappa}\,p - 2\,\kappa e^{-\frac{2p}{\sqrt{\kappa}}}, \tag{80}$$

where κ denotes a constant. Now equation (66) becomes

$$\frac{\partial^2 p}{\partial\alpha\,\partial\beta} = \sqrt{\kappa}\,\frac{4\,e^{-\frac{2p}{\sqrt{\kappa}}} - 1}{(1 + \alpha\beta)^2}. \tag{81}$$

* Vol. IV, p. 333.

In consequence of the identity

$$\frac{\partial^2}{\partial\alpha\,\partial\beta}\log(1+\alpha\beta)^2=\frac{2}{(1+\alpha\beta)^2},$$

the preceding equation is equivalent to

$$\frac{\partial^2}{\partial\alpha\,\partial\beta}\log(1+\alpha\beta)^2 e^{\frac{2p}{\sqrt{\kappa}}}=\frac{8}{e^{\frac{2p}{\sqrt{\kappa}}}(1+\alpha\beta)^2}.$$

If we put

$$e^{\theta}=e^{\frac{p}{\sqrt{\kappa}}}(1+\alpha\beta),$$

this equation takes the Liouville form

$$\frac{\partial^2\theta}{\partial\alpha\,\partial\beta}=4\,e^{-2\theta},$$

of which the general integral is

$$e^{-2\theta}=\frac{1}{4}\frac{A'B'}{(1+AB)^2},$$

where A and B are functions of α and β respectively, and the accents indicate differentiation with respect to these. Hence the general integral of (81) is

$$e^{\frac{p}{\sqrt{\kappa}}}=\frac{2(1+AB)}{\sqrt{A'B'}(1+\alpha\beta)},$$

and the linear element of S is

$$ds^2=du_1^2+2\left(u_1-v\sqrt{\kappa}-2\,\kappa e^{-\frac{2v}{\sqrt{\kappa}}}\right)dv^2. \tag{82}$$

If now, in addition to (80), we take

$$V=\frac{1}{2}e^{\frac{v}{\sqrt{\kappa}}},\qquad \int\frac{V}{V'}\,dv=v\sqrt{\kappa}-\frac{\kappa}{2},$$

the equations (79) take such a form that

$$(x+iy)\,x=-\kappa z. \tag{83}$$

Hence the surfaces with the linear element (82) are applicable to the imaginary paraboloid (83). The generator $x+iy=0$ of this paraboloid in the plane at infinity is tangent to the imaginary circle at the point $(x:y:z=1:i:0)$, which is a different point from that in which the plane at infinity touches the surface, that is, the point of intersection of the two generators.

Another interesting case is afforded when m in (73) has the value 2. Then $\overline{\omega}'(v) = -v^2$, and equation (71) becomes

$$ds^2 = du_1^2 + 2(u_1 - v^2)\, dv^2. \tag{84}$$

If we take $V = v/\sqrt{2\kappa}$, we obtain from equations (79)

$$x + iy = \sqrt{2\kappa}\, v, \qquad x - iy = \sqrt{\frac{2}{\kappa}}\, v\left(u_1 - \frac{v^2}{2}\right), \qquad z = u_1 - \frac{v^2}{2},$$

from which we find, by the elimination of u_1 and v,

$$(x + iy)z = \kappa(x - iy). \tag{85}$$

The generator $x + iy = 0$ in the plane at infinity on the paraboloid (85) is tangent to the imaginary circle at the point $(x:y:z = 1:i:0)$, just as in the case of the paraboloid (83), but the paraboloid (85) is tangent to the plane at infinity at the same point.

GENERAL EXAMPLES

1. A moving trihedral can be associated with a surface in an infinity of ways so that as the vertex of the trihedral describes a curve $u =$ const. the x-axis generates a ruled surface whose line of striction is this curve.

2. The tangents to the curves $v =$ const. on a surface at the points where these curves are met by an integral curve of the equation

$$\begin{Bmatrix} 11 \\ 2 \end{Bmatrix} du + \begin{Bmatrix} 12 \\ 2 \end{Bmatrix} dv = 0$$

form a ruled surface for which the latter curve is the line of striction.

3. If the ruled surface formed by an infinity of tangents to a surface S has the locus of the points of contact for its line of striction, this relation is unaltered by deformations of S.

4. Show that if D, D', D'' are the second fundamental coefficients of a surface with the linear element (53), the equation of the lines of curvature of the associated surface Σ is reducible to

$$\begin{vmatrix} D\,du_1 + D'\,dv_1 & D'\,du_1 + D''\,dv_1 \\ \dfrac{\rho''}{\rho'}\,du_1 & \rho\rho'\,dv_1 \end{vmatrix} = 0.$$

5. Show that the surface Σ associated by the method of Weingarten with a surface S applicable to a surface of revolution corresponds with parallelism of tangent planes to the surface S' complementary to S with respect to the deforms of the meridians; and that the lines of curvature on Σ and S' correspond.

6. Show that when ϕ has the form (57), the equation (51) is reducible to

$$(\rho_1 + p)(\rho_2 + p) = F(2q - p^2);$$

hence the determination of all the surfaces applicable to surfaces of revolution is equivalent to the determination of those surfaces Σ which are such that if M_1 and M_2 are the centers of principal curvature of Σ at a point M, and N is the projection of the origin O on the normal at M, the product $NM_1 \cdot NM_2$ is a function of ON.

7. Given any surface S applicable to a surface of revolution. Draw through a fixed point O segments parallel to the tangents to the deforms of the meridians and of lengths proportional to the radii of the corresponding parallels, and through the extremities of these segments draw lines parallel to the normals to S. Show that these lines form a normal congruence whose orthogonal surfaces Σ have the same spherical representation of their lines of curvature as S and are integral surfaces of the equation of Ex. 6.

8. Let S be a surface applicable to a surface of revolution and S' the surface complementary to S with respect to the deforms of the meridians; let also Σ and Σ' be surfaces associated with S and S' respectively after the manner of Ex. 7. Show that corresponding normals to Σ and Σ' are perpendicular to one another, and that the common perpendicular to these normals passes through the origin and is divided by it into two segments which are functions of one another.

9. Show that a surface determined by the equation

$$2q + \kappa + (\rho_1 + \rho_2)p + \rho_1\rho_2 = 0,$$

where κ is a constant, possesses the property that the sphere described on the segment of each normal between the centers of principal curvature with this segment for diameter cuts the sphere with center at the origin and of radius $\sqrt{\pm\kappa}$ in great circles, orthogonally, or passes through the origin, according as κ is positive, negative, or zero. These surfaces are called the *surfaces of Bianchi.*

10. Show that for the surfaces of Bianchi the function $\phi(p, q)$ is of the form

$$\phi = \sqrt{2q - p^2 + \kappa},$$

and that the linear element of the associated surface S applicable to a surface of revolution is

$$ds^2 = \left(\frac{1}{\psi^2} - \kappa\right)d\psi^2 + \psi^2 dp^2.$$

Show also that according as $\kappa = 0, > 0$, or < 0 the linear element of S is reducible to the respective forms

$$ds^2 = du^2 + e^{2u}dv^2,\quad ds^2 = \tanh^4 u\, du^2 + \operatorname{sech}^2 u\, dv^2,\quad ds^2 = \coth^4 u\, du^2 + \operatorname{csch}^2 u\, dv^2.$$

On account of this result and Ex. 10, p. 318, the surfaces of Bianchi are said to be of the *parabolic*, *elliptic*, or *hyperbolic type*, according as $\kappa = 0, > 0$, or < 0.

11. Let S be a pseudospherical surface with its linear element in the form (VIII, 32), and S_1 the Bianchi transform whose linear element is (VIII, 33). Find the coördinates $\bar{x}, \bar{y}, \bar{z}$ of the surface Σ associated with S_1 by the method of Weingarten, and show that by means of Ex. 8, p. 291, the expression for $\bar{x}$ is reducible to

$$\bar{x} = ae^{\frac{\xi}{a}}(\cos\theta X_1 + \sin\theta X_2) + \zeta X,$$

where X_1, X_2, X are the direction-cosines with respect to the x-axis of the tangents to the lines of curvature of S and of the normal to the latter.

12. Show that the surfaces S and Σ of Ex. 11 have the same spherical representation of their lines of curvature, that Σ is a surface of Bianchi of the parabolic type, and that consequently there is an infinity of these surfaces of the parabolic type which have the same spherical representation of their lines of curvature as a given pseudospherical surface S.

13. Show that if Σ_1 and Σ_2 are two surfaces of Bianchi of the parabolic type which have the same spherical representation of their lines of curvature, the locus of a point which divides in constant ratio the line joining corresponding points of Σ_1 and Σ_2 is a surface of Bianchi with the same representation of its lines of curvature, and that it is of the elliptic or hyperbolic type according as the point divides the segment internally or externally.

14. When S is a pseudospherical surface with its linear element in the form (VIII, 32), the coördinates $\bar{x}_1, \bar{y}_1, \bar{z}_1$ of the surface Σ determined by the method of Weingarten are reducible to

$$\bar{x}_1 = (ae^{-\frac{\xi}{a}} \cos\theta + \eta \sin\theta) X_1 + (ae^{-\frac{\xi}{a}} \sin\theta - \eta\cos\theta) X_2,$$

and analogous expressions for $\bar{y}_1$ and $\bar{z}_1$, where X_1, Y_1, Z_1; X_2, Y_2, Z_2 are the direction-cosines of the tangents to the lines of curvature of S. Show also that Σ has the same spherical representation of its lines of curvature as the surface S_1 with the linear element (VIII, 33).

15. Derive from the equations

$$\bar{x}X_1 + \bar{y}Y_1 + \bar{z}Z_1 = p, \qquad \bar{x}^2 + \bar{y}^2 + \bar{z}^2 = 2q,$$

by means of (44), (48), and (49), the equations

$$\sum \bar{x}\frac{\partial^2 x}{\partial u^2} = \Delta\frac{\partial^2\phi}{\partial q^2}, \qquad \sum \bar{x}\frac{\partial^2 x}{\partial u\,\partial v} = -\Delta\frac{\partial^2\phi}{\partial p\,\partial q}, \qquad \sum \bar{x}\frac{\partial^2 x}{\partial v^2} = \Delta\frac{\partial^2\phi}{\partial p^2}$$

where x, y, z are the coördinates of S.

16. Show that the equations for Σ similar to (IV, 27) are reducible to

$$\frac{\partial^2 x}{\partial u\,\partial v}du + \frac{\partial^2 x}{\partial v^2}dv + \bar{r}\left(\frac{\partial^2 x}{\partial u^2}du + \frac{\partial^2 x}{\partial u\,\partial v}dv\right) = 0,$$

and similar expressions in y and z. Derive therefrom (cf. Ex. 15) the equations

$$D'du + D''dv + \bar{r}(D\,du + D'dv) = 0,$$

$$\frac{\partial^2\phi}{\partial p\,\partial q}du - \frac{\partial^2\phi}{\partial p^2}dv - \bar{r}\left(\frac{\partial^2\phi}{\partial q^2}du - \frac{\partial^2\phi}{\partial p\,\partial q}dv\right) = 0,$$

where D, D', D'' are the second fundamental coefficients of S.

17. Show that the lines of curvature on Σ correspond to a conjugate system on S (cf. Ex. 16).

18. Show that for the surface Σ we have

$$\frac{\partial\bar{x}}{\partial p} = -\rho_1\rho_2\frac{\partial X_1}{\partial q}, \qquad \frac{\partial\bar{x}}{\partial q} = \frac{\partial X_1}{\partial p} - (\rho_1 + \rho_2)\frac{\partial X_1}{\partial q}.$$

19. Let S be the surface defined by (67) and S_1 the surface whose coördinates are

$$x_1 = x - u_1X_1, \qquad y_1 = y - u_1Y_1, \qquad z_1 = z - u_1Z_1.$$

Show that S_1 is an involute of S, that the curves $p = $ const. are geodesics on S and lines of curvature on S_1, and that the radii of principal curvature of S_1 are

$$\rho_1' = u_1, \qquad \rho_2' = -[u_1 + 2\bar{\omega}'(p)].$$

20. Show that when m in (73) is 0 or 1, the function p is the sum of two arbitrary functions of α and β respectively, that the linear element of S is

$$ds^2 = du_1^2 + 2\,u_1\,dv^2,$$

that S is an evolute of a minimal surface (cf. Ex. 19), and that the mean evolute of Σ is a point.

21. Show that when m in (73) is 2, the general integral of equation (74) is

$$p = f_1'(\alpha) + f_2'(\beta) - 2\frac{\beta f_1(\alpha) + \alpha f_2(\beta)}{1 + \alpha\beta},$$

where f_1 and f_2 are arbitrary functions of α and β respectively. Show also that the surface Σ is minimal (cf. § 151).

22. Show that the mean evolute of a surface of Goursat is a surface of Goursat homothetic to the given one.

23. Show that when $\overline{\omega}(p) = \frac{1}{2}\,ap^2$, then

$$p = a\log(1 + \alpha\beta) + f_1(\alpha) + f_2(\beta),$$

where f_1 and f_2 are arbitrary functions, that the linear element of S is

$$ds^2 = du_1^2 + 2\,(u_1 + av)\,dv^2,$$

and that the mean evolute of Σ is a sphere.

24. Show that the surfaces S of Ex. 23 are applicable to the surfaces of revolution S_0 whose equations are

$$x_0 = \alpha\bar{u}\cos\frac{\bar{v}}{\alpha}, \qquad y_0 = \alpha\bar{u}\sin\frac{\bar{v}}{\alpha}, \qquad z_0 = \int\sqrt{\bar{u}^2 - \alpha^2 - a^2}\,d\bar{u},$$

where α is an arbitrary constant. Show also that when $\alpha = ia$, S_0 is a paraboloid.

25. Show that when

$$\phi\,(p,\,q) = \left(q + \frac{m}{2}\right)^2 - mp^2,$$

the surfaces Σ are spherical or pseudospherical according as m is positive or negative; also that the surfaces S are applicable to the surface

$$x + iy = v, \qquad x - iy = \frac{v^2}{2} - \frac{u^2}{2\,m} - mv, \qquad z = u,$$

which is a paraboloid tangent to the plane at infinity at a point of the circle at infinity.

CHAPTER XI

INFINITESIMAL DEFORMATION OF SURFACES

152. General problem. The preceding chapters deal with pairs of isometric surfaces which are such that in order that one may be applied to the other a finite deformation is necessary. In the present chapter we shall be concerned with the infinitesimal deformations which constitute the intermediate steps in such a finite deformation.

Let x, y, z; x', y', z' respectively be the coördinates of a surface S and a surface S', the latter being obtained from the former by a very small deformation. If we put

$$(1) \qquad x' = x + \epsilon x_1, \qquad y' = y + \epsilon y_1, \qquad z' = z + \epsilon z_1,$$

where ϵ denotes a small constant and x_1, y_1, z_1 are determined functions of u and v, these functions are proportional to the direction-cosines of the line through corresponding points of S and S'. From these equations we have

$$dx'^2 + dy'^2 + dz'^2 = dx^2 + dy^2 + dz^2 + 2\epsilon(dx\,dx_1 + dy\,dy_1 + dz\,dz_1) + \epsilon^2(dx_1^2 + dy_1^2 + dz_1^2).$$

If the functions satisfy the condition

$$(2) \qquad dx\,dx_1 + dy\,dy_1 + dz\,dz_1 = 0,$$

corresponding small lengths on S and S' are equal to within terms of the second order in ϵ. When ϵ is taken so small that ϵ^2 may be neglected, the surface S' defined by (1) is said to arise from S by an *infinitesimal deformation* of the latter. In such a deformation each point of S undergoes a displacement along the line through it whose direction-cosines are proportional to x_1, y_1, z_1. These lines are called the *generatrices* of the deformation.

It is evident that the problem of infinitesimal deformation is equivalent to the solution of equation (2). Since x_1, y_1, z_1 are functions of u and v, they may be taken for the coördinates of a surface S_1. Equation (2) expresses the fact that the tangent to any curve on S is perpendicular to the tangent to the corresponding curve on S_1 at the homologous point. We say that in this case S and S_1 *correspond with orthogonality of corresponding linear elements*. And so we have:

The problem of the infinitesimal deformation of a surface S is equivalent to the determination of the surfaces corresponding to it with orthogonality of linear elements.

153. Characteristic function. We proceed to the determination of these surfaces S_1, and to this end replace equation (2) by the equivalent system

$$(3)\quad \sum \frac{\partial x}{\partial u}\frac{\partial x_1}{\partial u}=0, \quad \sum \frac{\partial x}{\partial v}\frac{\partial x_1}{\partial v}=0, \quad \sum \frac{\partial x}{\partial u}\frac{\partial x_1}{\partial v}+\sum \frac{\partial x}{\partial v}\frac{\partial x_1}{\partial u}=0.$$

Weingarten * replaced the last of these equations by the two

$$(4)\qquad \sum \frac{\partial x}{\partial u}\frac{\partial x_1}{\partial v}=\phi H, \qquad \sum \frac{\partial x}{\partial v}\frac{\partial x_1}{\partial u}=-\phi H,$$

thus defining a function ϕ, which Bianchi has called the *characteristic function;* as usual $H=\sqrt{EG-F^2}$.

If the first of equations (3) be differentiated with respect to v, and the second with respect to u, we have

$$\sum \frac{\partial x}{\partial u}\frac{\partial^2 x_1}{\partial u\,\partial v}+\sum \frac{\partial x_1}{\partial u}\frac{\partial^2 x}{\partial u\,\partial v}=0, \qquad \sum \frac{\partial x}{\partial v}\frac{\partial^2 x_1}{\partial u\,\partial v}+\sum \frac{\partial x_1}{\partial v}\frac{\partial^2 x}{\partial u\,\partial v}=0.$$

With the aid of these identities, of the formulas (V, 3), and of the Gauss equations (V, 7), the equations obtained by the differentiation of equations (4) with respect to u and v respectively are reducible to

$$\frac{\partial \phi}{\partial u}=\frac{D\sum X\frac{\partial x_1}{\partial v}-D'\sum X\frac{\partial x_1}{\partial u}}{H},$$

$$\frac{\partial \phi}{\partial v}=\frac{D'\sum X\frac{\partial x_1}{\partial v}-D''\sum X\frac{\partial x_1}{\partial u}}{H}.$$

* *Crelle*, Vol. C (1887), pp. 296–310.

Excluding the case where S is a developable surface, we solve these equations for $\sum X\frac{\partial x_1}{\partial u}$, $\sum X\frac{\partial x_1}{\partial v}$ and obtain

$$(5)\qquad \sum X\frac{\partial x_1}{\partial u}=\frac{D'\frac{\partial\phi}{\partial u}-D\frac{\partial\phi}{\partial v}}{KH},\qquad \sum X\frac{\partial x_1}{\partial v}=\frac{D''\frac{\partial\phi}{\partial u}-D'\frac{\partial\phi}{\partial v}}{KH},$$

where K denotes the total curvature of S. If we solve equations (3), (4), and (5) for the derivatives of x_1, y_1, z_1 with respect to u and v, we obtain

$$(6)\qquad \begin{cases} \dfrac{\partial x_1}{\partial u}=\dfrac{D\left(\phi\dfrac{\partial X}{\partial v}-X\dfrac{\partial\phi}{\partial v}\right)-D'\left(\phi\dfrac{\partial X}{\partial u}-X\dfrac{\partial\phi}{\partial u}\right)}{KH},\\[2ex] \dfrac{\partial x_1}{\partial v}=\dfrac{D'\left(\phi\dfrac{\partial X}{\partial v}-X\dfrac{\partial\phi}{\partial v}\right)-D''\left(\phi\dfrac{\partial X}{\partial u}-X\dfrac{\partial\phi}{\partial u}\right)}{KH}, \end{cases}$$

and similar expressions in y_1 and z_1. Hence, when the characteristic function is known, the surface S_1 can be obtained by quadratures. Our problem reduces therefore to the determination of ϕ.

If equations (5) be differentiated with respect to v and u respectively, and the resulting equations be subtracted from one another, we have

$$\frac{\partial}{\partial v}\left(\frac{D\frac{\partial\phi}{\partial v}-D'\frac{\partial\phi}{\partial u}}{KH}\right)+\frac{\partial}{\partial u}\left(\frac{D''\frac{\partial\phi}{\partial u}-D'\frac{\partial\phi}{\partial v}}{KH}\right)=\sum\frac{\partial x_1}{\partial v}\frac{\partial X}{\partial u}-\sum\frac{\partial x_1}{\partial u}\frac{\partial X}{\partial v}.$$

When the derivatives of X, Y, Z in the right-hand member are replaced by the expressions (V, 8), the above equation reduces to

$$(7)\qquad \frac{\partial}{\partial v}\left(\frac{D\frac{\partial\phi}{\partial v}-D'\frac{\partial\phi}{\partial u}}{KH}\right)+\frac{\partial}{\partial u}\left(\frac{D''\frac{\partial\phi}{\partial u}-D'\frac{\partial\phi}{\partial v}}{KH}\right)=\frac{2FD'-ED''-GD}{H}\phi.$$

Bianchi calls this the *characteristic equation.*

In consequence of (IV, 73, 74) equation (7) is reducible to

$$(8)\qquad \frac{\partial}{\partial v}\left(\frac{D\frac{\partial\phi}{\partial v}-D'\frac{\partial\phi}{\partial u}}{\mathcal{H}}\right)+\frac{\partial}{\partial u}\left(\frac{D''\frac{\partial\phi}{\partial u}-D'\frac{\partial\phi}{\partial v}}{\mathcal{H}}\right)=\frac{2\mathcal{F}D'-\mathcal{E}D''-\mathcal{G}D}{\mathcal{H}}\phi,$$

where $\mathcal{E}$, $\mathcal{F}$, $\mathcal{G}$ are the coefficients of the linear element of the spherical representation of S, namely

$$d\sigma^2 = \mathcal{E}\,du^2 + 2\,\mathcal{F}\,du\,dv + \mathcal{G}\,dv^2 \tag{9}$$

and

$$\mathcal{H} = \sqrt{\mathcal{E}\mathcal{G} - \mathcal{F}^2}.$$

By means of (V, 27) equation (8) is reducible to

$$\begin{aligned} D''\left(\frac{\partial^2\phi}{\partial u^2} - \begin{Bmatrix}11\\1\end{Bmatrix}'\frac{\partial\phi}{\partial u} - \begin{Bmatrix}11\\2\end{Bmatrix}'\frac{\partial\phi}{\partial v} + \mathcal{E}\phi\right) \\ + D\left(\frac{\partial^2\phi}{\partial v^2} - \begin{Bmatrix}22\\1\end{Bmatrix}'\frac{\partial\phi}{\partial u} - \begin{Bmatrix}22\\2\end{Bmatrix}'\frac{\partial\phi}{\partial v} + \mathcal{G}\phi\right) \\ - 2D'\left(\frac{\partial^2\phi}{\partial u\,\partial v} - \begin{Bmatrix}12\\1\end{Bmatrix}'\frac{\partial\phi}{\partial u} - \begin{Bmatrix}12\\2\end{Bmatrix}'\frac{\partial\phi}{\partial v} + \mathcal{F}\phi\right) = 0, \end{aligned} \tag{10}$$

where the Christoffel symbols are formed with respect to (9).

Since X, Y, Z are solutions of equations (V, 22), they are solutions of (10), and consequently also of equation (7). Therefore the latter equation may be written

$$\begin{aligned} \frac{\partial}{\partial v}\left[\frac{D\left(X\frac{\partial\phi}{\partial v} - \phi\frac{\partial X}{\partial v}\right) - D'\left(X\frac{\partial\phi}{\partial u} - \phi\frac{\partial X}{\partial u}\right)}{KH}\right] \\ + \frac{\partial}{\partial u}\left[\frac{D''\left(X\frac{\partial\phi}{\partial u} - \phi\frac{\partial X}{\partial u}\right) - D'\left(X\frac{\partial\phi}{\partial v} - \phi\frac{\partial X}{\partial v}\right)}{KH}\right] = 0. \end{aligned}$$

But this is the condition of integrability of equations (6). Hence we have the theorem:

Each solution of the characteristic equation determines a surface S_1 and consequently an infinitesimal deformation of S.

154. Asymptotic lines parametric. When the asymptotic lines on S are parametric, equation (10) is reducible, in consequence of (VI, 15), to

$$\frac{\partial^2\phi}{\partial u\,\partial v} + \frac{1}{2}\frac{\partial\log\rho}{\partial v}\frac{\partial\phi}{\partial u} + \frac{1}{2}\frac{\partial\log\rho}{\partial u}\frac{\partial\phi}{\partial v} + \mathcal{F}\phi = 0, \tag{11}$$

where

$$\frac{1}{\rho} = \sqrt{-K} = \frac{\mathcal{H}}{D'}.$$

If we put

$$\phi\sqrt{-\epsilon\rho} = \theta,$$

ϵ being $+1$ or -1 according as the curvature of S is positive or negative, equation (11) becomes

$$\frac{\partial^2\theta}{\partial u\,\partial v}=\left(\frac{1}{\sqrt{\rho}}\frac{\partial^2\sqrt{\rho}}{\partial u\,\partial v}-\mathcal{F}\right)\theta. \tag{12}$$

Since X, Y, Z are solutions of (11), the functions

$$\nu_1=X\sqrt{-\epsilon\rho},\qquad \nu_2=Y\sqrt{-\epsilon\rho},\qquad \nu_3=Z\sqrt{-\epsilon\rho}$$

are solutions of (12).

Now equations (6) may be put in the form

$$\frac{\partial x_1}{\partial u}=-\begin{vmatrix}\nu_1 & \theta\\ \dfrac{\partial\nu_1}{\partial u} & \dfrac{\partial\theta}{\partial u}\end{vmatrix},\qquad \frac{\partial x_1}{\partial v}=\begin{vmatrix}\nu_1 & \theta\\ \dfrac{\partial\nu_1}{\partial v} & \dfrac{\partial\theta}{\partial v}\end{vmatrix}. \tag{13}$$

The reader should compare these equations with the Lelieuvre formulas (§ 79), which give the expressions for the derivatives of the coördinates of S in terms of ν_1, ν_2, ν_3.

From these results it follows that any three solutions of an equation of the form

$$\frac{\partial^2\theta}{\partial u\,\partial v}=M\theta,$$

where M is any function of u and v, determine a surface S upon which the parametric curves are the asymptotic lines, and every other solution linearly independent of these three gives by quadratures an infinitesimal deformation of S

EXAMPLES

1. A necessary and sufficient condition that two surfaces satisfying the condition (2) be applicable is that they be minimal surfaces adjoint to one another.

2. If x, y, z and x_1, y_1, z_1 satisfy the condition (2), so also do ξ, η, ζ and ξ_1, η_1, ζ_1, the latter being given by

$$\begin{aligned}\xi&=a_1x+b_1y+c_1z+d_1, & x_1&=a_1\xi_1+a_2\eta_1+a_3\zeta_1+e_1,\\ \eta&=a_2x+b_2y+c_2z+d_2, & y_1&=b_1\xi_1+b_2\eta_1+b_3\zeta_1+e_2,\\ \zeta&=a_3x+b_3y+c_3z+d_3, & z_1&=c_1\xi_1+c_2\eta_1+c_3\zeta_1+e_3,\end{aligned}$$

where $a_1, a_2, \cdots, e_1, e_2, e_3$ are constants.

3. A necessary condition that the locus of the point (x_1, y_1, z_1) be a curve is that S be a developable surface. In this case any orthogonal trajectory of the tangent planes to S satisfies the condition.

4. Investigate the cases $\phi=0$ and $\phi=c$, where c is a constant different from zero.

5. If S_1 and S_1' correspond to S with orthogonality of linear elements, so also does the locus of a point dividing in constant ratio the line joining corresponding points on S_1 and S_1'.

155. Associate surfaces. The expressions in the parentheses of equation (10) differ only in sign from the second fundamental coefficients, D_0, D_0', D_0'', of the surface S_0 enveloped by the plane

$$Xx + Yy + Zz = \phi.* \tag{14}$$

Hence equation (10) may be written

$$D''D_0 + DD_0'' - 2D'D_0' = 0. \tag{15}$$

This is the condition that to the asymptotic lines upon either of the surfaces S, S_0 there corresponds a conjugate system on the other (§ 56). Bianchi applies the term *associate* to two surfaces whose tangent planes at corresponding points are parallel, and for which the asymptotic lines on either correspond to a conjugate system on the other. Since the converse of the preceding results are readily shown to be true, we have the theorem of Bianchi †:

When two surfaces are associate the expression for the distance from a fixed point in space to the tangent plane to one is the characteristic function for an infinitesimal deformation of the other.

Hence the problems of infinitesimal deformation and of the determination of surfaces associate to a given one are equivalent. We consider the latter problem.

Since the tangent planes to S and S_0 at corresponding points are parallel, we have

$$\frac{\partial x_0}{\partial u} = \lambda\frac{\partial x}{\partial u} - \mu\frac{\partial x}{\partial v}, \qquad \frac{\partial x_0}{\partial v} = \sigma\frac{\partial x}{\partial u} - \tau\frac{\partial x}{\partial v},$$

and similar equations in y_0 and z_0, where λ, μ, σ, τ are functions of u and v to be determined. ‡

If these equations be multiplied by $\frac{\partial X}{\partial u}, \frac{\partial Y}{\partial u}, \frac{\partial Z}{\partial u}$ and added, and likewise by $\frac{\partial X}{\partial v}, \frac{\partial Y}{\partial v}, \frac{\partial Z}{\partial v}$ and added, we obtain

$$\left\{\begin{array}{ll} D_0 = \lambda D - \mu D', & D_0'' = \sigma D' - \tau D'', \\ D_0' = \lambda D' - \mu D'' = \sigma D - \tau D', & \end{array}\right. \tag{16}$$

* Cf. § 67. † *Lezioni*, Vol. II, p. 9.

‡ The negative signs before μ and τ are taken so that subsequent results may have a suitable form.

where D_0, D_0', D_0'' are the second fundamental quantities for S_0. When these values are substituted in (15), we find

$$\lambda - \tau = 0. \tag{17}$$

Consequently the above equations reduce to

$$\frac{\partial x_0}{\partial u} = \lambda \frac{\partial x}{\partial u} - \mu \frac{\partial x}{\partial v}, \qquad \frac{\partial x_0}{\partial v} = \sigma \frac{\partial x}{\partial u} - \lambda \frac{\partial x}{\partial v}. \tag{18}$$

If we make use of the Gauss equations (V, 7), the condition of integrability of equations (18) is reducible to

$$A \frac{\partial x}{\partial u} + B \frac{\partial x}{\partial v} = 0,$$

where A and B are determinate functions. Since similar equations hold in y and z, both A and B must be identically zero. Calculating the expressions for these functions, we have the following equations to be satisfied by λ, μ, and σ:

$$\left\{\begin{aligned} &\frac{\partial \mu}{\partial v} - \frac{\partial \lambda}{\partial u} + \mu \begin{Bmatrix} 22 \\ 2 \end{Bmatrix} - 2\lambda \begin{Bmatrix} 12 \\ 2 \end{Bmatrix} + \sigma \begin{Bmatrix} 11 \\ 2 \end{Bmatrix} = 0, \\ &\frac{\partial \sigma}{\partial u} - \frac{\partial \lambda}{\partial v} + \mu \begin{Bmatrix} 22 \\ 1 \end{Bmatrix} - 2\lambda \begin{Bmatrix} 12 \\ 1 \end{Bmatrix} + \sigma \begin{Bmatrix} 11 \\ 1 \end{Bmatrix} = 0. \end{aligned}\right. \tag{19}$$

To these equations we must add

$$2\lambda D' - \mu D'' - \sigma D = 0, \tag{20}$$

obtained from the last of (16). The determination of the associate surfaces of a given surface referred to any parametric system requires the integration of this system of equations. Moreover, every set of solutions leads to an associate surface. We shall now consider several cases in which the parametric curves are of a particular kind.

156. Particular parametric curves. Suppose that S is a surface upon which the parametric curves form a conjugate system. We inquire under what conditions there exists an associate surface upon which also the corresponding curves form a conjugate system.

On this hypothesis we have, from (16),

$$\mu = \sigma = 0,$$

so that equations (19) reduce to

$$(21) \qquad \frac{\partial \log \lambda}{\partial u} = -2\begin{Bmatrix}12\\2\end{Bmatrix}, \qquad \frac{\partial \log \lambda}{\partial v} = -2\begin{Bmatrix}12\\1\end{Bmatrix},$$

which are consistent only when

$$(22) \qquad \frac{\partial}{\partial u}\begin{Bmatrix}12\\1\end{Bmatrix} = \frac{\partial}{\partial v}\begin{Bmatrix}12\\2\end{Bmatrix},$$

that is, when the point equation of S, namely

$$\frac{\partial^2\theta}{\partial u \partial v} = \begin{Bmatrix}12\\1\end{Bmatrix}\frac{\partial \theta}{\partial u} + \begin{Bmatrix}12\\2\end{Bmatrix}\frac{\partial \theta}{\partial v},$$

has equal invariants (cf. § 165).

Conversely, when condition (22) is satisfied, the function λ given by the quadratures (21) makes the equations

$$(23) \qquad \frac{\partial x_0}{\partial u} = \lambda\frac{\partial x}{\partial u}, \qquad \frac{\partial x_0}{\partial v} = -\lambda\frac{\partial x}{\partial v}$$

compatible, and thus the coördinates of an associate surface are obtained by quadratures. Hence we have the theorem of Cosserat*:

The infinitesimal deformation of a surface S is the same problem as the determination of the conjugate systems with equal point invariants on S.

Since the relation between S and S_0 is reciprocal and the parametric curves are conjugate for both surfaces, these curves on S_0 also have equal point invariants.

If S be referred to its asymptotic lines, the corresponding lines on S_0 form a conjugate system. In this case, as is seen from (16), λ is zero and equations (18) reduce to

$$(24) \qquad \frac{\partial x_0}{\partial u} = -\mu\frac{\partial x}{\partial v}, \qquad \frac{\partial x_0}{\partial v} = \sigma\frac{\partial x}{\partial u};$$

moreover, equations (19) become

$$(25) \qquad \begin{cases} \dfrac{\partial \mu}{\partial v} + \mu\begin{Bmatrix}22\\2\end{Bmatrix} + \sigma\begin{Bmatrix}11\\2\end{Bmatrix} = 0, \\[2ex] \dfrac{\partial \sigma}{\partial u} + \mu\begin{Bmatrix}22\\1\end{Bmatrix} + \sigma\begin{Bmatrix}11\\1\end{Bmatrix} = 0. \end{cases}$$

* *Toulouse Annales*, Vol. VII (1893), N. 60.

The solution of this system is the same problem as the integration of a partial differential equation of the second order, as is seen by the elimination of either unknown. When a solution of the former is obtained, the corresponding value of the other unknown is given directly by one of equations (25).

We make an application of these results to a ruled surface, which we suppose to be referred to its asymptotic lines. If the curves $v = \text{const.}$ are the generators, they are geodesics, and consequently (VI, 50)

$$\begin{Bmatrix} 11 \\ 2 \end{Bmatrix} = 0.$$

Now μ can be found by a quadrature. When this value is substituted in the second of equations (25), we have a linear equation in σ, and consequently σ also can be obtained by quadratures. Hence we have the theorem:

When the curved asymptotic lines on a ruled surface are known, its associate surfaces can be found by quadratures.

If S_0 were referred to its asymptotic lines, we should have equations similar to (24). These equations may be interpreted as follows:

The tangent to an asymptotic line on one of two associate surfaces is parallel to the direction conjugate to the corresponding curve on the other surface.

EXAMPLES

1. If two associate surfaces are applicable to one another, they are minimal surfaces.

2. Every surface of translation admits an associate surface of translation such that the generatrices of the two surfaces constitute the common conjugate system.

3. The surfaces associate to a sphere are minimal.

4. When the equations of the right helicoid are

$$x = u \cos v, \qquad y = u \sin v, \qquad z = av,$$

the characteristic function of any infinitesimal deformation is $\phi = (U+V)(u^2+a^2)^{-\frac{1}{2}}$, where U and V are arbitrary functions of u and v respectively. Find the surfaces S_1 and S_0, and show that the latter are molding surfaces.

5. If S_0 and S_0' are associate surfaces of a surface S, the locus of a point dividing in constant ratio the joins of corresponding points of S_0 and S_0' is an associate of S.

157. Relations between three surfaces S, S_1, S_0. Having thus discussed the various ways in which the problem of infinitesimal deformation may be attacked, we proceed to the consideration of other properties which are possessed by a set of three surfaces S, S_1, S_0.

We recall the differential equation

$$dx dx_1 + dy dy_1 + dz dz_1 = 0,$$

and remark that it may be replaced by the three

$$dx_1 = z_0 dy - y_0 dz, \quad dy_1 = x_0 dz - z_0 dx, \quad dz_1 = y_0 dx - x_0 dy, \tag{26}$$

if the functions x_0, y_0, z_0 are such a form that the conditions of integrability of equations (26) are satisfied. These conditions are

$$\frac{\partial y}{\partial u}\frac{\partial z_0}{\partial v} - \frac{\partial z}{\partial u}\frac{\partial y_0}{\partial v} = \frac{\partial y}{\partial v}\frac{\partial z_0}{\partial u} - \frac{\partial z}{\partial v}\frac{\partial y_0}{\partial u},$$

$$\frac{\partial z}{\partial u}\frac{\partial x_0}{\partial v} - \frac{\partial x}{\partial u}\frac{\partial z_0}{\partial v} = \frac{\partial z}{\partial v}\frac{\partial x_0}{\partial u} - \frac{\partial x}{\partial v}\frac{\partial z_0}{\partial u},$$

$$\frac{\partial x}{\partial u}\frac{\partial y_0}{\partial v} - \frac{\partial y}{\partial u}\frac{\partial x_0}{\partial v} = \frac{\partial x}{\partial v}\frac{\partial y_0}{\partial u} - \frac{\partial y}{\partial v}\frac{\partial x_0}{\partial u}.$$

If these equations be multiplied by $\frac{\partial x}{\partial u}$, $\frac{\partial y}{\partial u}$, $\frac{\partial z}{\partial u}$ respectively and added, and likewise by $\frac{\partial x}{\partial v}$, $\frac{\partial y}{\partial v}$, $\frac{\partial z}{\partial v}$, and by X, Y, Z, we obtain, by (IV, 2),

$$\sum X\frac{\partial x_0}{\partial u} = 0, \qquad \sum X\frac{\partial x_0}{\partial v} = 0, \tag{27}$$

$$\begin{vmatrix} X & Y & Z \\ \frac{\partial x}{\partial u} & \frac{\partial y}{\partial u} & \frac{\partial z}{\partial u} \\ \frac{\partial x_0}{\partial v} & \frac{\partial y_0}{\partial v} & \frac{\partial z_0}{\partial v} \end{vmatrix} = \begin{vmatrix} X & Y & Z \\ \frac{\partial x}{\partial v} & \frac{\partial y}{\partial v} & \frac{\partial z}{\partial v} \\ \frac{\partial x_0}{\partial u} & \frac{\partial y_0}{\partial u} & \frac{\partial z_0}{\partial u} \end{vmatrix}. \tag{28}$$

From the first two of these equations it follows that the locus of the point with coördinates x_0, y_0, z_0 corresponds to S with parallelism of tangent planes.

In order to interpret the last of these equations we recall from § 61 that

$$X = \frac{a}{\mathcal{H}}\frac{\partial(Y, Z)}{\partial(u, v)}, \qquad Y = \frac{a}{\mathcal{H}}\frac{\partial(Z, X)}{\partial(u, v)}, \qquad Z = \frac{a}{\mathcal{H}}\frac{\partial(X, Y)}{\partial(u, v)},$$

where a is ± 1 according as the curvature of the surface is positive or negative. If we substitute these values in the left-hand members of the following equations, and add and subtract $\frac{\partial x}{\partial u}\frac{\partial X}{\partial u}\frac{\partial X}{\partial v}$ and $\frac{\partial x}{\partial v}\frac{\partial X}{\partial u}\frac{\partial X}{\partial v}$ from these equations respectively, the resulting expressions are reducible to the form of the right-hand members

$$(29)\qquad \begin{cases} H\left(Y\dfrac{\partial z}{\partial u}-Z\dfrac{\partial y}{\partial u}\right)=a\left(D'\dfrac{\partial X}{\partial u}-D\dfrac{\partial X}{\partial v}\right), \\ H\left(Y\dfrac{\partial z}{\partial v}-Z\dfrac{\partial y}{\partial v}\right)=a\left(D''\dfrac{\partial X}{\partial u}-D'\dfrac{\partial X}{\partial v}\right). \end{cases}$$

By means of these and similar identities, equation (28) can be transformed into

$$D''\sum\frac{\partial x_0}{\partial u}\frac{\partial X}{\partial u}+D\sum\frac{\partial x_0}{\partial v}\frac{\partial X}{\partial v}-D'\left(\sum\frac{\partial x_0}{\partial u}\frac{\partial X}{\partial v}+\sum\frac{\partial x_0}{\partial v}\frac{\partial X}{\partial u}\right)=0.$$

Since this equation is equivalent to (15) because of (27), the quantities x_0, y_0, z_0 in (26) are the coördinates of S_0. Hence when a surface S_1 is known, the coördinates of the corresponding surface S_0 are readily found.

This result enables us to find another property of S_0 and S_1. If X_1, Y_1, Z_1 denote the direction-cosines of the normal to S_1, they are given by

$$X_1=\frac{1}{H_1}\frac{\partial(y_1, z_1)}{\partial(u, v)},\qquad Y_1=\frac{1}{H_1}\frac{\partial(z_1, x_1)}{\partial(u, v)},\qquad Z_1=\frac{1}{H_1}\frac{\partial(x_1, y_1)}{\partial(u, v)},$$

where $H_1=\sqrt{E_1G_1-F_1^2}$, E_1, F_1, G_1 being the coefficients of the linear element of S_1. If the values of the derivatives of x_1, y_1, z_1, as given by (26), be substituted in these expressions, we have, in consequence of (14),

$$(30)\qquad X_1=\frac{x_0H}{H_1}\phi,\qquad Y_1=\frac{y_0H}{H_1}\phi,\qquad Z_1=\frac{z_0H}{H_1}\phi.$$

As an immediate consequence we have the theorem:

A normal to S_1 is parallel to the radius vector of S_0 at the corresponding point.

By means of (30) we find readily the expressions for the second fundamental coefficients D_1, D_1', D_1'' of S_1. If we notice that

$$\sum x_0 \frac{\partial x_1}{\partial u} = 0, \qquad \sum x_0 \frac{\partial x_1}{\partial v} = 0,$$

and substitute the values from (6) and (30) in

$$D_1 = -\sum \frac{\partial x_1}{\partial u}\frac{\partial X_1}{\partial u}, \qquad D_1' = -\sum \frac{\partial x_1}{\partial u}\frac{\partial X_1}{\partial v} = -\sum \frac{\partial x_1}{\partial v}\frac{\partial X_1}{\partial u},$$

$$D_1'' = -\sum \frac{\partial x_1}{\partial v}\frac{\partial X_1}{\partial v},$$

we obtain

$$\text{(31)} \qquad \begin{cases} D_1 = \dfrac{\phi^2}{H_1 K}(DD_0' - D'D_0), \\ D_1' = \dfrac{\phi^2}{H_1 K}(DD_0'' - D'D_0') = \dfrac{\phi^2}{H_1 K}(D'D_0' - D''D_0), \\ D_1'' = \dfrac{\phi^2}{H_1 K}(D'D_0'' - D''D_0'). \end{cases}$$

From these expressions follow

$$\text{(32)} \qquad \begin{cases} D_1 D'' + D_1'' D - 2\,D_1' D' = 0, \\ D_1 D_0'' + D_1'' D_0 - 2\,D_1' D_0' = 0. \end{cases}$$

Combining this result with (15), we have:

The asymptotic lines upon any one of a group of three surfaces S, S_1, S_0 correspond to a conjugate system on the other two;

or, in other words:

The system of lines which is conjugate for any two of three surfaces S, S_1, S_0 corresponds to the asymptotic lines on the other.

If the curvature of S be negative, its asymptotic lines are real, and consequently the common conjugate system on S_1 and S_0 is real. If these lines be parametric, the second of equations (32) reduces to

$$D_1 D_0'' + D_1'' D_0 = 0.$$

As an odd number of the four quantities in this equation must be negative, either S_0 or S_1 has positive curvature and the other negative. Similar results follow if we begin with the assumption that S_1 or S_0 has negative curvature.

If the curvature of S be positive, the conjugate system common to it and S_1 is real (cf. § 56); consequently the asymptotic lines on S_0 are real, and the curvature of the latter is negative. But we saw that when the curvature of S_0 is negative, and of S positive, that of S_1 also is negative. Hence:

Given a set of three surfaces S, S_1, S_0; one and only one of them has positive curvature.

Suppose that S is referred to the conjugate system corresponding to asymptotic lines on S_0. The point equation of S is

$$\frac{\partial^2\theta}{\partial u\,\partial v}-\begin{Bmatrix}12\\1\end{Bmatrix}\frac{\partial\theta}{\partial u}-\begin{Bmatrix}12\\2\end{Bmatrix}\frac{\partial\theta}{\partial v}=0. \tag{33}$$

We shall prove that this is the point equation of S_1 also.

If we differentiate the equation

$$\frac{\partial x_1}{\partial u}=z_0\frac{\partial y}{\partial u}-y_0\frac{\partial z}{\partial u}$$

with respect to v, and make use of the fact that y and z are solutions of (33), we have, in consequence of (26),

$$\frac{\partial^2 x_1}{\partial u\,\partial v}=\left(\frac{\partial z_0}{\partial v}\frac{\partial y}{\partial u}-\frac{\partial y_0}{\partial v}\frac{\partial z}{\partial u}\right)+\begin{Bmatrix}12\\1\end{Bmatrix}\frac{\partial x_1}{\partial u}+\begin{Bmatrix}12\\2\end{Bmatrix}\frac{\partial x_1}{\partial v}.$$

But the expression in parenthesis is zero in consequence of equations similar to (24), and hence x_1 is a solution of (33).

Since the parametric curves on S_0 are its asymptotic lines, the spherical representation of S_0 and consequently of S must satisfy the condition

$$\frac{\partial}{\partial u}\begin{Bmatrix}12\\1\end{Bmatrix}'=\frac{\partial}{\partial v}\begin{Bmatrix}12\\2\end{Bmatrix}'.$$

Hence we have the theorem of Cosserat:

The problem of infinitesimal deformation of a surface is the same as the determination of the conjugate systems with equal tangential invariants upon the surface.

158. Surfaces resulting from an infinitesimal deformation. We pass to the consideration of the surface S' arising from an infinitesimal deformation of S. Its coördinates are given by

$$x'=x+\epsilon x_1,\qquad y'=y+\epsilon y_1,\qquad z'=z+\epsilon z_1, \tag{34}$$

where ϵ is a small constant whose powers higher than the first are neglected. Since the fundamental quantities of the first order for S', namely E', F', G', are equal to the corresponding ones for S, by means of (26) the expressions for the direction-cosines X', Y', Z' of the normal to S' are reducible to

$$X' = X + \epsilon(Yz_0 - Zy_0) \tag{35}$$

and similar expressions for Y' and Z'.

The derivatives of X' with respect to u and v are reducible by means of (29) to

$$\frac{\partial X'}{\partial u} = \frac{\partial X}{\partial u} + \epsilon\left(z_0\frac{\partial Y}{\partial u} - y_0\frac{\partial Z}{\partial u}\right) + \frac{\epsilon a}{\mathcal{H}}\left(D_0'\frac{\partial X}{\partial u} - D_0\frac{\partial X}{\partial v}\right),$$

$$\frac{\partial X'}{\partial v} = \frac{\partial X}{\partial v} + \epsilon\left(z_0\frac{\partial Y}{\partial v} - y_0\frac{\partial Z}{\partial v}\right) + \frac{\epsilon a}{\mathcal{H}}\left(D_0''\frac{\partial X}{\partial u} - D_0'\frac{\partial X}{\partial v}\right),$$

where a is ± 1 according as the curvature of S_0 is positive or negative. When these results are combined with (26) and (34), we obtain

$$\sum\frac{\partial x'}{\partial u}\frac{\partial X'}{\partial u} = \sum\frac{\partial x}{\partial u}\frac{\partial X}{\partial u} + \frac{\epsilon a}{\mathcal{H}}\sum\frac{\partial x}{\partial u}\left(D_0'\frac{\partial X}{\partial u} - D_0\frac{\partial X}{\partial v}\right)$$
$$+ \epsilon\sum\left[z_0\left(\frac{\partial X}{\partial u}\frac{\partial y}{\partial u} + \frac{\partial Y}{\partial u}\frac{\partial x}{\partial u}\right) - y_0\left(\frac{\partial X}{\partial u}\frac{\partial z}{\partial u} + \frac{\partial Z}{\partial u}\frac{\partial x}{\partial u}\right)\right].$$

The last expression is identically zero, as one sees by writing it out in full. From this and similar expressions for $\sum\frac{\partial x'}{\partial u}\frac{\partial X'}{\partial v}$, $\sum\frac{\partial x'}{\partial v}\frac{\partial X'}{\partial u}$, and $\sum\frac{\partial x'}{\partial v}\frac{\partial X'}{\partial v}$, the values for the second fundamental coefficients of S' can be given in the form

$$\left\{\begin{aligned} \mathcal{D} &= -\sum\frac{\partial x'}{\partial u}\frac{\partial X'}{\partial u} = D + \frac{\epsilon a}{\mathcal{H}}(D_0'D - D_0D'), \\ \mathcal{D}' &= D' + \frac{\epsilon a}{\mathcal{H}}(D_0''D - D_0'D') = D' + \frac{\epsilon a}{\mathcal{H}}(D_0'D' - D_0D''), \\ \mathcal{D}'' &= D'' + \frac{\epsilon a}{\mathcal{H}}(D_0''D' - D_0'D''). \end{aligned}\right. \tag{36}$$

We know that $\mathcal{H}$ is equal to $\pm HK$ according as the curvature of S is positive or negative (cf. § 60). Also, by § 157, one and only one of three surfaces S, S_1, S_0 has positive curvature. Recalling that a in the above formulas is ± 1 according as the curvature of S_0 is positive or negative, we can, in consequence of (31), write equations (36) in the form

$$\mathcal{D} = D \pm \epsilon\frac{H_1}{\phi^2 H}D_1, \quad \mathcal{D}' = D' \pm \epsilon\frac{H_1}{\phi^2 H}D_1', \quad \mathcal{D}'' = D'' \pm \frac{\epsilon H_1}{\phi^2 H}D_1'', \tag{37}$$

where the upper sign holds when S_1 has positive curvature.

From these equations it is seen that $\mathfrak{D}'$ and D' can be zero simultaneously only when D_1' is zero. Hence we have:

The unique conjugate system which remains conjugate in an infinitesimal deformation of a surface is the one corresponding to a conjugate system on S_1, or, what is the same thing, to the asymptotic lines on S_0.

In particular, in order that the curves of this conjugate system be the lines of curvature, it is necessary and sufficient that the spherical representation be orthogonal, and consequently that S_0 be a minimal surface (cf. § 55). From this it follows that the spherical representation of the lines of curvature of S is isothermal. Conversely, if a surface is of this kind, there is a unique minimal surface with the same representation of its asymptotic lines, and this surface can be found by quadratures. Hence the required infinitesimal deformation of the given surface can be effected by quadratures (26), and so we have the theorem of Weingarten *:

A necessary and sufficient condition that a surface admit an infinitesimal deformation which preserves its lines of curvature is that the spherical representation of the latter be isothermal; when such a surface is expressed in terms of parameters referring to its lines of curvature, the deformation can be effected by quadratures.

159. Isothermic surfaces. By means of the results of § 158 we obtain an important theorem concerning surfaces whose lines of curvature form an isothermal system. They are called *isothermic surfaces* (cf. Exs. 1, 3, p. 159).

From equations (23) it follows that if the common conjugate system on two associate surfaces is orthogonal for one it is the same for the other. In this case equation (22) reduces to

$$\frac{\partial^2}{\partial u\,\partial v}\log\frac{E}{G}=0,$$

of which the general integral is

$$\frac{E}{G}=\frac{U}{V},$$

where U and V are functions of u and v respectively. Hence the lines of curvature on S form an isothermal system (cf. § 41).

* *Sitzungsberichte der König. Akademie zu Berlin*, 1886.

If the parameters be isothermic and the linear element written

$$ds^2 = r(du^2 + dv^2),$$

it follows from (21) that

(37) $$\lambda = \frac{1}{r},$$

and equations (23) become

$$\frac{\partial x_0}{\partial u} = \frac{1}{r}\frac{\partial x}{\partial u}, \qquad \frac{\partial x_0}{\partial v} = -\frac{1}{r}\frac{\partial x}{\partial v}.$$

From these results we derive the following theorem of Bour * and Christoffel:

If the linear element of an isothermic surface referred to its lines of curvature be

$$ds^2 = r(du^2 + dv^2),$$

a second isothermic surface can be found by quadratures. It is associate to the given one, and its linear element is

$$ds_0^2 = \frac{1}{r}(du^2 + dv^2).$$

From equations (16) and (17) it follows that the equation of the common conjugate system (IV, 43) on two associate surfaces S, S_0 is reducible to

(38) $$\mu\, du^2 + 2\,\lambda\, dudv + \sigma\, dv^2 = 0.$$

The preceding results tell us that a necessary and sufficient condition that S be an isothermic surface is that there be a set of solutions of equations (19) such that (38) is the equation of the lines of curvature on S. Hence there must be a function ρ such that

$$\mu = \rho(ED' - FD), \quad 2\,\lambda = \rho(ED'' - GD), \quad \sigma = \rho(FD'' - GD')$$

satisfy equations (19).† Upon substitution we are brought to two equations of the form

$$\frac{\partial \log \rho}{\partial u} = \alpha, \qquad \frac{\partial \log \rho}{\partial v} = \beta,$$

where α and β are determinate functions of u and v. In order that S be isothermic, these functions must satisfy the condition

$$\frac{\partial \alpha}{\partial v} = \frac{\partial \beta}{\partial u}.$$

When it is satisfied, ρ and consequently μ, λ, σ are given by quadratures.

* *Journal de l'École Polytechnique*, Cahier 39 (1862), p. 118.

† Cf. Bianchi, Vol. II, p. 30.

Consider furthermore the form

$$H(\mu\, du^2 + 2\,\lambda\, du dv + \sigma\, dv^2). \tag{39}$$

From (37) it is seen that when the lines of curvature are parametric, this expression reduces to $2\, du dv$. Hence its curvature is zero (cf. V, 12), and consequently the curvature of (39) is zero. From § 135 it follows that this form is reducible to $du_1 dv_1$ by quadratures. Hence we have the theorem of Weingarten:

The lines of curvature upon an isothermic surface can be found by quadratures.

We conclude this discussion of isothermic surfaces with the proof of a theorem of Ribaucour. He introduced the term *limit surfaces* of a group of applicable surfaces to designate the members of the group whose mean curvature is a maximum or minimum. According to Ribaucour,

The limit surfaces of a group of applicable surfaces are isothermic.

In proving it we consider a member S of the group referred to its lines of curvature. Its mean curvature is given by $D/E + D''/G$. In consequence of equations (36) the mean curvature of a near-by surface is, to within terms of higher order,

$$\frac{\mathfrak{D}}{E} + \frac{\mathfrak{D}''}{G} = \frac{D}{E} + \frac{D''}{G} \pm \frac{\epsilon}{\sqrt{\gamma}} D_0' \left(\frac{D}{E} - \frac{D''}{G}\right).$$

A necessary and sufficient condition that the mean curvature of S be a maximum or minimum is consequently

$$D_0'\left(\frac{D}{E} - \frac{D''}{G}\right) = 0.$$

Excluding the case of the sphere for which the expression in parenthesis is zero, we have that D_0' is zero. Hence the common conjugate system of S and S_0 is composed of lines of curvature on the former, and therefore S is isothermic.

GENERAL EXAMPLES

1. If x, y, z and x_1, y_1, z_1 are the coördinates of two surfaces corresponding with orthogonality of linear elements, the coördinates of a pair of applicable surfaces are given by

$$\xi_1 = x + tx_1, \qquad \eta_1 = y + ty_1, \qquad \zeta_1 = z + tz_1,$$
$$\xi_2 = x - tx_1, \qquad \eta_2 = y - ty_1, \qquad \zeta_2 = z - tz_1,$$

where t is any constant.

2. If two surfaces are applicable, the locus of the mid-point of the line joining corresponding points admits of an infinitesimal deformation in which this line is the generatrix.

3. Whatever be the surface S, the characteristic equation (7) admits the solution $\phi = aX + bY + cZ$, where a, b, c are constants. Show that S_0 is the point (a, b, c) and that equations (26) become

$$x_1 = cy - bz + d, \qquad y_1 = az - cx + e, \qquad z_1 = bx - ay + f,$$

where d, e, f are constants; that consequently S_1 is a plane, and that the infinitesimal deformation is in reality an infinitesimal displacement.

4. Determine the form of the results of Exs. 1, 2, where ϕ has the value of Ex. 3.

5. Show that the first fundamental coefficients E_1, F_1, G_1 of a surface S_1 are of the form

$$E_1 = E\phi^2 + \frac{1}{H^2K^2}\left(D\frac{\partial\phi}{\partial v} - D'\frac{\partial\phi}{\partial u}\right)^2,$$

$$F_1 = F\phi^2 + \frac{1}{H^2K^2}\left(D\frac{\partial\phi}{\partial v} - D'\frac{\partial\phi}{\partial u}\right)\left(D'\frac{\partial\phi}{\partial v} - D''\frac{\partial\phi}{\partial u}\right),$$

$$G_1 = G\phi^2 + \frac{1}{H^2K^2}\left(D'\frac{\partial\phi}{\partial v} - D''\frac{\partial\phi}{\partial u}\right)^2.$$

6. Let Σ denote the locus of the point which bisects the segment of the normal to a surface S between the centers of principal curvature of the latter. In order that the lines on Σ corresponding to the lines of curvature on S shall form a conjugate system, it is necessary and sufficient that Σ correspond to a minimal surface with orthogonality of linear elements, and that the latter surface and S correspond with parallelism of tangent planes.

7. Show that when the spherical representation of the asymptotic lines of a surface S satisfies the condition

$$\frac{\partial}{\partial u}\begin{Bmatrix}11\\2\end{Bmatrix}' = \frac{\partial}{\partial v}\begin{Bmatrix}22\\1\end{Bmatrix}',$$

equations (25) admit two pairs of solutions which are such that $\mu = \sigma$ and $\mu = -\sigma$. On the two associate surfaces S_0, S_0' thus found by quadratures the parametric systems are isothermal-conjugate, and S_0 and S_0' are associates of one another.

8. Show that the equation of Ex. 7 is a necessary and sufficient condition that two surfaces associate to S be associate to one another.

9. Show that when the sphere is referred to its minimal lines, the condition of Ex. 7 is satisfied, and investigate this case.

10. On any surface associate to a pseudospherical surface the curves corresponding to the asymptotic lines of the latter are geodesics. A surface with a conjugate system of geodesics is called a *surface of Voss* (cf. § 170).

11. Determine whether minimal surfaces and the surfaces associate to pseudospherical surfaces are the only surfaces of Voss.

12. When the equations of a central quadric are in the form (VII, 35), the associate surfaces are given by

$$x_0 = \sqrt{a}\left[\int U(1-u^2)\,du + \int V(1-v^2)\,dv\right],$$

$$y_0 = 2\sqrt{b}\left[\int Uu\,du + \int Vv\,dv\right],$$

$$z_0 = i\sqrt{c}\left[\int U(1+u^2)\,du - \int V(1+v^2)\,dv\right],$$

where U and V are arbitrary functions of u and v respectively; hence the associates are surfaces of translation.

13. When the equations of a paraboloid are in the form

$$x = \sqrt{a}(u+v), \quad y = \sqrt{b}(u-v), \quad z = 2uv,$$

the associate surfaces are surfaces of translation whose generators are plane curves; their equations are

$$x_0 = \sqrt{a}(U+V), \quad y_0 = \sqrt{b}(V-U), \quad z_0 = 2\int uU'du + 2\int vV'dv,$$

where U and V are arbitrary functions of u and v respectively.

14. Show that a quadric admits of an infinitesimal deformation which preserves its lines of curvature, and determine the corresponding associate surface.

15. Since the relation between S and S_1 is reciprocal, there is a surface S_3 associate to S_1 which bears to S a relation similar to that of S_0 to S_1. Show that the asymptotic lines on S_0 and S_3 correspond, and that these surfaces are polar reciprocal with respect to the imaginary sphere $x^2 + y^2 + z^2 + 1 = 0$.

16. Since the relation between S and S_0 is reciprocal, there is a surface S_2 corresponding to S_0 with orthogonality of linear elements which bears to S a relation similar to that of S_1 to S_0. Show that the asymptotic lines on S_1 and S_2 correspond, that the coördinates of the latter are such that

$$x_1 - x_2 = yz_0 - zy_0, \quad y_1 - y_2 = zx_0 - xz_0, \quad z_1 - z_2 = xy_0 - yx_0,$$

and that the line joining corresponding points on S_1 and S_2 is tangent to both surfaces.

17. Show that if S_5 denotes the surface corresponding to S_3 with orthogonality of linear elements which is determined by S_1, associate to S_3, the surfaces S and S_5 are related to one another in a manner similar to S_1 and S_2 of Ex. 16.

18. Show that the surface S_4, which is the associate to S_2 determined by S_0, is the polar reciprocal of S with respect to the imaginary sphere $x^2 + y^2 + z^2 + 1 = 0$.

19. If we continue the process introduced in the foregoing examples, we obtain two sequences of surfaces

$$\begin{array}{llllllll} S, & S_1, & S_3, & S_5, & S_7, & S_9, & S_{11}, & \cdots, \\ S, & S_0, & S_2, & S_4, & S_6, & S_8, & S_{10}, & \cdots. \end{array}$$

Show that S_{11} and S_{10} are the same surface, likewise S_{12} and S_9, and that consequently there is a closed system of twelve surfaces; they are called the *twelve surfaces of Darboux*.

20. A necessary and sufficient condition that a surface referred to its minimal lines be isothermic is that

$$\frac{D}{D''} = \frac{U}{V},$$

where U and V are functions of u and v respectively.

21. A necessary and sufficient condition that the lines of curvature on an isothermic surface be represented on the sphere by an isothermal system is that

$$\frac{\rho_1}{\rho_2} = \frac{U}{V},$$

where U and V are functions of u and v respectively, the latter being parameters referring to the lines of curvature. Show that the parameters of the asymptotic lines on such a surface can be so chosen that $E = G$.

22. Show that an isothermic surface is transformed by an inversion into an isothermic surface.

23. If S_1 and S_2 are the sheets of the envelope of a family of spheres of two parameters, which are not orthogonal to a fixed sphere, and the points of contact of any sphere are said to correspond, in order that the correspondence be conformal, it is necessary that the lines of curvature on S_1 and S_2 correspond and that these surfaces be isothermic (cf. Ex. 15, Chap. XIII).

CHAPTER XII

RECTILINEAR CONGRUENCES

160. Definition of a congruence. Spherical representation. A two-parameter system of straight lines in space is called a *rectilinear congruence.* The normals to a surface constitute such a system; likewise the generatrices of an infinitesimal deformation of a surface (cf. § 152). Later we shall find that in general the lines of a congruence are not normal to a surface. Hence congruences of normals form a special class; they are called *normal congruences.* They were the first studied, particularly in investigations of the effects of reflection and refraction upon rays of light. The first purely mathematical treatment of general rectilinear congruences was given by Kummer in his memoir, Allgemeine Theorie der gradlinigen Strahlensysteme.* We begin our treatment of the subject with the derivation of certain of Kummer's results by methods similar to his own.

From the definition of a congruence it follows that its lines meet a given plane in such a way that through a point of the plane one line, or at most a finite number, pass. Similar results hold if a surface be taken instead of a plane; this surface is called the *surface of reference.* And so we may define a congruence analytically by means of the coördinates of the latter surface in terms of two parameters u, v, and by the direction-cosines of the lines in terms of these parameters. Thus, a congruence is defined by a set of equations such as

$$(1) \qquad \begin{cases} x=f_1(u, v), & y=f_2(u, v), & z=f_3(u, v), \\ X=\phi_1(u, v), & Y=\phi_2(u, v), & Z=\phi_3(u, v), \end{cases}$$

where the functions f and ϕ are analytic in the domain of u and v under consideration, and the functions ϕ are such that

$$X^2+Y^2+Z^2=1.$$

* *Crelle,* Vol. LVII (1860), pp. 189–230.

We make a representation of the congruence upon the unit sphere by drawing radii parallel to the lines of the congruence, and call it the *spherical representation* of the congruence. When we put

$$(2) \qquad \mathcal{E}=\sum\left(\frac{\partial X}{\partial u}\right)^2, \qquad \mathcal{F}=\sum\frac{\partial X}{\partial u}\frac{\partial X}{\partial v}, \qquad \mathcal{G}=\sum\left(\frac{\partial X}{\partial v}\right)^2,$$

the linear element of the spherical representation is

$$(3) \qquad d\sigma^2=\mathcal{E}du^2+2\mathcal{F}dudv+\mathcal{G}dv^2.$$

If we put

$$(4) \quad e=\sum\frac{\partial x}{\partial u}\frac{\partial X}{\partial u}, \quad f=\sum\frac{\partial x}{\partial v}\frac{\partial X}{\partial u}, \quad f'=\sum\frac{\partial x}{\partial u}\frac{\partial X}{\partial v}, \quad g=\sum\frac{\partial x}{\partial v}\frac{\partial X}{\partial v},$$

we have the second quadratic form

$$(5) \qquad \sum dxdX=e\,du^2+(f+f')\,dudv+g\,dv^2,$$

which is fundamental in the theory of congruences.

161. Normal congruences. Ruled surfaces of a congruence. If there be a surface S' normal to the congruence, the coördinates of S' are given by

$$(6) \qquad x'=x+tX, \qquad y'=y+tY, \qquad z'=z+tZ,$$

where t measures the distance from the surface of reference to S'. Since S' is normal to the congruence, we must have

$$(7) \qquad \sum Xd(x+tX)=0,$$

which is equivalent to

$$(8) \qquad \sum X\frac{\partial x}{\partial u}+\frac{\partial t}{\partial u}=0, \qquad \sum X\frac{\partial x}{\partial v}+\frac{\partial t}{\partial v}=0.$$

If these equations be differentiated with respect to v and u respectively, and the resulting equations be subtracted, we obtain

$$(9) \qquad f=f'.$$

Conversely, when this condition is satisfied, the function t given by the quadratures (8) satisfies equation (7). Since t involves an additive constant, equations (6) define a family of parallel surfaces normal to the congruence. Hence:

A necessary and sufficient condition for a normal congruence is that f and f' be equal.

The lines of the congruence which pass through a curve on the surface of reference S form a ruled surface. Such a curve, and consequently a ruled surface of the congruence, is determined by a relation between u and v. Hence a differential equation of the form

$$M\,du + N\,dv = 0 \tag{10}$$

defines a family of ruled surfaces of the congruence. We consider a line $l(u, v)$ of the congruence and the ruled surface Σ of this family upon which l is a generator; we say that Σ passes through l. We apply to Σ the results of §§ 103, 104.

If ds_0 denotes the linear element of the curve C in which Σ cuts the surface of reference, it follows from (VII, 54), (3), and (5) that the quantities a^2 and b for Σ have the values

$$\left\{\begin{aligned} a^2 &= \sum\left(\frac{dX}{ds_0}\right)^2 = \frac{d\sigma^2}{ds_0^2}, \\ b &= \sum\frac{dX}{ds_0}\frac{dx}{ds_0} = \frac{e\,du^2 + (f+f')\,du\,dv + g\,dv^2}{ds_0^2}. \end{aligned}\right. \tag{11}$$

From (VII, 58) we have that the direction-cosines λ, μ, ν of the common perpendicular to l and to the line l' of parameters $u + du$, $v + dv$, where dv/du is given by (10), have the values

$$\lambda = \left(Y\frac{dZ}{d\sigma} - Z\frac{dY}{d\sigma}\right), \quad \mu = \left(Z\frac{dX}{d\sigma} - X\frac{dZ}{d\sigma}\right), \quad \nu = \left(X\frac{dY}{d\sigma} - Y\frac{dX}{d\sigma}\right), \tag{12}$$

which, by means of (V, 31), are reducible to

$$\lambda = \frac{\left(\mathcal{E}\dfrac{\partial X}{\partial v} - \mathcal{F}\dfrac{\partial X}{\partial u}\right)du + \left(\mathcal{F}\dfrac{\partial X}{\partial v} - \mathcal{G}\dfrac{\partial X}{\partial u}\right)dv}{\sqrt{\mathcal{E}\mathcal{G} - \mathcal{F}^2}\cdot d\sigma}, \tag{13}$$

and similar expressions for μ and ν.

From (12) it follows that

$$\lambda\frac{dX}{d\sigma} + \mu\frac{dY}{d\sigma} + \nu\frac{dZ}{d\sigma} = 0.$$

Since $dX/d\sigma$, $dY/d\sigma$, $dZ/d\sigma$ are the direction-cosines of the tangent to the spherical representation of the generators of Σ, we have the theorem:

Given a ruled surface Σ of a congruence; let C be the curve on the unit sphere which represents Σ, and M the point of C corresponding to a generator L of Σ; the limiting position of the common perpendicular to L and a near-by generator of Σ is perpendicular to the tangent to C at M.

162. Limit points. Principal surfaces. By means of (VII, 62) and (12) we find that the expression for the shortest distance δ between l and l' is, to within terms of higher order,

$$\delta = \frac{ds_0^2}{d\sigma}\begin{vmatrix} \frac{dx}{ds_0} & \frac{dy}{ds_0} & \frac{dz}{ds_0} \\ X & Y & Z \\ \frac{dX}{ds_0} & \frac{dY}{ds_0} & \frac{dZ}{ds_0} \end{vmatrix} = \Sigma\lambda\, dx.$$

When the values (13) for λ, μ, ν are substituted in the right-hand member of this equation, the result is reducible to

$$\delta = \frac{1}{/\!\!\!f\, d\sigma}\begin{vmatrix} \mathcal{E}du + \mathcal{F}dv, & \mathcal{F}du + \mathcal{G}dv \\ e\, du + f\, dv, & f'du + g\, dv \end{vmatrix}. \tag{14}$$

If N denotes the point where this line of shortest distance meets l, the locus of N is the line of striction of Σ. Hence the distance of N from the surface S, measured along l, is given by (VII, 65); if it be denoted by r, we have, from (11),

$$r = -\frac{e\, du^2 + (f + f')\, du dv + g\, dv^2}{\mathcal{E}du^2 + 2\,\mathcal{F}du dv + \mathcal{G}dv^2}. \tag{15}$$

For the present* we exclude the case where the coefficients of the two quadratic forms are proportional. Hence r varies with the value of dv/du, that is, with the ruled surface Σ through l. If we limit our consideration to real surfaces Σ, the denominator is always positive, and consequently the quantity r has a finite maximum and minimum. In order to find the surfaces Σ for which r has these limiting values, we replace dv/du by t, and obtain

$$r = -\frac{e + (f + f')\, t + gt^2}{\mathcal{E} + 2\,\mathcal{F}t + \mathcal{G}t^2}. \tag{16}$$

If we equate to zero the derivative of the right-hand member with respect to t, we get

$$(\mathcal{E} + \mathcal{F}t)\left[\frac{1}{2}(f + f') + gt\right] - (\mathcal{F} + \mathcal{G}t)\left[e + \frac{1}{2}(f + f')\,t\right] = 0, \tag{17}$$

a quadratic in t. Since $\mathcal{E}\mathcal{G} - \mathcal{F}^2 > 0$, we may apply to this equation reasoning similar to that used in connection with equation (IV, 21),

* Cf. Ex. 1, § 171.

and thus prove that it has two real roots. The corresponding values of r follow from (16) when these values of t are substituted in the latter. Because of (17) the resulting equation may be written

$$\bar{r} = -\frac{e + \frac{1}{2}(f+f')\bar{t}}{\mathcal{E} + \mathcal{F}\bar{t}} = -\frac{\frac{1}{2}(f+f') + g\bar{t}}{\mathcal{F} + \mathcal{G}\bar{t}},$$

where $\bar{t}$ indicates a root of (17) and $\bar{r}$ the corresponding value of r.

When we write the preceding equations in the form

$$\left[\mathcal{E}\bar{r} + e\right] + \left[\mathcal{F}\bar{r} + \frac{1}{2}(f+f')\right]\bar{t} = 0,$$

$$\left[\mathcal{F}\bar{r} + \frac{1}{2}(f+f')\right] + \left[\mathcal{G}\bar{r} + g\right]\bar{t} = 0,$$

and eliminate $\bar{t}$, we obtain the following quadratic in $\bar{r}$:

$$(\mathcal{E}\mathcal{G} - \mathcal{F}^2)\bar{r}^2 + [g\mathcal{E} - (f+f')\mathcal{F} + e\mathcal{G}]\bar{r} + eg - \left(\frac{f+f'}{2}\right)^2 = 0. \tag{18}$$

If r_1 and r_2 denote the roots of this equation, we have

$$\left\{\begin{aligned} r_1 + r_2 &= \frac{(f+f')\mathcal{F} - g\mathcal{E} - e\mathcal{G}}{\mathcal{E}\mathcal{G} - \mathcal{F}^2}, \\ r_1 r_2 &= \frac{4\,eg - (f+f')^2}{4(\mathcal{E}\mathcal{G} - \mathcal{F}^2)}. \end{aligned}\right. \tag{19}$$

The points on l corresponding to these values of r are called its *limit points*. They are the boundaries of the segment of l upon which lie the feet of each perpendicular common to it and to a near-by line of the congruence. The ruled surfaces of the congruences which pass through l and are determined by equation (17) are called the *principal surfaces* for the line. There are two of them, and their tangent planes at the limit points are determined by l and by the perpendiculars of shortest distance at the limit points. They are called the *principal planes*.

In order to find other properties of the principal surfaces, we imagine that the parametric curves upon the sphere represent these surfaces. If equation (17) be written

$$\begin{vmatrix} \mathcal{E}\,du + \mathcal{F}\,dv, & \mathcal{F}\,du + \mathcal{G}\,dv \\ e\,du + \frac{1}{2}(f+f')\,dv, & \frac{1}{2}(f+f')\,du + g\,dv \end{vmatrix} = 0, \tag{20}$$

it is seen that a necessary and sufficient condition that the ruled surfaces $v=\text{const.}$, $u=\text{const.}$ be the principal surfaces, is

$$\frac{1}{2}(f+f')\mathcal{E}-e\mathcal{F}=0, \qquad g\mathcal{F}-\frac{1}{2}(f+f')\mathcal{G}=0.$$

From these it follows that since the coefficients of the two fundamental quadratic forms are not proportional, we must have

$$\mathcal{F}=0, \qquad f+f'=0. \tag{21}$$

From the first of these equations and the preceding theorem follows the result:

The principal surfaces of a congruence are represented on the sphere by an orthogonal system, and the two principal planes for each line are perpendicular to one another.

For this particular parametric system equation (13) reduces to

$$\lambda=\frac{\mathcal{E}\frac{\partial X}{\partial v}du-\mathcal{G}\frac{\partial X}{\partial u}dv}{\sqrt{\mathcal{E}\mathcal{G}}\,d\sigma}, \tag{22}$$

so that the direction-cosines λ_1, μ_1, ν_1 of the perpendicular whose foot is the limit point on l corresponding to $v=\text{const.}$ have the values

$$\lambda_1=\frac{1}{\sqrt{\mathcal{G}}}\frac{\partial X}{\partial v}, \qquad \mu_1=\frac{1}{\sqrt{\mathcal{G}}}\frac{\partial Y}{\partial v}, \qquad \nu_1=\frac{1}{\sqrt{\mathcal{G}}}\frac{\partial Z}{\partial v}.$$

Hence the angle ω between the lines with these direction-cosines and those with (22) is given by

$$\cos\omega=\sum\lambda\lambda_1=\frac{\sqrt{\mathcal{E}}\,du}{\sqrt{\mathcal{E}\,du^2+\mathcal{G}\,dv^2}}. \tag{23}$$

The values of r_1 and r_2 are now

$$r_1=-\frac{e}{\mathcal{E}}, \qquad r_2=-\frac{g}{\mathcal{G}},$$

so that with the aid of (23) equation (15) can be put in the form

$$r=r_1\cos^2\omega+r_2\sin^2\omega. \tag{24}$$

This is *Hamilton's equation.* We remark that it is independent of the choice of parameters.

163. Developable surfaces of a congruence. Focal surfaces. In order that a ruled surface be developable, it is necessary and sufficient that the perpendicular distance between very near generators be of the second or higher order. From (14) it follows that the ruled surfaces of a congruence satisfying the condition

$$\begin{vmatrix} \mathcal{E}\,du + \mathcal{F}\,dv, & \mathcal{F}\,du + \mathcal{G}\,dv \\ e\,du + f\,dv, & f'\,du + g\,dv \end{vmatrix} = 0 \tag{25}$$

are developable. Unlike equation (20), the values of dv/du satisfying this equation are not necessarily real. We have then the theorem:

Of all the ruled surfaces of a congruence through a line of it two are developable, but they are not necessarily real.

The normals to a real surface afford an example of a congruence with real developables; for, the normals along a line of curvature form a developable surface (§ 51). Since f and f' are equal in this case, equations (20) and (25) are equivalent. And, conversely, they are equivalent only in this case. Hence:

When a congruence is normal, and only then, the principal surfaces are developable.

When a ruled surface is developable its generators are tangent to a curve at the points where the lines of shortest distance meet them. Hence each line of a congruence is tangent to two curves in space, real or imaginary according to the character of the roots of equation (25). The points of contact are called the *focal points* for the line. By means of (25) we find that the values of r for these points are given by

$$\rho = -\frac{e\,du + f\,dv}{\mathcal{E}\,du + \mathcal{F}\,dv} = -\frac{f'\,du + g\,dv}{\mathcal{F}\,du + \mathcal{G}\,dv}.$$

If these equations be written in the form

$$(\mathcal{E}\rho + e)\,du + (\mathcal{F}\rho + f)\,dv = 0,$$
$$(\mathcal{F}\rho + f')\,du + (\mathcal{G}\rho + g)\,dv = 0,$$

and if du, dv be eliminated, we have

$$(\mathcal{E}\mathcal{G} - \mathcal{F}^2)\rho^2 + [g\mathcal{E} - (f + f')\mathcal{F} + e\mathcal{G}]\rho + eg - ff' = 0. \tag{26}$$

If ρ_1 and ρ_2 denote the roots of this equation, it follows that

$$(27)\quad \begin{cases} \rho_1+\rho_2=\dfrac{(f+f')\mathcal{F}-g\mathcal{E}-e\mathcal{G}}{\mathcal{E}\mathcal{G}-\mathcal{F}^2}, \\ \rho_1\rho_2=\dfrac{eg-ff'}{\mathcal{E}\mathcal{G}-\mathcal{F}^2}. \end{cases}$$

From (19) and (27) it is seen that

$$(28)\quad \begin{cases} r_1+r_2=\rho_1+\rho_2, \\ (r_1-r_2)^2-(\rho_1-\rho_2)^2=\dfrac{(f-f')^2}{(\mathcal{E}\mathcal{G}-\mathcal{F}^2)}. \end{cases}$$

These results may be interpreted as follows:

The mid-points of the two segments bounded respectively by the limit points and by the focal points coincide.

This point is called the *middle point* of the line and its locus the *middle surface* of the congruence.

The distance between the focal points is never greater than that between the limit points. They coincide when the congruence is normal.

Equation (24) may be written in the forms

$$\cos^2\omega=\frac{r-r_2}{r_1-r_2},\qquad \sin^2\omega=\frac{r_1-r}{r_1-r_2}.$$

Hence if ω_1 and ω_2 denote the values of ω corresponding to the developable surfaces, we have

$$\cos^2\omega_1=\frac{\rho_1-r_2}{r_1-r_2},\qquad \sin^2\omega_1=\frac{r_1-\rho_1}{r_1-r_2},$$

$$\cos^2\omega_2=\frac{\rho_2-r_2}{r_1-r_2}\qquad \sin^2\omega_2=\frac{r_1-\rho_2}{r_1-r_2}.$$

From these and the first of (28) it follows that

$$\cos^2\omega_1=\sin^2\omega_2,\qquad \sin^2\omega_1=\cos^2\omega_2,$$

so that

$$(29)\qquad \cos 2\,\omega_1+\cos 2\,\omega_2=0,$$

and consequently

$$(30)\qquad \omega_1+\omega_2=\frac{\pi}{2}\pm n\pi,$$

or

$$(31)\qquad \omega_1-\omega_2=\frac{\pi}{2}\pm n\pi,$$

where n denotes any integer. If the latter equation be true, the developable surfaces are represented on the sphere by an orthogonal system, as follows from the theorem at the close of § 161. But by § 34 the condition that equation (25) define an orthogonal system on the sphere is $f = f'$, that is, the congruence must be normal. Since in this case the principal surfaces are the developables, equation (30) as well as (31) is satisfied. Hence equation (30) is the general solution of (29).

The planes through l which make the angles ω_1, ω_2 with the principal plane $\omega = 0$ are called the *focal planes* for the line; they are the tangent planes to the two developable surfaces through the line. Incidentally we have proved the theorem:

A necessary and sufficient condition that the two focal planes for each line of a congruence be perpendicular is that the congruence be normal.

And from equation (30) it follows that

The focal planes are symmetrically placed with respect to the principal planes in such a way that the angles formed by the two pairs of planes have the same bisecting planes.

If θ denote the angle between the focal planes, then

$$\theta = \omega_2 - \omega_1 = \frac{\pi}{2} - 2\,\omega_1,$$

and

$$\sin\theta = \cos 2\,\omega_1 = \cos^2\omega_1 - \cos^2\omega_2 = \frac{\rho_1 - \rho_2}{r_1 - r_2}. \tag{32}$$

The loci of the focal points of a congruence are called its *focal surfaces*. Each line of the congruence touches both surfaces, being tangent to the edges of regression of the two developables through it. By reasoning similar to that employed in the discussion of surfaces of center (§ 74) we prove the theorem:

A congruence may be regarded as two families of developable surfaces. Each focal surface is touched by the developables of one family along their edges of regression and enveloped by those of the other family along the curves conjugate to these edges.

The preceding theorem shows that of the two focal planes through a line l one is tangent to the focal surface S_1 and the other is the osculating plane of the edge of regression on S_1 to which l is tangent; similar results hold for S_2. When the congruence is normal these planes are perpendicular, and consequently these edges of regression are geodesics on S_1 and S_2. Since the converse is true (§ 76), we have:

A necessary and sufficient condition that the tangents to a family of curves on a surface form a normal congruence is that the curves be geodesics.

EXAMPLES

1. If X, Y, Z are the direction-cosines of the normal to a minimal surface at the point (x, y, z), the line whose direction-cosines are Y, $-X$, Z and which passes through the point $(x, y, 0)$ generates a normal congruence.

2. Prove that the tangent planes to two confocal quadrics at the points of contact of a common tangent are perpendicular, and consequently that the common tangents to two confocal quadrics form a normal congruence.

3. Find the congruence of common tangents to the paraboloids

$$x^2 + y^2 = 2\,az, \qquad x^2 + y^2 = -2\,az,$$

and determine the focal surfaces.

4. If two ruled surfaces through a line L are represented on the sphere by orthogonal lines, their lines of striction meet L at points equally distant from the middle point.

5. In order that the focal planes for each line of a congruence meet under the same angle, it is necessary and sufficient that the osculating planes of the edges of regression of the developables meet the tangent planes to the focal surfaces under constant angle.

6. A necessary and sufficient condition that a surface of reference of a congruence be its middle surface is $g\mathcal{E} - (f + f')\mathcal{F} + e\mathcal{G} = 0$.

164. Associate normal congruences. If we put

$$\gamma = \sum X \frac{\partial x}{\partial u}, \qquad \gamma_1 = \sum X \frac{\partial x}{\partial v}, \tag{33}$$

equations (8) may be replaced by

$$t = c - \int \gamma\, du + \gamma_1 dv, \tag{34}$$

where c is a constant. Now equation (9) is equivalent to

$$\frac{\partial \gamma}{\partial v} = \frac{\partial \gamma_1}{\partial u}. \tag{35}$$

In consequence of this condition equation (34) may be written

$$t = c - \phi(u_1), \tag{36}$$

where u_1 is a function of u and v thus defined. If the orthogonal trajectories of the curves $u_1 =$ const. be taken as parametric curves $v_1 =$ const., it follows from (36) and from equations in u_1 and v_1 analogous to (33) and (34) that

$$\sum X \frac{\partial x}{\partial v_1} = 0.$$

From this result follows the theorem:

The lines of a normal congruence cut orthogonally the curves on the surface of reference at whose points t is constant.

If θ denotes the angle which a line of the congruence makes with the normal to the surface of reference at the point of intersection, we have

$$\sin\theta = \sum \frac{X}{\sqrt{E}} \frac{\partial x}{\partial u_1} = \frac{\phi'(u_1)}{\sqrt{E}}, \tag{37}$$

where the linear element of the surface is

$$ds^2 = E\, du_1^2 + G\, dv_1^2.$$

If S be taken for the surface of reference of a second congruence whose direction-cosines X_1, Y_1, Z_1 satisfy the conditions

$$\sum X_1 \frac{\partial x}{\partial u_1} = \phi_1'(u_1), \qquad \sum X_1 \frac{\partial x}{\partial v_1} = 0,$$

where $\phi_1(u_1)$ is any function whatever of u_1, this congruence is normal and t_1 has the value

$$t_1 = c_1 - \phi_1(u_1).$$

Since ϕ_1 is any function, there is a family of these normal congruences which we call the *associates* of the given congruence and of one another. Through any point of the surface of reference there passes a line of each congruence, and all of these lines lie in the plane normal to the curve $u_1 =$ const. through the point. Hence:

The two lines of two associate congruences through the same point of the surface of reference lie in a plane normal to the surface.

Combining with equation (37) a similar one for an associate congruence, we have

$$\frac{\sin\theta}{\sin\theta_1}=\frac{\phi'(u_1)}{\phi_1'(u_1)}=f(u_1). \tag{38}$$

Hence we have the theorem:

The ratio of the sines of the angles which the lines of two associate congruences make with the normal to their surface of reference is constant along the curves at whose points t is constant.

When in particular $f(u_1)$ in (38) is a constant, the former theorem and equation (38) constitute the laws of reflection and refraction of rays of light, according as the constant is equal to or different from minus one. And so we have the theorem of Malus and Dupin:

If a bundle of rays of light forming a normal congruence be reflected or refracted any number of times by the surfaces of successive homogeneous media, the rays continue to constitute a normal congruence.

By means of (37) equation (36) can be put in the form

$$t=c-\int\sqrt{E}\sin\theta\,du_1.$$

From this result follows the theorem of Beltrami *:

If a surface of reference of a normal congruence be deformed in such a way that the directions of the lines of the congruence with respect to the surface be unaltered, the congruence continues to be normal.

165. Derived congruences. It is evident that the tangents to the curves of any one-parameter family upon a surface S constitute a congruence. If these curves be taken for the parametric lines $v=$ const., and their conjugates for $u=$ const., the developables in one family have the curves $v=$ const. for edges of regression, and the developables of the other family envelop S along the curves $u=$ const. We may take S for the surface of reference. If S_1 be the other focal surface, the lines of the congruence are tangent to the curves $u=$ const. on S_1. The tangents to the curves $v=$ const. on S_1 form a second congruence of which S_1 is one focal surface, and the second surface S_2 is uniquely determined. Moreover, the

* *Giornale di matematiche*, Vol. II (1864), p. 281.

lines of the second congruence are tangent to the curves $u = \text{const.}$ on S_2. In turn we may construct a third congruence of tangents to the curves $v = \text{const.}$ on S_2. This process may be continued indefinitely unless one of these focal surfaces reduces to a curve, or is infinitely distant.

In like manner we get a congruence by drawing tangents to the curves $u = \text{const.}$ on S, which is one focal surface, and the other, S_{-1}, is completely determined. The tangents to the curves $u = \text{const.}$ on S_{-1} form still another, and so on. In this way we obtain a suite of surfaces

$$\cdots \quad S_{-2}, \quad S_{-1}, \quad S, \quad S_1, \quad S_2 \quad \cdots,$$

which is terminated only when a surface reduces to a curve, or its points are infinitely distant. Upon each of these surfaces the parametric curves form a conjugate system. The congruences thus obtained have been called *derived congruences* by Darboux.* It is clear that the problem of finding all the derived congruences of a given one reduces to the integration of the equation of its developables (25); for, when the developables are known we have the conjugate system on its focal surfaces.

In order to derive the analytical expressions for these results, we recall (§ 80) that the coördinates x, y, z of S are solutions of an equation of the form

$$\frac{\partial^2 \theta}{\partial u\, \partial v} + a \frac{\partial \theta}{\partial u} + b \frac{\partial \theta}{\partial v} = 0, \tag{39}$$

where a and b are determinate functions of u and v. If the coördinates of S_1 be denoted by x_1, y_1, z_1, they are given by

$$x_1 = x + \lambda_1 \frac{\partial x}{\partial u}, \qquad y_1 = y + \lambda_1 \frac{\partial y}{\partial u}, \qquad z_1 = z + \lambda_1 \frac{\partial z}{\partial u},$$

where $\lambda_1 \sqrt{E}$ measures the distance between the focal points. But as the lines of the congruence are tangent to the curves $u = \text{const.}$ on S_1, we must have

$$\frac{\partial x_1}{\partial v} = \mu_1 \frac{\partial x}{\partial u}, \qquad \frac{\partial y_1}{\partial v} = \mu_1 \frac{\partial y}{\partial u}, \qquad \frac{\partial z_1}{\partial v} = \mu_1 \frac{\partial z}{\partial u}, \tag{40}$$

* Vol. II, pp. 16–22.

where μ_1 is a determinate function of u and v. When the above value for x_1 is substituted in the first of these equations, the result is reducible, by means of (39), to

$$\left(\frac{\partial \lambda_1}{\partial v} - a\lambda_1 - \mu_1\right)\frac{\partial x}{\partial u} + (1 - b\lambda_1)\frac{\partial x}{\partial v} = 0.$$

Since the same equation is true for y and z, the quantities in parentheses must be zero, that is,

$$\lambda_1 = \frac{1}{b}, \qquad \mu_1 = \frac{\partial}{\partial v}\frac{1}{b} - \frac{a}{b}.$$

Hence the surface S_1 is defined by

$$(41) \qquad x_1 = x + \frac{1}{b}\frac{\partial x}{\partial u}, \qquad y_1 = y + \frac{1}{b}\frac{\partial y}{\partial u}, \qquad z_1 = z + \frac{1}{b}\frac{\partial z}{\partial u},$$

and equations (40) become

$$(42) \quad \frac{\partial x_1}{\partial v} = \left(\frac{\partial}{\partial v}\frac{1}{b} - \frac{a}{b}\right)\frac{\partial x}{\partial u}, \quad \frac{\partial y_1}{\partial v} = \left(\frac{\partial}{\partial v}\frac{1}{b} - \frac{a}{b}\right)\frac{\partial y}{\partial u}, \quad \frac{\partial z_1}{\partial v} = \left(\frac{\partial}{\partial v}\frac{1}{b} - \frac{a}{b}\right)\frac{\partial z}{\partial u}.$$

Proceeding in a similar manner, we find that S_{-1} is defined by the equations

$$(43) \qquad x_{-1} = x + \frac{1}{a}\frac{\partial x}{\partial v}, \qquad y_{-1} = y + \frac{1}{a}\frac{\partial y}{\partial v}, \qquad z_{-1} = z + \frac{1}{a}\frac{\partial z}{\partial v},$$

and that

$$\frac{\partial x_{-1}}{\partial u} = \left(\frac{\partial}{\partial u}\frac{1}{a} - \frac{b}{a}\right)\frac{\partial x}{\partial v},$$

and similar expressions in y_{-1} and z_{-1}.

From (41) and (43) it is seen that the surface S_1 or S_{-1} is at infinity, according as b or a is zero. When a and b are both zero, S is a surface of translation (§ 81). Hence the tangents to the generators of a surface of translation form two congruences for each of which the other focal surface is at infinity.

In order that S_1 be a curve, x_1, y_1, z_1 must be functions of u alone. From (42) it follows that the condition for this is

$$\frac{\partial}{\partial v}\frac{1}{b} = \frac{a}{b}.$$

In like manner the condition that S_{-1} be a curve is

$$\frac{\partial}{\partial u}\frac{1}{a} = \frac{b}{a}.$$

The functions h and k, defined by

$$h = \frac{\partial a}{\partial u} + ab, \qquad k = \frac{\partial b}{\partial v} + ab,$$

are called the invariants of the differential equation (39). Hence the above results may be stated:

A necessary and sufficient condition that the focal surface S_1 or S_{-1} be a curve is that the invariant k or h respectively of the point equation of S be zero.

166. Fundamental equations of condition. We have seen (§ 160) that with every congruence there are associated two quadratic differential forms. Now we shall investigate under what conditions two quadratic forms determine a congruence. We assume that we have two such forms and that there is a corresponding congruence. The tangents to the parametric curves on the surface of reference at a point are determined by the angles which they make with the tangents to the parametric curves of the spherical representation of the congruence at the corresponding point, and with the normal to the unit sphere. Hence we have the relations

$$(44) \qquad \begin{cases} \dfrac{\partial x}{\partial u} = \alpha \dfrac{\partial X}{\partial u} + \beta \dfrac{\partial X}{\partial v} + \gamma X, \\[2ex] \dfrac{\partial x}{\partial v} = \alpha_1 \dfrac{\partial X}{\partial u} + \beta_1 \dfrac{\partial X}{\partial v} + \gamma_1 X, \end{cases}$$

and similar equations in y and z, where α, β, γ; $\alpha_1, \beta_1, \gamma_1$ are functions of u and v. If we multiply these equations by $\dfrac{\partial X}{\partial u}, \dfrac{\partial Y}{\partial u}, \dfrac{\partial Z}{\partial u}$ respectively, and add; also by $\dfrac{\partial X}{\partial v}, \dfrac{\partial Y}{\partial v}, \dfrac{\partial Z}{\partial v}$ and by X, Y, Z; we obtain

$$e = \alpha \mathcal{E} + \beta \mathcal{F}, \qquad f = \alpha_1 \mathcal{E} + \beta_1 \mathcal{F}, \qquad \gamma = \sum X \frac{\partial x}{\partial u},$$

$$f' = \alpha \mathcal{F} + \beta \mathcal{G}, \qquad g = \alpha_1 \mathcal{F} + \beta_1 \mathcal{G}, \qquad \gamma_1 = \sum X \frac{\partial x}{\partial v},$$

from which we derive

$$(45) \qquad \begin{cases} \alpha = \dfrac{e\mathcal{G} - f'\mathcal{F}}{\mathcal{E}\mathcal{G} - \mathcal{F}^2}, & \beta = \dfrac{f'\mathcal{E} - e\mathcal{F}}{\mathcal{E}\mathcal{G} - \mathcal{F}^2}, & \gamma = \sum X \dfrac{\partial x}{\partial u}, \\[2ex] \alpha_1 = \dfrac{f\mathcal{G} - g\mathcal{F}}{\mathcal{E}\mathcal{G} - \mathcal{F}^2}, & \beta_1 = \dfrac{g\mathcal{E} - f\mathcal{F}}{\mathcal{E}\mathcal{G} - \mathcal{F}^2}, & \gamma_1 = \sum X \dfrac{\partial x}{\partial v}. \end{cases}$$

In order that equations (44) be consistent, we must have

$$(46)\qquad \frac{\partial}{\partial u}\left(\frac{\partial x}{\partial v}\right)=\frac{\partial}{\partial v}\left(\frac{\partial x}{\partial u}\right),$$

which, in consequence of equations (V, 22), is reducible to the form

$$R\frac{\partial X}{\partial u}+S\frac{\partial X}{\partial v}+TX=0,$$

where R, S, T are determinate functions. Since this equation must be satisfied by Y and Z also, we must have $R=0$, $S=0$, $T=0$. When the values of α, β, α_1, β_1, from (45), are substituted in these equations, we have

$$(47)\qquad \begin{cases} \dfrac{\partial e}{\partial v}-\dfrac{\partial f}{\partial u}-\begin{Bmatrix}12\\1\end{Bmatrix}'e+\begin{Bmatrix}11\\1\end{Bmatrix}'f-\begin{Bmatrix}12\\2\end{Bmatrix}'f'+\begin{Bmatrix}11\\2\end{Bmatrix}'g+\mathcal{F}\gamma-\mathcal{E}\gamma_1=0,\\ \dfrac{\partial f'}{\partial v}-\dfrac{\partial g}{\partial u}-\begin{Bmatrix}22\\1\end{Bmatrix}'e+\begin{Bmatrix}12\\1\end{Bmatrix}'f-\begin{Bmatrix}22\\2\end{Bmatrix}'f'+\begin{Bmatrix}12\\2\end{Bmatrix}'g+\mathcal{G}\gamma-\mathcal{F}\gamma_1=0,\\ \dfrac{\partial \gamma}{\partial v}-\dfrac{\partial \gamma_1}{\partial u}+f-f'=0. \end{cases}$$

Conversely, when we have a quadratic form whose curvature is $+1$, it may be taken as the linear element of the spherical representation of a congruence, which is determined by any set of functions e, f, f', g, γ, γ_1, satisfying equations (47). For, when these equations are satisfied, so also is (46), and consequently the coördinates of the surface of reference are given by the quadratures (44).

Incidentally we remark that when the congruence is normal, and the surface of reference is one of the orthogonal surfaces, the last of equations (47) is satisfied identically, and the first two reduce to the Codazzi equations (V, 27).

We apply these results to the determination of the congruences with an assigned spherical representation of their principal surfaces, and those with a given representation of their developables.

167. Spherical representation of principal surfaces and of developables. A necessary and sufficient condition that the principal surfaces of a congruence cut the surface of reference in the parametric lines is given by (21).

If we require that the surface of reference be the middle surface of the congruence, and if r denote half the distance between the limit points, we have, from (15),

$$e = -r\mathcal{E}, \qquad g = r\mathcal{G}. \tag{48}$$

When these values are substituted in (47), the first two become

$$\begin{cases} \gamma_1 = -\dfrac{1}{\mathcal{E}}\dfrac{\partial}{\partial v}(r\mathcal{E}) - \sqrt{\dfrac{\mathcal{G}}{\mathcal{E}}}\dfrac{\partial}{\partial u}\left(\dfrac{f}{\sqrt{\mathcal{E}\mathcal{G}}}\right), \\ \gamma = \dfrac{1}{\mathcal{G}}\dfrac{\partial}{\partial u}(r\mathcal{G}) + \sqrt{\dfrac{\mathcal{E}}{\mathcal{G}}}\dfrac{\partial}{\partial v}\left(\dfrac{f}{\sqrt{\mathcal{E}\mathcal{G}}}\right), \end{cases} \tag{49}$$

and the last is reducible to *

$$\begin{aligned} 2\frac{\partial^2 r}{\partial u\,\partial v} + \frac{\partial \log \mathcal{E}}{\partial v}\frac{\partial r}{\partial u} + \frac{\partial \log \mathcal{G}}{\partial u}\frac{\partial r}{\partial v} + \frac{\partial^2 \log \mathcal{E}\mathcal{G}}{\partial u\,\partial v}r \\ + \frac{\partial}{\partial u}\left[\sqrt{\frac{\mathcal{G}}{\mathcal{E}}}\frac{\partial}{\partial u}\left(\frac{f}{\sqrt{\mathcal{E}\mathcal{G}}}\right)\right] + \frac{\partial}{\partial v}\left[\sqrt{\frac{\mathcal{E}}{\mathcal{G}}}\frac{\partial}{\partial v}\left(\frac{f}{\sqrt{\mathcal{E}\mathcal{G}}}\right)\right] + 2f = 0. \end{aligned} \tag{50}$$

Moreover, equations (44) become

$$\frac{\partial x}{\partial u} = -r\frac{\partial X}{\partial u} - \frac{f}{\mathcal{G}}\frac{\partial X}{\partial v} + \gamma X, \qquad \frac{\partial x}{\partial v} = \frac{f}{\mathcal{E}}\frac{\partial X}{\partial u} + r\frac{\partial X}{\partial v} + \gamma_1 X, \tag{51}$$

where γ and γ_1 are given by (49); and similar equations in y and z. Our problem reduces, therefore, to the determination of pairs of functions r and f which satisfy (50). Evidently either of these functions may be chosen arbitrarily and the other is found by the solution of a partial differential equation of the second order. Hence any orthogonal system on the unit sphere serves for the representation of the principal surfaces of a family of congruences, whose equations involve three arbitrary functions.

In order that the parametric curves on the sphere represent the developables of a congruence, it is necessary and sufficient that

$$f'\mathcal{E} - e\mathcal{F} = 0, \qquad g\mathcal{F} - f\mathcal{G} = 0,$$

as is seen from (25). If the surface of reference be the middle surface, and ρ denotes half the distance between the focal points, it follows from (15) that

$$\rho = -\frac{e}{\mathcal{E}} = \frac{g}{\mathcal{G}}.$$

* Cf. Bianchi, Vol. I, p. 314.

Combining these equations with the above, we have

$$(52) \qquad e = -\rho\mathcal{E}, \qquad f = -f' = \rho\mathcal{F}, \qquad g = \rho\mathcal{G}.$$

When these values are substituted in the first two of equations (47) and the resulting equations are solved for γ and γ_1, we find

$$\gamma = \frac{\partial \rho}{\partial u} + 2\begin{Bmatrix}12\\2\end{Bmatrix}'\rho, \qquad \gamma_1 = -\frac{\partial \rho}{\partial v} - 2\begin{Bmatrix}12\\1\end{Bmatrix}'\rho,$$

and the last of equations (47) reduces to

$$(53) \quad \frac{\partial^2 \rho}{\partial u \partial v} + \begin{Bmatrix}12\\1\end{Bmatrix}'\frac{\partial \rho}{\partial u} + \begin{Bmatrix}12\\2\end{Bmatrix}'\frac{\partial \rho}{\partial v} + \left[\frac{\partial}{\partial u}\begin{Bmatrix}12\\1\end{Bmatrix}' + \frac{\partial}{\partial v}\begin{Bmatrix}12\\2\end{Bmatrix}' + \mathcal{F}\right]\rho = 0.$$

Each solution of this equation determines a congruence with the given representation of its developables,* and the middle surface is given by the quadratures

$$(54) \quad \begin{cases} \dfrac{\partial x}{\partial u} = -\rho\dfrac{\partial X}{\partial u} + \left(\dfrac{\partial \rho}{\partial u} + 2\begin{Bmatrix}12\\2\end{Bmatrix}'\rho\right)X, \\[2ex] \dfrac{\partial x}{\partial v} = \rho\dfrac{\partial X}{\partial v} - \left(\dfrac{\partial \rho}{\partial v} + 2\begin{Bmatrix}12\\1\end{Bmatrix}'\rho\right)X, \end{cases}$$

and similar expressions in y and z.

When the values (52) are substituted in (18) the latter becomes

$$(\mathcal{E}\mathcal{G} - \mathcal{F}^2)\bar{r}^2 - \rho^2\mathcal{E}\mathcal{G} = 0.$$

Consequently equation (32) reduces to

$$\sin\theta = \frac{2\rho}{2\bar{r}} = \frac{\sqrt{\mathcal{E}\mathcal{G} - \mathcal{F}^2}}{\sqrt{\mathcal{E}\mathcal{G}}}.$$

Referring to equation (III, 16), we have:

The angle between the focal planes of a congruence is equal to the angle between the lines on the sphere representing the corresponding developables.

This result is obtained readily from geometrical considerations.

168. Fundamental quantities for the focal surfaces. We shall make use of these results in deriving the expressions for the fundamental quantities of the focal surfaces S_1 and S_2, which are defined by

$$x_1 = x + \rho X, \qquad y_1 = y + \rho Y, \qquad z_1 = z + \rho Z,$$
$$x_2 = x - \rho X, \qquad y_2 = y - \rho Y, \qquad z_2 = z - \rho Z.$$

* This result is due to Guichard, *Annales de l'École Normale*, Ser. 3, Vol. VI (1889), pp. 342–344.

From these and (54) we get

$$(55)\quad \begin{cases} \dfrac{\partial x_1}{\partial u} = 2\left(\dfrac{\partial \rho}{\partial u} + \left\{ {12 \atop 2} \right\}' \rho\right) X, & \dfrac{\partial x_1}{\partial v} = 2\,\rho\left(\dfrac{\partial X}{\partial v} - \left\{ {12 \atop 1} \right\}' X\right), \\[2ex] \dfrac{\partial x_2}{\partial u} = -\,2\,\rho\left(\dfrac{\partial X}{\partial u} - \left\{ {12 \atop 2} \right\}' X\right), & \dfrac{\partial x_2}{\partial v} = -\,2\left(\dfrac{\partial \rho}{\partial v} + \left\{ {12 \atop 1} \right\}' \rho\right) X. \end{cases}$$

The coefficients of the linear elements of S_1 and S_2, as derived from these formulas, are

$$(56)\quad \begin{cases} E_1 = 4\left(\dfrac{\partial \rho}{\partial u} + \left\{ {12 \atop 2} \right\}' \rho\right)^2, & F_1 = -\,4\,\rho \left\{ {12 \atop 1} \right\}' \left(\dfrac{\partial \rho}{\partial u} + \left\{ {12 \atop 2} \right\}' \rho\right), \\[2ex] G_1 = 4\,\rho^2\left(\mathcal{G} + \left\{ {12 \atop 1} \right\}'^2\right), & H_1 = 4\,\rho\sqrt{\mathcal{G}}\left(\dfrac{\partial \rho}{\partial u} + \left\{ {12 \atop 2} \right\}' \rho\right), \end{cases}$$

and

$$(57)\quad \begin{cases} E_2 = 4\,\rho^2\left(\mathcal{E} + \left\{ {12 \atop 2} \right\}'^2\right), & F_2 = -\,4\,\rho \left\{ {12 \atop 2} \right\}' \left(\dfrac{\partial \rho}{\partial v} + \left\{ {12 \atop 1} \right\}' \rho\right), \\[2ex] G_2 = 4\left(\dfrac{\partial \rho}{\partial v} + \left\{ {12 \atop 1} \right\}' \rho\right)^2, & H_2 = 4\,\rho\sqrt{\mathcal{E}}\left(\dfrac{\partial \rho}{\partial v} + \left\{ {12 \atop 1} \right\}' \rho\right). \end{cases}$$

The direction-cosines of the normals to S_1 and S_2 denoted by X_1, Y_1, Z_1; X_2, Y_2, Z_2 respectively are found from the above equations and (V, 31) to have the values

$$X_1 = \frac{1}{H_1}\frac{\partial(y_1, z_1)}{\partial(u, v)} = \frac{Y\dfrac{\partial Z}{\partial v} - Z\dfrac{\partial Y}{\partial v}}{\sqrt{\mathcal{G}}} = \frac{1}{\sqrt{\mathcal{G}}\mathcal{H}}\left(\mathcal{F}\frac{\partial X}{\partial v} - \mathcal{G}\frac{\partial X}{\partial u}\right),$$

$$X_2 = \frac{1}{H_2}\frac{\partial(y_2, z_2)}{\partial(u, v)} = \frac{Z\dfrac{\partial Y}{\partial u} - Y\dfrac{\partial Z}{\partial u}}{\sqrt{\mathcal{E}}} = \frac{-1}{\sqrt{\mathcal{E}}\mathcal{H}}\left(\mathcal{E}\frac{\partial X}{\partial v} - \mathcal{F}\frac{\partial X}{\partial u}\right),$$

and similar expressions for Y_i and Z_i. If these equations be differentiated, and the resulting equations be reduced by means of (V, 22), they can be put in the form

$$\frac{\partial X_1}{\partial u} = \frac{\mathcal{H}}{\sqrt{\mathcal{G}}}\left[\frac{1}{\mathcal{G}}\left\{ {12 \atop 1} \right\}'\frac{\partial X}{\partial v} + X\right], \quad \frac{\partial X_1}{\partial v} = \frac{\mathcal{H}}{\mathcal{G}^{\frac{3}{2}}}\left\{ {22 \atop 1} \right\}'\frac{\partial X}{\partial v},$$

$$\frac{\partial X_2}{\partial u} = \frac{\mathcal{H}}{\mathcal{E}^{\frac{3}{2}}}\left\{ {11 \atop 2} \right\}'\frac{\partial X}{\partial u}, \qquad \frac{\partial X_2}{\partial v} = \frac{\mathcal{H}}{\sqrt{\mathcal{E}}}\left[\frac{1}{\mathcal{E}}\left\{ {12 \atop 2} \right\}'\frac{\partial X}{\partial u} + X\right].$$

From these expressions and (55) we obtain

$$(58)\quad \begin{cases} D_1 = -\sum \dfrac{\partial x_1}{\partial u}\dfrac{\partial X_1}{\partial u} = -\dfrac{2\mathcal{H}}{\sqrt{\mathcal{G}}}\left(\dfrac{\partial \rho}{\partial u} + \begin{Bmatrix}12\\2\end{Bmatrix}'\rho\right), \\ D_1' = -\sum \dfrac{\partial x_1}{\partial v}\dfrac{\partial X_1}{\partial u} = -\sum \dfrac{\partial x_1}{\partial u}\dfrac{\partial X_1}{\partial v} = 0, \\ D_1'' = -\sum \dfrac{\partial x_1}{\partial v}\dfrac{\partial X_1}{\partial v} = -\dfrac{2\mathcal{H}}{\sqrt{\mathcal{G}}}\begin{Bmatrix}22\\1\end{Bmatrix}'\rho, \end{cases}$$

and

$$(59)\quad D_2 = \frac{2\mathcal{H}}{\sqrt{\mathcal{E}}}\begin{Bmatrix}11\\2\end{Bmatrix}'\rho, \qquad D_2' = 0, \qquad D_2'' = \frac{2\mathcal{H}}{\sqrt{\mathcal{E}}}\left(\frac{\partial \rho}{\partial v} + \begin{Bmatrix}12\\1\end{Bmatrix}'\rho\right).$$

From the foregoing formulas we derive the following expressions for the total curvature of S_1 and of S_2:

$$(60)\quad K_1 = \frac{\mathcal{H}^2\begin{Bmatrix}22\\1\end{Bmatrix}'}{4\rho\mathcal{G}^2\left(\dfrac{\partial \rho}{\partial u} + \begin{Bmatrix}12\\2\end{Bmatrix}'\rho\right)}, \qquad K_2 = \frac{\mathcal{H}^2\begin{Bmatrix}11\\2\end{Bmatrix}'}{4\rho\mathcal{E}^2\left(\dfrac{\partial \rho}{\partial v} + \begin{Bmatrix}12\\1\end{Bmatrix}'\rho\right)}.$$

EXAMPLES

1. If upon a surface of reference S of a normal congruence the curves orthogonal to the lines of the congruence are defined by $\phi(u, v) = \text{const.}$, and θ denotes the angle between a line of the congruence and the normal to the surface at the point of meeting, then $\sin^2\theta = \Delta_1 F(\phi)$ where the differential parameter is formed with respect to the linear element of S. Show that θ is constant along a line $\phi = \text{const.}$ only when the latter is a geodesic parallel.

2. When in the point equation of a surface, namely

$$\frac{\partial^2\theta}{\partial u\,\partial v} + a\frac{\partial\theta}{\partial u} + b\frac{\partial\theta}{\partial v} = 0,$$

a or b is zero, the coördinates of the surface can be found by quadratures.

3. Find the derived congruences of the tangents to the parametric curves on a tetrahedral surface (Ex. 2, p. 267), and determine under what conditions the surface S_i or S_{-i} is a curve.

4. Find the equation of the type (39) which admits as solutions the quantities x_1, y_1, z_1 given by (41).

5. When a congruence consists of the tangents to the lines of curvature in one system on a surface, the focal distances are equal to the radii of geodesic curvature of the lines of curvature in the other system.

6. Let S be a surface referred to its lines of curvature, let s_1 and s_2 denote the arcs of the curves $v =$ const. and $u =$ const. respectively, r_1 and r_2 their radii of first curvature, and R_1 and R_2 their radii of geodesic curvature; for the second focal sheet Σ_1 of the congruence of tangents to the curves $v =$ const. the linear element is reducible to

$$d\sigma_1^2 = (dR_2 - ds_1)^2 + \frac{R_2^2}{r_1^2} ds_1^2;$$

hence the curves $s_1 =$ const. are geodesics.

7. Show that Σ_1 of Ex. 6 is developable when $r_1 = f(s_1)$, and determine the most general form of r_1 so that Σ_1 shall be developable.

8. Determine the condition which ρ must satisfy in order that the asymptotic lines on either focal surface of a congruence shall correspond to a conjugate system on the other, and show that in this case

$$K_1 K_2 = -\left(\frac{\sin\theta}{2\rho}\right)^4,$$

where θ denotes the angle between the focal planes.

9. In order that the focal surfaces degenerate into curves, it is necessary and sufficient that the spherical representation satisfy the conditions

$$\frac{\partial}{\partial u}\left\{ \begin{matrix} 12 \\ 1 \end{matrix} \right\}' = \frac{\partial}{\partial v}\left\{ \begin{matrix} 12 \\ 2 \end{matrix} \right\}' = \left\{ \begin{matrix} 12 \\ 1 \end{matrix} \right\}'\left\{ \begin{matrix} 12 \\ 2 \end{matrix} \right\}' - \mathcal{F}.$$

10. Show that the surfaces orthogonal to a normal congruence of the type of Ex. 9 are cyclides of Dupin.

11. A necessary and sufficient condition that the second sheet of the congruence of tangents to a family of curves on a surface S be developable is that the curves be plane.

169. Isotropic congruences. An *isotropic congruence* is one whose focal surfaces are developables with minimal edges of regression. In § 31 we saw that $H = 0$ is a necessary and sufficient condition that a surface be of this kind. Referring to (56) and (57), we see that we must have

$$\rho\sqrt{\mathcal{G}}\left(\frac{\partial \rho}{\partial u} + \left\{ \begin{matrix} 12 \\ 2 \end{matrix} \right\}' \rho\right) = 0, \qquad \rho\sqrt{\mathcal{E}}\left(\frac{\partial \rho}{\partial v} + \left\{ \begin{matrix} 12 \\ 1 \end{matrix} \right\}' \rho\right) = 0.$$

From (54) it is seen that if ρ were zero the middle surface would be a point, and from (55) that if the expressions in parentheses were zero the surfaces S_1 and S_2 would be curves. Consequently

$$\text{(61)} \qquad \mathcal{E} = \mathcal{G} = 0.$$

Conversely, if this condition be satisfied, S_1 and S_2 are isotropic developables. Hence an isotropic congruence is one whose developables are represented on the sphere by minimal lines.

In consequence of (61) we have, from (52),

$$e = g = 0,$$

and since $f + f'$ also is zero, it follows that

$$dx\,dX + dy\,dY + dz\,dZ = 0. \tag{62}$$

Therefore r is zero, so that all the lines of striction lie on the middle surface. Since (61) is a consequence of (62), we have the following theorem of Ribaucour,* which is sometimes taken for the definition of isotropic congruences:

All the lines of striction of an isotropic congruence lie on the middle surface; and, conversely, when all the lines of striction lie on the middle surface, the congruence is isotropic; moreover, the middle sur face corresponds to the spherical representation with orthogonality of linear elements.

Ribaucour has established also the following theorem: †

The middle envelope of an isotropic congruence is a minimal surface.

Since the minimal lines on the sphere are parametric, in order to prove this theorem it is only necessary to show that on the middle envelope, that is, the envelope of the middle planes, the corresponding lines form a conjugate system. If W denotes the distance of the middle plane from the origin, the condition necessary and sufficient that the parametric lines be conjugate is that W satisfy the equation

$$\frac{\partial^2 \theta}{\partial u\, \partial v} + \mathcal{F}\theta = 0. \tag{63}$$

By definition

$$W = \Sigma X x,$$

and with the aid of (V, 22) we find

$$\frac{\partial^2 W}{\partial u\, \partial v} = \frac{\partial^2 \rho}{\partial u\, \partial v} + \rho\mathcal{F} - W\mathcal{F}.$$

Since equation (53) reduces to $\dfrac{\partial^2 \rho}{\partial u\, \partial v} + \rho\mathcal{F} = 0$, the function W satisfies (63).

* Étude des Élassoïdes ou Surfaces à Courbure Moyenne Nulle, *Memoires Couronnés par l'Academie de Belgique*, Vol. XLIV (1881), p. 63.

† *L.c.*, p. 31.

170. Congruences of Guichard. Guichard * proposed and solved the problem:

To determine the congruences whose focal surfaces are met by the developables in the lines of curvature.

With Bianchi we call them *congruences of Guichard.*

We remark that a necessary and sufficient condition that a congruence be of this kind is that F_1 and F_2 of § 168 be zero. From (56) and (57) it is seen that this is equivalent to

(64) $$\begin{Bmatrix} 12 \\ 1 \end{Bmatrix}' = \begin{Bmatrix} 12 \\ 2 \end{Bmatrix}' = 0.$$

Comparing this result with § 78, we have the theorem:

A necessary and sufficient condition that the developables of a congruence meet the focal surfaces in their lines of curvature is that the congruence be represented on the sphere by curves representing also the asymptotic lines on a pseudospherical surface.

In this case the parameters can be so chosen that †

$$\mathcal{E} = \mathcal{G} = 1, \qquad \mathcal{F} = -\cos\omega,$$

where ω is a solution of

$$\frac{\partial^2 \omega}{\partial u\, \partial v} = \sin\omega.$$

In this case equation (53) is

(65) $$\frac{\partial^2 \rho}{\partial u\, \partial v} = \rho \cos\omega.$$

In particular, this equation is satisfied by X, Y, Z (V, 22). If we replace ρ by X in (54), we have

$$\frac{\partial x}{\partial u} = 0, \qquad \frac{\partial x}{\partial v} = 0;$$

consequently, for the congruence determined by this value of ρ, the middle surface is a plane.

From (55) it follows that the lines of the congruence are tangent to the lines of curvature $v = \text{const.}$ on S_1. Consequently they are

* *L.c.*, p. 346.

† This is the only real solution of (64).

parallel to the normals to one of the sheets of the evolute of S_1 (cf. § 74); call it Σ_1. Hence the conjugate system on Σ_1 corresponding to the lines of curvature on S_1 is represented on the sphere by the same lines as the developables of the congruence. Referring to (VI, 38), we see that condition (64) is equivalent to

$$\left\{ \begin{matrix} 11 \\ 2 \end{matrix} \right\} = \left\{ \begin{matrix} 22 \\ 1 \end{matrix} \right\} = 0,$$

where the Christoffel symbols are formed with respect to the linear element of Σ_1. But these are the conditions that the parametric curves on Σ_1 be geodesics (cf. § 85). Surfaces with a conjugate system of geodesics were studied by Voss,* and on this account are called *surfaces of Voss*. Since the converse of the above results is true, we have the following theorem of Guichard:

A necessary and sufficient condition that the tangents to the lines of curvature in one family of a surface form a congruence of Guichard is that one sheet of the evolute of the surface be a surface of Voss, and that the tangents constituting the congruence be those which are parallel to the normals to the latter.

If W_1 denotes the distance from the origin to the tangent plane to the surface of Voss Σ_1, then W_1 is a solution of equation (65) (cf. § 84). Hence $W_1 + \kappa\rho$ is a solution of this equation, provided κ be a constant. But since the tangent plane to Σ_1 passes through the corresponding point of S_1, the above result shows that a plane normal to the lines of the congruence, and which divides in constant ratio the segment between the focal points, envelopes a surface of Voss. In particular, we have the corollary:

The middle envelope of a congruence of Guichard is a surface of Voss.

171. Pseudospherical congruences. The lines joining corresponding points on a pseudospherical surface S and on one of its Bäcklund transforms S_1 (cf. § 120) constitute an interesting congruence. We recall that the distance between corresponding points is constant, and that the tangent planes to the two surfaces at these points meet under constant angle. From (32) it follows that the distance between the limit points also is constant.

* *Münchener Berichte*, Vol. XVIII (1888), pp. 95–102.

Conversely, when the angle between the focal planes of a congruence is constant, and consequently also the angle θ between the parametric lines on the sphere representing the developables, we have, from (V, 4),

$$\frac{1}{\mathcal{E}}\begin{Bmatrix}11\\2\end{Bmatrix}' + \frac{1}{\mathcal{G}}\begin{Bmatrix}12\\1\end{Bmatrix}' = 0, \quad \frac{1}{\mathcal{E}}\begin{Bmatrix}12\\2\end{Bmatrix}' + \frac{1}{\mathcal{G}}\begin{Bmatrix}22\\1\end{Bmatrix}' = 0. \tag{66}$$

Furthermore, if the distance between the focal points is constant, we have $\rho = a$, and by (60)

$$K_1 = K_2 = -\frac{\sin^2\theta}{4a^2}.$$

Hence the two focal surfaces have the same constant curvature. Congruences of this kind were first studied by Bianchi.* He called them *pseudospherical congruences.*

In order that the two focal surfaces of the congruence be Bäcklund transforms of one another, it is necessary that their lines of curvature correspond. It is readily found that for both surfaces the equation of these lines is reducible by means of (66) to

$$\mathcal{E}\begin{Bmatrix}12\\1\end{Bmatrix}'\begin{Bmatrix}12\\2\end{Bmatrix}' du^2 - \left[\mathcal{E}\mathcal{G} + \mathcal{E}\begin{Bmatrix}12\\1\end{Bmatrix}'^2 + \mathcal{G}\begin{Bmatrix}12\\2\end{Bmatrix}'^2\right] du\,dv$$
$$+ \mathcal{G}\begin{Bmatrix}12\\1\end{Bmatrix}'\begin{Bmatrix}12\\2\end{Bmatrix}' dv^2 = 0.$$

Moreover, the differential equation of the asymptotic lines on each surface is $\mathcal{E}du^2 - \mathcal{G}dv^2 = 0$. Hence we have the theorems:

On the focal surfaces of a pseudospherical congruence the lines of curvature correspond, and likewise the asymptotic lines.

The focal surfaces of a pseudospherical congruence are Bäcklund transforms of one another.

EXAMPLES

1. When the parameters of a congruence are any whatsoever, and likewise the surface of reference, a condition necessary and sufficient that a congruence be isotropic is

$$\frac{e}{\mathcal{E}} = \frac{f + f'}{2\mathcal{F}} = \frac{g}{\mathcal{G}}.$$

2. A necessary and sufficient condition that a congruence be isotropic is that the locus of two points on each line at an equal constant distance from the middle surface shall describe applicable surfaces.

3. Show that equation (65) admits $\frac{\partial \omega}{\partial u}$ and $\frac{\partial \omega}{\partial v}$ as solutions. Prove that in each case one of the focal surfaces is a sphere.

* *Annali*, Ser. 2, Vol. XV (1887), pp. 161–172; also *Lezioni*, Vol. I, pp. 323, 324.

4. Determine all the congruences of Guichard for which one of the focal surfaces is a sphere.

5. When a surface is referred to its lines of curvature, a necessary and sufficient condition that the tangents to the curves $v = \text{const.}$ shall form a congruence of Guichard is

$$\frac{\partial}{\partial u}\left(\frac{1}{\sqrt{E}}\frac{\partial\sqrt{G}}{\partial u}\right) = 0.$$

6. Determine the surfaces which are such that the tangents to the lines of curvature in each system form a congruence of Guichard.

172. W-congruences. We have just seen that the asymptotic lines on the focal surfaces of a pseudospherical congruence correspond; the same is true in the case of the congruences of normals to a W-surface (cf. § 124). For this reason all congruences possessing this property are called *W-congruences*. We shall derive other properties of these congruences.

The condition that asymptotic lines correspond, namely

$$D_1 D_2'' = D_1'' D_2,$$

takes the following form in consequence of (58) and (59):

$$\left(\frac{\partial \rho}{\partial u} + \begin{Bmatrix} 12 \\ 2 \end{Bmatrix}' \rho\right)\left(\frac{\partial \rho}{\partial v} + \begin{Bmatrix} 12 \\ 1 \end{Bmatrix}' \rho\right) = \rho^2 \begin{Bmatrix} 11 \\ 2 \end{Bmatrix}' \begin{Bmatrix} 22 \\ 1 \end{Bmatrix}'.$$

Hence from (60) it follows that a necessary and sufficient condition for a W-congruence is

$$K_1 K_2 = \left(\frac{\sin\theta}{2\rho}\right)^4. \tag{67}$$

In order to obtain an idea of the analytical problem involved in the determination of W-congruences, we suppose that we have two surfaces S, $\bar{S}$ referred to their asymptotic lines, and inquire under what conditions the lines joining corresponding points on the surfaces are tangent to them. We assume that the coördinates of the surfaces are defined * by means of the Lelieuvre formulas (cf. § 79), thus:

$$\left\{\begin{aligned}
\frac{\partial x}{\partial u} &= \begin{vmatrix} \nu_2 & \nu_3 \\ \dfrac{\partial \nu_2}{\partial u} & \dfrac{\partial \nu_3}{\partial u} \end{vmatrix}, & \frac{\partial x}{\partial v} &= -\begin{vmatrix} \nu_2 & \nu_3 \\ \dfrac{\partial \nu_2}{\partial v} & \dfrac{\partial \nu_3}{\partial v} \end{vmatrix}, \\
\frac{\partial \bar{x}}{\partial u} &= \begin{vmatrix} \bar{\nu}_2 & \bar{\nu}_3 \\ \dfrac{\partial \bar{\nu}_2}{\partial u} & \dfrac{\partial \bar{\nu}_3}{\partial u} \end{vmatrix}, & \frac{\partial \bar{x}}{\partial v} &= -\begin{vmatrix} \bar{\nu}_2 & \bar{\nu}_3 \\ \dfrac{\partial \bar{\nu}_2}{\partial v} & \dfrac{\partial \bar{\nu}_3}{\partial v} \end{vmatrix},
\end{aligned}\right. \tag{68}$$

* Cf. Guichard, *Comptes Rendus*, Vol. CX (1890), pp. 126–127.

and similar equations in $\bar{y}$, $\bar{z}$, y, and z. The functions ν_1, ν_2, ν_3; $\bar{\nu}_1$, $\bar{\nu}_2$, $\bar{\nu}_3$ respectively are solutions of equations of the form

$$\frac{\partial^2\theta}{\partial u\,\partial v}=\lambda\theta,\qquad \frac{\partial^2\bar{\theta}}{\partial u\,\partial v}=\bar{\lambda}\bar{\theta}, \tag{69}$$

and they are such that

$$\nu_1^2+\nu_2^2+\nu_3^2=\alpha,\qquad \bar{\nu}_1^2+\bar{\nu}_2^2+\bar{\nu}_3^2=\bar{\alpha}, \tag{70}$$

where α and $\bar{\alpha}$ are defined by

$$K=-\frac{1}{\alpha^2},\qquad \bar{K}=-\frac{1}{\bar{\alpha}^2}. \tag{71}$$

Since ν_1, ν_2, ν_3 and $\bar{\nu}_1$, $\bar{\nu}_2$, $\bar{\nu}_3$ are proportional to the direction-cosines of the normals to S and $\bar{S}$, the condition that the lines joining corresponding points be tangent to the surfaces S and $\bar{S}$ is

$$\nu_1(\bar{x}-x)+\nu_2(\bar{y}-y)+\nu_3(\bar{z}-z)=0,$$

$$\bar{\nu}_1(\bar{x}-x)+\bar{\nu}_2(\bar{y}-y)+\bar{\nu}_3(\bar{z}-z)=0.$$

Hence

$$\frac{\bar{x}-x}{\nu_2\bar{\nu}_3-\nu_3\bar{\nu}_2}=\frac{\bar{y}-y}{\nu_3\bar{\nu}_1-\nu_1\bar{\nu}_3}=\frac{\bar{z}-z}{\nu_1\bar{\nu}_2-\nu_2\bar{\nu}_1}=m,$$

where m denotes a factor of proportionality. In order to find its value, we notice that from these equations follow the relations

$$\begin{aligned}(2\,\rho)^2=\Sigma(\bar{x}-x)^2&=m^2\Sigma(\nu_2\bar{\nu}_3-\nu_3\bar{\nu}_2)^2\\ &=m^2\left[\alpha\bar{\alpha}-(\Sigma\nu_1\bar{\nu}_1)^2\right]=m^2\alpha\bar{\alpha}\sin^2\theta,\end{aligned}$$

where θ denotes the angle between the focal planes. If this value of $2\,\rho$ and the values of K and $\bar{K}$ from (71) be substituted in (67), it is found that $m^4=1$. We take $m=1$, thus fixing the signs of $\bar{\nu}_1$, $\bar{\nu}_2$, $\bar{\nu}_3$, and the above equations become

$$\bar{x}-x=\nu_2\bar{\nu}_3-\nu_3\bar{\nu}_2,\quad \bar{y}-y=\nu_3\bar{\nu}_1-\nu_1\bar{\nu}_3,\quad \bar{z}-z=\nu_1\bar{\nu}_2-\nu_2\bar{\nu}_1. \tag{72}$$

If the first of these equations be differentiated with respect to u, the result is reducible by (68) to

$$(\bar{\nu}_3-\nu_3)\left(\frac{\partial\bar{\nu}_2}{\partial u}+\frac{\partial\nu_2}{\partial u}\right)=(\bar{\nu}_2-\nu_2)\left(\frac{\partial\bar{\nu}_3}{\partial u}+\frac{\partial\nu_3}{\partial u}\right).$$

Proceeding in like manner with the others, and also differentiating with respect to v, we are brought to

$$(73)\qquad \begin{cases} \dfrac{\partial}{\partial u}(\bar{\nu}_i + \nu_i) = k(\bar{\nu}_i - \nu_i), \\[2ex] \dfrac{\partial}{\partial v}(\bar{\nu}_i - \nu_i) = l(\bar{\nu}_i + \nu_i), \end{cases} \qquad (i = 1, 2, 3)$$

where l and k are factors of proportionality to be determined.

If the first of these equations be differentiated with respect to v, and in the reduction we make use of the second and of (69), we find

$$\left(\bar{\lambda} - kl - \frac{\partial k}{\partial v}\right)\bar{\nu}_i + \left(\lambda - kl + \frac{\partial k}{\partial v}\right)\nu_i = 0.$$

In like manner, if the second of the above equations be differentiated with respect to u, we obtain

$$\left(\bar{\lambda} - kl - \frac{\partial l}{\partial u}\right)\bar{\nu}_i - \left(\lambda - kl + \frac{\partial l}{\partial u}\right)\nu_i = 0.$$

Since these equations are true for $i = 1, 2, 3$, the quantities in parentheses must be zero. This gives

$$\bar{\lambda} = \frac{\partial k}{\partial v} + kl, \qquad \lambda = -\frac{\partial k}{\partial v} + kl, \qquad \frac{\partial l}{\partial u} = \frac{\partial k}{\partial v}.$$

In accordance with the last we put

$$k = \frac{\partial}{\partial u}\log\frac{1}{\theta_1}, \qquad l = \frac{\partial}{\partial v}\log\frac{1}{\theta_1},$$

and the others become

$$\bar{\lambda} = \theta_1 \frac{\partial^2}{\partial u\,\partial v}\left(\frac{1}{\theta_1}\right), \qquad \lambda = \frac{1}{\theta_1}\frac{\partial^2 \theta_1}{\partial u\,\partial v}.$$

Hence equations (69) may be written

$$\frac{\partial^2\theta}{\partial u\,\partial v} = \frac{1}{\theta_1}\frac{\partial^2\theta_1}{\partial u\,\partial v}\theta, \qquad \frac{\partial^2\bar{\theta}}{\partial u\,\partial v} = \theta_1\frac{\partial^2}{\partial u\,\partial v}\left(\frac{1}{\theta_1}\right)\bar{\theta},$$

from which it follows that θ_1 is a solution of the first of equations (69) and $1/\theta_1$ of the second. Moreover, equations (73) may now be written in the form

$$(74)\qquad \frac{\partial}{\partial u}(\theta_1\bar{\nu}_i) = -\begin{vmatrix} \theta_1 & \nu_i \\ \dfrac{\partial\theta_1}{\partial u} & \dfrac{\partial\nu_i}{\partial u} \end{vmatrix}, \qquad \frac{\partial}{\partial v}(\theta_1\bar{\nu}_i) = \begin{vmatrix} \theta_1 & \nu_i \\ \dfrac{\partial\theta_1}{\partial v} & \dfrac{\partial\nu_i}{\partial v} \end{vmatrix}.$$

Hence if θ_1 be a known solution of the first of equations (69), we obtain by quadratures three functions $\bar{\nu}_i$, which lead by the quadratures (68) to a surface $\bar{S}$. The latter is referred to its asymptotic lines and the joins of corresponding points on S and $\bar{S}$ are tangent to the latter. And so we have:

If a surface S be referred to its asymptotic lines, and the equations of the surface be in the Lelieuvre form, each solution of the corresponding equation

$$\frac{\partial^2 \theta}{\partial u\, \partial v} = \lambda \theta$$

determines a surface $\bar{S}$, found by quadratures, such that S and $\bar{S}$ are the focal surfaces of a W-congruence.

Comparing (74) with (XI, 13), we see that if we put

$$x_1 = \theta_1 \bar{\nu}_1, \qquad y_1 = \theta_1 \bar{\nu}_2, \qquad z_1 = \theta_1 \bar{\nu}_3,$$

the locus of the point (x_1, y_1, z_1) corresponds to S with orthogonality of linear elements. Hence $\bar{\nu}_1, \bar{\nu}_2, \bar{\nu}_3$ are proportional to the direction-cosines of the generatrices of an infinitesimal deformation of S, so that we have:

Each focal surface of a W-congruence admits of an infinitesimal deformation whose generatrices are parallel to the normals to the other focal surface.

Since the steps in the preceding argument are reversible, we have the theorem:

The tangents to a surface which are perpendicular to the generatrices of an infinitesimal deformation of the latter constitute a W-congruence of the most general kind; and the normals to the other surface are parallel to the generatrices of the deformation.

173. Congruences of Ribaucour. In his study of surfaces corresponding with orthogonality of linear elements Ribaucour considered the congruence formed by the lines through points on one surface parallel to the normals to a surface corresponding with the former in this manner. Bianchi * calls such a congruence a *congruence of Ribaucour*, and the second surface the *director surface.*

In order to ascertain the properties of such a congruence, we recall the results of § 153. Let S_1 be taken for the surface of

* Vol. II, p. 17.

reference, and draw lines parallel to the normals to S. If the latter be referred to its asymptotic lines, it follows from (XI, 6) that

$$e=\sum \frac{\partial x_1}{\partial u}\frac{\partial X}{\partial u}=-\frac{D'\mathcal{E}\phi}{HK}, \qquad f=\sum \frac{\partial x_1}{\partial v}\frac{\partial X}{\partial u}=\frac{D'\mathcal{F}\phi}{HK},$$

$$f'=\sum \frac{\partial x_1}{\partial u}\frac{\partial X}{\partial v}=-\frac{D'\mathcal{F}\phi}{HK}, \qquad g=\sum \frac{\partial x_1}{\partial v}\frac{\partial X}{\partial v}=\frac{D'\mathcal{G}\phi}{HK}.$$

Since these values satisfy the conditions

$$f'\mathcal{E}-e\mathcal{F}=0, \qquad g\mathcal{F}-f\mathcal{G}=0,$$

the ruled surfaces $u=\text{const.}$, $v=\text{const.}$ are the developables. And since also $\rho_1+\rho_2$ is equal to zero, S_1 is the middle surface of the congruence. But the parametric curves on S_1 form a conjugate system when the asymptotic lines on S are parametric. Hence we have the theorem:

The developable surfaces of a congruence of Ribaucour cut the middle surface in a conjugate system.

Guichard * proved that this property is characteristic of congruences of Ribaucour. In order to obtain this result, we differentiate the first of equations (54) with respect to v, and in the reduction make use of the fact that X and ρ satisfy equations (V, 22) and (53) respectively. This gives

$$\frac{\partial^2 x_1}{\partial u\,\partial v}=\left(\frac{\partial \log\rho}{\partial v}+\begin{Bmatrix}12\\1\end{Bmatrix}'\right)\frac{\partial x_1}{\partial u}+\left(\frac{\partial \log\rho}{\partial u}+\begin{Bmatrix}12\\2\end{Bmatrix}'\right)\frac{\partial x_1}{\partial v}$$
$$+\left(\frac{\partial}{\partial v}\begin{Bmatrix}12\\2\end{Bmatrix}'-\frac{\partial}{\partial u}\begin{Bmatrix}12\\1\end{Bmatrix}'\right)\rho X.$$

From this and similar equations in y_1 and z_1 it follows that a necessary and sufficient condition that the parametric curves form a conjugate system is

$$\frac{\partial}{\partial u}\begin{Bmatrix}12\\1\end{Bmatrix}'=\frac{\partial}{\partial v}\begin{Bmatrix}12\\2\end{Bmatrix}'.$$

When this condition is satisfied by a system of curves on the sphere, they represent the asymptotic lines on a unique surface S, whose coördinates are given by the quadratures (VI, 14)

$$\frac{\partial x}{\partial u}=\frac{D'}{\mathcal{H}^2}\left(\mathcal{F}\frac{\partial X}{\partial u}-\mathcal{E}\frac{\partial X}{\partial v}\right), \qquad \frac{\partial x}{\partial v}=\frac{D'}{\mathcal{H}^2}\left(-\mathcal{G}\frac{\partial X}{\partial u}+\mathcal{F}\frac{\partial X}{\partial v}\right),$$

* *Annales L'École Normale*, Ser. 3, Vol. VI (1889), pp. 344, 345.

and similar expressions for y and z. Combining these equations with (54), we find that

$$\sum \frac{\partial x_1}{\partial u}\frac{\partial x}{\partial u}=0, \qquad \sum \frac{\partial x_1}{\partial u}\frac{\partial x}{\partial v}+\sum \frac{\partial x_1}{\partial v}\frac{\partial x}{\partial u}=0, \qquad \sum \frac{\partial x_1}{\partial v}\frac{\partial x}{\partial v}=0.$$

Hence S and S_1 correspond with orthogonality of linear elements, and the normals to the former are parallel to the lines of the congruence. Hence:

A necessary and sufficient condition that the developables of a congruence cut the middle surface in a conjugate system is that their representation be that also of the asymptotic lines of a surface, in which case the latter and the middle surface correspond with orthogonality of linear elements.

EXAMPLES

1. When the coördinates of the unit sphere are in the form (III, 35), the parametric curves are asymptotic lines. Find the W-congruences for which the sphere is one of the focal sheets.

2. Let $\nu_i = f_i(u) + \phi_i(v)$, where f_i and ϕ_i are functions of u and v respectively, and $i = 1, 2, 3$, be three solutions of the first of equations (69), in which case $\lambda = 0$, and let θ_1 in (74) be unity. Show that for the corresponding W-congruence the middle surface is a surface of translation with the generatrices $u =$ const., $v =$ const., that the functions f_i and ϕ_i are proportional to the direction-cosines of the binormals to these generatrices, and that the intersections of the osculating planes of these generatrices are the lines of the congruence.

3. Show that isotropic congruences and congruences of Guichard are congruences of Ribaucour.

4. A necessary and sufficient condition that a congruence of Ribaucour be normal is that the spherical representation of its developables be isothermic.

5. The normals to quadrics and to the cyclides of Dupin constitute congruences of Ribaucour.

6. When the middle surface of a congruence is plane, the congruence is of the Ribaucour type.

7. Show that the congruence of Ribaucour, whose director surface is a right helicoid, is a normal congruence, and that the normal surfaces are molding surfaces.

8. Show that a necessary and sufficient condition that a congruence of Ribaucour be normal is that the director surface be minimal.

GENERAL EXAMPLES

1. Through each line of a congruence there pass two ruled surfaces of the congruence whose lines of striction lie on the middle surface; their equation is

$$\frac{g\mathcal{E}-(f+f')\mathcal{F}+e\mathcal{G}}{2(\mathcal{E}\mathcal{G}-\mathcal{F}^2)}=\frac{e\,du^2+(f+f')\,du\,dv+g\,dv^2}{\mathcal{E}du^2+2\mathcal{F}du\,dv+\mathcal{G}dv^2};$$

they are called the *mean ruled surfaces* of the congruence.

2. Show that the mean ruled surfaces of a congruence are represented on the sphere by an orthogonal system of real lines, and that their central planes (§ 105) bisect the angles between the focal planes. Let $u =$ const., $v =$ const. be the mean ruled surfaces and develop a theory analogous to that in § 167.

3. If the two focal surfaces of a congruence intersect, the intersection is the envelope of the edges of regression of the two families of developable surfaces of the congruence.

4. If a congruence consists of the lines joining points on two twisted curves, the focal planes for a line of the congruence are determined by the line and the tangent to each curve at the point where the curve is met by the line.

5. In order that the lines which join the centers of geodesic curvature of the curves of an orthogonal system on a surface shall form a normal congruence, it is necessary and sufficient that the corresponding radii of geodesic curvature be functions of one another, or that the curves in one family have constant geodesic curvature.

6. Let S be a surface whose lines of curvature in one system are circles; let C denote the vertex of the cone circumscribing S along a circle, and L the corresponding generator of the envelope of the planes of the circles; a necessary and sufficient condition that the lines through the points C and the corresponding lines L form a normal congruence is that the distance from C to the points of the corresponding circle shall be the same for every circle; if this distance be denoted by a, the radius of the sphere is given by

$$R^2 = R'^2(R^2 + a^2),$$

where the accent indicates differentiation with respect to the arc of the curve of centers of the spheres.

7. Let S be a surface referred to its lines of curvature, C_1 and C_2 the centers of principal normal curvature at a point, G_1 and G_2 the centers of geodesic curvature of the lines of curvature at this point; a necessary and sufficient condition that the line joining C_2 and G_1 form a normal congruence is that ρ_2 be a function of ρ_{gu}, or that one of these radii be a constant.

8. Let S be a surface of the kind defined in Ex. 6; the cone formed by the normals to the surface at points of a circle A is tangent to the second sheet of the evolute of S in a conic Γ (cf. § 132). Show that the lines through points of Γ and the vertex C of the cone which circumscribes S along A generate a normal congruence, and that C lies in the plane of Γ.

9. Given an isothermal orthogonal system on the sphere for which the linear element is

$$d\sigma^2 = \lambda^2 (du^2 + dv^2):$$

on each tangent to a curve $v =$ const. lay off the segment of length λ measured from the point of contact, and through the extremity of the segment draw a line parallel to the radius of the sphere at the point of contact. Show that this congruence is isotropic.

10. When a congruence is isotropic and its direction-cosines are of the form (III, 35), equation (53) reduces to

$$\frac{\partial^2 \rho}{\partial u\, \partial v} = -\frac{2\rho}{(1 + uv)^2},$$

Show that the general integral is

$$\rho = 2\left[u\phi(v) - vf(u)\right](1 + uv)^{-1} + f' - \phi',$$

where f and ϕ are arbitrary functions of u and v respectively. Find the equations of the middle surface.

11. Show that the intersections of the planes

$$(1-u^2)x + i(1+u^2)y + 2uz + 4f(u) = 0,$$
$$(1-v^2)x - i(1+v^2)y + 2vz + 4\phi(v) = 0$$

constitute an isotropic congruence, for which these are the focal planes; that the locus of the mid-points of the lines joining points on the edges of regression of the developables enveloped by these planes is the minimal surface which is the middle envelope of the congruence, by finding the coördinates of the point in which the tangent plane to this surface meets the intersection of the above planes.

12. Show that the middle surface of an isotropic congruence is the most general surface which corresponds to a sphere with orthogonality of linear elements, and that the corresponding associate surface in the infinitesimal deformation of the sphere is the minimal surface adjoint to the middle envelope.

13. Find the surface associate to the middle surface of an isotropic congruence when the surface corresponding to the latter with orthogonality of linear elements is a sphere, and show that it is the polar reciprocal, with respect to the imaginary sphere $x^2 + y^2 + z^2 + 1 = 0$, of the minimal surface adjoint to the middle envelope of the congruence.

14. The lines of intersection of the osculating planes of the generatrices of a surface of translation constitute a W-congruence of which the given surface is the middle surface; if the generatrices be curves of constant torsion, equal but of opposite sign, the congruence is normal to a W-surface of the type (VIII, 72).

15. If the points of a surface S be projected orthogonally upon any plane A, and if, after the latter has been rotated about any line normal to it through a right angle, lines be drawn through points of A parallel to the corresponding normals to S, these lines form a congruence of Ribaucour.

16. A necessary and sufficient condition that the tangents to the curves $v = \text{const.}$ on a surface, whose point equation is (VI, 26), shall form a congruence of Ribaucour is

$$\frac{\partial a}{\partial u} - \frac{\partial b}{\partial v} + \frac{\partial^2 \log b}{\partial u\, \partial v} = 0.$$

17. Show that the tangents to each system of parametric curves on a surface form congruences of Ribaucour when the point equation is

$$\frac{\partial^2 \theta}{\partial u\, \partial v} + U_1 V_1' \frac{\partial \theta}{\partial u} + U_1' V_1 \frac{\partial \theta}{\partial v} = 0,$$

where U_1 and V_1 are functions of u and v respectively, and the accents indicate differentiation.

18. Show that if the parametric curves on a surface S form a conjugate system, and the tangents to the curves of each family form a congruence of Ribaucour, the same is true of the surfaces S_1 and S_{-1}, which together with S constitute the focal surfaces of the two congruences.

19. Show that the parameter of distribution p of the ruled surface of a congruence, determined by a value of dv/du, is given by

$$p = \frac{1}{\mathcal{H}\, d\sigma^2} \begin{vmatrix} \mathcal{E}du + \mathcal{F}dv, & \mathcal{F}du + \mathcal{G}dv \\ e\,du + f\,dv, & f'du + g\,dv \end{vmatrix}.$$

20. Show that the mean ruled surfaces (cf. Ex. 1) of a congruence are characterized by the property that for these surfaces the parameter of distribution has the maximum and minimum values.

21. If S and S_0 are two associate surfaces, and through each point of one a line be drawn parallel to the corresponding radius vector of the other, the developables of the congruence thus formed correspond to the common conjugate system of S and S_0.

22. In order that two surfaces S and S_0 corresponding with parallelism of tangent planes be associate surfaces, it is necessary and sufficient that for the congruence formed by the joins of corresponding points M and M_0 of these surfaces the developables cut S and S_0 in their common conjugate system, and that the focal points M and M_0 form a harmonic range.

23. In order that a surface S be isothermic, it is necessary and sufficient that there exist a congruence of Ribaucour of which S is the middle surface, such that the developables cut S in its lines of curvature.

CHAPTER XIII

CYCLIC SYSTEMS

174. General equations of cyclic systems. The term *congruence* is not restricted to two-parameter systems of straight lines, but is applied to two-parameter systems of any kind of curves. Darboux * has made a study of these general congruences and Ribaucour † has considered congruences of plane curves. Of particular interest is the case where these curves are circles. Ribaucour has given the name *cyclic systems* to congruences of circles which admit of a one-parameter family of orthogonal surfaces. This chapter is devoted to a study of cyclic systems.

We begin with the general case where the planes of the circles envelop a nondevelopable surface S. We associate with the latter a moving trihedral (§ 68), and for the present assume that the parametric curves on the surface are any whatever.

As the circles lie in the tangent planes to S, the coördinates of a point on one of them with respect to the corresponding trihedral are of the form

$$a + R\cos\theta, \quad b + R\sin\theta, \quad 0, \tag{1}$$

where a, b are the coördinates of the center, R the radius, and θ the angle which the latter to a given point makes with the moving x-axis.

In § 69 we found the following expressions for the projections of a displacement of a point with respect to the moving axes:

$$\tag{2} \left\{\begin{array}{l} dx + \xi\,du + \xi_1 dv + (q\,du + q_1 dv)\,z - (r\,du + r_1 dv)\,y, \\ dy + \eta\,du + \eta_1 dv + (r\,du + r_1 dv)\,x - (p\,du + p_1 dv)\,z, \\ dz \qquad\qquad\qquad\quad + (p\,du + p_1 dv)\,y - (q\,du + q_1 dv)\,x, \end{array}\right.$$

* Vol. II, pp. 1–10; also Eisenhart, Congruences of Curves, *Transactions of the Amer. Math. Soc.*, Vol. IV (1903), pp. 470–488.

† Mémoire sur la theorie générale des surfaces courbes, *Journal des Mathématiques*, Ser. 4, Vol. VII (1891), § 117 *et. seq.*

where the translations ξ, ξ_1, η, η_1 and the rotations $p, q, r; p_1, q_1, r_1$ satisfy the conditions

$$(3)\quad \begin{cases} \dfrac{\partial p}{\partial v}-\dfrac{\partial p_1}{\partial u}=qr_1-rq_1, & \dfrac{\partial \xi}{\partial v}-\dfrac{\partial \xi_1}{\partial u}=\eta r_1-\eta_1 r, \\ \dfrac{\partial q}{\partial v}-\dfrac{\partial q_1}{\partial u}=rp_1-pr_1, & \dfrac{\partial \eta}{\partial v}-\dfrac{\partial \eta_1}{\partial u}=\xi_1 r-\xi r_1, \\ \dfrac{\partial r}{\partial v}-\dfrac{\partial r_1}{\partial u}=pq_1-qp_1, & p\eta_1-p_1\eta=q\xi_1-q_1\xi. \end{cases}$$

When the values (1) are substituted in (2) the latter are reducible to

$$(4)\quad \begin{cases} A\,du+A_1dv+\cos\theta\, dR-(d\theta+r\,du+r_1dv)R\sin\theta, \\ B\,du+B_1dv+\sin\theta\, dR+(d\theta+r\,du+r_1dv)R\cos\theta, \\ (p\,du+p_1dv)(b+R\sin\theta)-(q\,du+q_1dv)(a+R\cos\theta), \end{cases}$$

where we have put, for the sake of brevity,

$$(5)\quad \begin{cases} A=\dfrac{\partial a}{\partial u}+\xi-rb, & B=\dfrac{\partial b}{\partial u}+\eta+ra, \\ A_1=\dfrac{\partial a}{\partial v}+\xi_1-r_1b, & B_1=\dfrac{\partial b}{\partial v}+\eta_1+r_1a. \end{cases}$$

The conditions that

$$\frac{\partial}{\partial u}\left(\frac{\partial a}{\partial v}\right)=\frac{\partial}{\partial v}\left(\frac{\partial a}{\partial u}\right), \qquad \frac{\partial}{\partial u}\left(\frac{\partial b}{\partial v}\right)=\frac{\partial}{\partial v}\left(\frac{\partial b}{\partial u}\right)$$

are reducible, by means of (3), to

$$(6)\quad \begin{cases} \dfrac{\partial A}{\partial v}-\dfrac{\partial A_1}{\partial u}=r_1B-rB_1-b(pq_1-p_1q), \\ \dfrac{\partial B}{\partial v}-\dfrac{\partial B_1}{\partial u}=-r_1A+rA_1+a(pq_1-p_1q). \end{cases}$$

The direction-cosines of the tangent to the given circle at the point (1) are

$$(7)\qquad -\sin\theta, \quad \cos\theta, \quad 0.$$

Hence the condition that the locus of the point, as u and v vary, be orthogonal to the circle is that the sum of the expressions (4) multiplied respectively by the quantities (7) be zero. This gives

$$(8)\quad R\,d\theta+(B\cos\theta-A\sin\theta+Rr)du+(B_1\cos\theta-A_1\sin\theta+Rr_1)dv=0.$$

In order that the system of circles be normal to a family of surfaces this equation must admit of a solution involving a parameter. Since it is of the form

$$R\,d\theta + U\,du + V\,dv = 0, \tag{9}$$

the condition that such an integral exist is that the equation

$$R\left(\frac{\partial U}{\partial v} - \frac{\partial V}{\partial u}\right) + U\left(\frac{\partial V}{\partial \theta} - \frac{\partial R}{\partial v}\right) + V\left(\frac{\partial R}{\partial u} - \frac{\partial U}{\partial \theta}\right) = 0 \tag{10}$$

be satisfied identically.* For equation (8) this condition is reducible to

$$\left\{\begin{aligned} &\sin\theta\left[\frac{\partial R}{\partial v}A - \frac{\partial R}{\partial u}A_1 + Rb(pq_1 - p_1q)\right] \\ &\qquad - \cos\theta\left[\frac{\partial R}{\partial v}B - \frac{\partial R}{\partial u}B_1 - Ra(pq_1 - p_1q)\right] \\ &\qquad + (AB_1 - A_1B) + R^2(pq_1 - p_1q) = 0. \end{aligned}\right. \tag{11}$$

In order that this equation be satisfied identically, the expressions in the brackets must be zero. If they are not zero, it is possible that the two solutions of this equation will satisfy (8), and thus determine two surfaces orthogonal to the congruence of circles. Hence we have the theorem of Ribaucour:

If the circles of a congruence are normal to more than two surfaces, they form a cyclic system.

The equations of condition that the system be cyclic are consequently

$$\left\{\begin{aligned} &\frac{\partial R}{\partial v}A - \frac{\partial R}{\partial u}A_1 + Rb(pq_1 - p_1q) = 0, \\ &\frac{\partial R}{\partial v}B - \frac{\partial R}{\partial u}B_1 - Ra(pq_1 - p_1q) = 0, \\ &(AB_1 - A_1B) + R^2(pq_1 - p_1q) = 0. \end{aligned}\right. \tag{12}$$

The total curvature of S is given by (cf. § 70)

$$K = \frac{pq_1 - p_1q}{\xi\eta_1 - \xi_1\eta}.$$

*Murray, *Differential Equations*, p. 137. New York, 1897; also Forsyth, *Differential Equations*, p. 257. London, 1888.

From this and (5) it is seen that equations (12) involve only functions relating to the linear element of S and to the circle. Hence we have the theorem of Ribaucour:

If the envelope of the planes of the circles of a cyclic system be deformed in any manner without disturbing the size or position of the circles relative to the point of contact, the congruence of circles continues to form a cyclic system.

Furthermore, if we put

$$t = \tan\frac{\theta}{2},$$

equation (8) assumes the Riccati form,

$$dt + (a_1 t^2 + a_2 t + a_3)\,du + (b_1 t^2 + b_2 t + b_3)\,dv = 0,$$

where the a's and b's are functions of u and v. Recalling a fundamental property of such equations (§ 14), we have:

Any four orthogonal surfaces of a cyclic system meet the circles in four points whose cross-ratio is constant.

Since by hypothesis S is nondevelopable, equations (12) may be replaced by

$$\text{(13)}\qquad \begin{cases} R\dfrac{\partial R}{\partial u} = Aa + Bb, \\[2mm] R\dfrac{\partial R}{\partial v} = A_1 a + B_1 b, \\[2mm] AB_1 - A_1 B + R^2(\xi\eta_1 - \xi_1\eta)\,K = 0. \end{cases}$$

By (5) the first two of these equations are reducible to

$$\text{(14)}\qquad \begin{cases} \dfrac{\partial}{\partial u}(R^2 - a^2 - b^2) = 2(a\xi + b\eta), \\[2mm] \dfrac{\partial}{\partial v}(R^2 - a^2 - b^2) = 2(a\xi_1 + b\eta_1). \end{cases}$$

The condition of integrability of these equations is

$$\text{(15)}\qquad \frac{\partial a}{\partial v}\xi + \frac{\partial b}{\partial v}\eta - \frac{\partial a}{\partial u}\xi_1 - \frac{\partial b}{\partial u}\eta_1 = r(a\eta_1 - b\xi_1) - r_1(a\eta - b\xi).$$

Instead of considering this equation, we introduce a function ϕ by the equation

$$\text{(16)}\qquad 2\phi = R^2 - a^2 - b^2,$$

and determine the condition which ϕ must satisfy. We take for a and b the expressions obtained by solving (14); that is

$$(17)\qquad a=\frac{\eta_1\dfrac{\partial\phi}{\partial u}-\eta\dfrac{\partial\phi}{\partial v}}{\xi\eta_1-\xi_1\eta},\qquad b=\frac{-\xi_1\dfrac{\partial\phi}{\partial u}+\xi\dfrac{\partial\phi}{\partial v}}{\xi\eta_1-\xi_1\eta}.$$

Now the equation (15) vanishes identically, and the only other condition to be satisfied is the last of (13); this, by the substitution of these values of a, b, R, becomes a partial differential equation in ϕ of the form

$$(18)\qquad \frac{\partial^2\phi}{\partial u^2}\frac{\partial^2\phi}{\partial v^2}-\left(\frac{\partial^2\phi}{\partial u\,\partial v}\right)^2+J\frac{\partial^2\phi}{\partial u^2}+L\frac{\partial^2\phi}{\partial u\,\partial v}+M\frac{\partial^2\phi}{\partial v^2}+N=0,$$

where J, L, M, N denote functions of ϕ, ξ, $\cdots r_1$, and their derivatives of the first order. Conversely, each solution of this equation gives a cyclic system whose circles lie in the tangent planes to S.

EXAMPLES

1. Let S be a surface of revolution defined by (III, 99), and let T be the trihedral whose x-axis is tangent to the curve $v=$ const. Determine the condition which the function $\psi(u)$ must satisfy in order that the quantities a, b in (1) may have the values

$$a=\frac{\psi(u)}{\sqrt{1+\phi'^2(u)}},\qquad b=\frac{1}{u},$$

and determine also the expression for R.

2. A necessary and sufficient condition that all the circles of a cyclic system whose planes envelop a nondevelopable surface shall have the same radius, is that the planes of the circles touch their envelope S at the centers of the circles, and that S be pseudospherical.

3. Let S be a surface referred to an orthogonal system of lines, and let T be the trihedral whose x-axis is tangent to the curve $v=$ const. With reference to the trihedral the equations of a curve in the tangent plane are of the form

$$x=\rho\cos\theta,\qquad y=\rho\sin\theta,\qquad z=0,$$

where in general ρ is a function of θ, u, and v. Show that the condition that there be a surface orthogonal to these curves is that there exist a relation between θ, u, and v which satisfies the equation

$$\left[\rho^2+\left(\frac{\partial\rho}{\partial\theta}\right)^2\right]d\theta+\left(\frac{\partial\rho}{\partial\theta}\frac{\partial\rho}{\partial u}+\xi\cos\theta\frac{\partial\rho}{\partial\theta}-\rho\xi\sin\theta+\rho^2 r\right)du$$
$$+\left(\frac{\partial\rho}{\partial\theta}\frac{\partial\rho}{\partial v}+\eta_1\sin\theta\frac{\partial\rho}{\partial\theta}+\rho\eta_1\cos\theta+\rho^2 r_1\right)dv=0.$$

When this condition is satisfied by a function θ which involves an arbitrary constant, there is an infinity of normal surfaces. In this case the curves are said to form a *normal congruence*.

4. When the surface enveloped by the planes of the curves of a normal congruence of plane curves is deformed in such a way that the curves remain invariably fixed to the surface, the congruence continues to be normal.

175. Cyclic congruences. The axes of the circles of a cyclic system constitute a rectilinear congruence which Bianchi * has called a *cyclic congruence*. In order to derive the properties of this congruence and further results concerning cyclic systems, we assume that the parametric curves on S correspond to the developables of the congruence.

The coördinates of the focal points of a line of the congruence with reference to the corresponding trihedral are of the form $a, b, \rho_1; a, b, \rho_2$. On the hypothesis that the former are the coördinates of the focal point for the developable $v = \text{const.}$ through the line, we have, from (2),

$$\frac{\partial a}{\partial u} + \xi + q\rho_1 - rb = 0, \qquad \frac{\partial b}{\partial u} + \eta - p\rho_1 + ra = 0.$$

Proceeding in like manner with the other point, we obtain a pair of similar equations. All of these equations may be written in the abbreviated form

$$(19) \qquad A + q\rho_1 = 0, \quad B - p\rho_1 = 0, \quad A_1 + q_1\rho_2 = 0, \quad B_1 - p_1\rho_2 = 0,$$

in consequence of (5). When these values are substituted in the last of equations (13), it is found that

$$(20) \qquad R^2 = -\rho_1\rho_2.$$

Hence the lines joining a point on the circle to the focal points are perpendicular. If we put

$$2\rho = \rho_1 - \rho_2, \qquad 2\delta = \rho_1 + \rho_2,$$

thus indicating by 2ρ the distance between the focal points, and by δ the distance between the center of the circle and the mid-point of the line of the congruence, we find that

$$R^2 + \delta^2 = \rho^2.$$

We replace this équation by the two

$$(21) \qquad \delta = \rho\cos\sigma, \qquad R = \rho\sin\sigma,$$

thus defining a function σ. Now we have

$$\rho_1 = \rho(\cos\sigma + 1), \qquad \rho_2 = \rho(\cos\sigma - 1),$$

so that equations (19) may be written

$$(22) \qquad \begin{cases} A = -q\rho(\cos\sigma + 1), & B = p\rho(\cos\sigma + 1), \\ A_1 = -q_1\rho(\cos\sigma - 1), & B_1 = p_1\rho(\cos\sigma - 1). \end{cases}$$

* Vol. II, p. 161.

By means of (5) equation (15) can be put in the form

$$A_1\xi - A\xi_1 + B_1\eta - B\eta_1 = 0.$$

When the values (22) are substituted in this equation, it becomes

$$\cos\sigma\,(p\eta_1 - p_1\eta - q\xi_1 + q_1\xi) + (p_1\eta - q_1\xi + p\eta_1 - q\xi_1) = 0.$$

Since by (3) the expression in the first parenthesis is zero, the same is true of the second, and so we have

$$p\eta_1 - q\xi_1 = 0, \qquad p_1\eta - q_1\xi = 0.$$

But these are the conditions (V, 67) that the parametric curves on S form a conjugate system. Hence we have the theorem of Ribaucour:

On the envelope of the planes of the circles of a cyclic system the curves corresponding to the developables of the associated cyclic congruence form a conjugate system.

176. Spherical representation of cyclic congruences. When the expressions (22) are substituted in (6), we obtain

$$q_1\frac{\partial}{\partial u}[\rho(\cos\sigma - 1)] - q\frac{\partial}{\partial v}[\rho(\cos\sigma + 1)]$$
$$= \rho\left(\frac{\partial q}{\partial v} + \frac{\partial q_1}{\partial u} + r_1 p + r p_1\right) - b\,(pq_1 - p_1 q),$$

$$p_1\frac{\partial}{\partial u}[\rho(\cos\sigma - 1)] - p\frac{\partial}{\partial v}[\rho(\cos\sigma + 1)]$$
$$= \rho\left(\frac{\partial p}{\partial v} + \frac{\partial p_1}{\partial u} - r_1 q - r q_1\right) - a\,(pq_1 - p_1 q).$$

Since $pq_1 - p_1 q \neq 0$ unless S is developable, the preceding equations may be replaced by

$$(23)\qquad \begin{cases} \dfrac{\partial}{\partial u}[\rho(\cos\sigma - 1)] = 2\rho\begin{Bmatrix}12\\2\end{Bmatrix}' + (aq - bp), \\[2ex] \dfrac{\partial}{\partial v}[\rho(\cos\sigma + 1)] = -2\rho\begin{Bmatrix}12\\1\end{Bmatrix}' + (aq_1 - bp_1), \end{cases}$$

where the Christoffel symbols are formed with respect to

$$(24)\qquad (p\,du + p_1\,dv)^2 + (q\,du + q_1\,dv)^2,$$

the linear element of the spherical representation of S.

When in like manner we substitute in the first two of equations (13), taking

$$R^2 = \rho^2 \sin^2\sigma = \rho^2(1-\cos^2\sigma),$$

we obtain

$$(1-\cos\sigma)\frac{\partial\rho}{\partial u} - \frac{\rho\cos\sigma}{1+\cos\sigma}\frac{\partial}{\partial u}\cos\sigma = pb - qa,$$

$$(1+\cos\sigma)\frac{\partial\rho}{\partial v} - \frac{\rho\cos\sigma}{1-\cos\sigma}\frac{\partial}{\partial v}\cos\sigma = -p_1 b + q_1 a.$$

From these equations and (23) we find

$$(25)\qquad \begin{cases} \dfrac{\partial}{\partial u}\cos\sigma = 2(\cos\sigma+1)\begin{Bmatrix}12\\2\end{Bmatrix}', \\[2ex] \dfrac{\partial}{\partial v}\cos\sigma = 2(\cos\sigma-1)\begin{Bmatrix}12\\1\end{Bmatrix}', \end{cases}$$

and

$$(26)\qquad \begin{cases} (\cos\sigma-1)\dfrac{\partial\rho}{\partial u} = -2\rho\cos\sigma\begin{Bmatrix}12\\2\end{Bmatrix}' + qa - pb, \\[2ex] (\cos\sigma+1)\dfrac{\partial\rho}{\partial v} = -2\rho\cos\sigma\begin{Bmatrix}12\\1\end{Bmatrix}' + q_1 a - p_1 b. \end{cases}$$

The condition of integrability of equations (25) is reducible to

$$(27)\quad \left(\frac{\partial}{\partial u}\begin{Bmatrix}12\\1\end{Bmatrix}' - \frac{\partial}{\partial v}\begin{Bmatrix}12\\2\end{Bmatrix}'\right)\cos\sigma = \frac{\partial}{\partial u}\begin{Bmatrix}12\\1\end{Bmatrix}' + \frac{\partial}{\partial v}\begin{Bmatrix}12\\2\end{Bmatrix}' - 4\begin{Bmatrix}12\\1\end{Bmatrix}'\begin{Bmatrix}12\\2\end{Bmatrix}'.$$

If the expression for $\cos\sigma$ obtained from this equation be substituted in (25), we find two conditions upon the curves on the sphere in order that they may represent the developables of a cyclic congruence. A particular case is that in which (27) is identically satisfied, when the two conditions are

$$(28)\qquad \frac{\partial}{\partial u}\begin{Bmatrix}12\\1\end{Bmatrix}' = \frac{\partial}{\partial v}\begin{Bmatrix}12\\2\end{Bmatrix}' = 2\begin{Bmatrix}12\\1\end{Bmatrix}'\begin{Bmatrix}12\\2\end{Bmatrix}'.$$

It is now our purpose to show that if any system of curves on the sphere satisfies either set of conditions, all the congruences whose developables are thus represented on the sphere are cyclic.

We assume that the sphere is referred to such a system and that we have a solution ρ of

$$(29)\quad \frac{\partial^2\rho}{\partial u\,\partial v} + \begin{Bmatrix}12\\1\end{Bmatrix}'\frac{\partial\rho}{\partial u} + \begin{Bmatrix}12\\2\end{Bmatrix}'\frac{\partial\rho}{\partial v} + \left(\frac{\partial}{\partial u}\begin{Bmatrix}12\\1\end{Bmatrix}' + \frac{\partial}{\partial v}\begin{Bmatrix}12\\2\end{Bmatrix}' + \mathcal{F}\right)\rho = 0.$$

By the method of § 167, or that hereinafter explained, we find the middle surface of the congruence. Then we take the point on each line at the distance $-\rho\cos\sigma$ from the mid-point as the center of the circle of radius $\rho\sin\sigma$ and for which the line is the axis. These circles form a cyclic system, as we shall show.

In the first place we determine the middle surface with reference to a trihedral of fixed vertex, whose z-axis coincides with the radius of the sphere parallel to the line of the congruence and whose x- and y-axes are any whatever. If x_0, y_0, z_0 denote the coördinates of the mid-point of a line with reference to the corresponding trihedral, the coördinates of the focal points are

$$x_0, \quad y_0, \quad z_0+\rho; \qquad x_0, \quad y_0, \quad z_0-\rho.$$

From (2) it is seen that if these points correspond to the developables $v=$ const. and $u=$ const. respectively, we must have

$$\frac{\partial x_0}{\partial u}+q(z_0+\rho)-ry_0=0, \qquad \frac{\partial y_0}{\partial u}+rx_0-p(z_0+\rho)=0,$$

$$\frac{\partial x_0}{\partial v}+q_1(z_0-\rho)-r_1y_0=0, \qquad \frac{\partial y_0}{\partial v}+r_1x_0-p_1(z_0-\rho)=0.$$

Since $pq_1-p_1q\neq 0$, the conditions of integrability of these equations can be put in the form

$$(30)\qquad \left\{\begin{aligned} &\frac{\partial z_0}{\partial u}-\frac{\partial \rho}{\partial u}-2\rho\begin{Bmatrix}12\\2\end{Bmatrix}'+(py_0-qx_0)=0,\\ &\frac{\partial z_0}{\partial v}+\frac{\partial \rho}{\partial v}+2\rho\begin{Bmatrix}12\\1\end{Bmatrix}'+(p_1y_0-q_1x_0)=0.\end{aligned}\right.$$

It is readily found that the condition of integrability of these equations is reducible to (29).

It will be to our advantage to have also the coördinates of the point of contact of the plane of the circle with its envelope S. If $\bar{x}$, $\bar{y}$, $z_0-\rho\cos\sigma$ denote these coördinates with reference to the above trihedral, it follows from (2) that

$$\frac{\partial}{\partial u}(z_0-\rho\cos\sigma)+p\bar{y}-q\bar{x}=0,$$

$$\frac{\partial}{\partial v}(z_0-\rho\cos\sigma)+p_1\bar{y}-q_1\bar{x}=0.$$

if these equations be subtracted from the respective ones of (30), the results are reducible, by means of (25), to

$$(\cos\sigma - 1)\frac{\partial\rho}{\partial u} + 2\rho\cos\sigma\begin{Bmatrix}12\\2\end{Bmatrix}' + p(y_0 - \bar{y}) - q(x_0 - \bar{x}) = 0,$$

$$(\cos\sigma + 1)\frac{\partial\rho}{\partial v} + 2\rho\cos\sigma\begin{Bmatrix}12\\1\end{Bmatrix}' + p_1(y_0 - \bar{y}) - q_1(x_0 - \bar{x}) = 0,$$

which are the same as (26). For, the quantities $x_0 - \bar{x}$, $y_0 - \bar{y}$ are the coördinates of the center of the circle with reference to the trihedral parallel to the preceding one and with the corresponding point on S for vertex.

If, then, we have a solution σ of (25) and ρ of (29), the corresponding values of a and b given by (26) satisfy (22), since the latter are the conditions that the parametric curves on the sphere represent the developables of the congruence. However, we have seen that when the values (22) are substituted in (12), we obtain equations reducible to (25) and (26). Hence the circles constructed as indicated above form a cyclic system.

Since equations (25) admit only one solution (27) unless the condition (28) is satisfied, we have the theorem:

With each cyclic congruence there is associated a unique cyclic system unless it is at the same time a congruence of Ribaucour, in which case there is an infinity of associated cyclic systems.

Recalling the results of § 141, we have the theorem of Bianchi *:

When the total curvature of a surface referred to its asymptotic lines is of the form

$$K = -\frac{1}{[\phi(u) + \psi(v)]^2},$$

it is the surface generatrix of a congruence of Ribaucour which is cyclic in an infinity of ways, and these are the only cyclic congruences with an infinity of associated cyclic systems.

In this case the general solution of equations (25) is

$$\text{(31)} \qquad \cos\sigma = \frac{\psi - \phi + a}{\phi + \psi},$$

where a is an arbitrary constant.

* Vol. II, p. 165.

177. Surfaces orthogonal to a cyclic system. In this section we consider the surfaces S_1 orthogonal to the circles of a cyclic system. Since the direction-cosines of the normals to the surfaces with reference to the moving trihedral in § 174 are $-\sin\theta$, $\cos\theta$, 0, the spherical representation of these surfaces is given by the point whose coördinates are these with respect to a trihedral of fixed vertex parallel to the above trihedral. From (2) we find that the expressions for the projections of a displacement of this point are

$$-\cos\theta\,(d\theta + r\,du + r_1\,dv),$$
$$-\sin\theta\,(d\theta + r\,du + r_1\,dv),$$
$$(p\,du + p_1\,dv)\cos\theta + (q\,du + q_1\,dv)\sin\theta.$$

Moreover, by means of (8), (21), (22), we obtain the identity

$$\sin\sigma\,(d\theta + r\,du + r_1\,dv) = -(1+\cos\sigma)(p\cos\theta + q\sin\theta)\,du + (1-\cos\sigma)(p_1\cos\theta + q_1\sin\theta)\,dv. \tag{32}$$

Hence the linear element of the spherical representation of S_1 is

$$d\sigma_1^2 = \frac{2}{1-\cos\sigma}(p\cos\theta + q\sin\theta)^2\,du^2 + \frac{2}{1+\cos\sigma}(p_1\cos\theta + q_1\sin\theta)^2\,dv^2. \tag{33}$$

Since the parametric curves on the sphere form an orthogonal system, the parametric curves on the surface are the lines of curvature, if they form an orthogonal system. In order to show that this condition is satisfied, we first reduce the expressions (4) for the projections of a displacement of a point on S_1, by means of (21), (22), (25), (26), and (32), to

$$\begin{cases} \cos\theta\sin\sigma\left(\dfrac{C\,du}{1-\cos\sigma} - \dfrac{D\,dv}{1+\cos\sigma}\right), \\ \sin\theta\sin\sigma\left(\dfrac{C\,du}{1-\cos\sigma} - \dfrac{D\,dv}{1+\cos\sigma}\right), \\ C\,du + D\,dv, \end{cases} \tag{34}$$

where we have put

$$\begin{cases} C = p(b + R\sin\theta) - q(a + R\cos\theta), \\ D = p_1(b + R\sin\theta) - q_1(a + R\cos\theta). \end{cases} \tag{35}$$

Hence the linear element of S_1 is

$$(36)\qquad ds_1^2 = 2\frac{C^2 du^2}{1-\cos\sigma} + 2\frac{D^2 dv^2}{1+\cos\sigma},$$

from which it is seen that the parametric curves on S_1 form an orthogonal system, and consequently are the lines of curvature.

Furthermore, it is seen from (34) that the tangents to the curves $v = \text{const.}$, $u = \text{const.}$ make with the plane of the circle the respective angles

$$(37)\qquad \tan^{-1}\left(\frac{1-\cos\sigma}{\sin\sigma}\right), \qquad \tan^{-1}\left(-\frac{1+\cos\sigma}{\sin\sigma}\right).$$

But it follows from (21) that the lines joining a point on the circumference of a circle to the focal points of its axis make the angles (37) with the radius to the point. Hence we have:

The lines of curvature on a surface orthogonal to a cyclic system correspond to the developables of the congruence of axes of the circles, and the tangents to the two lines of curvature through a point of the surface meet the corresponding axis in its focal points.

178. Normal cyclic congruences. Since the developables of a cyclic congruence correspond to a conjugate system on the envelope S of the planes of the circles, this system consists of the lines of curvature when the congruence is normal, and only in this case (cf. § 83). If, under these conditions, we take two of the edges of the trihedral tangent to the lines of curvature, we have

$$(38)\qquad \xi_1 = \eta = 0, \qquad p = q_1 = 0,$$

and equations (25) become

$$\frac{\partial}{\partial u}\log\cos\frac{\sigma}{2} = \frac{\partial}{\partial u}\log p_1, \qquad \frac{\partial}{\partial v}\log\sin\frac{\sigma}{2} = \frac{\partial}{\partial v}\log q.$$

By a suitable choice of parameters we have

$$p_1 = \cos\frac{\sigma}{2}, \qquad q = -\sin\frac{\sigma}{2},$$

so that if we put $\omega = -\sigma/2$, the linear element of the sphere is

$$(39)\qquad d\sigma^2 = \sin^2\omega\, du^2 + \cos^2\omega\, dv^2.$$

Comparing this result with (§ 119), we have the theorem:

The normals to a surface Σ with the same spherical representation of its lines of curvature as a pseudospherical surface constitute the only kind of normal cyclic congruences.

Since the surface Σ and the envelope S of the planes of the circles have the same representation of their lines of curvature, the tangents to the latter at corresponding points on the two surfaces are parallel. Hence with reference to a trihedral for Σ parallel to the trihedral for S the coördinates of a point on the circle are $R\cos\theta$, $R\sin\theta$, μ, where μ remains to be determined and θ is given by (32), which can be put in the form

$$(40) \qquad \frac{\partial\theta}{\partial u}+\frac{\partial\omega}{\partial v}=\cos\omega\sin\theta, \qquad \frac{\partial\theta}{\partial v}+\frac{\partial\omega}{\partial u}=-\sin\omega\cos\theta.$$

If we express, by means of (2), the condition that all displacements of this point be orthogonal to the line whose direction-cosines are $-\sin\theta$, $\cos\theta$, 0, the resulting equation is reducible, by means of (40), to

$$\sin\theta(R\cos\omega-\mu\sin\omega-\xi)\,du \\ -\cos\theta(R\sin\omega+\mu\cos\omega-\eta_1)\,dv=0.$$

Hence the quantities in parentheses are zero, from which we obtain

$$(41) \qquad R=\xi\cos\omega+\eta_1\sin\omega, \qquad \mu=-\xi\sin\omega+\eta_1\cos\omega.$$

When, in particular, Σ is a pseudospherical surface of curvature $-1/a^2$, we have (VIII, 22)

$$\xi=a\cos\omega, \qquad \eta_1=a\sin\omega,$$

so that $R=a$ and $\mu=0$. Hence the circles are of constant radius and the envelope of their planes is the locus of their centers (cf. Ex. 2, § 174). Conversely, when the latter condition is satisfied, it follows from (13) that R is constant. Moreover, in this case ρ_1 and ρ_2, as defined in § 175, are the principal radii of the surface, which by (20) is pseudospherical. When these values are substituted in (36) and (33), it is found that the linear element of each orthogonal surface is

$$ds_1^2=a^2(\cos^2\theta\,du^2+\sin^2\theta\,dv^2),$$

and of its spherical representation

$$(42) \qquad d\sigma_1^2=\sin^2\theta\,du^2+\cos^2\theta\,dv^2.$$

Hence these orthogonal surfaces are the transforms of Σ by means of the Bianchi transformation (§ 119).

The expression (42) is the linear element of the spherical representation of the surfaces orthogonal to the circles associated with any surface Σ, whether it be pseudospherical or not, whose spherical representation is given by (39). Since these orthogonal surfaces have this representation of their lines of curvature, they are of the same kind as Σ. We have thus for all surfaces with the same representation of their lines of curvature as pseudospherical surfaces, a transformation into similar surfaces of which the Bianchi transformation is a particular case; we call it a generalized Bianchi transformation.*

179. Cyclic systems for which the envelope of the planes of the circles is a curve. We consider now the particular cases which have been excluded from the preceding discussion, and begin with that for which the envelope S of the planes of the circles is a curve C.

We take the moving trihedral such that its xy-plane, as before, is that of the circle, and take the x-axis tangent to C. If s denotes the arc of the latter, we have

$$ds = \xi\, du + \xi_1 dv, \qquad \eta = \eta_1 = 0,$$

and by (3)

$$r\xi_1 - r_1\xi = 0, \qquad q\xi_1 - q_1\xi = 0. \tag{43}$$

From (14), (15), and (16) it follows that a and ϕ are functions of s, so that these equations may be replaced by

$$R^2 = a^2 + b^2 + 2\,\phi(s). \tag{44}$$

If the parametric curves on the sphere represent the developables of the congruence, the conditions (19) must hold. But from (5), (15), and (43) we obtain

$$A\xi_1 - A_1\xi = 0.$$

If the values from (19) be substituted in this equation, we have, from (43),

$$\rho_1 - \rho_2 = 0.$$

Hence the focal surfaces coincide. If we put

$$\rho = \rho_1 = \rho_2$$

in (19) and substitute in the last of (12), we obtain

$$(\rho^2 + R^2)(pq_1 - p_1q) = 0.$$

*Cf. *American Journal*, Vol. XXVI (1905), pp. 127–132.

The vanishing of $pq_1 - p_1 q$ is the condition that there be a single infinity of planes, which case we exclude for the present. Hence $\rho = \pm iR$; that is, the developables of the cyclic congruence are imaginary.

Instead of retaining as parametric curves those representing the developables, we make the following choice. We take the arc of C for the parameter u; consequently $\xi = 1$, $\xi_1 = 0$. Since $\eta = \eta_1 = 0$ also, we have, from (3),

$$q_1 = r_1 = 0, \qquad \frac{\partial p}{\partial v} - \frac{\partial p_1}{\partial u} = 0;$$

hence we may choose the parameter v so that $p = 0$, $p_1 = 1$. From (3) it follows, furthermore, that

$$\frac{\partial q}{\partial v} = r, \qquad \frac{\partial r}{\partial v} = -q,$$

of which the general integral is

$$q = U_1 \cos v + U_2 \sin v, \qquad r = -U_1 \sin v + U_2 \cos v,$$

where U_1 and U_2 are arbitrary functions of u. From (5) we have

$$A = \phi''(u) + 1 - rb, \qquad A_1 = 0, \qquad B_1 = \frac{\partial b}{\partial v},$$

so that the third of equations (12) is reducible by (44) to

$$(45) \qquad \frac{(\phi'' + 1)\dfrac{\partial b}{\partial v}}{(b^2 + \phi'^2 + 2\phi)^{\frac{3}{2}}} = \frac{\partial}{\partial v} \frac{U_1 \sin v - U_2 \cos v}{(b^2 + \phi'^2 + 2\phi)^{\frac{1}{2}}}.$$

Hence if we take for a any function of u denoted by $\phi'(u)$, equation (45) gives b, and R follows directly from (44).

180. Cyclic systems for which the planes of the circles pass through a point. If the planes of the circles of a cyclic system pass through a point O, we take it for the origin and for the vertex of a moving trihedral whose z-axis is parallel to the axis of the circle under consideration. In this case equations (14) may be replaced by

$$(46) \qquad R^2 = a^2 + b^2 - c,$$

where c denotes a constant. But this is the condition that all the circles are orthogonal to a sphere with center at O, or cut it in diametrically opposite points, or pass through O, according as c is positive, negative, or zero. Hence we have the theorem:

If the planes of the circles of a cyclic system pass through a point. the circles are orthogonal to a sphere with its center at the point, or meet the sphere in opposite points, or pass through the center.

From geometrical considerations we see that the converse of this theorem is true.

When c in (46) is zero all the circles pass through O. Then by (21) we have

$$a = -\rho \sin\sigma \cos\theta, \qquad b = -\rho \sin\sigma \sin\theta, \tag{47}$$

and equations (26) become

$$\left\{\begin{aligned} &(\cos\sigma - 1)\frac{\partial \log\rho}{\partial u} = -2\cos\sigma\left\{\begin{matrix}12\\2\end{matrix}\right\}' + \sin\sigma\,(p\sin\theta - q\cos\theta), \\ &(\cos\sigma + 1)\frac{\partial \log\rho}{\partial v} = -2\cos\sigma\left\{\begin{matrix}12\\1\end{matrix}\right\}' + \sin\sigma\,(p_1\sin\theta - q_1\cos\theta). \end{aligned}\right. \tag{48}$$

These equations are obtained likewise when we substitute the values (47) in equations (22) and reduce by means of (25) and (32). Because of (22) the function ρ given by (26) is a solution of (29), and therefore ρ given by (48) is a solution. But the solution θ of (32) involves a parameter. Hence we have the theorem of Bianchi*.

Among all the cyclic congruences with the same spherical representation of their developables there are an infinity for which the circles of the associated cyclic system pass through a point.

If we take the line through O and the center of the circle for the x-axis of the trihedral, equation (11) must admit of the solution $\theta = \pi$, and consequently must be of the form

$$\sin\theta L + (\cos\theta + 1)M = 0.$$

In order that this equation admit of a solution other than π, both L and M must be zero and the system cyclic. We combine this result with the preceding theorem to obtain the following:

A two-parameter family of circles through a point and orthogonal to any surface constitute a cyclic system, and the most general spherical representation of the developables of a cyclic congruence is afforded by the representation of the axes of such a system of circles.†

* Vol. II, p. 169.

† Bianchi, Vol. II, p. 170.

We consider finally the case where the planes of the circles depend upon a single parameter. If we take for moving axes the tangent, principal normal, and binormal of the edge of regression of these planes and its arc for the parameter u, we have

$$\xi = 1, \qquad \xi_1 = \eta = \eta_1 = 0, \qquad p_1 = q_1 = r_1 = 0;$$

and comparing (V, 50) with (2), we see that

$$p = -\frac{1}{\tau}, \qquad q = 0, \qquad r = \frac{1}{\rho},$$

where ρ and τ are the radii of first and second curvature of the edge of regression. Now

$$A = \frac{\partial a}{\partial u} + 1 - \frac{b}{\rho}, \qquad A_1 = \frac{\partial a}{\partial v}, \qquad B = \frac{\partial b}{\partial u} + \frac{a}{\rho}, \qquad B_1 = \frac{\partial b}{\partial v}.$$

The equations (12) reduce to two. One of the functions a, b may be chosen arbitrarily; then the other and R can be obtained by the solution of partial differential equations of the first order.

EXAMPLES

1. Show that a congruence of Ribaucour whose surface generator is the right helicoid is cyclic, and determine the cyclic systems.

2. A congruence of Guichard is a cyclic congruence, and the envelope of the planes of the circles of each associated cyclic system is a surface of Voss.

3. The surface generator of a cyclic congruence of Ribaucour is an associate surface of the planes of the circles of each associated cyclic system.

4. If S is a surface whose lines of curvature have the same spherical representation as a pseudospherical surface, and S_1 is a transform of S resulting from a generalized Bianchi transformation (§ 178), the tangents to the lines of curvature of S_1 pass through the centers of principal curvature of S.

5. When the focal segment of each line of a cyclic congruence is divided in constant ratio by the center of the circle, the envelope of the planes of the circles is a surface of Voss.

6. The circles of the cyclic system whose axes are normal to the surface Σ, defined in Ex. 11, p. 370, pass through a point, and the surfaces orthogonal to the circles are surfaces of Bianchi of the parabolic type.

7. If the spheres with the focal segments of the lines of a congruence for diameters pass through a point, the congruence is cyclic, and the circles pass through the point.

8. Show that the converse of Ex. 7 is true.

GENERAL EXAMPLES

1. Determine the normal congruences of Ribaucour which are cyclic.

2. If the envelope of the planes of the circles of a cyclic system is a surface of Voss whose conjugate geodesic system corresponds to the developables of the associated cyclic congruence, any family of planes cutting the focal segments in constant ratio and perpendicular to them envelop a surface of Voss.

3. A necessary and sufficient condition that a congruence be cyclic is that the developables have the same spherical representation as the conjugate lines of a surface which remain conjugate in a deformation of the surface. If the developables of the congruence are real, the deforms of the surface are imaginary.

4. The planes of the cyclic systems associated with a cyclic congruence of Ribaucour touch their respective envelopes in such a way that the points of contact of all the planes corresponding to the same line of the congruence lie on a straight line.

5. If the spheres described on the focal segments of a congruence as diameters cut a fixed sphere orthogonally or in great circles, the congruence is cyclic and the circles cut the fixed sphere orthogonally or in diametrically opposite points.

6. If one draws the circles which are normal to a surface S and which cut a fixed sphere S_0 in diametrically opposite points or orthogonally, the spheres described on the focal segments of the congruence of axes as diameters cut S_0 in great circles or orthogonally.

7. Determine the cyclic systems of equal circles whose planes envelop a developable surface.

8. Let Σ_1 be the surface defined in Ex. 14, p. 371, and let S_0 be the sphere with center at the origin and radius r. Draw the circles which are normal to Σ_1 and which cut S_0 orthogonally or in diametrically opposite points. Show that the cyclic congruence of the axes of these circles is a normal congruence, and that the coördinates of the normal surfaces are of the form

$$x = \left[\frac{1}{2a}\left\{a^2 e^{-\frac{\xi}{a}} - (\eta^2+\kappa)\, e^{\frac{\xi}{a}}\right\}\cos\theta + \eta\sin\theta\right]X_1$$
$$+\left[\frac{1}{2a}\left\{a^2 e^{-\frac{\xi}{a}} - (\eta^2+\kappa)\, e^{\frac{\xi}{a}}\right\}\sin\theta - \eta\cos\theta\right]X_2 + tX,$$

where κ is equal to $-r^2$ or $+r^2$, according as the circles cut S_0 orthogonally or in diametrically opposite points, and where t is given by

$$dt = \left[\frac{1}{2a}\left\{a^2 e^{-\frac{\xi}{a}} - (\eta^2+\kappa)\, e^{\frac{\xi}{a}}\right\}\cos\theta + \eta\sin\theta\right]\sin\omega\, du$$
$$-\left[\frac{1}{2a}\left\{a^2 e^{-\frac{\xi}{a}} - (\eta^2+\kappa)\, e^{\frac{\xi}{a}}\right\}\sin\theta - \eta\cos\theta\right]\cos\omega\, dv.$$

9. Show that the surfaces of Ex. 8 are surfaces of Bianchi which have the same spherical representation of their lines of curvature as the pseudospherical surface S referred to in Ex. 14, p. 371.

10. Show that the surfaces orthogonal to the cyclic system of Ex. 8 are surfaces of Bianchi of the parabolic type.

11. Let S be a surface referred to an orthogonal system, and let T be the trihedral whose x-axis is tangent to the curve $v = \text{const.}$ The equations

$$x = \rho(1 + \cos\theta), \qquad y = 0, \qquad z = \rho\sin\theta$$

define a circle normal to S. Show that the necessary and sufficient conditions that the circles so defined form a cyclic system are

$$\xi\frac{\partial\rho}{\partial v} + \rho\eta_1 r = 0, \qquad \rho(pr_1 - p_1 r) - q_1\left(\xi + \frac{\partial\rho}{\partial u}\right) + q\frac{\partial\rho}{\partial v} = 0.$$

12. A necessary and sufficient condition that a cyclic system remain cyclic when an orthogonal surface S is deformed is that S be applicable to a surface of revolution and that

$$\rho = \frac{1}{\phi}\left(c - \int\phi\, du\right),$$

where c is a constant and the linear element of S is $ds^2 = du^2 + \phi^2(u)\,dv^2$ (cf. Ex. 11).

13. Determine under what conditions the lines of intersection of the planes of the circles of a cyclic system and the tangent planes to an orthogonal surface form a normal congruence.

14. Let Σ_1 and Σ_2 be two surfaces orthogonal to a cyclic system, and let M_1 and M_2 be the points of intersection of one of the circles with Σ_1 and Σ_2. Show that the normals to Σ_1 and Σ_2 at the points M_1 and M_2 meet in a point M equidistant from these points, and show that Σ_1 and Σ_2 constitute the sheets of the envelope of a two-parameter family of spheres such that the lines of curvature on Σ_1 and Σ_2 correspond.

15. Let S be the surface of centers of a two-parameter family of spheres of variable radius R, and let Σ_1 and Σ_2 denote the two sheets of the envelope of these spheres. Show that the points of contact M_1 and M_2 of a sphere with these sheets are symmetric with respect to the tangent plane to S at the corresponding point M. Let S be referred to a moving trihedral whose plane $y = 0$ is the plane M_1MM_2, and let the parametric curves be tangent to the x- and y-axes respectively. Show that if σ denotes the angle which the radius MM_1 makes with the x-axis of the trihedral, the lines of curvature on Σ_1 are given by

$$\begin{aligned}\xi\sin\sigma(\sin\sigma p - r\cos\sigma)\,du^2 + \eta_1\left(q_1 - \frac{\partial\sigma}{\partial v}\right)dv^2 \\ +\left[\eta_1\left(q - \frac{\partial\sigma}{\partial u}\right) - \xi\sin\sigma(\cos\sigma r_1 - p_1\sin\sigma)\right]du\,dv = 0.\end{aligned}$$

16. Find the condition that the lines of curvature on Σ_1 and Σ_2 of Ex. 15 correspond, and show that in this case these curves correspond to a conjugate system on S.

17. Show that the circles orthogonal to two surfaces form a cyclic system, provided that the lines of curvature on the two surfaces correspond.

18. Let S be a pseudospherical surface with the linear element (VIII, 22), the lines of curvature being parametric, and let A be a surface with the same spherical representation of its lines of curvature as S; furthermore, let A_1 denote the envelope of the plane which makes the constant angle σ with the tangent plane at a point M of A and meets this plane in a line L, which forms with the tangent to the curve $v = \text{const.}$ at M an angle θ defined by equations (VIII, 35). If M_1

denotes the point of contact of this plane, we drop from M_1 a perpendicular on L, meeting the latter in N. Show that if λ and μ denote the lengths MN and NM_1, they are given by

$$\lambda = (\sqrt{E}\cos\omega + \sqrt{G}\sin\omega)\sin\sigma, \qquad \mu = (-\sqrt{E}\sin\omega + \sqrt{G}\cos\omega)\sin\sigma,$$

where E and G are the first fundamental coefficients of A.

19. Show that when the surface A in Ex. 18 is the pseudospherical surface S, then A_1 is the Bäcklund transform S_1 of S by means of the functions (θ, σ), and that when A is other than S the lines of curvature on the four surfaces S, A, S_1, A_1 correspond, and the last two have the same spherical representation.

20. Show that as θ is given all values satisfying equations (VIII, 35) for a given σ, the locus of the point M_1, defined in Ex. 18, is a circle whose axis is normal to the surface A at M.

21. Show that when A in Ex. 18 is a surface of Bianchi of the parabolic type (Ex. 11, p. 370) the surfaces A_1 are of the same kind, whatever be σ.

CHAPTER XIV

TRIPLY ORTHOGONAL SYSTEMS OF SURFACES

181. Triple system of surfaces associated with a cyclic system. Let S_1 be one of the surfaces orthogonal to a cyclic system, and let its lines of curvature be parametric. The locus Σ_1 of the circles which meet S_1 in the line of curvature $v = \text{const.}$ through a point M is a surface which cuts S_1 orthogonally. Hence, by Joachimsthal's theorem (§ 59), the line of intersection is a line of curvature for Σ_1. In like manner, the locus Σ_2 of the circles which meet S_1 in the line of curvature $u = \text{const.}$ through M cuts S_1 orthogonally, and the curve of intersection is a line of curvature on Σ_2 also. Since the developables of the associated cyclic congruence correspond to the lines of curvature on all of the orthogonal surfaces, each of the latter is met by Σ_1 and Σ_2 in a line of curvature of both surfaces. At each point of the circle through M the tangent to the circle is perpendicular to the line of curvature $v = \text{const.}$ on Σ_1 through the point and to $u = \text{const.}$ on Σ_2. Hence the circle is a line of curvature for both Σ_1 and Σ_2, and these surfaces cut one another orthogonally along the circle.

Since there is a surface Σ_1 for each curve $v = \text{const.}$ on S_1 and a surface Σ_2 for each $u = \text{const.}$, the circles of a cyclic system and the orthogonal surfaces may be looked upon as a system of three families of surfaces such that through each point in space there passes a surface of each family. Moreover, each of these three surfaces meets the other two orthogonally, and each curve of intersection is a line of curvature on both surfaces. We have seen (§ 96) that the confocal quadrics form such a system of surfaces, and another example is afforded by a family of parallel surfaces and the developables of the congruence of normals to these surfaces.

When three families of surfaces are so constituted that through each point of space there passes a surface of each family and each of the three surfaces meets the other two orthogonally, they are said to form a *triply orthogonal system.* In the preceding examples the curve of intersection of any two surfaces is a line of curvature for both. Dupin showed that this is a property of all triply orthogonal systems. We shall prove this theorem in the next section.

182. General equations. Theorem of Dupin. The simplest example of an orthogonal system is afforded by the planes parallel to the coördinate planes. The equations of the system are

$$x = u_1, \qquad y = u_2, \qquad z = u_3,$$

where u_1, u_2, u_3 are parameters. Evidently the values of these parameters corresponding to the planes through a point are the rectangular coördinates of the point. In like manner, the surfaces of each family of any triply orthogonal system may be determined by a parameter, and the values of the three parameters for the three surfaces through a point constitute the *curvilinear coördinates* of the point. Between the latter and the rectangular coördinates there obtain equations of the form

$$(1) \qquad x = f_1(u_1, u_2, u_3), \qquad y = f_2(u_1, u_2, u_3), \qquad z = f_3(u_1, u_2, u_3),$$

where the functions f are analytic in the domain considered. An example of this is afforded by formulas (VII, 8), which define space referred to a system of confocal quadrics.

In order that the system be orthogonal it is necessary and sufficient that these functions satisfy the three conditions

$$(2) \qquad \sum \frac{\partial x}{\partial u_1}\frac{\partial x}{\partial u_2} = 0, \qquad \sum \frac{\partial x}{\partial u_2}\frac{\partial x}{\partial u_3} = 0, \qquad \sum \frac{\partial x}{\partial u_3}\frac{\partial x}{\partial u_1} = 0.$$

Any one of the surfaces $u_i =$ const. is defined by (1) when u_i is given this constant value.

By the *linear element of space at a point* we mean the linear element at the point of any curve through it. This is

$$ds^2 = dx^2 + dy^2 + dz^2,$$

which, in consequence of (2), may be written in the parametric form

$$(3) \qquad ds^2 = H_1^2\, du_1^2 + H_2^2\, du_2^2 + H_3^2\, du_3^2,$$

where

$$(4) \qquad H_1^2 = \sum \left(\frac{\partial x}{\partial u_1}\right)^2, \qquad H_2^2 = \sum \left(\frac{\partial x}{\partial u_2}\right)^2, \qquad H_3^2 = \sum \left(\frac{\partial x}{\partial u_3}\right)^2.$$

As thus defined, the functions H_1, H_2, H_3 are real and we shall assume that they are positive.

From (3) we have at once the linear element of any of the surfaces of the system. For instance, the linear element of a surface $u_3 =$ const. is

$$ds_3^2 = H_1^2 du_1^2 + H_2^2 du_2^2.$$

Now we shall find that the second quadratic forms of these surfaces are expressible in terms of the functions H and their derivatives.

If X_i, Y_i, Z_i denote the direction-cosines of the normals to the surfaces $u_i =$ const., we have

$$X_i = \frac{1}{H_i}\frac{\partial x}{\partial u_i}, \qquad Y_i = \frac{1}{H_i}\frac{\partial y}{\partial u_i}, \qquad Z_i = \frac{1}{H_i}\frac{\partial z}{\partial u_i}. \tag{5}$$

We choose the axes such that

$$\begin{vmatrix} X_1 & Y_1 & Z_1 \\ X_2 & Y_2 & Z_2 \\ X_3 & Y_3 & Z_3 \end{vmatrix} = +1. \tag{6}$$

In consequence of (5) the second fundamental coefficients of a surface $u_i =$ const. are defined by

$$D_i = \frac{1}{H_i}\sum \frac{\partial x}{\partial u_i}\frac{\partial^2 x}{\partial u_\kappa^2}, \quad D_i' = \frac{1}{H_i}\sum \frac{\partial x}{\partial u_i}\frac{\partial^2 x}{\partial u_\kappa \partial u_l}, \quad D_i'' = \frac{1}{H_i}\sum \frac{\partial x}{\partial u_i}\frac{\partial^2 x}{\partial u_l^2},$$

where i, κ, l take the values 1, 2, 3 in cyclic order, and the sign Σ refers to the summation of terms in x, y, z, as formerly. In order to evaluate these expressions we differentiate equations (2) with respect to u_3, u_1, u_2 respectively. This gives

$$\sum \frac{\partial x}{\partial u_1}\frac{\partial^2 x}{\partial u_2 \partial u_3} + \sum \frac{\partial x}{\partial u_2}\frac{\partial^2 x}{\partial u_1 \partial u_3} = 0,$$

$$\sum \frac{\partial x}{\partial u_2}\frac{\partial^2 x}{\partial u_1 \partial u_3} + \sum \frac{\partial x}{\partial u_3}\frac{\partial^2 x}{\partial u_1 \partial u_2} = 0,$$

$$\sum \frac{\partial x}{\partial u_3}\frac{\partial^2 x}{\partial u_1 \partial u_2} + \sum \frac{\partial x}{\partial u_1}\frac{\partial^2 x}{\partial u_2 \partial u_3} = 0.$$

If each of these equations be subtracted from one half of the sum of the three, we have

$$\sum \frac{\partial x}{\partial u_1}\frac{\partial^2 x}{\partial u_2 \partial u_3} = 0, \quad \sum \frac{\partial x}{\partial u_2}\frac{\partial^2 x}{\partial u_3 \partial u_1} = 0, \quad \sum \frac{\partial x}{\partial u_3}\frac{\partial^2 x}{\partial u_1 \partial u_2} = 0;$$

consequently $D_i' = 0$.

If the first and third of (2) be differentiated with respect to u_2 and u_3 respectively, and the second and third of (4) with respect to u_1, we have

$$\sum \frac{\partial x}{\partial u_1}\frac{\partial^2 x}{\partial u_2^2} = -\sum \frac{\partial x}{\partial u_2}\frac{\partial^2 x}{\partial u_1 \partial u_2} = -H_2\frac{\partial H_2}{\partial u_1},$$

$$\sum \frac{\partial x}{\partial u_1}\frac{\partial^2 x}{\partial u_3^2} = -\sum \frac{\partial x}{\partial u_3}\frac{\partial^2 x}{\partial u_1 \partial u_3} = -H_3\frac{\partial H_3}{\partial u_1}.$$

Hence we have

$$D_1 = -\frac{H_2}{H_1}\frac{\partial H_2}{\partial u_1}, \qquad D_1'' = -\frac{H_3}{H_1}\frac{\partial H_3}{\partial u_1}.$$

Proceeding in like manner, we find the expressions for the other D's, which we write as follows:

$$D_i = -\frac{H_\kappa}{H_i}\frac{\partial H_\kappa}{\partial u_i}, \qquad D_i' = 0, \qquad D_i'' = -\frac{H_l}{H_i}\frac{\partial H_l}{\partial u_i}, \tag{7}$$

where i, κ, l take the values 1, 2, 3 in cyclic order. From the second of these equations and the fact that the parametric system on each surface is orthogonal, follows the theorem of Dupin:

The surfaces of a triply orthogonal system meet one another in lines of curvature of each.

183. Equations of Lamé. By means of these results we find the conditions to be satisfied by H_1, H_2, H_3, in order that (3) may be the linear element of space referred to a triply orthogonal system of surfaces. For each surface the Codazzi and Gauss equations must be satisfied. When the above values are substituted in these equations, we find the following six equations which it is necessary and sufficient that the functions H satisfy:

$$\frac{\partial^2 H_i}{\partial u_\kappa \partial u_l} = \frac{1}{H_\kappa}\frac{\partial H_\kappa}{\partial u_l}\frac{\partial H_i}{\partial u_\kappa} + \frac{1}{H_l}\frac{\partial H_l}{\partial u_\kappa}\frac{\partial H_i}{\partial u_l}, \tag{8}$$

$$\frac{\partial}{\partial u_i}\left(\frac{1}{H_i}\frac{\partial H_\kappa}{\partial u_i}\right) + \frac{\partial}{\partial u_\kappa}\left(\frac{1}{H_\kappa}\frac{\partial H_i}{\partial u_\kappa}\right) + \frac{1}{H_l^2}\frac{\partial H_i}{\partial u_l}\frac{\partial H_\kappa}{\partial u_l} = 0, \tag{9}$$

where i, κ, l take the values 1, 2, 3 in cyclic order. These are the *equations of Lamé*, being named for the geometer who first deduced them.*

* *Leçons sur les coordonnées curvilignes et leurs diverses applications*, pp. 73–79. Paris, 1859.

For each of the surfaces there is a system of equations of the form (V, 16). When the values from (7) are substituted in these equations we have

$$(10)\qquad \begin{cases} \dfrac{\partial X_i}{\partial u_i} = -\dfrac{1}{H_\kappa}\dfrac{\partial H_i}{\partial u_\kappa}X_\kappa - \dfrac{1}{H_l}\dfrac{\partial H_i}{\partial u_l}X_l, \\ \dfrac{\partial X_i}{\partial u_\kappa} = \dfrac{1}{H_i}\dfrac{\partial H_\kappa}{\partial u_i}X_\kappa, \qquad \dfrac{\partial X_i}{\partial u_l} = \dfrac{1}{H_i}\dfrac{\partial H_l}{\partial u_i}X_l. \end{cases}$$

Recalling the results of § 65, we have that each set of solutions of equations (8), (9) determine a triply orthogonal system, unique to within a motion in space. In order to obtain the coördinates of space referred to this system, we must find nine functions X_i, Y_i, Z_i which satisfy (10) and

$$\Sigma X_i^2 = 1, \qquad \Sigma X_i X_\kappa = 0. \qquad (i \neq \kappa)$$

Then the coördinates of space are given by quadratures of the form

$$x = \int H_1 X_1\, du_1 + H_2 X_2\, du_2 + H_3 X_3\, du_3.$$

If $\rho_{i\kappa}$ denotes the principal radius of a surface $u_i =$ const. in the direction of the curve of parameter u_κ, we have, from (7),

$$(11)\qquad \frac{1}{\rho_{i\kappa}} = -\frac{1}{H_i H_\kappa}\frac{\partial H_\kappa}{\partial u_i}.$$

Let ρ_1 denote the radius of first curvature of a curve of parameter u_1. In accordance with § 49 we let $\overline{\omega}_1$ and $\overline{\omega}_1 - \pi/2$ denote the angles which the tangents to the curves of parameter u_3 and u_2 respectively through the given point make, in the positive sense, with the positive direction of the principal normal of the curve of parameter u_1. Hence, by (IV, 16), we have

$$(12)\qquad \frac{\cos\overline{\omega}_1}{\rho_1} = \frac{1}{\rho_{31}}, \qquad \frac{\sin\overline{\omega}_1}{\rho_1} = \frac{1}{\rho_{21}}.$$

From these equations and similar ones for curves of parameter u_2 and u_3, we deduce the relations

$$(13)\qquad \frac{1}{\rho_i^2} = \frac{1}{\rho_{\kappa i}^2} + \frac{1}{\rho_{li}^2}, \qquad \tan\overline{\omega}_i = \frac{\rho_{li}}{\rho_{\kappa i}},$$

where i, κ, l take the values 1, 2, 3 in cyclic order. Moreover, since the parametric curves are lines of curvature, it follows from (§ 59) that the torsion of a curve of parameter u_i is

$$\frac{1}{\tau_i} = \frac{1}{H_i}\frac{\partial \overline{\omega}_i}{\partial u_i}. \tag{14}$$

184. Triple systems containing one family of surfaces of revolution. Given a family of plane curves and their orthogonal trajectories; if the plane be revolved about a line of the plane as an axis, the two families of surfaces of revolution thus generated, and the planes through the axis, form a triply orthogonal system. We inquire whether there are any other triple systems containing a family of surfaces of revolution.

Suppose that the surfaces $u_3 = \text{const.}$ of a triple system are surfaces of revolution, and that the curves $u_2 = \text{const.}$ upon them are the meridians. Since the latter are geodesics, we must have

$$\frac{\partial H_1}{\partial u_2} = 0. \tag{15}$$

From (8) it follows that either

$$\frac{\partial H_1}{\partial u_3} = 0, \quad \text{or} \quad \frac{\partial H_3}{\partial u_2} = 0.$$

In the first case it follows from (11) that $1/\rho_{31} = 0$. Consequently, the surfaces of revolution $u_3 = \text{const.}$ are developables, that is, either circular cylinders or circular cones. Furthermore, from (15) and (11), we have $1/\rho_{21} = 0$, so that the surfaces $u_2 = \text{const.}$ also are developables, and in addition we have, from (13), that $1/\rho_1 = 0$, that is, the curves of parameter u_1 are straight lines and consequently the surfaces $u_1 = \text{const.}$ are parallel. The latter are planes when the surfaces $u_3 = \text{const.}$ are cylinders, and surfaces with circular lines of curvature when $u_3 = \text{const.}$ are circular cones. Conversely, from the theorem of Darboux (§ 187) and from § 132, it follows that any system of circular cylinders with parallel generators, or any family of circular cones whose axes are tangent to the locus of the vertex, leads to a triple system of the kind sought.

We consider now the second case, namely

$$\frac{\partial H_1}{\partial u_2} = 0, \qquad \frac{\partial H_3}{\partial u_2} = 0.$$

From (11) we find that $1/\rho_{21}=0$, and $1/\rho_{23}=0$; consequently the surfaces $u_2=$ const. are planes. Since these are the planes of the meridians, it follows that the axes of the surfaces coincide, and consequently the case cited at the beginning of this section is the only one for nondevelopable surfaces.

185. Triple systems of Bianchi and of Weingarten. In § 119 it was found that all the Bianchi transforms of a given pseudospherical surface are pseudospherical surfaces of the same total curvature, and that they are the orthogonal surfaces of a cyclic system of circles of constant radius. Hence the totality of these circles and surfaces constitutes a triply orthogonal system, such that the surfaces in one family are pseudospherical. As systems of this sort were first considered by Ribaucour (cf. § 119), they are called the *triple systems of Ribaucour*. We proceed to the consideration of all triple systems such that the surfaces of one family are pseudospherical. These systems were first studied by Bianchi,* and consequently Darboux † has called them the *systems of Bianchi.*

From § 119 it follows that the parameters of the lines of curvature of a pseudospherical surface of curvature $-1/a^2$ can be so chosen that the linear element takes the form

$$ds^2=\cos^2\omega\, du^2+\sin^2\omega\, dv^2, \tag{16}$$

where ω is a solution of the equation

$$\frac{\partial^2\omega}{\partial u^2}-\frac{\partial^2\omega}{\partial v^2}=\frac{\sin\omega\cos\omega}{a^2}. \tag{17}$$

In this case the principal radii are given by

$$\frac{1}{\rho_1}=-\frac{\tan\omega}{a},\qquad \frac{1}{\rho_2}=\frac{\cot\omega}{a}. \tag{18}$$

In general the total curvature of the pseudospherical surfaces of a system of Bianchi varies with the surfaces. If the surfaces $u_3=$ const. are the pseudospherical surfaces, we may write the curvature in the form $-1/U_3^2$, where U_3 is a function of u_3 alone.

* *Annali*, Ser. 2, Vol. XIII (1885), pp. 177–234; Vol. XIV (1886), pp. 115–130; *Lezioni*, Vol. II, chap. xxvii.

† *Leçons sur les systèmes orthogonaux et les coordonnées curvilignes*, pp. 308–323. Paris, 1898.

In accordance with (11) and (18) we put

$$(19)\qquad \begin{cases} \dfrac{1}{\rho_{31}} = -\dfrac{1}{H_3 H_1}\dfrac{\partial H_1}{\partial u_3} = \dfrac{\tan\omega}{U_3}, \\ \dfrac{1}{\rho_{32}} = -\dfrac{1}{H_3 H_2}\dfrac{\partial H_2}{\partial u_3} = -\dfrac{\cot\omega}{U_3}. \end{cases}$$

If these values of $\dfrac{\partial H_1}{\partial u_3}$ and $\dfrac{\partial H_2}{\partial u_3}$ be substituted in equations (8) for i equal to 1 and 2 respectively, we obtain

$$\frac{1}{H_1}\frac{\partial H_1}{\partial u_2} = -\tan\omega\,\frac{\partial \omega}{\partial u_2}, \qquad \frac{1}{H_2}\frac{\partial H_2}{\partial u_1} = \cot\omega\,\frac{\partial \omega}{\partial u_1}.$$

From these equations we have, by integration,

$$(20)\qquad H_1 = \phi_{13}\cdot\cos\omega, \qquad H_2 = \phi_{23}\cdot\sin\omega,$$

where ϕ_{13} and ϕ_{23} are functions independent of u_2 and u_1 respectively. We shall show that both of them are independent of u_3.

When the values of H_1 and H_2 from (20) are substituted in (19), we have respectively

$$(21)\qquad \begin{cases} H_3 = U_3\cot\omega\left(\tan\omega\,\dfrac{\partial\omega}{\partial u_3} - \dfrac{\partial \log \phi_{13}}{\partial u_3}\right), \\ H_3 = U_3\tan\omega\left(\cot\omega\,\dfrac{\partial\omega}{\partial u_3} + \dfrac{\partial \log \phi_{23}}{\partial u_3}\right). \end{cases}$$

From these equations it follows that

$$(22)\qquad \cot\omega\,\frac{\partial \log\phi_{13}}{\partial u_3} + \tan\omega\,\frac{\partial \log\phi_{23}}{\partial u_3} = 0.$$

Hence, unless ϕ_{13} and ϕ_{23} are independent of u_3, $\tan\omega$ is equal to the ratio of a function of u_1 and u_3 and of a function of u_2 and u_3.

We consider the latter case and study for the moment a particular surface $u_3 = c$. By the change of parameters

$$\phi_{13}(u_1, c)\,du_1 = du, \qquad \phi_{23}(u_2, c)\,du_2 = dv,$$

the linear element of the surface reduces to (16), and (22) becomes

$$\tan\omega = \frac{U}{V},$$

where U and V are functions of u and v respectively. When this value is substituted in (17), we obtain

$$(23)\qquad \left(\frac{U''}{U} + \frac{V''}{V}\right)(U^2+V^2) = \frac{U^2+V^2}{a^2} + 2\,U'^2 + 2\,V'^2.$$

If this equation be differentiated successively with respect to u and v, we find

$$\left(\frac{U''}{U}\right)'\frac{1}{UU'}+\left(\frac{V''}{V}\right)'\frac{1}{VV'}=0,$$

unless U' or V' is equal to zero. From this it follows that

$$\left(\frac{U''}{U}\right)'=4\,\kappa UU', \qquad \left(\frac{V''}{V}\right)'=-4\,\kappa VV',$$

where κ denotes a constant. Integrating, we have

$$U''=2\,\kappa U^3+\alpha U, \qquad V''=-2\,\kappa V^3+\beta V,$$

α and β being constants, and another integration gives

$$U'^2=\kappa U^4+\alpha U^2+\gamma, \qquad V'^2=-\kappa V^4+\beta V^2+\delta.$$

When these expressions are substituted in (23), we find

$$\left(\beta-\alpha-\frac{1}{a^2}\right)U^2+\left(\alpha-\beta-\frac{1}{a^2}\right)V^2-2\,\gamma-2\,\delta=0.$$

This condition can be satisfied only when the curvature is zero. Hence U' or V' must be zero, that is, ω must be a function of u or v alone. In this case the surface is a surface of revolution. In accordance with § 184 a triple system of Bianchi arises from an infinity of pseudospherical surfaces of revolution with the same axis.

When exception is made of this case, the functions ϕ_{13} and ϕ_{23} in (20) are independent of u_3. Hence the parameters of the systems may be chosen so that we have

$$H_1=\cos\omega, \qquad H_2=\sin\omega, \qquad H_3=U_3\frac{\partial\omega}{\partial u_3}. \tag{24}$$

When these values are substituted in the six equations (8), (9), they reduce to the four equations

$$\left\{\begin{aligned}
&\frac{\partial^2\omega}{\partial u_1^2}-\frac{\partial^2\omega}{\partial u_2^2}-\frac{\sin\omega\cos\omega}{U_3^2}=0,\\
&\frac{\partial^3\omega}{\partial u_1\partial u_2\partial u_3}-\cot\omega\frac{\partial\omega}{\partial u_1}\frac{\partial^2\omega}{\partial u_2\partial u_3}+\tan\omega\frac{\partial\omega}{\partial u_2}\frac{\partial^2\omega}{\partial u_1\partial u_3}=0,\\
&\frac{\partial}{\partial u_1}\left(\frac{1}{\cos\omega}\frac{\partial^2\omega}{\partial u_1\partial u_3}\right)-\frac{1}{U_3}\frac{\partial}{\partial u_3}\left(\frac{\sin\omega}{U_3}\right)-\frac{1}{\sin\omega}\frac{\partial\omega}{\partial u_2}\frac{\partial^2\omega}{\partial u_2\partial u_3}=0,\\
&\frac{\partial}{\partial u_2}\left(\frac{1}{\sin\omega}\frac{\partial^2\omega}{\partial u_2\partial u_3}\right)+\frac{1}{U_3}\frac{\partial}{\partial u_3}\left(\frac{\cos\omega}{U_3}\right)+\frac{1}{\cos\omega}\frac{\partial\omega}{\partial u_1}\frac{\partial^2\omega}{\partial u_1\partial u_3}=0.
\end{aligned}\right. \tag{25}$$

Darboux has inquired into the generality of the solution of this system of equations, and he has found that the general solution involves five arbitrary functions of a single variable. We shall not give a proof of this fact, but refer the reader to the investigation of Darboux.*

We turn to the consideration of the particular case where the total curvature of all the pseudospherical surfaces is the same, which may be taken to be -1 without any loss of generality. As triple systems of this sort were first discussed by Weingarten, we follow Bianchi in calling them *systems of Weingarten*. Of this kind are the triple systems of Ribaucour.

For this case we have $U_3=1$, so that the linear element of space is

$$ds^2 = \cos^2\omega\, du_1^2 + \sin^2\omega\, du_2^2 + \left(\frac{\partial\omega}{\partial u_3}\right)^2 du_3^2. \tag{26}$$

Since the second of equations (25) may be written in either of the forms

$$\left\{\begin{aligned} \frac{\partial}{\partial u_1}\left(\frac{1}{\sin\omega}\frac{\partial^2\omega}{\partial u_2\,\partial u_3}\right) &= -\frac{1}{\cos\omega}\frac{\partial\omega}{\partial u_2}\frac{\partial^2\omega}{\partial u_1\,\partial u_3}, \\ \frac{\partial}{\partial u_2}\left(\frac{1}{\cos\omega}\frac{\partial^2\omega}{\partial u_1\,\partial u_3}\right) &= \frac{1}{\sin\omega}\frac{\partial\omega}{\partial u_1}\frac{\partial^2\omega}{\partial u_2\,\partial u_3}, \end{aligned}\right. \tag{27}$$

if we put

$$\Phi = \left(\frac{1}{\cos\omega}\frac{\partial^2\omega}{\partial u_1\,\partial u_3}\right)^2 + \left(\frac{1}{\sin\omega}\frac{\partial^2\omega}{\partial u_2\,\partial u_3}\right)^2 - \left(\frac{\partial\omega}{\partial u_3}\right)^2,$$

it follows from the last two of (25) and from (27) that

$$\frac{\partial\Phi}{\partial u_1} = 0, \qquad \frac{\partial\Phi}{\partial u_2} = 0.$$

Hence Φ is a function of u_3 alone. But by changing the parameter u_3, an operation which will not affect the form of (26), we can give Φ a constant value, say c. Consequently we have

$$\left(\frac{1}{\cos\omega}\frac{\partial^2\omega}{\partial u_1\,\partial u_3}\right)^2 + \left(\frac{1}{\sin\omega}\frac{\partial^2\omega}{\partial u_2\,\partial u_3}\right)^2 - \left(\frac{\partial\omega}{\partial u_3}\right)^2 = c. \tag{28}$$

Bianchi has shown † that equation (28) and the first of (25) are equivalent to the system (25), when $U_3=1$. Consequently the problem of the determination of triple systems of Weingarten is the problem of finding common solutions of these two equations.

* *L.c.*, pp. 313, 314; Bianchi, Vol. II, pp. 531, 532.

† Vol. II, p. 550.

EXAMPLES

1. Show that the equations

$$x = r\cos u\cos v, \qquad y = r\cos u\sin v, \qquad z = r\sin u$$

define space referred to a triply orthogonal system.

2. A necessary and sufficient condition that the surfaces $u_3 =$ const. of a triply orthogonal system be parallel is that H_3 be a function of u_3 alone. What are the other surfaces $u_1 =$ const., $u_2 =$ const.?

3. Two near-by surfaces $u_3 =$ const. intercept equal segments on those orthogonal trajectories of the surfaces $u_3 =$ const. which pass through a curve $H_3 =$ const. on the former; on this account the curves $H_3 =$ const. on the surfaces $u_3 =$ const. are called *curves of equidistance*.

4. Let the surfaces $u_3 =$ const. of a triple system be different positions of the same pseudosphere, obtained by translating the surface in the direction of its axis. Determine the character of the other surfaces of the system.

5. Derive the following results for a triple system of Weingarten:

$$\Delta_1 \frac{\partial \omega}{\partial u_3} = c + \left(\frac{\partial \omega}{\partial u_3}\right)^2, \qquad \rho_g = -\frac{\sqrt{c + \left(\dfrac{\partial \omega}{\partial u_3}\right)^2}}{\dfrac{\partial \omega}{\partial u_3}},$$

where the differential parameter is formed with respect to the linear element of a surface $u_3 =$ const., and ρ_g is the radius of geodesic curvature of a curve $\dfrac{\partial \omega}{\partial u_3} =$ const. on this surface. Show that the curves of equidistance on the surfaces $u_3 =$ const. are geodesic parallels of constant geodesic curvature.

6. Show that when c in (28) is equal to zero, the first curvature $1/\rho_3$ of the curves of parameter u_3 is constant and equal to unity; that equations similar to (12) become

$$\frac{\partial^2 \omega}{\partial u_2 \partial u_3} = -\sin\omega\cos\bar{\omega}_3 \frac{\partial \omega}{\partial u_3}, \qquad \frac{\partial^2 \omega}{\partial u_1 \partial u_3} = -\cos\omega\sin\bar{\omega}_3 \frac{\partial \omega}{\partial u_3};$$

that if we put $\theta = \dfrac{\pi}{2} - \bar{\omega}_3$, the last two of equations (25), where $U_3 = 1$, may be written

$$\frac{\partial \theta}{\partial u_1} + \frac{\partial \omega}{\partial u_2} = \sin\theta\cos\omega, \qquad \frac{\partial \theta}{\partial u_2} + \frac{\partial \omega}{\partial u_1} = -\cos\theta\sin\omega;$$

and that

$$\frac{\partial^2 \theta}{\partial u_1^2} - \frac{\partial^2 \theta}{\partial u_2^2} = \sin\theta\cos\theta, \qquad \left(\frac{1}{\cos\theta}\frac{\partial^2 \theta}{\partial u_1 \partial u_3}\right)^2 + \left(\frac{1}{\sin\theta}\frac{\partial^2 \theta}{\partial u_2 \partial u_3}\right)^2 = \left(\frac{\partial \theta}{\partial u_3}\right)^2.$$

When $c = 0$ in (28) the system is said to be of *constant curvature*.

7. A necessary and sufficient condition that the curves of parameter u_3 of a system of Weingarten be circles is that $\bar{\omega}_3$ be independent of u_3. In this case (cf. Ex. 6) the surfaces $u_3 =$ const. are the Bianchi transforms of the pseudospherical surface with the linear element

$$ds^2 = \cos^2\theta\, du^2 + \sin^2\theta\, dv^2.$$

186. Theorem of Ribaucour. The following theorem is due to Ribaucour*:

Given a family of surfaces of a triply orthogonal system and their orthogonal trajectories; the osculating circles to the latter at their points of meeting with any surface of the family form a cyclic system.

In proving this theorem we first derive the conditions to be satisfied by a system of circles orthogonal to a surface S so that they may form a cyclic system. Let the lines of curvature on S be parametric and refer the surface to the moving trihedral whose x- and y-axes are tangent to the curves $v=$ const., $u=$ const. We have (V, 63)

$$\xi_1 = \eta = p = q_1 = 0. \tag{29}$$

If ϕ denotes the angle which the plane of the circle through a point makes with the corresponding xz-plane, θ the angle which the radius to a point P of the circle makes with its projection in the xy-plane, and R the radius of the circle, the coördinates of P with reference to the moving axes are

$$x = R(1+\cos\theta)\cos\phi, \quad y = R(1+\cos\theta)\sin\phi, \quad z = R\sin\theta.$$

Moreover, the direction-cosines of the tangent to the circle at P are

$$-\sin\theta\cos\phi, \qquad -\sin\theta\sin\phi, \qquad \cos\theta.$$

If we express the condition that every displacement of P must be at right angles to this line, we have, from (29) and (V, 51),

$$d\theta - \left[\sin\theta\left(\frac{1}{R}\frac{\partial R}{\partial u} + \frac{\xi\cos\phi}{R}\right) + q\cos\phi(1+\cos\theta)\right]du$$
$$-\left[\sin\theta\left(\frac{1}{R}\frac{\partial R}{\partial v} + \frac{\eta_1\sin\phi}{R}\right) - p_1\sin\phi(1+\cos\theta)\right]dv = 0.$$

The condition that this equation admit an integral is reducible to

$$\left[\frac{\partial}{\partial u}\left(\frac{\eta_1\sin\phi}{R}\right) - \frac{\partial}{\partial v}\left(\frac{\xi\cos\phi}{R}\right)\right]\sin\theta + \left[\frac{\sin\phi\cos\phi}{R}(\xi p_1 + \eta_1 q)\right.$$
$$\left. - R\frac{\partial}{\partial u}\left(\frac{\sin\phi}{R}p_1\right) - R\frac{\partial}{\partial v}\left(\frac{\cos\phi}{R}q\right)\right](1+\cos\theta) = 0.$$

Hence, as remarked before (§ 174), if there are three surfaces orthogonal to a system of circles, the system is cyclic.

* *Comptes Rendus*, Vol. LXX (1870), pp. 330–333.

The condition that it be cyclic is

$$
(30)\qquad \begin{cases} \dfrac{\partial}{\partial u}\left(\dfrac{\eta_1 \sin\phi}{R}\right) - \dfrac{\partial}{\partial v}\left(\dfrac{\xi\cos\phi}{R}\right) = 0, \\ \dfrac{\sin\phi\cos\phi}{R^2}(\xi p_1 + \eta_1 q) - \dfrac{\partial}{\partial u}\left(\dfrac{\sin\phi}{R}p_1\right) - \dfrac{\partial}{\partial v}\left(\dfrac{\cos\phi}{R}q\right) = 0. \end{cases}
$$

Since the principal radii of S are given by

$$
(31)\qquad \frac{1}{\rho_1} = -\frac{q}{\xi}, \qquad \frac{1}{\rho_2} = \frac{p_1}{\eta_1},
$$

the second of equations (30) reduces to the first when S is a sphere or a plane. Hence we have incidentally the theorem:

A two-parameter system of circles orthogonal to a sphere and to any other surface constitute a cyclic system.

We return to the proof of the theorem of Ribaucour and apply the foregoing results to the system of osculating circles of the curves of parameter u_3 of an orthogonal system at their points of intersection with a surface $u_3 = \text{const.}$

From equations similar to (12) we have, by (11),

$$
\frac{\cos\phi}{R} = -\frac{1}{H_1 H_3}\frac{\partial H_3}{\partial u_1}, \qquad \frac{\sin\phi}{R} = -\frac{1}{H_2 H_3}\frac{\partial H_3}{\partial u_2},
$$

and the equations analogous to (31) are

$$
\frac{1}{\rho_{31}} = -\frac{q}{H_1} = -\frac{1}{H_1 H_3}\frac{\partial H_1}{\partial u_3}, \qquad \frac{1}{\rho_{32}} = \frac{p_1}{H_2} = -\frac{1}{H_2 H_3}\frac{\partial H_2}{\partial u_3}.
$$

When these values are substituted in equations (30) the first vanishes identically, likewise the second, in consequence of equations (8). Hence the theorem of Ribaucour is proved.*

187. Theorems of Darboux. The question naturally arises whether any family of surfaces whatever forms part of a triply orthogonal system. This question will be answered with the aid of the following theorem of Darboux,† which we establish by his methods:

A necessary and sufficient condition that two families of surfaces orthogonal to one another admit of a third family orthogonal to both is that the first two meet one another in lines of curvature.

* For a geometrical proof the reader is referred to Darboux, *l.c.*, p. 77. † *L.c.*, pp. 6–8.

Let the two families of surfaces be defined by

$$(32) \qquad \alpha(x, y, z) = a, \qquad \beta(x, y, z) = b,$$

where a and b are the parameters. The condition of orthogonality is

$$(33) \qquad \frac{\partial \alpha}{\partial x}\frac{\partial \beta}{\partial x} + \frac{\partial \alpha}{\partial y}\frac{\partial \beta}{\partial y} + \frac{\partial \alpha}{\partial z}\frac{\partial \beta}{\partial z} = 0.$$

In order that a third family of surfaces exist orthogonal to the surfaces of the other families, there must be a function $\gamma(x, y, z)$ satisfying the equations

$$\frac{\partial \alpha}{\partial x}\frac{\partial \gamma}{\partial x} + \frac{\partial \alpha}{\partial y}\frac{\partial \gamma}{\partial y} + \frac{\partial \alpha}{\partial z}\frac{\partial \gamma}{\partial z} = 0, \qquad \frac{\partial \beta}{\partial x}\frac{\partial \gamma}{\partial x} + \frac{\partial \beta}{\partial y}\frac{\partial \gamma}{\partial y} + \frac{\partial \beta}{\partial z}\frac{\partial \gamma}{\partial z} = 0.$$

If dx, dy, dz denote the projections on the axes of a displacement of a point on one of the surfaces $\gamma = \text{const.}$, we must have

$$\begin{vmatrix} dx & dy & dz \\ \frac{\partial \alpha}{\partial x} & \frac{\partial \alpha}{\partial y} & \frac{\partial \alpha}{\partial z} \\ \frac{\partial \beta}{\partial x} & \frac{\partial \beta}{\partial y} & \frac{\partial \beta}{\partial z} \end{vmatrix} = 0.$$

This equation is of the form (XIII, 9). The condition (XIII, 10) that it admit of an integral involving a parameter is

$$\sum \frac{\partial(\alpha, \beta)}{\partial(y, z)} \left[\frac{\partial \beta}{\partial x}\frac{\partial^2 \alpha}{\partial z^2} + \frac{\partial \alpha}{\partial z}\frac{\partial^2 \beta}{\partial x \partial z} - \frac{\partial \alpha}{\partial x}\frac{\partial^2 \beta}{\partial z^2} - \frac{\partial \beta}{\partial z}\frac{\partial^2 \alpha}{\partial x \partial z} \right.$$
$$\left. - \frac{\partial \beta}{\partial y}\frac{\partial^2 \alpha}{\partial x \partial y} - \frac{\partial \alpha}{\partial x}\frac{\partial^2 \beta}{\partial y^2} + \frac{\partial \alpha}{\partial y}\frac{\partial^2 \beta}{\partial x \partial y} + \frac{\partial \beta}{\partial x}\frac{\partial^2 \alpha}{\partial y^2} \right] = 0,$$

where Σ indicates the sum of the three terms obtained by permuting x, y, z in this expression. If we add to this equation the identity

$$\sum \frac{\partial(\alpha, \beta)}{\partial(y, z)} \left[\frac{\partial \alpha}{\partial x} \sum \frac{\partial^2 \beta}{\partial x^2} - \frac{\partial \beta}{\partial x} \sum \frac{\partial^2 \alpha}{\partial x^2} \right] = 0,$$

the resulting equation may be written in the form

$$(34) \qquad \begin{vmatrix} \frac{\partial \alpha}{\partial x} & \frac{\partial \beta}{\partial x} & \delta\left(\alpha, \frac{\partial \beta}{\partial x}\right) - \delta\left(\beta, \frac{\partial \alpha}{\partial x}\right) \\ \frac{\partial \alpha}{\partial y} & \frac{\partial \beta}{\partial y} & \delta\left(\alpha, \frac{\partial \beta}{\partial y}\right) - \delta\left(\beta, \frac{\partial \alpha}{\partial y}\right) \\ \frac{\partial \alpha}{\partial z} & \frac{\partial \beta}{\partial z} & \delta\left(\alpha, \frac{\partial \beta}{\partial z}\right) - \delta\left(\beta, \frac{\partial \alpha}{\partial z}\right) \end{vmatrix} = 0,$$

where, for the sake of brevity, we have introduced the symbol $\delta(\theta, \phi)$, defined by

$$\delta(\theta, \phi) = \frac{\partial \theta}{\partial x}\frac{\partial \phi}{\partial x} + \frac{\partial \theta}{\partial y}\frac{\partial \phi}{\partial y} + \frac{\partial \theta}{\partial z}\frac{\partial \phi}{\partial z}.$$

If equation (33) be differentiated with respect to x, the result may be written

$$\delta\left(\alpha, \frac{\partial \beta}{\partial x}\right) + \delta\left(\beta, \frac{\partial \alpha}{\partial x}\right) = 0.$$

Consequently equation (34) is reducible to

$$\text{(35)} \qquad \begin{vmatrix} \frac{\partial \alpha}{\partial x} & \frac{\partial \beta}{\partial x} & \delta\left(\alpha, \frac{\partial \beta}{\partial x}\right) \\ \frac{\partial \alpha}{\partial y} & \frac{\partial \beta}{\partial y} & \delta\left(\alpha, \frac{\partial \beta}{\partial y}\right) \\ \frac{\partial \alpha}{\partial z} & \frac{\partial \beta}{\partial z} & \delta\left(\alpha, \frac{\partial \beta}{\partial z}\right) \end{vmatrix} = 0,$$

which is therefore the condition upon α and β in order that the desired function γ exist.

A displacement along a curve orthogonal to the surfaces $\alpha = \text{const.}$ is given by

$$\frac{dx}{\frac{\partial \alpha}{\partial x}} = \frac{dy}{\frac{\partial \alpha}{\partial y}} = \frac{dz}{\frac{\partial \alpha}{\partial z}}.$$

Such a curve lies upon a surface $\beta = \text{const.}$, and since, by (35), it satisfies the condition

$$\begin{vmatrix} dx & \frac{\partial \beta}{\partial x} & d\frac{\partial \beta}{\partial x} \\ dy & \frac{\partial \beta}{\partial y} & d\frac{\partial \beta}{\partial y} \\ dz & \frac{\partial \beta}{\partial z} & d\frac{\partial \beta}{\partial z} \end{vmatrix} = 0,$$

it is a line of curvature on the surface (cf. Ex. 3, p. 247). Hence the curves of intersection of the surfaces $\alpha = \text{const.}$, $\beta = \text{const.}$, being the orthogonal trajectories of the above curves, are lines of curvature on the surfaces $\beta = \text{const.}$ And by Joachimsthal's theorem (§ 59) they are lines of curvature on the surfaces $\alpha = \text{const.}$ also. Having thus established the theorem of Darboux, we are in a position to answer the question at the beginning of this section.

Given a family of surfaces $\alpha = \text{const.}$; the lines of curvature in one family form a congruence of curves which must admit a family of orthogonal surfaces, if the surfaces $\alpha = \text{const.}$ are to form part of an orthogonal system. If this condition is satisfied, then, according to the theorem of Darboux, there is a third family of surfaces which together with the other two form an orthogonal system.

If X_1, Y_1, Z_1 denote the direction-cosines of the tangents to the lines of curvature in one family on the surfaces $\alpha = \text{const.}$, the analytical condition that there be a family of surfaces orthogonal to these curves is that the equation

$$X_1 dx + Y_1 dy + Z_1 dz = 0$$

admit an integral involving a parameter. The condition for this is

$$(36) \qquad X_1\left(\frac{\partial Z_1}{\partial y} - \frac{\partial Y_1}{\partial z}\right) + Y_1\left(\frac{\partial X_1}{\partial z} - \frac{\partial Z_1}{\partial x}\right) + Z_1\left(\frac{\partial Y_1}{\partial x} - \frac{\partial X_1}{\partial y}\right) = 0.$$

In order to find X_1, Y_1, Z_1 we remark that since they are the direction-cosines of the tangents to a line of curvature we must have

$$\frac{\partial X}{\partial x} X_1 + \frac{\partial X}{\partial y} Y_1 + \frac{\partial X}{\partial z} Z_1 = \lambda X_1,$$

and similar equations in Y, Z, where the function λ is a factor of proportionality to be determined and X, Y, Z are the direction-cosines of the normal to the surface $\alpha = \text{const.}$ Hence, if the surfaces are defined by $\alpha = \text{const.}$, the functions X_1, Y_1, Z_1 are expressible in terms of the first and second derivatives of α, and so equation (36) is of the third order in these derivatives. Therefore we have the theorem of Darboux*:

The determination of all triply orthogonal systems requires the integration of a partial differential equation of the third order.

Darboux has given the name *family of Lamé* to a family of surfaces which forms part of a triply orthogonal system.

188. Transformation of Combescure. We close our study of triply orthogonal surfaces with an exposition of the *transformation of Combescure*,† by means of which from a given orthogonal system others can be obtained such that the normals to the surfaces of one system are parallel to the normals to the corresponding surfaces of the other system at corresponding points.

* *L.c.*, p. 12. † *Annales de l'École Normale Superieure*, Vol. IV (1867), pp. 102–122.

We make use of a set of functions $\beta_{i\kappa}$, introduced by Darboux * in his development of a similar transformation in space of n dimensions. By definition

$$\beta_{i\kappa} = \frac{1}{H_i}\frac{\partial H_\kappa}{\partial u_i}. \qquad (i \neq \kappa)$$

In terms of these functions equations (8), (9) are expressible in the form

$$(37) \qquad \frac{\partial \beta_{i\kappa}}{\partial u_l} = \beta_{il}\beta_{l\kappa}, \qquad \frac{\partial \beta_{i\kappa}}{\partial u_i} + \frac{\partial \beta_{\kappa i}}{\partial u_\kappa} + \beta_{li}\beta_{l\kappa} = 0,$$

and formulas (10) become

$$(38) \qquad \frac{\partial X_i}{\partial u_i} = -\beta_{\kappa i}X_\kappa - \beta_{li}X_l, \qquad \frac{\partial X_i}{\partial u_\kappa} = \beta_{ik}X_\kappa.$$

Equations (37), (38) are the necessary and sufficient conditions that the expression

$$X_1 H_1\, du_1 + X_2 H_2\, du_2 + X_3 H_3\, du_3$$

be an exact differential. From their form it is seen that if we have another set of functions H_1', H_2', H_3' satisfying the six conditions

$$(39) \qquad \beta_{i\kappa} = \frac{1}{H_i'}\frac{\partial H_\kappa'}{\partial u_i},$$

where the functions $\beta_{i\kappa}$ have the same values as for the given system, the expression

$$X_1 H_1'\, du_1 + X_2 H_2'\, du_2 + X_3 H_3'\, du_3,$$

and similar ones in Y, Z, are exact differentials, and so by quadratures we obtain an orthogonal system possessing the desired property.

In order to ascertain the analytical character of this problem, we eliminate H_2' and H_3' from equations (39) and obtain the three equations

$$\frac{\partial^2 H_1'}{\partial u_1 \partial u_2} = \frac{1}{\beta_{21}}\frac{\partial \beta_{21}}{\partial u_1}\frac{\partial H_1'}{\partial u_2} + \beta_{21}\beta_{12}H_1',$$

$$\frac{\partial^2 H_1'}{\partial u_2 \partial u_3} = \frac{\beta_{31}\beta_{23}}{\beta_{21}}\frac{\partial H_1'}{\partial u_2} + \frac{\beta_{21}\beta_{32}}{\beta_{31}}\frac{\partial H_1'}{\partial u_3},$$

$$\frac{\partial^2 H_1'}{\partial u_3 \partial u_1} = \frac{1}{\beta_{31}}\frac{\partial \beta_{31}}{\partial u_1}\frac{\partial H_1'}{\partial u_3} + \beta_{31}\beta_{13}H_1'.$$

The general integral of a system of equations of this kind involves three arbitrary functions each of a single parameter u_i. When one

* *L.c.*, p. 161.

has an integral, the corresponding values of H_2', H_3' are given directly by (39). Hence we have the theorem:

*With every triply orthogonal system there is associated an infinity of others, depending upon three arbitrary functions, such that the normals to the surfaces of any two systems at corresponding points are parallel.**

EXAMPLES

1. In every system of Weingarten for which c in (28) is zero, the system of circles osculating the curves of parameter u_3 at points of a surface $u_3 =$ const. form a system of Ribaucour (§ 185).

2. If the orthogonal trajectories of a family of Lamé are twisted curves of the same constant first curvature, the surfaces of the family are pseudospherical surfaces of equal curvature.

3. Every triply orthogonal system which is derived from a cyclic system by a transformation of Combescure possesses one family of plane orthogonal trajectories.

4. If the orthogonal trajectories of a family of Lamé are plane curves, the cyclic system of circles osculating these trajectories at the points of any surface of the family may be obtained from the given system by a transformation of Combescure.

5. Determine the triply orthogonal systems which result from the application of the transformation of Combescure to a system of Ribaucour (§ 185).

GENERAL EXAMPLES

1. If an inversion by reciprocal radii (§ 80) be effected upon a triply orthogonal system, the resulting system will be of the same kind.

2. Determine the character of the surfaces of the system obtained by an inversion from the system of Ex. 1, § 185, and show that all the curves of intersection are circles.

3. Establish the existence of a triply orthogonal system of spheres.

4. A necessary and sufficient condition that the asymptotic lines correspond on the surfaces $u_3 =$ const. of a triply orthogonal system is that there exist a relation of the form

$$\phi_1 H_1^2 + \phi_2 H_2^2 + \phi_3 = 0,$$

where ϕ_1, ϕ_2, ϕ_3 are functions independent of u_3.

5. When the condition of Ex. 4 is satisfied, those orthogonal trajectories of the surfaces $u_3 =$ const. which pass through points of an asymptotic line on a surface $u_3 =$ const. constitute a surface S which meets the surfaces $u_3 =$ const. in asymptotic lines of the latter and geodesics on S.

6. Show that the asymptotic lines correspond on the pseudospherical surfaces of a triple system of Bianchi.

7. Show that there exist triply orthogonal systems for which the surfaces in one family, say $u_3 =$ const., are spherical, and that the parameters can be chosen so that

$$H_1 = \cosh\theta, \qquad H_2 = \sinh\theta, \qquad H_3 = U_3 \frac{\partial\theta}{\partial u_3}.$$

Find the equations of Lamé for this case.

8. Every one-parameter family of spheres or planes is a family of Lamé.

* Cf. Bianchi, Vol. II, p. 494.

9. In order to obtain the most general triply orthogonal system for which the surfaces in one family are planes, one need construct an orthogonal system of curves in a plane and allow the latter to roll over a developable surface, in which case the curves generate the other surfaces. When the developable is given, the determination of the system reduces to quadratures.

10. Show that the most general triply orthogonal system for which one family of Lamé consists of spheres passing through a point can be found by quadratures.

11. Show that a family of parallel surfaces is a family of Lamé.

12. Show that the triply orthogonal systems for which the curves of parameter u_3 are circles passing through a point can be found without quadrature.

13. By means of Ex. 6, § 185, show that for a system of Weingarten of constant curvature the principal normals to the curves of parameter u_3 at the points of meeting with a surface $u_3 =$ const. form a normal pseudospherical congruence, and that the surfaces complementary to the surfaces $u_3 =$ const. and their orthogonal trajectories constitute a system of Weingarten of constant curvature.

14. By means of Ex. 13 show that for a triple system arising from a system of Weingarten of constant curvature by a transformation of Combescure the osculating planes of the curves $u_3 =$ const., at points of a surface $u_3 =$ const., envelop a surface S of the same kind as this surface $u_3 =$ const.; and these surfaces S and their orthogonal trajectories constitute a system of the same kind as the one resulting from the Combescure transformation of the given system of Weingarten.

15. Show that a necessary condition that the curves of parameter u_1 of a triple system of Bianchi be plane is that ω satisfy also the conditions

$$\frac{\partial \omega}{\partial u_2} = \phi_{23} \sin \omega, \qquad \frac{\partial \omega}{\partial u_1} = \phi_{13} \sin \omega,$$

where ϕ_{23} and ϕ_{13} are independent of u_1 and u_2 respectively (cf. Ex. 5, p. 317). Show that if ϕ_{13} and ϕ_{23} satisfy the conditions

$$\left(\frac{\partial \phi_{13}}{\partial u_1}\right)^2 = \phi_{13}^4 + 2\,a\phi_{13}^2 + b, \qquad \left(\frac{\partial \phi_{23}}{\partial u_2}\right)^2 = \left(\phi_{23}^2 + \frac{1}{U_3^2}\right)^2 + 2\,a\left(\phi_{23}^2 + \frac{1}{U_3^2}\right) + b,$$

where a and b are constants and U_3 is an arbitrary function of u_3, the function ω, given by

$$\cos \omega = \frac{\dfrac{\partial \phi_{23}}{\partial u_2} - \dfrac{\partial \phi_{13}}{\partial u_1}}{\phi_{13}^2 - \phi_{23}^2 - \dfrac{1}{U_3^2}},$$

determines a triply orthogonal system of Bianchi of the kind sought.

16. When $U_3 = 1$ and ω is independent of u_2, the first and fourth of equations (25) may be replaced by

$$\frac{\partial \omega}{\partial u_1} = \sin \omega.$$

Show that for a value of ω satisfying this condition and the other equations (25) the expressions

$$H_1 = \cos \omega \left(\int \frac{\phi_3\, du_3}{\sin \omega} + \phi_1\right) - \int \frac{\phi_3 \cos \omega}{\sin \omega} du_3 + \phi_1',$$

$$H_2 = \sin \omega \left(\int \frac{\phi_3\, du_3}{\sin \omega} + \phi_1\right) - \int \phi_3\, du_3 + \phi_2,$$

$$H_3 = \left(\int \frac{\phi_3\, du_3}{\sin \omega} + \phi_1\right) \frac{\partial \omega}{\partial u_3},$$

where ϕ_1, ϕ_2, ϕ_3 are functions of u_1, u_2, u_3 respectively, and the accent indicates differentiation, define a triply orthogonal system for which the surfaces $u_3 = \text{const.}$ are molding surfaces.

17. Under what conditions do the functions

$$H_1 = U_2 \frac{\partial \omega}{\partial u_2}, \quad H_2 = -\left(\frac{U_2 \dfrac{\partial^2 \omega}{\partial u_2^2}}{\dfrac{\partial \omega}{\partial u_2}} + U_2' \right), \quad H_3 = -\frac{U_2 U_3}{\sin \omega} \frac{\partial^2 \omega}{\partial u_2 \partial u_3},$$

where U_2 and U_3 are functions of u_2 and u_3 respectively, determine a triply orthogonal system arising from a triple system of Bianchi by a transformation of Combescure? Show that in this case the surfaces $u_2 = \text{const.}$ are spheres of radius U_2, and that the curves of parameter u_2 in the system of Bianchi are plane or spherical.

18. Prove that the equations

$$x = A(u_1 - a)^{m_1}(u_2 - a)^{m_2}(u_3 - a)^{m_3},$$
$$y = B(u_1 - b)^{m_1}(u_2 - b)^{m_2}(u_3 - b)^{m_3},$$
$$z = C(u_1 - c)^{m_1}(u_2 - c)^{m_2}(u_3 - c)^{m_3},$$

where A, B, C, a, b, c, m_i are constants, define space referred to a triple system of surfaces, such that each surface is cut by the surfaces of the other two families in a conjugate system.

19. Given a surface S and a sphere Σ; the circles orthogonal to both constitute a cyclic system; hence the locus of a point upon these circles which is in constant cross-ratio with the points of intersection with S and Σ is a surface S_1 orthogonal to the circles; S_1 may be looked upon as derived from S by a contact transformation which preserves lines of curvature; such a transformation preserves planes and spheres.

20. When S of Ex. 19 is a cyclide of Dupin, so are the surfaces S_1, and also the surface which is the locus of the circles which meet S in any line of curvature; hence all of these surfaces form a triple system of cyclides of Dupin.

21. Given three functions U_i defined by

$$U_i = m_i u_i^2 + 2 n_i u_i + p_i, \qquad (i = 1, 2, 3)$$

where m_i, n_i, p_i are constants satisfying the conditions

$$\Sigma m_i = 0, \qquad \Sigma n_i = 0, \qquad \Sigma p_i = 0;$$

and given also the function

$$N = \alpha_1(u_2 - u_3)\sqrt{U_1} + \alpha_2(u_3 - u_1)\sqrt{U_2} + \alpha_3(u_1 - u_2)\sqrt{U_3}$$
$$+ \beta \Sigma m_i u_i + \gamma(p_1 u_2 u_3 + p_2 u_3 u_1 + p_3 u_1 u_2),$$

where α_i, β, γ are constants; determine under what condition the functions

$$H_1 = \frac{u_2 - u_3}{N\sqrt{U_1}}, \qquad H_2 = \frac{u_3 - u_1}{N\sqrt{U_2}}, \qquad H_3 = \frac{u_1 - u_2}{N\sqrt{U_3}}$$

determine a triply orthogonal system. Show that all of the surfaces are isothermic, and that they are cyclides of Dupin.

22. Determine whether there exist triply orthogonal systems of minimal surfaces.

INDEX

The numbers refer to pages. References to an author and his contributions are made in the form of the first Bianchi paragraph, whereas when a proper name is part of a title the reference is given the form as in the second Bianchi paragraph.

* References to asymptotic lines, geodesics, lines of curvature, etc., on particular kinds of surfaces are listed under the latter.

* **For references such as Equations of Codazzi, see Codazzi.**

* See footnote, p. 467.

* See footnote, p. 467.

* This reference is to nondevelopable ruled surfaces. For developable ruled surfaces, see Developables.
† For references such as Surface of Bianchi, see Bianchi.
‡ Surfaces of center of certain surfaces are referred to under these surfaces.

* For references such as Transformation of Bäcklund, see Bäcklund.

DOVER PHOENIX EDITIONS

An Introduction to the Approximation of Functions, Theodore J. Rivlin. Unabridged republication of the 1969 edition. 160pp. 49554-X

An Essay on the Foundations of Geometry, Bertrand Russell. Unabridged republication of the 1897 edition. 224pp. 49555-8

Elements of Number Theory, I. M. Vinogradov. Unabridged republication of the first edition, 1954. 240pp. 49530-2

Asymptotic Expansions for Ordinary Differential Equations, Wolfgang Wasow. Unabridged republication of the 1976 corrected, slightly enlarged reprint of the original 1965 edition. 384pp. 49518-3

Physics

Semiconductor Statistics, J. S. Blakemore. Unabridged, corrected, and slightly enlarged republication of the 1962 edition. 141 illustrations. xviii+318pp. 49502-7

Wave Propagation in Periodic Structures, L. Brillouin. Unabridged republication of the 1946 edition. 131 illustrations. 272pp. 49556-6

The Conceptual Foundations of the Statistical Approach in Mechanics, Paul and Tatiana Ehrenfest. Unabridged republication of the 1959 edition. 128pp. 49504-3

The Analytical Theory of Heat, Joseph Fourier. Unabridged republication of the 1878 edition. 20 figures. 496pp. 49531-0

States of Matter, David L. Goodstein. Unabridged republication of the 1975 edition. 154 figures. 4 tables. 512pp. 49506-X

The Principles of Mechanics, Heinrich Hertz. Unabridged republication of the 1900 edition. 320pp. 49557-4

Thermodynamics of Small Systems, Terrell L. Hill. Unabridged and corrected republication in one volume of the two-volume edition published in 1963–1964. 32 illustrations. 408pp. 6½ x 9¼. 49509-4

Theoretical Physics, A. S. Kompaneyets. Unabridged republication of the 1961 edition. 56 figures. 592pp. 49532-9

Quantum Mechanics, H. A. Kramers. Unabridged republication of the 1957 edition. 14 figures. 512pp. 49533-7

The Theory of Electrons, H. A. Lorentz. Unabridged reproduction of the 1915 edition. 9 figures. 352pp. 49558-2

The Principles of Physical Optics, Ernst Mach. Unabridged republication of the 1926 edition. 279 figures. 10 portraits. 336pp. 49559-0

The Scientific Papers of James Clerk Maxwell, James Clerk Maxwell. Unabridged republication of the 1890 edition. 197 figures. 39 tables. Total of 1,456pp.
Volume I (640pp.) 49560-4; *Volume II* (816pp.) 49561-2

Vectors and Tensors in Crystallography, Donald E. Sands. Unabridged and corrected republication of the 1982 edition. xviii+228pp. 49516-7

Principles of Mechanics and Dynamics, Sir William (Lord Kelvin) Thompson and Peter Guthrie Tait. Unabridged republication of the 1912 edition. 168 diagrams. Total of 1,088pp. *Volume I* (528pp.) 49562-0; *Volume II* (560pp.) 49563-9

Treatise on Irreversible and Statistical Thermophysics: An Introduction to Nonclassical Thermodynamics, Wolfgang Yourgrau, Alwyn van der Merwe, and Gough Raw. Unabridged, corrected republication of the 1966 edition. xx+268pp. 49519-1

Engineering

Principles of Aeroelasticity, Raymond L. Bisplinghoff and Holt Ashley. Unabridged, corrected republication of the original 1962 edition. xi+527pp. 49500-0

Statics of Deformable Solids, Raymond L. Bisplinghoff, James W. Mar, and Theodore H. H. Pian. Unabridged and corrected Dover republication of the edition published in 1965. 376 illustrations. xii+322pp. 6½ x 9¼. 49501-9